Binders for Durable and Sustainable Concrete

Modern Concrete Technology Series

Series Editors
Arnon Bentur
Faculty of Civil and Environmental Engineering
Technion-Israel Institute of Technology

Sidney Mindess
Department of Civil Engineering
University of British Columbia

A series of books presenting the state-of-the-art in concrete technology

Concrete in the Maritime Environment
P. K. Mehta

Concrete in Hot Environments
I. Soroka

Durability of Concrete in Cold Climates
M. Pigeon & R. Pleau

High Performance Concrete
P. C. Aïtcin

Steel Corrosion in Concrete
A. Bentur, S. Diamond and N. Berke

Optimization Methods for Material Design of Cement-based Composites
Edited by A. Brandt

Special Inorganic Cements
I. Odler

Concrete Mixture Proportioning
F. de Larrard

Sulfate Attack on Concrete
J. Skalny, J. Marchand & I. Odler

Determination of Pore Structure Parameters
K. K. Aligizaki

Fundamentals of Durable Reinforced Concrete
M. G. Richardson

Aggregates in Concrete
M. G. Alexander & S. Mindess

Diffusion of Chloride in Concrete
E. Poulsen & L. Mejlbro

Fibre Reinforced Cementitious Composites 2nd edition
A. Bentur & S. Mindess

Binders for Durable and Sustainable Concrete

Pierre-Claude Aïtcin

CRC Press is an imprint of the
Taylor & Francis Group, an **informa** business
A TAYLOR & FRANCIS BOOK

CRC Press
Taylor & Francis Group
6000 Broken Sound Parkway NW, Suite 300
Boca Raton, FL 33487-2742

First issued in paperback 2019

ISBN-13: 978-0-415-38588-6 (hbk)
ISBN-13: 978-0-367-86412-5 (pbk)

Typeset in Sabon by
RefineCatch Limited, Bungay, Suffolk

British Library Cataloguing in Publication Data
A catalogue record for this book is available from the British Library

Library of Congress Cataloging in Publication Data
Aïtcin, Pierre-Claude, 1938–
Binders for durable and sustainable concrete /
Pierre-Claude Aïtcin.
p. cm.—(Modern concrete technology series)
Includes bibliographical references and index.
1. Concrete—Additives. 2. Binders (Materials) I. Title.
II. Series: Modern concrete technology series (E & FN Spon); 15.
III. Series.
TP884.A3A36 2006
666′.893—dc22
2006006233

Visit the Taylor & Francis Web site at
http://www.taylorandfrancis.com

and the CRC Press Web site at
http://www.crcpress.com

To my grandchildren

'We don't inherit the world from our ancestors, we borrow it from our children[1]*'*

Old African saying included by Léopold Sédar Senghor, a past President of Sénégal in his introductory speech at the Académie française

[1] and grandchildren

Science, deprived of any effective control, wanders along in vain imaginings, while industry, deprived of a specific direction, is immobilized in aimless empirical experimentation

Henri Le Chatelier, 'Recherches expérimentales sur la constitution des mortiers hydrauliques' (1904)
(*Free translation by Christine Couture*)

The great end of life is not knowledge but action

Thomas Huxley

Contents

Foreword xix
Preface xxi
Acknowledgements xxiii
Note from the author xxvii

1 **Introduction** 1
1.1 *Concrete: the most widely used construction material* 1
1.2 *Concrete helps satisfy a fundamental need* 3
1.3 *The strengths and weaknesses of concrete* 10
1.4 *Presentation of the different chapters of the book* 14
References 15

2 **Binders and concrete of yesterday** 16
2.1 *Introduction* 16
2.2 *Gypsum, one of the first binders* 20
2.3 *The use of lime* 25
2.4 *Hydraulic binders based on pozzolans* 27
2.5 *The discovery of Portland cement* 30
2.6 *The development of modern concrete* 36
2.7 *Conclusion* 39
References 39

3 **The hydraulic binders and concrete industries at the beginning of the twenty-first century** 44
3.1 *Evolution of cement consumption during the twentieth century* 44
3.2 *The new ecological policy: from Portland cement to hydraulic binders* 52
3.3 *The structure of the cement industry during the second half of the twentieth century* 54
3.4 *The law of the market* 56

3.5 *Technological regulation of the cement industry* 58
3.6 *International associations* 60
3.7 *Technical and educational societies* 60
3.8 *Research in the area of cement and concrete* 60
3.9 *Scientific literature* 61
3.10 *Scientific meetings* 62
3.11 *Conclusion* 62
References 63

4 The chemical and phase composition of Portland cement 64
4.1 *The chemical composition of Portland cement: its importance, its limitations* 64
4.2 *The phase composition of Portland cement* 70
4.3 *Simplified chemical notations* 74
4.4 *Bogue potential composition* 74
4.5 *The ionic nature of the mineral phases found in Portland cement clinker* 76
4.6 *Amorphous solids: glasses* 78
4.7 *The CaO–SiO_2 and SiO_2–CaO–Al_2O_3 phase diagrams* 80
4.7.1 Introduction 80
4.7.2 The CaO–SiO_2 phase diagram 81
4.7.3 The CaO–SiO_2–Al_2O_3 phase diagram 83
4.8 *Making of Portland cement clinker* 85
4.9 *Conclusion* 88
References 88

5 Production of Portland cement (This chapter was written in collaboration with J.-C. Weiss) 90
5.1 *Historical background of Portland cement production* 90
5.2 *Manufacturing processes* 106
5.3 *Selection and preparation of the raw materials* 112
5.4 *The fuels* 116
5.5 *Schematic representation of a cement kiln* 117
5.5.1 Kiln with a pre-heater 117
5.5.2 Two-pier short kiln equipped with a precalcinator 119
5.5.3 Long kiln with three piers equipped with a precalcinator 121
5.6 *Circulation of volatiles* 121
5.6.1 Chlorine cycle 122
5.6.2 Sulphur and alkali cycles 123
5.7 *Microscopical examination of some industrial clinkers* 124
5.7.1 Examination of Figure 5.26 124
5.7.2 Examination of Figure 5.27 124

5.7.3 Examination of Figure 5.28 124
5.7.4 Examination of Figure 5.29 125
5.7.5 Examination of Figure 5.30 125
5.7.6 Examination of Figure 5.31 126
5.7.7 Examination of Figure 5.32 127
5.8 *Clinker storage 128*
5.9 *The addition of calcium sulphate 129*
5.10 *Final grinding 133*
5.11 *Quality control 135*
5.11.1 Chemical analysis 135
5.11.2 Physical properties 136
5.11.3 Mechanical properties 137
5.12 *Conditioning and shipping 138*
5.13 *Evolution of the characteristics of Portland cement 140*
5.14 *The ecological impact of Portland cement manufacturing 142*
5.15 *Does an ideal Portland cement exist? 143*
5.16 *Conclusion 144*
References 144

6 Portland cement hydration 146

6.1 *Introduction 146*
6.2 *The scientific complexity of a simple technological phenomenon 148*
6.3 *How to approach the hydration reaction 151*
6.4 *Portland cement and gypsum 153*
6.4.1 Controlling the hydration of C_3A 153
6.4.2 The true nature of the calcium sulphate found in Portland cement 156
6.4.3 The SO_3 content of clinker 159
6.5 *Direct observation of the structural modifications occurring during cement hydration 161*
6.6 *Indirect observations of the physico-chemical and thermodynamic changes occurring during Portland cement hydration 166*
6.6.1 Variation of the electrical conductivity 167
6.6.2 Portland cement setting 168
6.6.3 Development of the heat of hydration 168
6.6.4 Volumetric variations associated with hydration reaction 172
6.6.4.1 Le Chatelier's observations 173
6.6.4.2 Autogenous shrinkage 174
6.7 *Portland cement hydration 176*
6.7.1 Some experimental data on hydration reaction 176

6.7.2 Jensen's and Hansen's graphical representation of hydration reaction 177
6.7.2.1 Case of cement paste having a water/binder ratio equal to 0.60 181
6.7.2.2 Case of cement paste having a water/binder ratio equal to 0.42 182
6.7.2.3 Case of cement paste having a water/binder ratio equal to 0.30 182
6.7.2.4 Case of cement paste having a water/binder ratio equal to 0.60 that is hydrating in the presence of an external source of water 182
6.7.2.5 Case of cement paste having a water/binder ratio of 0.42 that is hydrating in the presence of an external source of water 183
6.7.2.6 Case of cement paste having a water/binder ratio of 0.30 that is hydrating in the presence of an external source of water 183
6.7.3 Amount of external water necessary to avoid self-desiccation of the cement paste (using Power's definition of self-desiccation) 184
6.7.4 Volumetric changes during the hardening of concrete 186
6.7.5 Influence of the temperature in the kinetics of hydration reaction 187
6.8 Sequential description of hydration reaction 188
6.8.1 First stage: mixing period 189
6.8.2 Second stage: dormant period 189
6.8.3 Third stage: setting and acceleration of hydration 190
6.8.4 Fourth stage: hardening 192
6.8.5 Fifth stage: slow-down period 193
6.8.6 Schematic representation of a hydrated cement paste at twenty-eight days 193
6.9 Linking hydration and mechanical strength 194
6.10 Structural models of C-S-H 195
6.11 The origin of the cohesive forces 198
6.12 Modelling hydration reaction 200
6.13 Conclusion 200
References 201

7 Admixtures **206**
7.1 Historical background 207
7.2 An artificially complicated terminology 209

7.3 *Dispersing agents 213*
7.3.1 The dispersion of cement particles 213
7.3.2 Lignosulphonates 218
7.3.3 Superplasticizers 221
7.3.3.1 Polynaphthalenes (PNS) 222
7.3.3.2 Hydration in the presence of polysulphonate 224
7.3.3.3 The crucial role of calcium sulphate 228
7.3.3.4 The compatibility between polysulphonate and cement 229
7.3.3.5 The limits of the testing of the rheology of cement paste 231
7.3.3.6 Some typical cases 234
7.3.3.7 Selection of a superplasticizer 240
7.3.3.8 Superplasticizer dosage 248
7.3.3.9 Practical advice 249
7.4 *Admixtures that modify hydration kinetics 251*
7.4.1 Accelerators 251
7.4.2 Set accelerators 253
7.4.3 Retarders 254
7.4.4 Cocktails of dispersing agent: accelerator or dispersing-agent retarder 256
7.5 *Admixtures that react with one of the by-products of hydration reaction 257*
7.6 *Air-entraining agents 258*
7.7 *Other types of admixtures 260*
7.7.1 Colloidal agents 260
7.7.2 Water-repellent admixtures 261
7.7.3 Shrinkage-reducing admixtures 261
7.7.4 Anti-freeze admixtures 262
7.7.5 Latexes 262
7.7.6 Foaming agents 263
7.7.7 Corrosion inhibitors 263
7.8 *Conclusion 264*
References 266

8 Cementitious materials other than Portland cement: supplementary cementitious materials, mineral components Portland cement additions 273
8.1 *Terminology 273*
8.2 *Blast furnace slag 276*
8.2.1 Fabrication process 276
8.2.2 Slag hydration 282
8.2.3 Effect of slag on the main characteristics of concrete 283

8.3 Pozzolans 285
8.3.1 Fly ashes 287
8.3.2 Silica fume 293
8.3.3 Calcined clays and shales 299
8.3.4 Rice husk ash 299
8.3.5 Natural pozzolans 300
8.3.6 Diatomaceous earth 303
8.3.7 Perlite 304
8.4 Fillers 305
8.5 Producing blended cement or adding cementitious materials directly in the concrete mixer? 306
8.6 Effects of cementitious materials on the principal characteristics of concrete 307
8.7 Blended cements 309
8.8 Conclusion 310
References 310

9 Special Portland cements and other types of hydraulic binder 314
9.0 Introduction 314
9.1 Special Portland cements 314
9.1.1 White Portland cement 315
9.1.2 Buff cement 315
9.1.3 Oil well cements 316
9.1.4 Shrinkage compensating cements 317
9.1.5 Regulated set cements 317
9.1.6 Masonry cement 318
9.1.7 Air-entrained Portland cements 318
9.1.8 Low alkali Portland cements 319
9.1.9 Microcements 320
9.2 Aluminous cements 320
9.3 Calcium sulphoaluminate cements 325
9.3.1 Introduction 325
9.3.2 Composition and hydration of calcium sulphoaluminate cements 325
9.3.3 Applications of calcium sulphoaluminate cements 326
9.4 Other types of cementitious systems 327
9.5 Other types of cements 328
9.5.1 Sorel cement 328
9.5.2 Oxisulphate magnesium-based cement 328
9.5.3 Other cements 328
9.6 Conclusion 329
References 329

10 The art and science of high-performance concrete 332
10.1 Introduction 332
10.2 What is high-performance concrete? 333
10.3 Water/cement or water/binder ratio? 333
10.4 Concrete as a composite material 334
10.5 Making high-performance concrete 335
10.6 Temperature rise 335
10.7 Shrinkage 341
10.8 Curing 343
10.9 Durability 347
10.9.1 General matters 347
10.9.2 Durability in a marine environment 348
10.9.2.1 Nature of the aggressive action 348
10.9.2.2 Chemical attack on concrete 349
10.9.2.3 Physical attack 350
10.9.2.4 Mechanical attack 350
10.9.2.5 Conclusion 351
10.10 Freeze-thaw resistance 351
10.11 The fire resistance of HPC 352
10.11.1 The fire in the Channel Tunnel 353
10.11.2 The Düsseldorf Airport fire 355
10.11.3 Spalling of concrete under fire conditions 355
10.11.4 The Britl-Euram HITECO BE-1158 Research Project 355
10.12 The future of HPCs 356
10.13 Some HPC structures 357
10.13.1 Monuments in HPC 357
10.13.2 Skyscrapers 358
10.13.2.1 The Petronas Towers in Kuala Lumpur 360
10.13.2.2 High-rise buildings in Montreal 361
10.13.2.3 High-rise buildings in Canada 361
10.13.2.4 A very special American high-rise building 361
10.13.3 HPC bridges 362
10.13.3.1 Joigny Bridge in France 364
10.13.3.2 Small bridges 364
10.13.3.3 Portneuf Bridge 365
10.13.3.4 Montée St-Rémi viaduct 370
10.13.3.5 The Normandie Bridge in France 370
10.13.3.6 The Confederation Bridge in Canada 371
10.13.4 Hibernia offshore platform 373
10.13.5 Miscellaneous uses of HPC 375
10.13.5.1 Reconstruction of an entrance of a McDonald's restaurant 375
10.13.5.2 Piglet farm 377

10.13.6 Special HPCs 378
10.13.6.1 Self-compacting HPCs 378
10.13.6.2 Roller-compacted HPC 379
10.14 Reactive powder concrete 379
10.14.1 Reactive powder concrete concept 383
10.14.1.1 Increase of homogeneity 383
10.14.1.2 Increase of compactness 384
10.14.1.3 Improvement of the microstructure by thermal treatment 385
10.14.1.4 Improvement of the ductility of reactive powder concrete 385
10.14.2 The Sherbrooke pedestrian bikeway 387
10.14.3 Fabrication 388
10.14.3.1 Phase I: fabrication of the confined post tensioned diagonals 388
10.14.3.2 Phase II: construction of the deck and of the lower beam 388
10.14.3.3 Phase III: curing 389
10.14.3.4 Phase IV: transportation to the site 389
10.14.4 Erection 389
10.14.4.1 Phase I: assembling the prefabricated elements 389
10.14.4.2 Phase II: installation of each half of the bridge 391
10.14.5 The past and the future 391
References 395

11 The development of the cement and concrete industries within a sustainable development policy 397
11.1 Introduction 397
11.2 How to lower the environmental impact of concrete 400
11.2.1 Reducing the water/binder ratio 400
11.2.2 Increasing the service life of concrete structures 402
11.2.3 Using concretes having a lower cement content 403
11.2.3.1 High-performance concrete having a low heat of hydration 403
11.2.3.2 High volume fly ash and slag concretes 404
11.2.4 Recycling concrete 404
11.3 The manufacturing of hydraulic binders presenting a more energy-efficient and ecological performance than present Portland cement 405
11.3.1 Decreasing the energetic content of clinker 405
11.3.1.1 Decreasing the clinkering temperature through the use of mineralizers 406

11.3.2 Producing a belitic clinker 406
11.3.3 Using a source of lime other than limestone 407
11.3.4 Use of mineral components 408
11.4 Would it be possible to eliminate Portland cement? 409
11.5 Conclusion 410
References 411

12 Cements of yesterday and today, concretes of tomorrow 413
12.1 Introduction 413
12.2 Concrete: the most widely used construction material in the world 414
12.3 Progress achieved by the cement industry in recent years 415
12.4 The emergence of a science of concrete 416
12.4.1 Recent progress achieved in the field of chemical admixtures 416
12.4.2 Progress achieved in observing the microstructure and in understanding the nanostructure of concretes 419
12.5 Cements of yesterday and today 420
12.5.1 Evolution of their characteristics 420
12.5.2 Standards 421
12.5.3 Cement admixture/compatibility 421
12.6 Concretes of yesterday and today 423
12.6.1 A commodity product or a niche product 423
12.6.2 Strength or durability 425
12.6.3 The race for more MPa 426
12.7 The concrete of tomorrow in a sustainable development perspective 427
12.7.1 The ecological impact of concrete 428
12.7.2 The binders of tomorrow 429
12.7.3 The admixtures of tomorrow 429
12.7.4 The concrete of tomorrow 430
12.8 The development of the concrete industry and the cement industry in the twenty-first century 430
12.9 Conclusion 431
References 432

13 My vision of clinkers and binders 433
13.1 Introduction 433
13.2 My vision of clinkers and binders 433
13.3 The ideal Portland cement 435
13.4 Perverse effects of C_3A 436
13.4.1 Perverse effects of C_3A on the rheology of concrete 436

13.4.2 Perverse effects of C_3A on the compatibility and robustness of polysulphonate-based superplasticizers and water reducers 437
13.4.3 The perverse effect of C_3A on the durability of concrete 438
13.5 Making concrete with an ASTM Type V cement 439
13.6 Conclusion 440
References 441

Appendix I How Vicat prepared his artificial lime (Mary 1862) 442
Appendix II AD 1824. . . . . .No. 5022: artificial stone: Aspdin's specification 444
Appendix III And if the first North American natural cement was made in 1676 in Montreal 446
Appendix IV The SAL and KOSH treatments 447
Appendix V Determination of the Bogue potential composition 448
Appendix VI Example of a very simple binary diagram 452
Appendix VII Ternary diagrams 453
Appendix VIII Ternary phase diagrams 460
Appendix IX Influence of the alkalis on the nature and morphology of hydration products 466
Appendix X Colouring concrete 473
Appendix XI Relevant ASTM standards 478

Index 482

Foreword

Modern concretes constitute a sophisticated family of materials. Portland cement itself is a complicated material, manufactured by first burning an intimate mixture of limestone and clay or shale in a kiln at temperatures in the range of 1400°C to 1500°C, and then intergrinding the resulting clinker with gypsum. When used in concrete, it is now commonly combined with one or more supplementary cementitious materials, including, but not limited to, fly ash, silica fume, blast furnace slag, metakaolinite, and limestone powder. In addition, modern concretes often contain one or more chemical admixtures, such as air-entraining agents, superplasticizers, retarders, corrosion-inhibiting chemicals, and so on. We are thus dealing with a very complex, and not completely understood, binder system. Logically, then, one would expect that cement and concrete technology would fall under the purview of chemical engineers, but for historical reasons this is almost never the case. Rather, most of the people involved in the cement and concrete industries are civil engineers, who are usually not well equipped to understand the chemistry of the concrete system, or to appreciate the fundamental nature of concrete. Unfortunately, they most often see concrete as a 'black box' whose properties can be defined completely by its elastic modulus (E), its compressive strength (f'_c) and perhaps its Poisson's ratio (v).

Pierre-Claude Aïtcin has gone a long way toward addressing the problem. This book, which has grown out of the notes he prepared for his students, is a distillation of his almost forty years of experience in teaching, research and practice. It is perhaps the only book on the chemistry of modern concrete binders written by a civil engineer for civil engineers. It focuses not on the minutiae of the chemical reactions, but on the relevance of these reactions to the behavior of concrete. There is a particular focus on the durability of concrete, which is central to ensuring the sustainability of the cement and concrete industries.

Unlike most books on cement and concrete, this one has a certain charm. When he first asked me to read the initial draft of this book, I agreed only with reluctance, assuming that this would be another dry recitation of facts and chemical equations. However, once I started to read it, I found that I could hardly put it down. As in his earlier book, *High Performance Concrete*, Pierre-Claude has managed to personalize the material, and his passion for the subject comes through very clearly. Indeed, it is his passionate commitment to making the very best use of the materials available to us that underlies everything in this book. I now look forward eagerly to his promised next book on *Durable and Sustainable Concrete*.

Sidney Mindess
June, 2006

Preface

Presently the cement industry is evolving relatively quickly, which is quite unusual for a commodity industry known for its resistance to change. This evolution results from the implementation of the Kyoto Protocol, the focus given to sustainability, globalization and, more recently, for a third time, a drastic increase in the price of fuels. It takes such tremors to see such a commodity industry engage itself in radical moves in a short period of time. It seems that the time of the comfortable slow path of change is over.

Portland cement no longer reigns as the absolute king of the cement market; it has to share its prevalent position with all kinds of blended cements made with more or less well known but often ill defined supplementary cementitious materials. Blended cements are now being promoted after a long period during which, in some parts of the world, they were tolerated or ignored, if not strongly resisted.

Performance standards are becoming the rule for testing and specifying Portland cement and blended cements. Personally I strongly support this more competitive approach, but it is time to raise the real question: are we sure that we are presently appropriately testing the right performance indicators when we are testing cements? Almost forty years of experience have taught me that this is not always true. It would be a pity to forget this very important aspect of cements for users that are facing such costly field problems.

There are many books written on the subject of Portland cement and concrete. Many conferences, symposia, colloquia have been held on these themes, so why write another book on the subject? Initially, this was not written as a book; it was a personal compilation of different notes that I was giving to the graduate students of the Department of Civil Engineering of the Faculty of Engineering of the University of Sherbrooke. About six years ago, I start thinking seriously about my retirement and the idea of transforming these notes into a book came to my mind. As any professor, I was convinced that the efforts carried out in synthesizing the material and in assembling these notes could be of some help to colleagues, students, practioners, specifiers and users. I was encouraged by the relative success of my first book, *High Performance*

Concrete, which has also been published in French, Portuguese, Czech and (in the near future) Spanish and Chinese.

At that time, I did not realize how much effort would be necessary to transform French lecture notes into an English book, this is why it took me so long. Now that it is finished, it is time to start writing my third book, *Durable and Sustainable Concrete*, in spite of the fact that I do not believe that I am a serious masochist!

This book is essentially the result of the work of others; only a very small part of it is from my own research; it is primarily a work of synthesis. It has been written essentially to pass on my experience, accumulated for almost forty years in the domain of concrete and hydraulic binders, to future generations, hoping that it could be useful. I propose a very personal vision of hydraulic binders, the one of a lover of concrete. Too often books on Portland cement and blended cements have been written by very knowledgeable scientists or by people from the industry, but not very often by concrete users. I have tried to write a book on Portland cement and blended cements from the perspective of a user. It contains some very personal ideas that are not shared by everybody, and in such cases I clearly identify my personal ideas in the text.

Presently, I am convinced that the optimization of Portland cement and blended cement is not always done with the optimization of concrete properties in mind. This is the next major change that will have to occur. This change will have to be carried out taking into consideration the tremendous advantages brought by admixtures.

I think that it is necessary to offer such a work of synthesis at the moment when the industry is changing so much. I hope that this book will help people in Universities, in government, in municipal agencies and in the field, to better understand the science and technology supporting the best use of the wonderful materials that are Portland cement clinker and blended cements. We must learn to build durable and sustainable concrete structures with these (not so new) binders.

Acknowledgements

Writing the acknowledgements is the most delicate part of a book. A book is essentially the result of the work of one or several authors as indicated on its cover, but it is also the result of the hidden help from colleagues, collaborators, students and friends. It would have been very simple to write: I thank all of the people who in one way or another helped me to write this book. It would have been short but not so sweet. It would have solved the problem of forgetting nobody. However when thinking about it, such a brief sentence does not do justice to some contributions that were more important than others. I think that it is fairer to establish a certain hierarchy in the help that I received with the risk that I could hurt some susceptibilities. Memory becomes less and less trustful when advancing in age and when the writing of such a book covered so many years. In fact it started many years ago as a note book written for the graduate students of the Department of Civil Engineering of the Faculty of Engineering of the University of Sherbrooke.

The transformation of a notebook into a book necessitates the help of many persons. First I would like to thank Christine Couture and Lise Morency who typed and retyped the different chapters, as well as Gilles Breton who had the responsibility of computerizing all the figures, tables and photos. My knowledge of computers is too rudimentary to have accomplished such a task by myself.

The passage from French into English was done in two steps. First, Christine Couture transformed my broken English into a more acceptable English, and finally Sidney Mindess made it more understandable both from a language and also a technical point of view. I had to answer a number of queries from him in order to clarify the sense of my sentences and my thinking. Sidney has the advantage of understanding, and even guessing, the thinking of a person who is used to think and express himself in French. This wonderful work improved the content of this book, and I want to thank him very specially.

Several of my former students and postdoctoral fellows will recognize some of the curves, tables and figures that I borrowed from their work at the University of Sherbrooke. I would like to thank them very warmly. I would like to mention some of them more particularly: Philippe Pinsonneault,

Gilles Chanvillard, Moussa Baalbaki, Pierre-Claver Nkinamubanzi and Nikola Petrov that helped me a number of times. I would like also to thank my co-worker Irène Kelsey-Lévesque for the beautiful scanning electron microscope photos that she took when she worked with me in the concrete research group. Before leaving our research group she compiled the best ones in a compendium, a true piece of art. Some of these pictures adorn this book. I will also not forget to thank my colleagues Arezki Tagnit-Hamou, Richard Gagné and Kamal Khayat not only for their direct contribution to this book but also for their hidden ones, as the result of discussions on the different subjects of this book. Their experience was very useful in clarifying my thoughts in many cases.

There were also colleagues others than those at the University of Sherbrooke who helped me. First I would like to thank Micheline Moranville-Regourd, a long time friend of mine, who taught me so much about the crystallography of Portland cement clinker and about Portland cement hydration. There is also Jean-Pierre Ollivier, who took the time to read the French draft. Most of his numerous suggestions were followed because they improved the message I wanted to pass along. All this help resulted in a strengthening of my thoughts. I would like also to thank J. Stark, B. Möser and A. Eckart, who allowed me to reproduce a part of their work in the Appendices.

There are also numerous friends working in the industry who gave me advice so that I could take advantage of their practical experience. I would like to thank particularly Jean-Claude Weiss, André Lerat, Juri Gebauer and Jean-Claude Roumain from Holcim and Pierre Richard and Gunnar Idorn, two particularly good friends, unfortunately deceased today. I would not like to forget the Lafarge, Holcim Ciment Québec and Vicat cement companies and the following equipment companies: Polysius, Loesche GMbH and KHD Humboldt Wedag, who allowed me to reproduce some of their equipment to illustrate this book.

I would like also to thank particularly Steven Kosmatka from the Portland Cement Association and Alain Capmas from l'Association Technique de l'Industrie des Liants Hydrauliques de France for their invaluable help.

I would like to thank some professional and amateur photographers who helped me to illustrate this book: Heiko Wittenborn, Alain Boily, Gilles Jodoin, Ben Zimmernan, Philippe Pattyn, Philippe Pinsonneault as well as Sheila Saralegi-Anabeitia who gave me the permission to reproduce the magistral piece of Art of Calatrava: the Tenerife Auditorium. I thank also the Museum of Art of Milwaukee that allows me to reproduce another major work of Calatrava. Finally I would extend my deepest thanks to the Portland Cement Association and more particularly to William J. Burns who allowed me to draw at will on its photo bank to illustrate the two first chapter of my book.

I would like to thank the American Concrete Institute, Cement and Concrete Research, the American Ceramic Society, Prentice-Hall, The MIT Press, and

Eyrolles, who allowed me to reproduce some of the figures that adorn this book.

Finally, I would like to thank Madeleine for her patience and encouragement when we worked on the two most exacting part of the editing of this book: the collection of the permissions to reproduce the numerous photos and figures that I took from others, and the compilation of the index.

Thank you very much to all of the others whom I forget to mention specifically.

Pierre-Claude Aïtcin
Sherbrooke, May 2006

Note from the author

This book is neither theoretical nor technological; it does not pretend to compete with general or specialized books published on the subject by Lea (1956), by Baron and Sauterey (1982), by Taylor (1997), and by Hewlet (1998) to which many references will be made.

This book also does not pretend cover in minute details all the knowledge accumulated in the domain of Portland cement and hydraulic binders. Its chief objective is to review the main physical, chemical, and thermodynamic principles that governs the fabrication and the use of Portland cement and hydraulic binders. The author is convinced that a better fundamental understanding of Portland cement and hydraulic binders is absolutely necessary to make better concrete.

As written by professor Jorg Schlaich (1987): 'One cannot design and work a material which one does not know and understand thoroughly. Therefore, design quality starts with education'.

Moreover, a better understanding of hydraulic binders is becoming very important to cope with the new integrated vision that is developing in the construction industry towards sustainable development. It is time to make more durable and more ecological concrete, and to do so, it is necessary to have a full understanding of the characteristics and the properties of hydraulic binders.

In fact, this book has been written from the notes of a graduate course given at the Université de Sherbrooke during the past twenty years. Of course, the selection of what has been considered as essential knowledge is subjective, but it has been always guided by the objective of bridging the present knowledge between Portland cement and concrete, between hydraulic binders and the organic molecules that are presently used to highlight or correct a particular strength or weakness of Portland cement and hydraulic binders.

In order to better understand the present interaction of the cement and concrete industries the author has found it useful to show how human beings passed from the first binders used in the Antiquity, and even during Prehistory to modern binders and to the 'smart' concretes that we are presently using.

Of course, our knowledge on hydraulic binders and concrete will continue to progress, but in the future this progression will be the result of a scientific approach rather than the fruit of fortuitous discoveries as in the past. In fact, as a result of a constant research effort starting with the pioneering work of Smeaton, Vicat, Le Chatelier, and continued by Powers and his team at the Portland Cement Association, our scientific knowledge of Portland cement, hydraulic binders and concrete has progressed. However, these efforts have highlighted that Portland cement, hydraulic binders, and concrete are very complex materials, that obey the laws of physics chemistry and thermodynamics, without forgetting the laws of the market.

References

Baron, J. and Ollivier, J.-P. (1992), *La Durabilité des bétons*, Paris: Presses de l'École Nationale des Ponts et Chaussées.

Baron, J. and Sauterey, R. (1982), *Le béton hydraulique*, Paris: Presses de l'École Nationale des Ponts et Chaussées.

Hewlett, P.C. (ed.) (1998), *Lea's Chemistry of Cement and Concrete*, 4th edn, London: Arnold.

Lea, F.M. (1956), *The Chemistry of Cement and Concrete*, New York: St Martin's Press.

Schlaich, J. (1987), 'Quality and economy', *Concrete Structures for the Future, Proceedings of IABSE Symposium, Versailles*, Paris: IABSE-AIPGIVBH.

Taylor, H.F.W. (1997), *Cement Chemistry*, 2nd edn, London: Thomas Telford.

Chapter 1

Introduction

1.1 Concrete: the most widely used construction material

It might seem surprising to start the introduction of a book on hydraulic binders by a chapter exclusively devoted to concrete, in which only a few sentences deal with Portland cement and hydraulic binders. This has been done deliberately to point out that Portland cement and hydraulic binders are made almost exclusively to make concrete.

It is unfortunate that, for historical as well as economical reasons that will be presented in Chapter 3, the world of Portland cement and hydraulic binder producers and the world of concrete users have evolved in such an unintegrated way. Of course, presently some vertical integration is occurring; cement companies are increasingly becoming involved in the concrete business and not only in the promotion of concrete while, at the same time, some international concrete companies are becoming cement producers. It is even foreseeable that in the very near future, cement companies will realize that they have to take more interest in the elaboration of construction codes that govern the use of concrete and not only in Portland cement and hydraulic binder standards.

In the year 2000, more than 1.5 billion tonnes of cement were produced to make, on average, nearly 1 cubic metre of concrete per capita (Figure 1.1). Easy to make, technologically simple and inexpensive to produce, concrete has been the construction material of the twentieth century. However, in spite of the great efforts in research done up to now and that will continue to be done, concrete faces a fundamental weakness: the lack of communication in a very fragmented industry lacking a vision of the strength and future of its own material. Too often, concrete is still specified in terms of its twenty-eight-day compressive strength, without taking into account the environmental conditions under which it will have to fulfil its structural function. As a consequence, many concrete structures are prematurely failing, projecting a bad image of concrete to the public (Figure 1.2)

Moreover, concrete is too often mistreated during its placing and curing, so that in the field, an excellent concrete can be instantaneously transformed into a

Figure 1.1 Annual concrete consumption per capita. Bon appétit!

poor concrete. Finally, the philosophy of the *lowest* bidder which is prevalent worldwide in awarding contracts in the construction industry does not promote the cause of good concrete and, more importantly, its adequate use. On too few occasions have contracts been awarded to the *best* bidder rather than to the *lowest* bidder.

Fortunately, concrete is presently reshaping its image due to recent technological developments such as the use of superplasticizers, silica fume, sophisticated admixtures and the increasing use of different types of high-performance concrete (Figure 1.3). Moreover, concrete has diversified into very specialized applications, for instance, self-compacting concretes, underwater concretes, high-performance roller-compacted concretes and reactive powder concretes. Modern concrete is becoming a sophisticated construction material, designed for specific applications so that it can be used more easily and at a more economical environmental cost in spite of an increased unit price. The increased added value of these modern concretes makes them more interesting for all the players involved in the construction industry.

In spite of the fact that the construction industry is a very competitive industry, concrete will have a future if it evolves from essentially a low-cost commodity product, as it is essentially now, to a very efficient and ecological material having a high added value. The strength of concrete will have to be built on its technological simplicity as well as on a scientific approach to

Figure 1.2 Examples of concretes unadapted to their environment: (a) seawater attack (courtesy of Rusty Morgan); (b) freezing and thawing cycles in the presence of de-icing salts (courtesy of Daniel Vézina); (c) de-icing salt attack (courtesy of John Bickley); (d) Alkali-aggregate reaction (courtesy of Benoît Fournier).

improve its performance. The final objective of this evolution is to improve the quality and durability of concrete structures while minimizing their ecological impact.

1.2 Concrete helps satisfy a fundamental need

It is said that human beings have three fundamental needs: food, clothing and dwelling: the history of building materials is closely linked to the last need. As long as human beings were nomads, their major concern regarding dwellings was lightness. Therefore, they lived in caves, natural or temporary shelters or tents. Tents are still used by necessity by some nomads or by modern people who want to experience the type of life our ancestors lived.

(a) (b) (c)

Figure 1.3 Some outstanding concrete structures: (a) the Grande Arche de la Défense in Paris (courtesy of Pierre Richard); (b) Confederation Bridge in Canada (courtesy of Gamil Tadros, photo by Alain Boily); (c) Tenerife Auditorium by Calatrava (courtesy of Auditorio de Tenerife, photo by Jose Ramon Oller).

From a structural point of view, a tent is a remarkable stucture: the material and the ropes work in pure tension while the poles work in compression, and, therefore, it is impossible to design a lighter structure. However, this lightness becomes a problem when passing from static to dynamic conditions under the effect of a strong wind or of a storm. Moreover, staying under a tent when it is raining or snowing rapidly becomes very uncomfortable. Therefore, when human beings started to settle down because they had succeeded in domesticating animals and learned how to grow cereals, they started to build more solid and more comfortable permanent shelters. Eventually, the house became a sign of social status.

It is surprising to see how human beings have been creative in matters of housing: they use all kinds of materials to build their houses: branches, leaves, stones (Figure 1.4), reeds (Figure 1.5), earth (adobe) (Figure 1.6) and even compacted snow (Figure 1.7).

However, it quickly became evident that the use of natural materials had its own limitations as far as durability, comfort and social status were concerned, which is why human beings started developing composite materials such as dried clay bricks reinforced with straw (the Bible, Gen. 11: 4–9, Exod. 5: 7–19). They started to fill the interstices between dry stones or pebbles with clay (Figure 1.8) or even a mixture of sand and bitumen. However, in this case too, it was realized that these first composite materials and corresponding constructions had their limitations.

It was discovered, most probably by chance, that limestone and gypsum, two very common superficial rocks, could be easily transformed into binders after being heated and mixed with water. These first binders were used to build

Figure 1.4 Borie in Périgord. A dry stone shepherd shelter (courtesy of Philippe Pattyn).

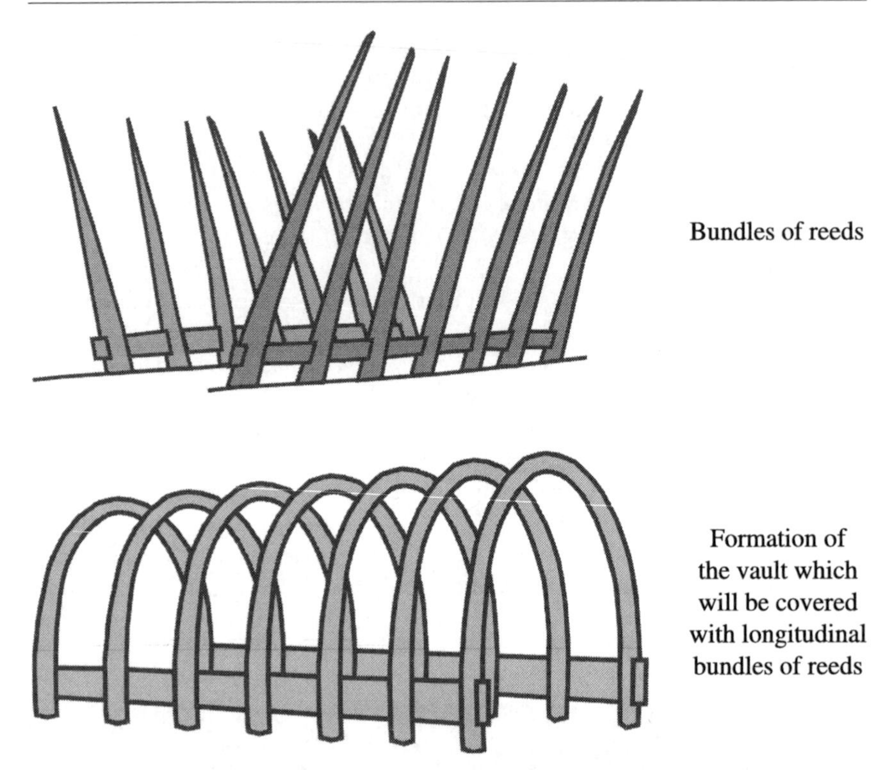

Figure 1.5 Principle of the construction of a dwelling made from reeds in the marsh area of the Tigris and Euphrates in Iraq. A picture of such a house can be seen in the article 'The Adventures of Marco Polo' published in the May 2001 issue of *National Geographic* (p. 12).

stronger and more durable floors and walls (Figure 1.9). Several thousands of years later, these two binders are still used in our houses, as will be seen in the next chapter.

A great step forward was made when it was found that certain natural materials could react at ambient temperature with lime to make a solid material able to harden under water. Such mixtures were used to build roads, bridges, aqueducts and harbours. Several thousands of years later, these materials, which are called pozzolanic materials, are still used, and will be used more extensively in the future due to recent technological developments and a new philosophy of sustainable development that is starting to guide the development of our societies. The Pantheon in Rome, one of the most remarkable and durable concrete structures ever built, illustrates the structural possibilities of such a simple combination of building materials (Figure 1.10). For 1900 years, the Pantheon has defied time and history, in spite of the fact that it was built with a 10 MPa concrete (Mark 1990). A high compressive strength is not

Figure 1.6 The earth house of Gilles Chanvillard in Lyon area in France.

Figure 1.7 An igloo is built with blocks of compacted snow, not with ice blocks. Photo by Heiko Wittenborn.

Figure 1.8 Improving the tightness of a borie with clay joints (courtesy of Philippe Pattyn).

Figure 1.9 The Jauregi farmhouse built by one of my ancestors before 1365 in the Basque province of Xuberoa in France. Initially the house was built with pebbles of a nearly mountain stream jointed with a lime mortar. The lime–Portland cement roughcast was added more recently. The house resisted a nearby strong earthquake in the Pyrenees, three patched cracks are still visible.

Figure 1.10 The Pantheon in Rome (courtesy of Paulo Helene).

mandatory to build a durable structure when construction details have been carefully designed and adequately executed.

It was necessary to wait almost 2000 years to see the emergence of a new building material with the discovery of Portland cement. Around 1820, for the first time in history, an artificial (non-natural) building material was developed: Portland cement, a material that had to undergo a complex physicochemical and pyrotechnical process to become a binder. This artificial material does not return to its original form when it has reacted with water (at least on a human time scale). It can be used to improve the tightness of joints (Figure 1.11). Of course, on a geological time scale, in the environmental conditions prevailing on Earth, the final and stable form of a concrete made with Portland cement and aggregate is: limestone, clay, silica and calcium sulphate, that is, the raw materials used to make it.

Figure 1.11 Improving the durability of the joints of a stone house with mortar joints (courtesy of Philippe Pattyn).

Figure 1.12 The lotus-flower-shaped Bahia temple in India.

1.3 The strengths and weaknesses of concrete

Concrete definitely presents technological advantages: it can be made from local inexpensive materials and it can be cast in any shape (Figures 1.12 to 1.15). Concrete has a good compressive strength, it does not rot, it is not much affected by humidity, it does not burn (Figure 1.16), and it is not attacked by insects (but it can be attacked by certain bacteria, as will be seen later).

Figure 1.13 The Sagrada Familia in Barcelona (photo by Isabelle Aïtcin).

Moreover, when concrete is well proportioned, adequately mixed, transported, placed and cured, it becomes a durable construction material in most environmental conditions. Concrete technology is simple: it consists essentially in thoroughly mixing a fine powder with aggregates, water and admixtures and in compacting this freshly mixed material into forms where it takes its final hardened shape and strength within less than a day. We will see later that technological progress has produced certain modern concretes that do not even need to be compacted because they are self-compacting.

Concrete also presents some weaknesses: it is weak in tension, it is heavy, it is not volumetrically stable because it shrinks and creeps or sometimes swells. Moreover, concrete must be properly cured to reach its full potential as a structural material, and its durability can be impaired in severe environmental conditions, usually acidic conditions (Baron and Ollivier 1992). Today, some of

Figure 1.14 The use of high-performance concrete in the Sagrada Familia (photo by Isabelle Aïtcin).

these weaknesses have been partially overcome with reinforcing bars, fibres, admixtures and lightweight aggregates.

As concrete's advantages outweigh its weaknesses, it is not surprising that concrete is presently the most widely used material after water. As already mentioned, every year more than 1.5 billion tonnes of cement are produced (Syndicat Français de l'Industrie Cimentière 1998). From this huge amount of cement, around 6 billion cubic metres of concrete are made. If we suppose an average cement content of 250 kilograms per cubic metre, this represents practically one cubic metre of concrete per person. As one cubic metre of concrete weighs 2.5 tonnes, each human being consumes on average 2.5 tonnes of concrete per year. According to *Paris-Match*, this weight represents twice the amount of food and liquid consumed by each French person in 1956 (Théron 1998). Moreover, concrete can be made from the Equator to the Poles, in the desert, on the sea, under water and even on the Moon (ACI Committee 125). It is so successful a material that it can be considered as the basis on which our present civilization has been built.

Figure 1.15 Milwaukee Art Museum by Calatrava (courtesy of the Milwaukee Art Museum).

Figure 1.16 Concrete and gypsum are the two materials that best resist fire (courtesy of the Portland Cement Association).

1.4 Presentation of the different chapters of the book

In the next chapter, it will be shown how, during the past centuries and even the past millennia, human beings have progressed in the mastering of present hydraulic binders. This brief historical review will show how a series of fortuitous discoveries, sometimes combined with the scientific approach of people like Smeaton, Vicat, Aspdin, and many others, resulted in the discovery of the artificial binder that is Portland cement.

Before presenting Portland cement's chemical and phase composition, data on past and present Portland cement and hydraulic binders consumption will be presented to show how this consumption is presently satisfied by the cement and concrete industries. The challenges that these industries had to face, and will have to face in the near future, are also presented.

The principle behind the making of Portland cement will then be exposed, to show both the complexity of Portland cement manufacturing as well as its theoretical simplicity. The environmental consequences of the production of Portland cement will be briefly reviewed, in order to present the different measures that should be adopted to make Portland cement a more ecological material. As human beings cannot live without Portland cement, it is important that they use it appropriately, within a perspective of sustainable development.

The different approaches that can be taken to explain the phenomenon called hydration are presented. In particular, it will be shown that, in spite of all the progress made, this apparently very simple phenomenon is not yet fully understood in all its minute details, even although more than 1.5 billion tonnes of Portland cement are used all over the world every year.

It will also be shown that over the years, a certain number of molecules, usually organic molecules, that could be used to modify the development or the consequences of the hydration reaction, were discovered. These molecules, known under the general term of admixtures, play an essential role during the fabrication of not only usual concrete, but also of the more complex concretes used in very specialized applications.

The author is convinced that the future of Portland cement and hydraulic binders in general is closely linked to progress achieved in understanding the interaction between certain organic molecules, cement particles and the ionic species found in hydraulic cement pastes. The era of fortuitous discoveries is likely finished in this domain. The latest admixtures that have appeared on the market were the result of a scientific approach: a new science of admixtures is developing. Moreover, a too often ignored aspect of the use of admixtures will be highlighted in the near future: their ecological incidence. In particular, it will be shown that the use of a small amount of very specific polymers results in savings in natural resources (cement, aggregate, fuel) and in a decrease in carbon dioxide emissions, a greenhouse-effect gas.

The use of the materials known as supplementary cementitious material or mineral components or mineral additives will be reviewed. In most parts of the world, their use will increase in the very near future to limit the amount of greenhouse gas emissions and to save natural resources. It will be explained why in the present economical context, the use of mineral additives has not increased beyond an approximate 35 to 45 per cent limit in certain countries, that are most advanced in the use of these materials, such as Belgium, the Netherlands and Germany, who for historical reasons have used these materials for a longer time than many other countries.

High performance and reactive powder concretes will then be presented in order to show how the present knowledge of Portland cement, mineral additives and admixtures can be used to develop these new types of performing concretes. It will be seen that a science of concrete is developing which takes advantage of new experimental techniques such as electron microscopy, nuclear magnetic resonance, advanced mathematical modelling or even fractal geometry.

A special chapter has been devoted to presenting, in detail, the environmental impact of the fabrication of Portland cement and concrete and how this environmental impact can be reduced. The Portland cement and concrete industries have already done great efforts to become greener industries, but more can be done. Moreover, the beneficial environmental aspects of the Portland cement and concrete industries, which are too often ignored or not publicized enough, will be presented. Many toxic wastes are presently eliminated in cement kilns or neutralized when encapsulated in concrete.

Finally, the type of clinker I have been dreaming of throughout my career will be presented. It is not a revolutionary clinker: it is simply a clinker with a low C_3A content and enough rapidly soluble alkali sulphate to be compatible with polysulphonate superplasticizers.

References

ACI (American Concrete Institute) Committee 125, Lunar concrete.

Baron, J. and Ollivier, J.-P. (1992) *La Durabilité des bétons*, Paris: Presses de l'École Nationale des Ponts et Chaussées.

Mark, R. (1990) *Light, Wind and Structure: The Mystery of Master Builders*, Cambridge, Mass. MIT. Press.

Syndicat Français de l'Industrie Cimentière (1998) '*Ciment 1998*', Paris: Syndicat Français de l'Industrie Cimentiere.

Théron, R. (1998) *50 ans de Paris Match, 1949–1998*, Paris: Filipacchi.

Chapter 2

Binders and concrete of yesterday

2.1 Introduction

Many people still believe that concrete was invented by the Romans. Of course, some concrete structures built by the Romans are still in pretty good shape and even in some cases still in use, such as the Pantheon and Coliseum in Rome, the Pont du Gard in France, the Aqueduct of Segovia in Spain, Hadrian's Wall in England and the Cologne aqueduct in Germany (Figures 2.1 to 2.5). Archaeologists, however, have discovered more and more proof that even before the Romans, the Greeks (Koui and Ftikos 1998) and Phoenicians (Baronio et al. 1997) used concrete on a smaller scale (Papadakis and Venuat 1966).

The first use of some 'protoconcrete' dates back to the Neolithic age, at a time when human beings lived essentially in natural caves carved in limestone cliffs. Based on carbon 14 dating techniques, Felder-Casagrande et al. (1997) found that mortars used in Nevali Çore in Turkey were made 22600 years ago. Excavation work done in Jericho, the first fortified town of the Neolithic age, has brought to light a 7000-year-old floor (Malinowski and Garfinkel 1991). In Jiftah El, another 180-m^2 concrete floor was accidentally discovered by a bulldozer during construction work. From these discoveries, Malinoswki and Garfinkel (1991) conclude that concrete was discovered during the Neolithic age. According to Kingery (1980), a well-known ceramicist, the use of lime preceded the use of pottery.

According to these authors, the discovery of the binding properties of lime was probably discovered when fires were used to heat caves or to cook. As quick lime hydrates very easily in the presence of water and hardens in air, it was not too difficult to discover the binding properties of lime.

Based a carbon-14 dating, Thornton (1996) reports that one of the oldest uses of concrete in central Europe can be found in Lepenski Vir near the Danube River, where a 250-mm thick floor was built some 5600 years before Christ. This concrete was a mixture of red lime, gravel and sand.

The use of lime was not limited only to the Middle East area, the shores of the Mediterranean Sea (Bonen et al. 1995; Güleç and Tulun 1997) or Europe. Thornton (1996) mentions that lime was used in 3000 BC in Dadiwan, Qinan

Figure 2.1 Inside view of the Pantheon in Rome (courtesy of M. Collins).

county, in the Chinese province of Ganow, to build a floor. This concrete was composed of sand, stone, pottery rubble and bones.

However, not everyone is yet convinced, as claimed by Davidovits (1987) and Morris (1991), that Egyptians used concrete blocks to build their pyramids. These two researchers state that the interior blocks of the pyramids were made of concrete and that only the external blocks were made of limestone. According to Davidovits, the concrete used in the pyramids was made from a special binder, which he calls a geopolymer, obtained by mixing silt from the Nile river, limestone powder and turquoise chrysocolla mine tailings from the Mount Sinai area. The binding properties of this geopolymer were based on an alkaline activation that results in the formation of zeolithic phases such as analcine $Na_2O \cdot Al_2O_3 \cdot 4SiO_2 \cdot 2H_2O$ or phillipsite $3CaO \cdot Al_2O_3 \cdot 10SiO_2 \cdot 12H_2O$.

Figure 2.2 The Pont du Gard (courtesy of Michael Collins). The Pont du Gard was erected during the first century BC, it is a part of the 50-kilometre long aqueduct that used to bring water to Nîmes. It is formed of three superimposed arch bridges containing six, eleven and thirty-five arches. The Pont du Gard was built as solid rock. The central arch is 48 metres high and 48 metres wide. In 1743 the States of Languedoc built a bridge at the base of the Pont du Gard.

According to Davidovits, this technology disappeared with the depletion of the Mount Sinai mines. This vision of the construction of the pyramids has been strongly opposed by Campbell and Folk (1991).

The discovery of the binding properties of these geopolymers was patented by Davidovits, and the American cement producer LONESTAR bought a licence to produce it. LONESTAR began a vigorous marketing campaign to promote a revolutionary new cement named PYRAMENT, a word made from the first syllables of *pyramid* and the last of *cement*. The only problem for LONESTAR was that it has been impossible to produce an industrial binder with such a prestigious name that could compete with ordinary Portland cement.

It is certain that it is the Romans who in antiquity developed the use of concrete in all kinds of applications throughout their empire (Figure 2.6). However, after the fall of the Roman Empire under the pressure of barbarians attacks, it was necessary to wait until practically the end of the eighteenth and

Figure 2.3 Segovia aqueduct in Spain.

Figure 2.4 Hadrian's Wall, North of England (courtesy of Elizabeth Neville).

Figure 2.5 Section of a Roman aqueduct from Köln (Germany) (courtesy of the Portland Cement Association).

the beginning of the nineteenth centuries to see new interest in hydraulic binders and concrete. This new interest resulted in the patenting of Portland cement by Joseph Aspdin, an English mason (Blezard 1998).

2.2 Gypsum, one of the first binders

In spite of the fact that gypsum was not the first artificial binder used, it is convenient to start this historical review with it, because from a chemical and processing point of view, it is the simplest processed building material. According to Stark and Wicht (1999), the binding properties of gypsum were discovered 10000 to 20000 years ago. It is difficult to give a more precise date because of the lack of durability of gypsum in a humid environment. The first

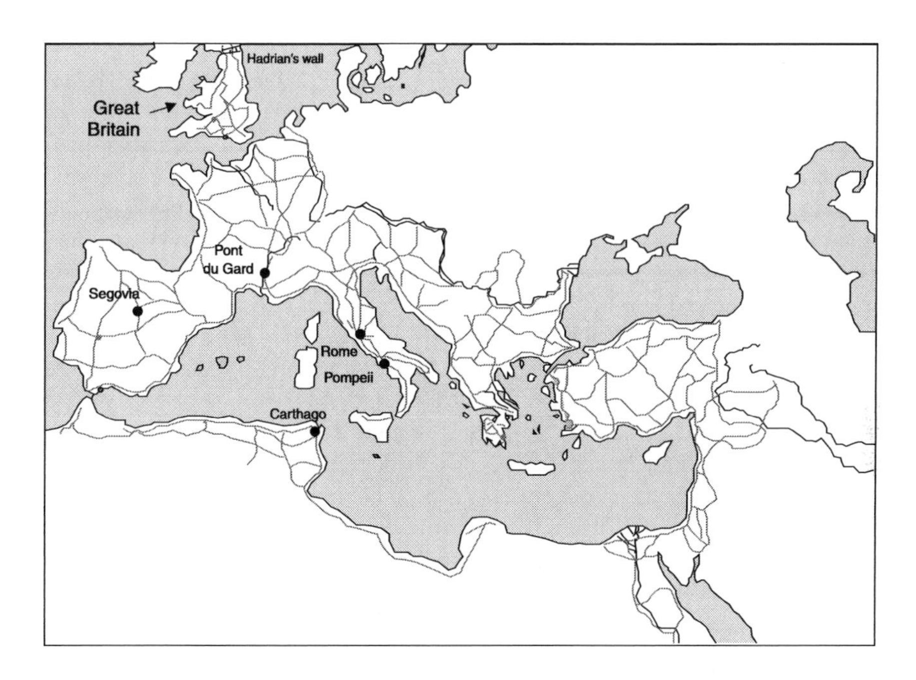

Figure 2.6 The Roman and American highway networks.

use of gypsum that can be dated precisely is in a fresco found in Çatal Hüyük in Anatolia (Turkey), created 9000 years before Christ (Torres and Emeric 1999). Archaeological excavations done recently in Israel have unearthed a gypsum floor in Jericho near the Sea of Galilee that was built 6000 to 7000 years BC. Moreover, several references are made to gypsum in Sumerian and Babylonian cuneiform texts.

Many archaeological studies have demonstrated that Egyptians used gypsum as a binder in mortars (Clarke and Engelbach 1930; Lucas and Harris 1962; Papadakis and Venuat 1966; Ghorab et al. 1986; Ragai et al. 1987; Ragai 1988a and 1988b; Regourd et al. 1988; Ragai 1989; Martinet 1991, 1992; Martinet et al. 1992; Blezard 1998). Snell and Snell (2000) state that in certain cases, gypsum mortars were only used as a lubricant between large stone blocks to facilitate their final placing.

All of these archaeological finds have demonstrated that Egyptians used different types of binders made from more or less pure gypsum. These binders covered a wide range, from the present calcium hemihydrate (plaster of Paris) to dry-bone gypsum, with all kinds of intermediate forms of soluble anhydrite. According to Martinet (1992) and Martinet et al. (1992), all the mortars used during the pharaonic period were made of calcium sulphate in which some admixtures were used (eggs, milk, wine, beer, animal blood, etc.). According to these authors, the first gypsums used were overburned (300°C), but later ones were prepared, like present gypsum, at temperatures between 120 and 160°C.

Egyptian gypsum technology was borrowed by the Cretans and the Greeks before reaching Rome (Stark and Wicht 1999). However, it seems that gypsum technology disappeared for a while after the fall of the Roman Empire. Significant use of gypsum resumed during the eleventh century in monasteries and later on in Roman and Gothic cathedrals (Luxán et al. 1995). During the thirteenth century, a French royal charter mentions the exploitation of eighteen different gypsum quarries in the Paris area (Chanvillard 1999). After the great London fire of 1666, in France Louis XIV signed a decree obliging architects to protect all wood construction with gypsum so that they could resist fire (Chanvillard 1999; Torres and Emeric 1999).

In 1768, Lavoisier presented the results of the first scientific work on gypsum at the French Académie des Sciences. This first scientific study led to the major scientific work done by Le Chatelier (Chanvillard 1999; Torres and Emeric, 1999) that explained the hardening process that transforms hemihydrate (plaster of Paris) back into gypsum.

Gypsum was a very successful construction material during the Baroque and rococco periods in the eighteenth century and its 'Belle Époque' occurred between 1900 and 1928 (Torres and Emeric 1999).

It is easy to make gypsum (Fritsch 1923). Gypsum deposits are found in many places at ground surface because they resulted from the evaporation of lakes or interior seas (geologists qualify them as evaporitic formations). They are quite common in the Middle East where the first human settlements were established (Antolini 1977).

Gypsum dehydration involves a very low heating up to 120 to 160°C, and its rehydration occurs in a matter of minutes at ambient temperature (Foucault 1977; Murat 1977). Figure 2.7 illustrates the gypsum cycle.

As a construction material, gypsum presents a significant advantage due to its good fire resistance, which is why it is presently used worldwide in the very convenient form of gypsum board (Kuntze 1987; Torres and Emeric 1999). Gypsum board is a particularly performing composite material which combines the good tensile strength of paper and the low compressive strength of gypsum to make a very handy and versatile construction material.

From a production point of view, gypsum is an ideal material than can be formed easily in automated production lines (Anonymous 1982; Olejnik 1999). In a modern gypsum plant (Figure 2.8), 12.5- to 16-mm thick gypsum boards are produced at a speed of 15 km/h. The gypsum board hardens in four minutes on a 300-m long production line. The boards are then cut to the desired

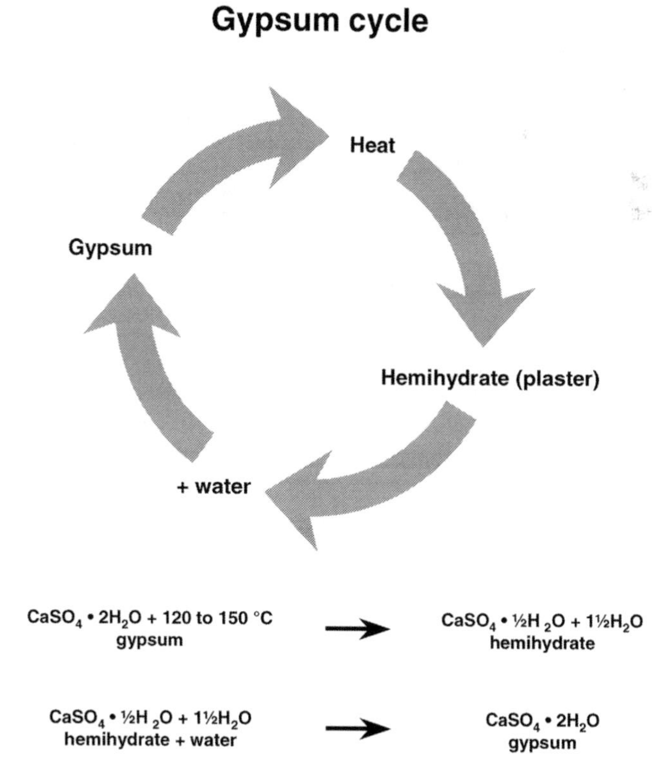

Figure 2.7 Gypsum cycle.

Figure 2.8 A fully automated gypsum factory (courtesy of PABCO in Las Vegas, photo by Ben Zimmerman).

length and dried for 40 minutes to eliminate non-stoechiometric water. In the USA, when the construction industry is running full speed, a gypsum plant can operate 360 days per year, 24 hours per day, under the supervision of only six persons for two production lines. To these six persons, four more are responsible for shipping. Every week there is a four-hour shutdown to carry out necessary maintenance.

Presently, gypsum is also used as a decorative material. Gypsum, however, has a weakness as a construction material: it loses almost all its resistance under humid conditions due to its partial dissolution. This is why the most ancient uses of gypsum are found inside Egyptian temples and tombs, where gypsum was well protected from humidity in the nearly desert climate of Egypt.

2.3 The use of lime

As previously mentioned, according to Kingery (1980), lime technology came long before pottery. The latest archaeological discoveries show that the use of lime as a binder was discovered during the Neolithic period, 10 000 to 20 000 years ago (Papadakis and Venuat 1966; Felder-Casagrande et al. 1997). Malinowski and Garfinkel (1991) also share this view. The mastering of the firing of clays dates back only to 4000 years BC. Very old lime mortars have been found in Crete (Blezard 1998); Cyprus, near Lanarca, and in Athens (Draffin 1943).

However, the use of lime was not widespread. For example, in Egypt, lime was only used under Ptolemy, 300 years before Christ, when the Greeks brought lime technology with them (Martinet et al. 1992). It is possible that it is the lack of combustible materials that explains this very late use of lime (Deloye 1996; Blezard 1998), because Egypt was not deprived of limestone.

The use of lime and concrete was not restricted to the Eastern Mediterranean or the Middle East. Small lime masonry walls were built 200 years before Christ in Bali in Indonesia (Deloye 1996), and Rivera-Villareal and Krayer (1996) have recently shown that Pre-Columbian Americans used lime as a binder.

During the Middle Ages, it seems that lime technology regressed, as has been found by studying lime mortars from different English castles built during the ninth, tenth and eleventh centuries. Lime was poorly calcinated and badly mixed (Bogue 1952). However, more recent work has shown that this was not always the case (Adams et al. 1993).

Lime preparation did not evolve much until vertical kilns were used. These vertical kilns were alternatively loaded from the top by layers of limestone and wood or wood charcoal. Lime was recovered at the bottom of the kiln (Deloye 1996).

Quite early, it was found that lime could be 'diluted' with sand to produce a handy mortar to assemble irregular stone blocks or pebbles (Draffin 1943; Deloye 1996; Blezard 1998; Koui and Ftikos 1998). Hydrated lime processing was for a long time a quite controversial subject: should lime be slaked just before its use or would it be better to slake it as soon as it is produced and keep it under water until use? During the construction of the Palace of Versailles, Loriot and de La Faye disagreed on the subject, Loriot saying that lime had to be slaked only just before its use, while de La Faye thought that lime should be slaked as soon as possible and kept under water, as recommended by Faujas de Saint-Fond (1778).

Progressively, masons started to differentiate between fat limes obtained from pure limestone and lean limes produced with impure limestone. Lean limes were not as easy to use, but they were found to be more durable (Draffin 1943; Bogue 1952; Blezard 1998). During the construction of the Eddystone Lighthouse, J. Smeaton conducted a survey of the binding qualities of the different limes available in England to find the one having the highest strength when immersed in seawater. This was the first known systematic work on improving the quality and durability of a masonry work.

Producing lime is quite simple as it involves only the heating of limestone blocks up to 850°C, a temperature easily obtained by firing wood (Figure 2.9) (Candlot 1906). However, when compared to the firing of gypsum, it is more energy demanding, which is why it was only used very late in the Middle East and Egypt where they lacked fuel, except for Phoenicia and Greece where wooded areas were easily accessible. To harden, hydrated lime needs the CO_2 present in the atmosphere, which is why in the centre of recently unearthed walls in Pompei, uncarbonated lime has been found (Adam 1995).

Lime is a binder which needs air to be transformed back into calcium carbonate. One of the main drawbacks of lime as a binder is that it loses part of its strength when it comes into contact with water due to partial dissolution of calcium carbonate. This strength loss is not as important as that of gypsum, but it is significant enough to limit the use of lime to construction. The durability of roads, bridges, pillars and wharfs built with lime mortars is essentially a function of the environmental conditions to which they are exposed.

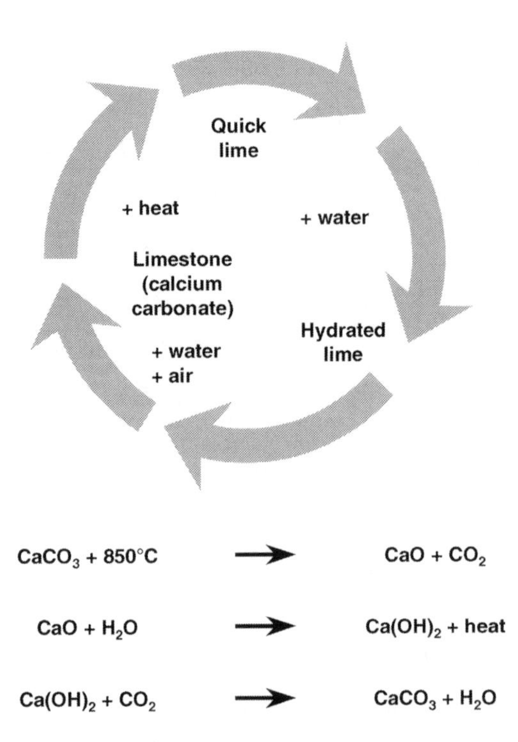

Figure 2.9 Calcium carbonate cycle.

Presently, lime is still used as a construction material by masons, who mix it half-and-half with Portland cement to make their mortars.

2.4 Hydraulic binders based on pozzolans

Although pozzolanic mortars and concretes are only very specific lime mortars, they played a key role as construction materials, because they were the first hydraulic binders that hardened under water. Pozzolanic binders represent a technological breakthrough, because they do not lose strength when in contact with water. When pozzolanic mortars or concrete are well compacted they can reach, in a matter of days, a compressive strength of 10 MPa (Mark 1987).

Recent archaeological discoveries tend to prove that pozzolanic materials were known during the Neolithic period (Malinowski and Garfinkel 1991), but this technology was only exploited by the Phoenicians on a small scale and later on by the Romans on a much larger scale (Papadakis and Venuat 1966).

The Phoenicians, and perhaps other people before them, discovered that fired clay wastes could be used to make particularly performing mortars (Figure 2.10). As for the Greeks, they mixed lime with the volcanic tuff of the Island of Thera, presently known as Santorin earth. The Romans used a red to

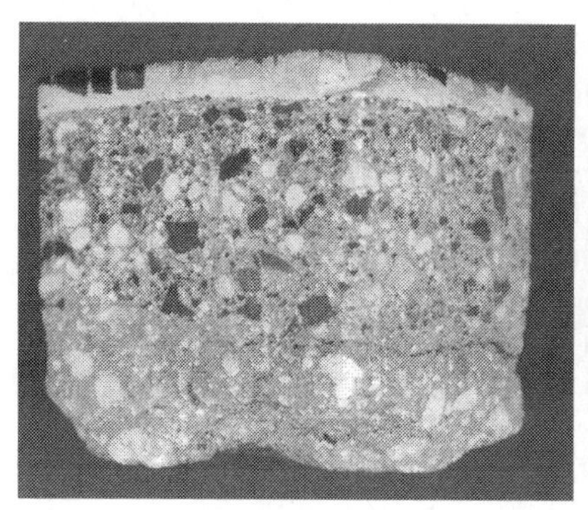

Figure 2.10 Piece of Carthaginean floor, 2,500 years old. This piece of floor is composed of three layers: a layer of clay treated with coarse lumps of hydrated lime, a layer of a mortar made from pieces of burnt clay tiles or amphoras mixed with lime lumps smaller than in the previous layer. These two layers are topped with a mosaic of small cubes of marbles bound by a hardened slurry of fine grained lime. The mosaic represented a rooster made of thirteen marbles of different colours. These marbles came from different quarries located all over the Roman Empire. Adam (1995) lists the principal quarries exploited by the Romans and the colour of each marble. This floor was demolished by a back-hoe operator who was looking for a treasure that could have been hidden under this floor.

purple volcanic tuff found in abundance in the Baie area near modern Naples, in a city called Puteoli, whose name in modern Italian is Puzzoli (Vitruvius 1979). It is the modern name of this city that has given us the name pozzolan. A pozzolan is a natural material that reacts with hydrated lime at ambient temperature to form insoluble calcium silicate hydrate, similar to those produced when hydrating Portland cement (Blezard 1998; Sabbioni et al. 1998).

As these mortars hardened under water, they were used to build roads, bridges and to line aqueducts. The Romans understood very well the economic and military potential of this material and they used it all over their empire, wherever they found pozzolanic materials (Papadakis and Venuat 1966; Malinowski 1982; Hill 1984; Rassineux et al. 1989; Thomassin and Rassineux 1992; Puertas et al. 1994 a and 1994 b; Steiger 1995; Adam 1995; Harries 1995; Lamprecht 1995; Sabbioni et al. 1998). For example, an aqueduct near Cologne in Germany was built using a local trass. The linings of the Pont du Gard and Segovia aqueduct are made of a pozzolanic mortar. In these two cases, it is important to note that pozzolanic mortars were not used to build the aqueduct but rather as a lining of the water canal, because pozzolanic material had to be imported over a long distance. It was cheaper to have limestone blocks squared by slaves or by soldiers to build the support of these aqueducts. However, in some areas where pozzolanic materials were more abundant, they were used to build arches and domes (Favre and de Castro 2000).

The discoveries made in Herculanum and Pompeii, two Roman cities in the Bay of Naples destroyed by an eruption of Mount Vesuvius, are particularly interesting because these two cities were under reconstruction due to their partial destruction by an earlier earthquake. These two cities were so suddenly submerged by several metres of ashes on 25 August 79 BC that some of the people died as they performed their daily tasks.

Roman concretes were dry concretes that were compacted with rams (Adam 1995; Lessing and Varone 1995). Pliny, cited by Bogue (1952), mentions that the floors of cisterns had to be compacted with iron rams. Rondelet (1805) concluded that the longevity of Roman mortars was essentially linked to the energy used to compact them. It was 2000 years before Féret and Abrams rediscovered the importance of voids on concrete compressive strength. It is also known that Roman masons mixed lime with egg whites and animal blood (Venuat 1984), and it is now possible to explain why blood could be used as a dispersent (or water reducer) when making a mortar or a concrete.

It would be unfair to end this paragraph on pozzolanic mortars and concretes without presenting the Pantheon in greater details. In many ways, this concrete structure is still remarkable. It has been studied under different angles and has been the subject of numerous papers (Mark and Hutchinson 1986; Mark 1987; Mark 1990; Adam 1995). Le Corbusier (1958), the famous Swiss architect and engineer, was a great admirer of the Pantheon.

Although very few domes older than the Pantheon still exist, it seems that the largest one ever built before the Pantheon was a dome in a bath complex in the Bay of Naples; its diameter was only 21.5 m, that is a little less than half the diameter of the Pantheon.

For centuries, no architect or engineer built a dome larger than the Pantheon. Michelangelo designed the dome of St Peters in Rome with a diameter of 42 m and it was necessary to rapidly reinforce its base with an iron chain to stop its premature cracking. It was only in 1889 that the span of the Pantheon's dome was exceeded during the construction of a glass and iron arch in the Galerie des Machines (span of 113 m) at the Paris Universal Exhibition. Presently, concrete pre-stressing has resulted in the construction of larger arches or vaults, the largest free span concrete roof presently in existence is the CNIT in La Défense, in Paris. It has a free span of 219 m (Anonymous 1958).

A finite element study conducted on the Pantheon's dome has shown the great importance of the step construction one can see on the outside of the dome. From a stress point of view, most of the concrete is working in compression, except in very few areas where 0.06 to 0.13 MPa tensile stresses are found (Mark and Hutchinson 1986). As far as the waffle pattern seen in the internal face of the vault is concerned, it has been found that it served only a decorative function, as it is too superficial to reduce significantly the stresses within the vault. Mathematical models have also been used to study the effects of the sun on the dimensional variations that can be induced in the dome.

From a material point of view, the Pantheon's dome is also remarkable. It was built using six concretes with different unit masses (Figure 2.11). The foundations were built using a 2200 kg/m^3 concrete, that is almost the unit mass of a modern concrete. The walls that were covered with decorative clay bricks

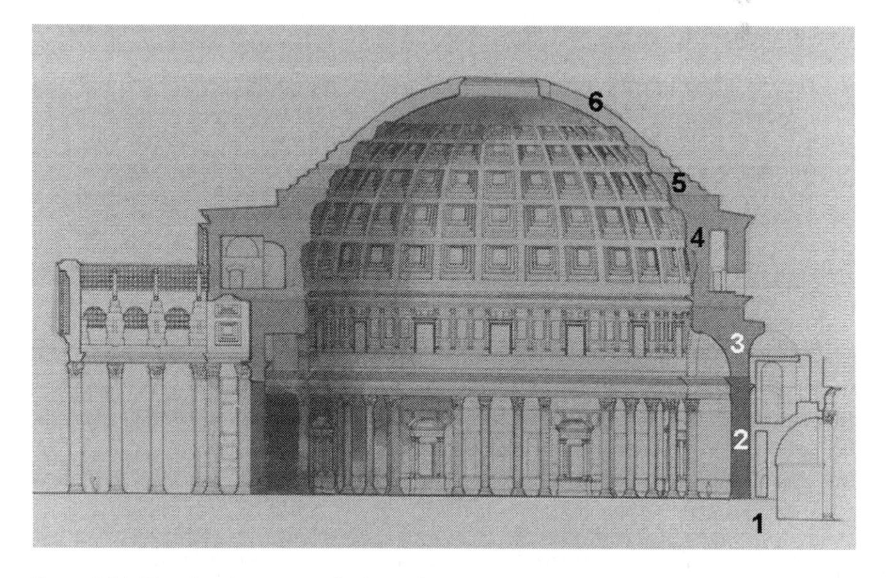

Figure 2.11 The Pantheon was built with six concretes having different unit masses (courtesy of M. Collins 2001, with permission of ACI).

were built using a lighter concrete having a unit mass of 1750 kg/m^3. At the base of the dome, the concrete unit mass is 1600 kg/m^3 and at the top, near the oculus, it is only 1350 kg/m^3. Pumice was used as an aggregate to lighten the structure.

For Mark and Hutchinson (1986), the use of concretes having different unit masses was the key to success in the construction of so huge a dome, especially when compared to a dome built with clay bricks. Only Christopher Wren succeeded in designing a lighter dome when he built St Paul's Cathedral in London, because he had subdivided the construction of the roof into three parts, each having a particular architectural or structural function (Figure 2.12). If the same constructive techniques used by Wren had been applied to the Pantheon, it would have been possible to reduce the average thickness of the Pantheon's dome to 0.65 m instead of its actual thickness of 1.5 metres, but this takes nothing away from the merit of the Pantheon's builders.

2.5 The discovery of Portland cement

It is surprising that the great construction activity that occurred during the so-called Roman and Gothic periods did not result in any resurrection of concrete as a building material, because the use of lime was not totally lost. Lime and gypsum were only used as support for frescoes and not very often in mortars.

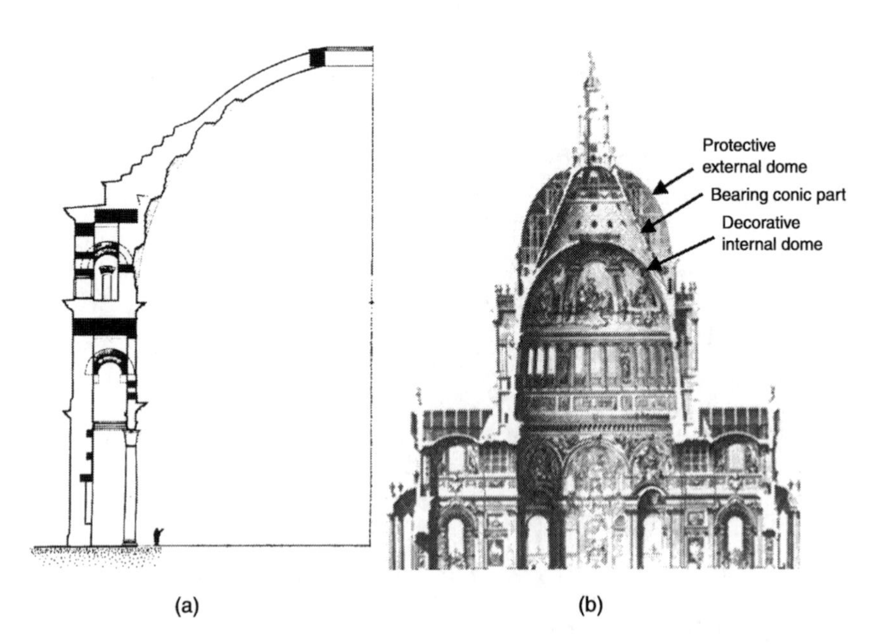

Figure 2.12 Comparison of the structural concept of two famous domes: the monolithic Pantheon (a) and the one of St Paul's Cathedral in London (b) (Mark 1990, courtesy of McGraw-Hill Companies).

Time was not a concern when building a cathedral, and stone-cutters were numerous and probably badly paid.

It is only at the end of the eighteenth century that a certain interest in the use of lime reappeared. As previously mentioned, the first known scientific work done on lime mortars was carried out by John Smeaton, who was hired to rebuild the Eddystone Lighthouse in Plymouth, England (Figures 2.13 and 2.14) (Bogue 1952; Papadakis and Venuat 1966). Smeaton found that usual lime mortars lacked time to harden before being immersed by high tide and exposed to the action of waves, therefore, he decided to compare the hardening properties of the different limes produced in the south of England. He found that some mortars made from impure limestone rich in clay hardened faster and were more durable than those made with pure limestone. Smeaton was aware that the Romans used to make durable concrete by mixing pozzolanic

Figure 2.13 Eddystone Lighthouse, rebuilt on land of Plymouth Hoe (courtesy of Elizabeth Neville).

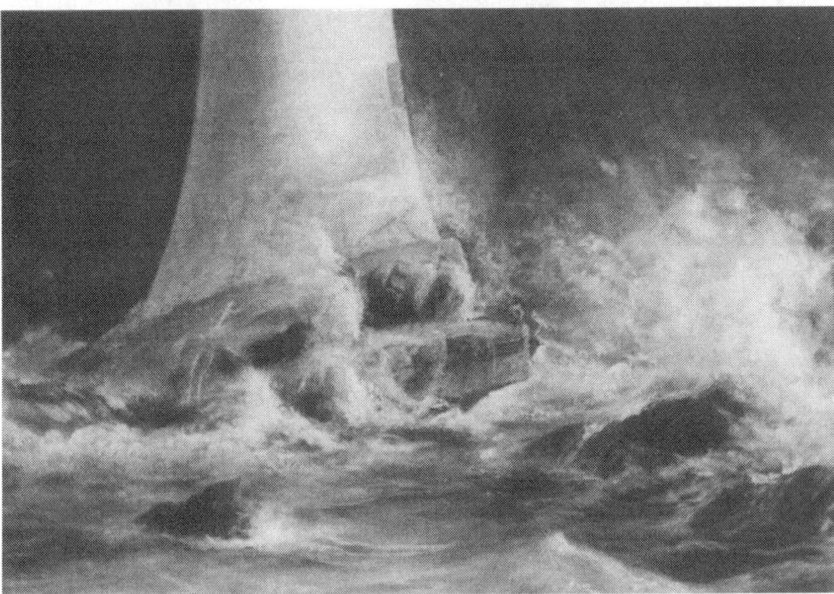

Figure 2.14 Eddystone lighthouse painted by Anton Melbye. Anton Melbye stayed at Napoleon III's court in Paris and taught painting to Camille Pissarro (courtesy of G. Idorn).

materials with lime, and therefore he decided to import pozzolans from Civita Vecchia in Italy to strengthen his mortars. All his work and observations were consigned in a report written in 1791 (Draffin 1943; Bogue 1952; Blezard 1998). In order to make the base of the lighthouse stronger, he gave it the shape of an oak (Chanvillard 1999). He selected granite blocks for the exterior of the lighthouse wall and Portland limestone for its inner part. Portland limestone had a great reputation as a construction material in England at that time; it is the stone that was used to build Westminster Bridge (Chanvillard 1999).

In 1882, the Eddystone lighthouse was replaced by a more modern and higher lighthouse, but considering its historical character, it was decided to dismantle it stone by stone and rebuild it on the shore at Plymouth Hoe, where it can be admired today. At low tide, near the modern lighthouse, it is possible to see the base of the Smeaton lighthouse on the ocean bottom.

Smeaton was not the only European to take an interest in lime. For example, in Sweden, Bergmann wrote that it was necessary that lime contain some manganese to harden under water. This finding was opposed by Collet-Descotils of the École des Mines in Paris, who showed that it was more important that the lime contained small particles of silica (Bogue 1952). Loriot, Faujas de St Fond, Guyton de Morveau, General Treussart, Berthier, and Chaptal all studied lime. Saussure in Switzerland, Higgins in England and Gersdorf and Fuchs in Germany also took an interest in lime (Blezard 1998).

In 1796, the first hydraulic binder was patented in England by Parker, who called his binder Roman cement (Papadakis and Venuat 1966). This Roman cement was in fact a natural cement made by burning a clayish limestone containing approximately 30 per cent clay. The first plant to produce such a binder was built in 1810 (Papadakis and Venuat 1966). Another patent was later granted to Dobbs in 1811. In 1810, the Science Society of the Netherlands organized a contest aimed at finding an explanation for the fact that the lime produced from limestone was far superior to the lime produced by calcining shells. It was a chemistry professor from Berlin University by the name of John who won the contest medal. He proved that it is was the presence of clay, silica and iron oxide that improved the quality of certain limestones (Blezard 1998).

Vicat's work, that began in 1812, constitutes a real milestone in the determination of the optimum proportions of clay and limestone in producing an hydraulic lime (Appendix I). The results of his research were published in 1818 and 1828. Vicat was unable to provide the exact proportions of silica, alumina and iron oxide that a limestone should contain to make a strong hydraulic binder, he only indicated upper and lower limits: one fifth and one fourth. He also confirmed Smeaton's findings that the best limestones were those containing clay. The first hydraulic lime plant was built in Nemours, France in 1818.

Concommittently with these first works on lime, Parker in England and Lesage in Boulogne-sur-Mer in France began firing beach pebbles at a high temperature to produce hard nodules, which were ground. The powder obtained hardened under water. In 1812, Frost took a patent on a so-called Britannic cement which was very similar to the binder produced by Parker and Lesage.

Vicat took an interest in limestone marls and characterized them according to their hydraulicity index (Papadakis and Venuat 1966). He found that when limestone contains more than 27 to 30 per cent clay, it is quite difficult to calcine it to produce lime; he found that it behaved instead as a natural cement that hardened as fast as calcium hemihydrate.

In spite of the fact that the first hydraulic limes and natural cements could have irregular properties due to the lack of comprehensive knowledge regarding the control of their ideal composition, their optimum firing temperature and their grinding, hydraulic limes and natural cements started being used in Europe, in Russia and in the USA. In particular, natural cement was used during the construction of the Lake Erie canal in 1818.

On 21 October 1824, Joseph Aspdin, an English mason, applied for the first patent on an improved cement he called Portland cement, because when it had hardened, its colour was like that of the well-known limestone of Portland island (Appendix II). Careful reading of Aspdin's patent shows that he definitely never produced what we presently call Portland cement (Bogue 1952; Blezard 1998). In 1911, at 101 years of age, Johnson wrote that Aspdin's Portland cement looked no more like modern cement than lime looks like cheese (Bogue 1952). It is only around 1850 that William Aspdin, Joseph Aspdin's son, succeeded in producing what Blezard (1998) calls a proto-Portland cement when he fired a mixture of limestone and clay at a sufficiently high temperature to produce tricalcium silicate.

In France, the first major plant to produce hydraulic lime was built in 1830 in Le Teil, in the South of France, by Pavin de Lafarge. At the beginning, in 1839, this plant had only two 5-metre-high vertical kilns that produced 3000 tonnes of lime. In 1865, the Le Teil plant received an order for 120000 tonnes for the construction of the Suez canal (Papadakis and Venuat 1966).

Around 1860, the hard nodules found when sieving slacked lime were ground. These nodules called 'grappier' were ground between hard stone wheels (Papadakis and Venuat 1966) in Lafarge's Le Teil plant to make a slow hardening hydraulic cement.

The binders used evolved from lime to hydraulic lime, from hydraulic lime to natural cement and from natural cement to Portland cement. This technological evolution was the result of a mixture of scientific work, fortuitous discoveries, luck, and intuition (Table 2.1). To consider Joseph Aspdin as the inventor of modern Portland cement is less and less an accepted fact. Blezard (1998) even contests the use of the name Portland, because according to him, Higgins and Smeaton had previously compared their hardened hydraulic lime to the limestone of Portland island. However, it is Joseph Aspdin who, for the first time, associated the words Portland and cement in a patent.

Table 2.1 Chronology of Portland cement discovery

1749	**Holbeig** mentions the fabrication of a 'roman cement' in Bornholm Island in Baltic Sea (Idorn 1997) (hydraulic lime).
1791	**John Smeaton** presents his conclusions on the reconstruction of Eddystone lighthouse (hydraulic lime).
1796	**James Parker** patents a roman cement – a natural cement made from the calcination of nodules found in clay limestone. Lesage produces a similar cement at Boulogne-sur-Mer (natural cement).
1811	**Edgar Dobbs** patents a cement obtained by calcining a mixture of limestone and clay (natural cement).
1818	**Louis Vicat** found the proportions of limestone and clay to produce 'chaux factice' following a scientific process (proto-Portland cement*).
1822	**James Frost** patents a Britannic cement (natural cement).
1824	**Joseph Aspdin** patents the fabrication of Portland cement. According to Blezard it is only an hydraulic lime (proto-Portland cement*).
1840–1850	**William Aspdin** and **Johnson** make for the first time a calcium silicate cement (meso-Portland cement*).

* Expression proposed by Blezard (1998)

The first French Portland cement plant was built in 1840 in Boulogne-sur-Mer. In 1871, 500 barrels (85 tonnes) of English Portland cement were shipped to the USA. English exports of Portland cement increased to a peak of 3 millions barrels (510 tonnes) in 1896, after which these exports began to decrease because local US production of natural cement increased. Cement barrels were often used to ballast ships (Dumez and Jeunemaître 2000).

In 1898, French production of Portland cement amounted to 400 000 tonnes, produced in sixteen cement plants. In 1905, this amount increased to 550 000 tonnes, to which 400 000 tonnes of natural cement and 60 000 tonnes of slag should be added (Papadakis and Venuat 1966).

In the USA, the first cement plant was built in 1865 in Copley, Pennsylvania. However, it was only between 1890 and 1899 that, for the first time, the production of Portland cement was greater than the production of natural cement in the USA.

In 1900, the world production of Portland cement was in the order of 10 million tonnes, as will be seen in the next chapter, and the production of natural cement was already marginal (2 per cent of cement production).

It is only in 1900 that the first rotary cement kilns replaced vertical kilns and that the first ball mills were used. In 1901, the Atlas cement company in Pennsylvania was operating fifty-one rotary kilns having a diameter of 2 metres and a length of 20 metres. Each kiln produced 25 tonnes of clinker per day (Papadakis and Venuat 1966). It is far from the 10 000 tonnes per day that some kilns presently produce, but the use of the rotary kiln to produce Portland cement was a decisive move toward an increase in mechanization.

Around the year 1900, the French chemist Giron, who was director of the Atlas cement plant in the USA, had the idea of adding, gypsum during clinker grinding to control Portland cement setting and hardening (Draffin 1943). This is the last major chemical innovation of the manufacturing process of Portland cement. From a technological and marketing point of view, it is a very important improvement, since by adding a small amount of gypsum it became possible to control the rheology, setting and hardening of Portland cement.

Of course, many other technological and mechanical innovations transformed the production of Portland cement into a very efficient and economical process. For example, around 1950, the dry process was developed, then precalcination was introduced around 1970 (Pliskin 1993) and more recently the use of alternative fuels. However, the fabrication of Portland cement remains a very simple pyrotechnical process, at least in principle: it consists in firing an adequately proportioned mixture of lime, silica, alumina and iron oxide at a temperature of about 1450°C in order to transform these four major oxides into the four active phases that constitute the essence of Portland cement.

For the first time in the history of construction materials, it was possible to manufacture an artificial binder able to harden under water, with which it is possible to produce a hardened artificial rock of any desired shape, stronger than many natural rocks. However, when looking at Portland cement production, it can be seen that this is not a cyclic process as it is for gypsum and lime, it is a linear process that transforms a well-proportioned mixture of clay and limestone into a multiphase material (Figure 2.15) that does not return back to its initial form once it hardens. However, as shown in Figure 2.16, if the time scale is changed to a geological time frame, most of the present concretes will one day end up as a mixture of limestone, clay and gypsum, which are the stable forms of the different mineral oxides used to make it.

2.6 The development of modern concrete

Isambard Brunel was one of the first engineers to use reinforced concrete during the construction of a tunnel under the Thames in 1828. In 1875, Lambot built a small boat displayed at the Paris Exhibition (Figure 2.17). In 1865, Joseph Monier patented concrete flower pots, reinforced with a steel mesh (Bosc 2000). However, it is Coignet (Steiger 1993) who can be considered as the real father of

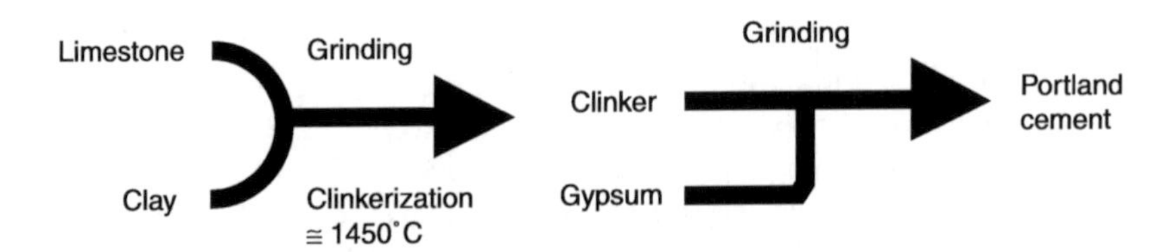

Figure 2.15 Principle of Portland cement manufacturing.

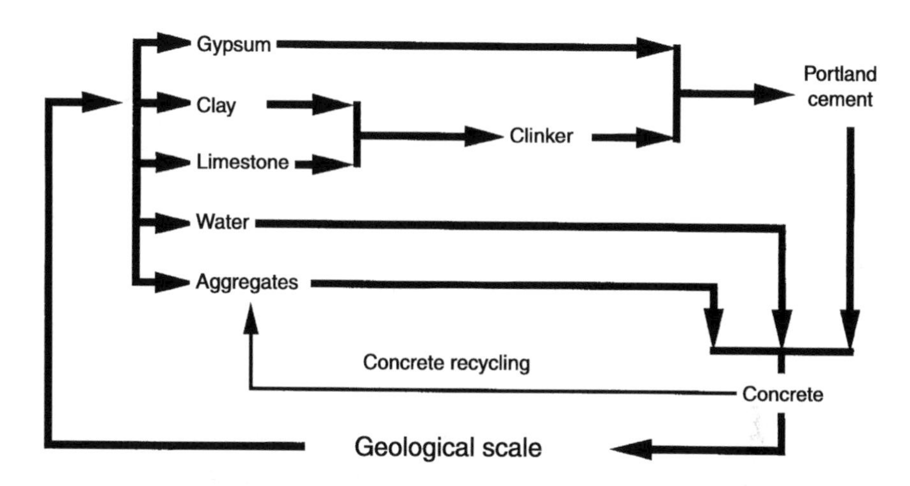

Figure 2.16 Durability of concrete on a geological scale.

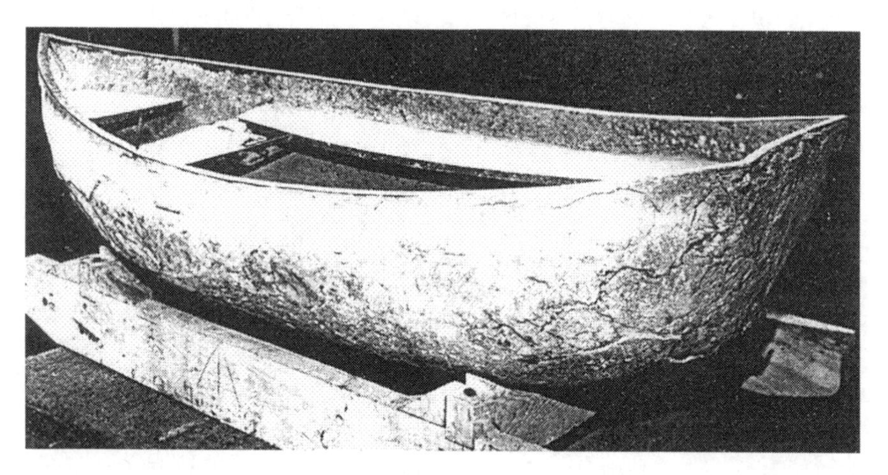

Figure 2.17 Lambot's concrete row boat (*Bâtir*, 45 (1955): 9). Photograph by H. Baranger.

modern concrete. In spite of the fact that Coignet was aware that concrete could be made with fat lime, he always used natural cement or Portland cement to make his concretes which he qualified as 'agglomerated'. Coignet recommended using the least water possible to obtain a viscous paste after a prolonged mixing period. The placing of agglomerated concrete required a strong compaction effort, layer by layer, in order to finally obtain a very dense material before its chemical hardening.

Coignet took an interest in the cleanliness of the sand; he recommended the use of sands containing less than 5 to 6 per cent clay; beyond this limit he recommended the washing of the sand. Moreover, he found that the gravel used to make his agglomerated concrete did not need to contain particles larger than a pea and should also not contain particles coarser than a hazelnut, the final objective being to minimize intergranular voids.

Coignet developed a special helicoidal mixer which he called a 'malaxator' that continuously mixed the ingredients of his agglomerated concrete. Finally, Coignet proposed to build a reinforced concrete bridge over the Seine, but his suggestion was passed over for a steel bridge construction, judged safer at that time.

Reinforced concrete structures started to be built in Europe under the pressure of engineers such as Hennebique, Coignet and Tedesco in France, and Koenen and Wayss in Germany. Concrete tensile strength was already being neglected in their calculations. Along with these theoretical calculations, some experimental work began in France, Germany and the USA, so that the first reinforced arch bridge was built in Copenhagen, Denmark in 1879 (Anonymous 1949) and one of the first bridges in reinforced concrete was built by Monier (Figure 2.18).

It is only quite recently that Freyssinet thought to prestress concrete (Freyssinet 1966). This simple technology actually combines the structural advantages of concrete and steel. Although it took a long time to see this revolutionary technology take off, it must be admitted that it had a significant

Figure 2.18 The first bridge in reinforced concrete: the Chazelet bridge by Monier (courtesy of J.-L. Bosc).

impact on the use of concrete, not only from a mechanical point of view, but also from an aesthetic point of view.

2.7 Conclusion

This has been my version of the history of hydraulic binders and cement. Other historical versions have already been written, others will be written in the future. Certainly if I had a better knowledge of the German, Spanish, Scandinavian, Russian, Chinese and Japanese languages, I would have written a quite different version, at least in its modern part. However, the essential purpose of this chapter was not to present a chauvinistic approach to Portland cement, hydraulic binders and the history of concrete but rather to show how both simple and complex materials were developed over centuries as a consequence of fortuitous discoveries, patient observation, trial and error and, in some cases, through a scientific approach.

The discovery of Portland cement was not a pure coincidence but, rather, the result of a slow evolution; its manufacturing process is now well under control. However, as with any other human activity, there is still room for improvement as far as the performance, consistency and ecological impact of Portland cement are concerned. The search for improvement of the manufacturing process and performance of Portland cement is one of the challenges that the cement industry will have to face in the years to come, as will be seen later.

For readers interested in a more complete historical background of the evolution of hydraulic binders through the ages, the following references are recommended: Davis 1924; Draffin 1943; Bogue 1952; Bealy and Duffy 1987; Blezard 1998; Idorn 1997. These versions are slightly different in some details essentially because they do not necessarily focus on the same developments.

References

Adam, J.-P. (1995) *La Construction romaine: Matériaux et techniques*, Paris: Les Grands Manuels Picards.

Adams, J., Dollimore, D., and Griffiths, D.L. (1993) 'Thermal Analytical Investigation of Ancient Mortars from Gothic Churches', Proceedings of the Tenth International Congress on Thermal Analysis, *Journal of Thermal Analysis*, 40 (1): 275–84.

Anonymous (1949) 'Cent ans de béton armé', *Travaux*, 194: 34–41.

—— (1958) 'Visite du CNIT (Centre National des Industries et des Techniques', *Travaux*, September: 886–7.

—— (1982) *Le Plâtre: Physico-chimie, fabrication et emploi*, Paris: Eyrolles.

Antolini, P. (1977) 'The Geology of Calcium Sulphates: Gypsum and Anhydrite', Colloque International de la RILEM, St-Rémy-les-Chevreuses, pp. 1–25.

Baronio, G, Binda, L., and Lombardini, N. (1997) 'The Role of Brick Pebbles and Dust Inconglomerates Based on Hydrated Lime and Crushed Bricks', *Construction and Building Materials*, 11 (1): 33–40.

Bealy, M. and Duffy, J.J. (1987) 'Concrete facts', *ASTM Standardization News*, September, pp. 28–31.
Blezard, R.G. (1998) 'The History of Calcareous Cements', in P.C. Hewlett (ed.) *Lea's Chemistry of Cement and Concrete*, London: Arnold, p. 6–23.
Bogue, R.H. (1952) *La Chimie du ciment Portland*, Paris: Eyrolles.
Bonen, D., Tasdemir, M.A., and Sarkar, S.L. (1995) 'The Evolution of Cementitious Materials through History', *Material Research Society Symposium, Proceedings* 370: pp. 159–68.
Bosc, J.-L. (2000) 'Du ciment armé au béton armé: la contribution oubliée de Joseph Monier', *Annales du Bâtiment et des Travaux Publics*, 1, February: 37–43.
Campbell, D.H. and Folk, R.L. (1991) 'The Ancient Egyptian Pyramids: Concrete or Rock?' *Concrete International*, 13 (8): 28–39.
Candlot, E. (1906) *Ciments et chaux hydrauliques: fabrication, propriétés, emploi*, Paris: La Librairie Polytechnique, Ch. Béranger.
Chanvillard, G. (1999) *Le Matériau béton: connaissances générales*, Lyon: ENTPE ALÉAS.
Clarke, S. and Engelbach, R. (1930) *Ancient Egyptian Masonry: The Building Craft*, Oxford: Oxford University Press.
Davidovits, J. (1987) 'Ancient and Modern Concretes: What Is the Real Difference', *Concrete International*, 9 (12): 23–35.
Davis, A.C. (1924) *A Hundred Years of Portland Cement*, London: Concrete Publications.
de Bélidor, B.F. (1788) *Architecture hydraulique*, Paris.
de la Faye (1777) *Recherches sur la préparation que les Romains donnoient à la chaux dont ils se servoient pour leurs constructions, et sur la composition et l'emploi de leurs mortiers*, Paris.
Deloye, F.-X. (1996) 'La Chaux à travers les âges', *Bulletin des laboratoires des Ponts et Chaussées*, 201, (January–February): 94–8.
Draffin, J.O. (1943) 'A Brief History of Lime, Cement, Concrete, and Reinforced Concrete', *University of Illinois Bulletin*, 40 (45, 29 June): 5–38.
Dumez, H. and Jeunemaître, A. (2000) *Understanding and Regulating the Market at a Time of Globalization: The Case of the Cement Industry*, Basingstoke: Macmillan Press.
Duriez, M. and Arrambide, J. (1960) *Nouveau traité de matériaux de construction*, Paris: Dunod.
Faujas de Saint-Fond (1778) *Recherches sur la pozzolane sur la théorie de la chaux et sur la cause de la dureté du mortier*, Grenoble and Paris.
Favre, R. and de Castro, J. (2000) 'La Pérennité de l'arc dans la construction', *Revue IAS Ingénieurs et Architectes Suisses*, 13 (July 5).
Felder-Casagrande, S., Wiedemann, H.G., and Reller, A. (1997) 'The Calcination of Limestone: Studies on the Past, the Present and the Future of a Crucial Industrial Process', *Journal of Thermal Analysis*, 49: 971–8.
Foucault, M. (1977) 'Le plâtre', *Colloque International de la RILEM*, St-Rémy-les-Chevreuses, pp. 271–84.
Freyssinet, E. (1966) *Un demi-siècle de technique française de la précontrainte*, Vol. I, Paris: Éditions Science et Industrie.
Fritsch, J. (1923) *Le Plâtre: Fabrication, propriétés, applications*, Paris: Librairie Générale Scientifique et Industrielle Desforges.

Ghorab, H.Y., Ragai, J., and Antar, A. (1986) 'Surface and Bulk Properties of Ancient Egyptian Mortars. Part I: X-ray Diffraction Studies', *Cement and Concrete Research*, 16 (6): 813–22.

Gourdin, P. (1984) 'La Cimenterie', *Revue Usine Nouvelle*, May: 146–55.

Güleç, A. and Tulun, T. (1997) 'Physico-Chemical and Petrographical Studies of Old Mortars and Plasters of Anatolia', *Cement and Concrete Research*, 27 (2): 227–34.

Harries, K.A. (1995) 'Concrete Construction in Early Rome', *Concrete International*, 17 (1, January): 58–62.

Hill, D. (1984) *A History of Engineering in Classical and Medieval Times*, La Salle, Ill.: Open Court.

Idorn, G.M. (1997) *Concrete Progress from Antiquity to the Third Millenium*, London: Thomas Telford.

Kingery, W.D. (1980) 'Social Needs and Ceramics Technology', *Ceramic Bulletin*, 59 (6): 598–600.

Koui, M. and Ftikos, C. (1998) 'The Ancient Kamirian Water Storage Tank: A Proof of Concrete Technology and Durability for Three Milleniums', *Matériaux et Constructions*, 31 (November): 623–7.

Kuntze, R.A. (1987) 'Gypsum: Its History and Use in Fire Rated Construction', *Construction Canada*, (1, January): 14–15.

Lamprecht, H.-O. (1995) 'Rationalisation grâce au béton romain', *Betonwerk+ Fertigteil-Technik*, (4): 124–33.

Le Corbusier (1958) *Vers une architecture*, Paris: Éditions Vincent, Fréal et Cie.

Leduc, E. (1925) *Chaux et ciments*, Paris: Librairie J.-B. Baillière et Fils.

Lessing, E. and Varone, A. (1995) *Pompéi*, Paris: Pierre Terrail.

Lucas, A. and Harris, J.R. (1962) *Ancient Egyptian Materials and Industries*, London: Edward Arnold.

Luxán, M.P., Dorrego, F., and Laborde, A. (1995) 'Ancient Gypsum Mortars from St. Engracia (Saragoza, Spain): Characterization, Identification of Additives and Treatments', *Cement and Concrete Research*, 25 (8): 1755–65.

Malinowski, R. (1979) 'Concretes and Mortars in Ancient Aqueducts', *Concrete International*, 1 (1): 66–76.

Malinowski, R. (1982) 'Durable Ancient Mortars and Concretes', *Nordic Concrete Research*, 1: 19.1–19.22.

Malinowski, R. and Garfinkel, Y. (1991) 'Prehistory of Concrete', *Concrete International*, 13 (3, March): 62–8.

Mark, R. and Hutchinson, P. (1986) 'On the Structure of the Roman Pantheon', *The Art Bulletin*, 68 (1, March): 24–34.

Mark, R. (1987) 'Reinterpreting Ancient Roman Structure', *American Scientist*, 75 (2 March–April): 142–9.

Mark, R. (1990) *Light, Wind and Structure: The Mystery of the Master Builders*, New York McGraw-Hill.

Martinet, G. (1991) 'Les Mortiers d'époque pharaonique à Karnak', *Bulletin de Liaison des Laboratoires des Ponts et Chaussées*, 172 (March–April): 157.

Martinet, G. (1992) *Grès et mortiers du temple d'Amon à Karnak (Haute Égypte): Étude des altérations, aide à la restauration*, Paris: Laboratoire central des Ponts et Chaussées.

Martinet, G. Deloye, F.-X., and Golvin, J.-C. (1992) 'Charactérisation des mortiers pharaoniques du temple d'Amon à Karnak', *Bulletin de Liaison des Laboratoires des Ponts et Chaussées*, 181 (September–October): 39–45.

Mary, M. (2000) *Louis Vicat invente le ciment artificiel*, Lonrai: Ciments Vicat. First published 1903.

Morris, M. (1991) 'The Cast-in-Place Theory of Pyramid Construction', *Concrete International*, 13 (8): 29, 39–44.

Murat, M. (1977) 'Structure, cristallographie et réactivité des sulphates de calcium', Colloque International de la RILEM, St-Rémy-les-Chevreuses, pp. 59–172.

Olejnik, R. (1999) 'Moderne Anwendungen von Gipsprodukten', *ZKG International*, 12: 649–53.

Papadakis, M., Venuat, M., and Vandamme, J. (1970) *Industrie de la chaux, du ciment et du plâtre*, Paris: Dunod.

Papadakis, M. and Venuat, M. (1966) *Fabrication et utilisation des liants hydrauliques*, n.p.

Pliskin, L. (1993) *La Fabrication du ciment*, Paris: Eyrolles.

Puertas, F., Blanco-Varela, T., and Palomo, A. (1994a) 'Stuccos and Roman Concretes of the Baelo Claudia city (Cádiz): Characterization and Causes of Decay', *Materiales de construcción*, 44 (236): 15–29.

Puertas, F., Blanco-Varela, M.T., Palomo, A., Ortega-Calvo, J.J., Ariño, X., and Saiz-Jimenez, C. (1994b) 'Decay of Roman and Repair Mortars in Mosaics from Italica, Spain', *The Science of Total Environment*, Amsterdam: Elsevier, Vol. 153, pp. 23–31.

Ragai, J. (1988a) 'Surface and Bulk Properties of Ancient Egyptian Mortars. Part III: X-ray Diffraction Studies (b)', *Cement and Concrete Research*, 18 (1): 9–17.

Ragai, J. (1988b) 'Surface and Bulk Properties of Ancient Egyptian Mortars. Part IV: Thermal Studies', *Cement and Concrete Research*, 18 (2): 179–84.

Ragai, J. (1989) 'Surface and Bulk Properties of Ancient Egyptian Mortars. Part V', *Cement and Concrete Research*, 19 (1): 42–6.

Ragai, J., Ghorab, H.Y., and Antar, A. (1987) 'Surface and Bulk Properties of Ancient Egyptian Mortars. Part II: Adsorption and Infrared Studies', *Cement and Concrete Research*, 17 (1): 12–21.

Rassineux, F., Petit, J.-C., and Meunier, A. (1989) 'Ancient Analogues of Modern Cement: Calcium Hydrosilicate in Mortars and Concretes from Gallo-Roman Thermal Baths of Western France', *Journal of the American Ceramic Society*, 92 (6, June): 1026–3102.

Regourd, M., Kerisel, J., Deletie, P., and Haguenauer, B. (1988) 'Microstructure of Mortars from Three Egyptian Pyramids', *Cement and Concrete Research*, 18 (1): 81–90.

Rivera-Villareal, R. and Krayer, S. (1996) 'Ancient Structural Concrete in Mesoamerica', *Concrete International*, 18 (6): 67–70.

Rondelet, J. (1805) *L'Art de bâtir*, Paris.

Sabbioni, C., Riontino, C. Zappia, G., and Ghedini, N. (1998) 'Characterization of Ancient Hydraulic Mortars on Roman Monuments', International Series on Advances in Architecture, *Proceedings of the 1997 5th International Conference on Structural Studies, Repairs and Maintenance of Historical Buildings*, Ashurst: Computational Mechanics, pp. 245–254.

Sabbioni, C., Riontino, C. Zappia, G. and Ghedini, N. (1998) *Structural Studies: Repairs and Maintenance of Historical Buildings*, Southampton: Wessex Institute.

Snell, L.M. and Snell, B.G. (2000) 'The Early Roots of Cement', *Concrete International*, 22 (2): 83–5.

Stark, J. and Wicht, B. (1999) 'Zur Historie des gipses', *ZKG International*, 10: 527–33.

Steiger, R.W. (1993) 'In the Beginning There Was – Béton Aggloméré', *Concrete Construction*, 38 (8, August): 567–71.

Steiger, R.W. (1995) 'Roads of the Roman Empire', *Concrete Construction*, 40 (11): 949–53.

Thomassin, J.H. and Rassineux, F. (1992) 'Ancient Analogues of Cement-Based Materials: Stability of Calcium Silicate Hydrates', Applied Geochemistry, Laboratoire des Matériaux et Géologie Environnementale, Poitiers.

Thornton, G. (1996) *Cast in Concrete: Concrete Construction in New Zealand 1850–1939*, Auckland: Reed Books.

Torres, F. and Emeric, F. (1999) *Lafarge plâtres histoires pour l'avenir*, Paris: J.P. de Monza.

Venuat, M. (1984) *Adjuvants et traitements*, Chatillons-sous-Bagneux, n.p.

Vitruvius, (1979) *Les Dix Livres d'architecture de Vitruve* translated by C. Perreault in 1684, Paris: P. Mardaga.

Chapter 3

The hydraulic binders and concrete industries at the beginning of the twenty-first century

3.1 Evolution of cement consumption during the twentieth century

Chapter 2 could have ended with a presentation of the Portland cement and hydraulic binders industries at the beginning of the twenty-first century, but it is deemed preferable to devote a full chapter to this subject. The importance of this presentation would have unbalanced the reading of Chapter 2. Moreover, it is very important to insist specifically on the more recent developments that have occurred in the past thirty years in order to better understand today's cement and concrete industries: why they have reached such a development, and as will be seen later in Chapter 12, what kind of future they face. In this last case, as it is always risky to predict the future, we would rather concentrate our analysis on major trends that can already be guessed rather than attempt to make precise predictions.

As already mentioned in Chapter 1, the production of Portland cement in the year 2000 was slightly greater than 1.5 billion tonnes, while it was only about 10 millions tonnes in 1900. However, during the twentieth century, cement consumption did not increase steadily. Table 3.1 and Figure 3.1a show that cement consumption started to increase significantly at the beginning of the 1960s in industrialized countries, then in the past fifteen to twenty years in South-East Asian countries (Table 3.2). As seen in Figure 3.1b, the total production of Portland cement has almost doubled during the past twenty years. Cement consumption has also doubled on almost all continents (Table 3.3).

When comparing cement production between 1900 and 2000, it can be seen that the use of Portland cement developed at the same time as an increase in the standard of living during the last century. In fact, in most countries, cement consumption can be linked to the gross domestic product (GDP) per capita. Figure 3.2 presents the world cement consumption per capita.

Presently, it is difficult to make projections concerning the long-term consumption of cement throughout the world, because the future of the cement and concrete industries is closely linked not only to demographic development, that is to birth rate figures (Figure 3.3) (Wilson 2000) but also to the rapidity

Table 3.1 Progression of the world cement production during the twentieth century

Year	*Millions of tonnes*	*Source*
1900	10	Estimation
1906	13	Candlot (1906)
1913	39	Davis (1924)
1924	54	Davis (1924)
1938	86	Bogue (1952)
1948	102	Gourdin (1984)
1955	215	ATILH
1965	430	ATILH
1973	717	Gourdin (1984)
1980	850	Gourdin (1984)
1990	1,140	CEMBUREAU
1995	1,440	CEMBUREAU
1998	1,520	CEMBUREAU

Table 3.2 Cement production by country in millions of tonnes (source: CEMBUREAU)

Country	*1987*	*1992*	*1997*	*2004*
China	180	308	493	935
Japan	71	89	92	72
United States	71	74	84	95
India	37	56	81	137
South Korea	26	47	60	56
Brazil	26	24	38	35
Thailand	10	22	37	37
Turkey	22	29	36	41
Italy	37	41	34	46
Germany	23	33	31	33
Mexico	22	27	30	35
Indonesia	n.a.*	19	28	30
Taiwan	15	20	21	19
France	23	21,5	18	22

* n.a.: not available

with which a sustainable development policy will be implemented in all human activities.

Instead of looking at general statistics, the consumption of cement can be analysed country by country. It is evident that there are great disparities in cement consumption throughout the world (Table 3.2). The three main producers of cement are China, Japan and the USA. The weight of China in the worldwide production of cement is closely tied to its demographic weight. However, India which has a population approaching, or greater than, 1 billion, is presently using less cement than Japan whose population is eight times smaller.

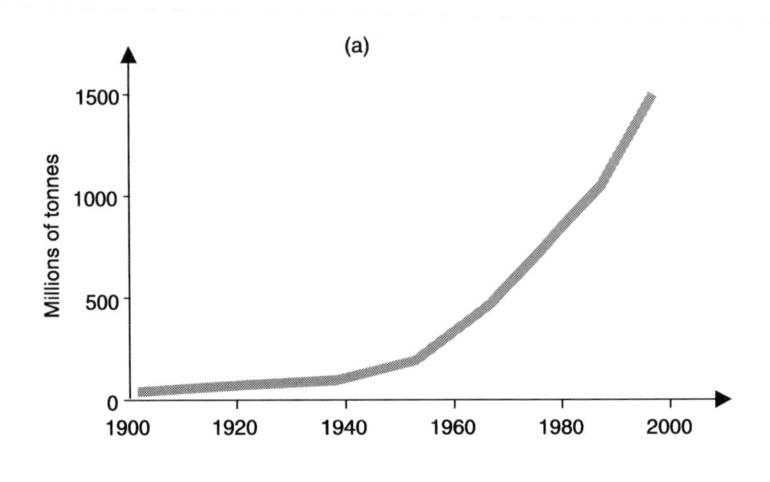

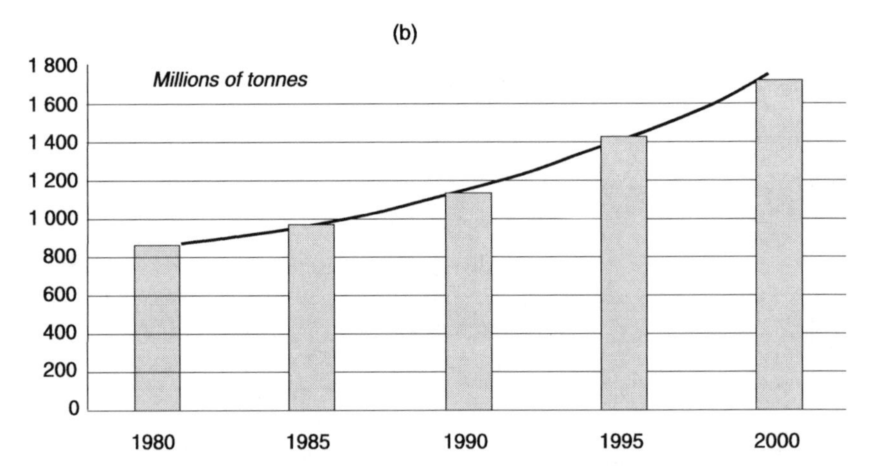

Figure 3.1 Progression of worldwide Portland cement production:
(a) during the twentieth century,
(b) during the last twenty years of the twentieth century.

When the per-capita consumption of cement is calculated country by country, as shown in Table 3.4, it is seen that cement consumption varies greatly. In 1993, Luxembourg and Portugal were the champions of cement consumption in Europe and in the world. The strong cement consumption in Portugal can be explained by the fact that Portugal has a large infrastructure programme underway in order to adjust its development to that of most European countries. The strong consumption of cement in Luxembourg can be related to the construction of European Union administrative and banking infrastructures. Luxembourg is becoming, in a certain way, an important crossroad in Europe.

Table 3.3 Cement production by continent in millions of tonnes (source: CEMBUREAU)

Millions of tonnes	*1924*	*1938*	*1996*	*2004*
Europe	25.00	52.0	285	296
Russia	–	–	–	71
America	25.00	22.0	195	229
Asia	2.75	9.5	450	516
China			495	928
Oceania	0.60	1.0	15	10
Africa	0.35	1.5	60	91
Total	53.70	86.0	1 500	2 141

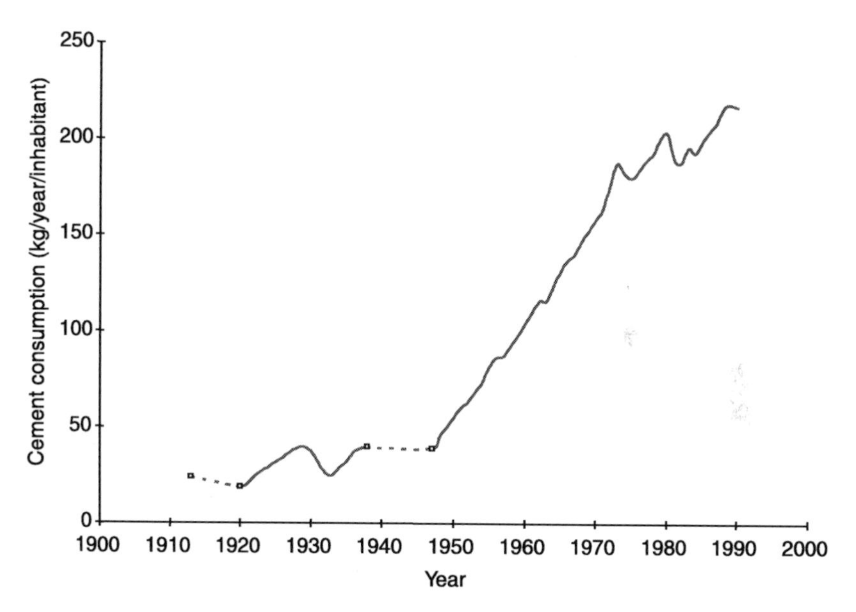

Figure 3.2 Increase of the average per-capita consumption of cement during the twentieth century (source: CEMBUREAU).

On the contrary, cement consumption in the poorest countries is still very low. For example, in 1998, cement consumption in Madagascar was only 14 kg/year/per capita (and only 7 kg/year/person in 1995), which is very far from the 1 100 kg/year/per capita used by Luxembourg at the same period (Weiss 2000).

Cement consumption is directly linked to the level of development of today's societies. Cement consumption in 1993 and 1996 has been plotted as a function of GDP for different countries in Figure 3.4. The curves obtained have a more or less parabolic shape with their concavity turned downwards. In these figures

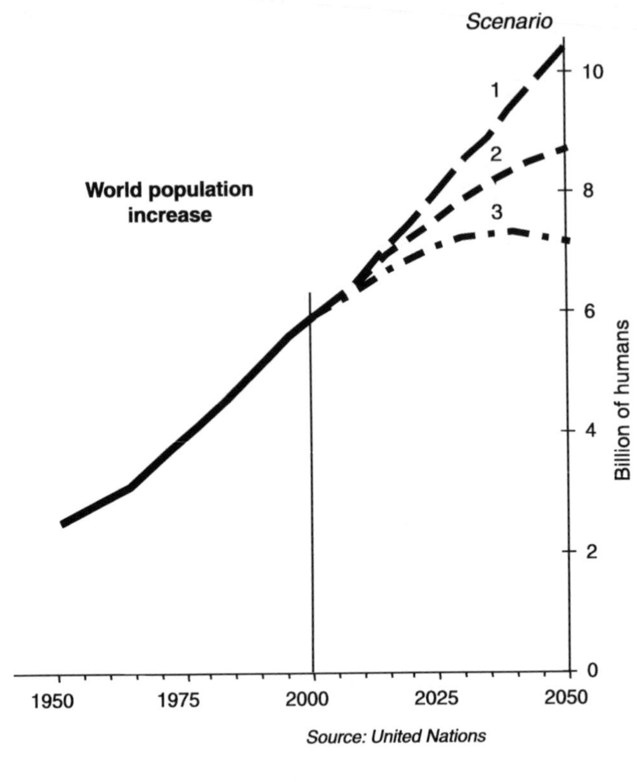

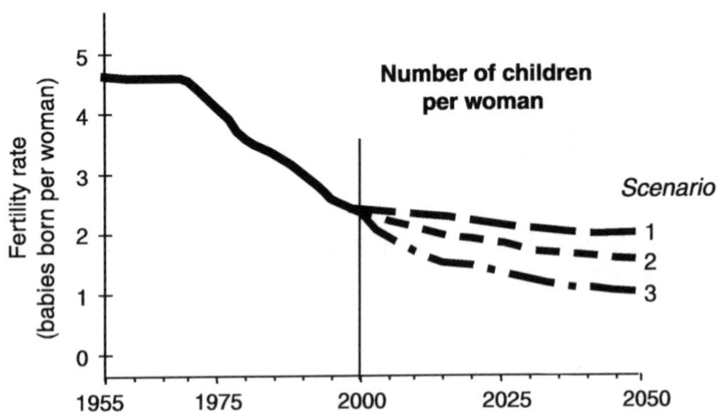

Figure 3.3 Different scenarios of worldwide population increase at the beginning of the twenty-first century (UN source).

Table 3.4 Per-capita consumption of cement, in kg (source: CEMBUREAU)

Country	*1987*	*1992*	*1997*	*2004*
Luxembourg	895	1 247	1 123	1 221
Portugal	564	769	948	867
Greece	604	739	716	963
Spain	522	666	681	1166
Ireland	345	409	628	1000
Japan	584	656	622	453
Austria	595	669	605	565
Italy	522	770	593	795
Belgium	415	579	566	557
Switzerland	738	658	528	569
Germany	380	455	419	353
China	168	263	388	712
Netherlands	334	344	355	313
United States	342	292	347	409
France	404	376	320	366
Denmark	311	241	270	296
United Kingdom	260	209	217	216

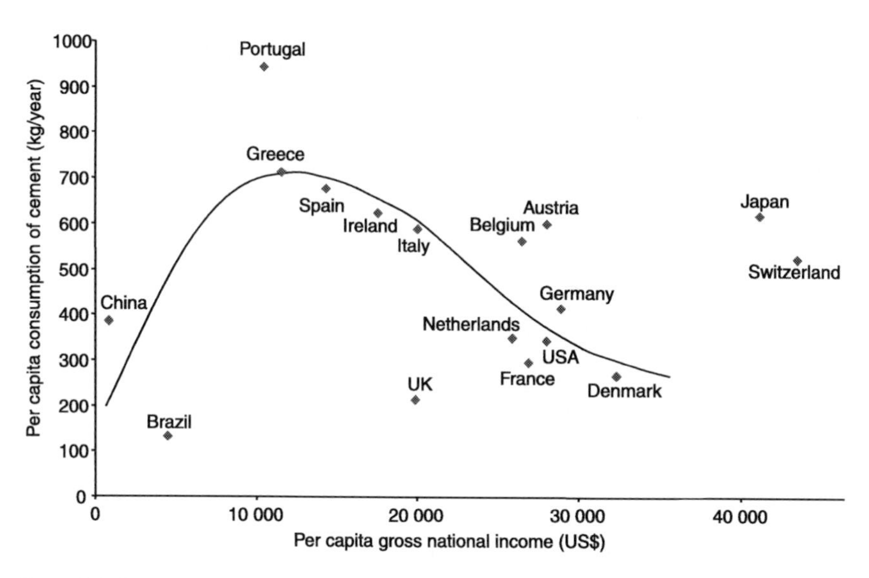

Figure 3.4 Correlation between the per-capita consumption of cement and the per-capita gross national income in 1997.

it is seen that as long as the GDP is lower than US$1000 per year, the per-capita cement consumption is lower than 100 kg/year. Human beings are more concerned with satisfying their two basic needs to feed and clothe themselves than with building strong houses. It is only when this critical US$1000 GDP is reached that people start to build small houses using concrete and concrete blocks. At the same time, it is observed that governments have enough money to begin to invest in infrastructure works.

When a GDP of about US$10 000 is reached, cement consumption begins to decrease, most people have decent dwellings, major infrastructure works have been completed, and only secondary work and maintenance work have to be done. This trend can be observed in Europe and America as well as in Asia.

Though cement consumption presently tends to be decreasing in industrialized countries, it is, on the contrary, increasing rapidly in many underdeveloped countries. Tables 3.3 and 3.4 show a great difference in the cement consumption in different countries towards the end of the twentieth century.

A large country like Brazil consumes only 135 kg of cement/year/per capita, but in this specific case, statistics hide great variations in cement consumption within the country. A San Paulista uses on average more cement than a poor countryman from the Nordeste province. Even in São Paulo or in Rio de Janeiro, the people living in fashionable neighbourhoods use more cement than the poor people living in favelas. Such examples could be multiplied in many countries, even within the richer ones.

Cement and concrete companies operating worldwide are not too preoccupied with the future, at least for the first half of the twenty-first century. However, these companies know that they will have to move their main activities from developed countries to developing countries. They will build cement plants in politically stable countries, grinding units in less sure countries, or perhaps build only a terminal in the least stable countries. Portland cement will flow from industrialized countries to developing countries (Dumez and Jeunemaître 2000).

Of course, cement consumption depends on many other factors than the per-capita GDP. Economic factors such as the availability of competing construction materials (wood, steel, aluminum, plastics, composite materials, glass, etc.), geographical factors (flat or mountainous countries), demographical factors (density of population, degree of urbanization), climatic factors (Belgians who spend more time in their homes in winter need larger houses and apartments than the French or the Spanish), cultural factors (Anglo-Saxon architects and engineers prefer to design with steel, while their Latin counterparts prefer concrete and their Scandinavian counterparts prefer wood). Many other factors of a historical, commercial, political and fiscal nature influence the use of cement and concrete in a country.

The main objective of the cement and concrete industries is to satisfy the needs of the construction industry. It is not the cement and concrete industries that force people to consume cement and concrete; people are using more concrete to improve their standard of living. Some people see in cement and concrete the essential elements supporting the improvement in the standard of living experienced over the past fifty years in industrialized countries.

At the dawn of the twenty-first century, it is difficult to make a precise description of the great cement companies operating on a worldwide basis, because, as in many other industries, there is a strong consolidation movement, but, as in old traditional heavy industries, this concentration has been carried out at a relatively slow pace. However, very recently, the concentration movement has accelerated somewhat to establish a new order among the bigger players (Dumez and Jeunemaître 2000).

In the ready-mix concrete business, the situation is more complex and confused. The ready-mix industry is an industry of proximity that obeys local market laws and is still very traditional. In some places, it is the tradition for cement companies to be involved in the ready-mix business, in others not, but as a general trend, it is observed worldwide that cement companies are getting more and more involved in the marketing of concrete. This trend should continue; after all, what can be done with cement other than make concrete?

Multinational ready-mix groups were developing very fast until recently. This was the case of the English group RMC which is the biggest ready-mix producer in France and of the Australian group Pioneer. These groups were buying cement companies before being bought by the three big cement companies. This globalization phenomenon by vertical integration will increase in the years to come.

Why is there a such great difference in the business approach of two very similar industries interested in the same product: cement? It is a matter of market scale and initial investment. The construction cost of a modern cement plant is in the order of US$200 to $300 million for a typical production capacity of 1 million tonnes per year, that is, 2000 to 2500 tonnes/per day. This investment pays off over a certain period of time, therefore, it can be done only in a stable market, not only from an economic point of view but also from a political point of view. It is also necessary to benefit from good maintenance services in electronics and computerized controls, since modern cement plants are highly automated and run by expert systems. Any shutdown due to a breakdown in the electronic or computer system has dramatic economic consequences.

However, a clear distinction between the cement and concrete industries should be made in the near future, because in order to maintain their level of business in industrialized countries, cement companies will have to get involved in activities related to the transformation of cement into concrete and even in construction activities. In fact, in the near future, more and more infrastructure works will be done in a Built Own Operate and Transfer (BOOT) mode, because of the disinvolvement of governments in huge socio-economic investments. Moreover, the new user-payer philosophy will become the rule for the development of intrastructures. The Confederation Bridge in Canada, linking Prince Edward Island to Mainland Canada, was built as a BOOT project by a consortium composed of French, Dutch and Canadian companies. This consortium made the feasibility studies, built the bridge and will manage toll perception and maintenance of the bridge for thirty-five years, after which the bridge will become the property of the Canadian Government. There are many other projects in the world that have been conducted in such a way and many more to be built.

If the cement industry does not adopt such an approach for the development of infrastructure works, highways will be built with asphalt, bridges with steel and buildings with glass and composite materials.

3.2 The new ecological policy: from Portland cement to hydraulic binders

It is not the purpose of this section to present blended cements (their detailed presentation will be made in Chapter 8), but rather to show when, how and why these binders appeared on the market.

The first blended cements were slag cements, produced as early as 1901 in Germany. Since Prüssing's work on slag, in 1888, it was known that by quenching slag it was possible to use it to replace a part of Portland cement, without decreasing the twenty-eight day compressive strength (Papadakis and Venuat 1966; Papadakis et al. 1970). As it will be seen in Chapter 8, slag needs to be activated to develop its binding properties. This activation can be obtained by Portland cement, gypsum, alkalis and alkali sulphates.

In 1900, German Portland cement producers did not see favorably the arrival of a competitive product on the market, and they started refusing to sell clinker to the metallurgical industry. The steel industry had no other choice than to build its own Portland cement clinker kilns. The confrontation developed until 1909, when the first two blended cements containing 30 and 70 per cent of slag were put on the market. After that initial stand-off, both industries decided to cooperate to establish the first standards on slag-blended cement.

Slag cement was produced rapidly in Belgium and in the Netherlands. In France, it was not until 1918 that the first slag cements started to be used (Papadakis et al. 1970). Slag cement was used in the lining of the Paris Metro and was accepted for marine works in 1930.

Supersulphated cement was invented by H. Kühl in 1908. Its production started in 1914 in Germany, and in 1932 in France and in Belgium (Papadakis and Venuat 1966). Supersulfated cement contains only 5 per cent of cement, 80 per cent of slag and 15 per cent of calcium sulphate.

In the USA, Portland cement producers succeeded in delaying the introduction of slag on the market until recently, in spite of the existence of standards on slag blended cement. Consequently, in some areas in Canada and in the USA, slag was, and still is, sold separately directly to concrete producers.

Pozzolanic materials have been more or less blended with Portland cement for a long time in some Mediterranean countries: in Italy, Greece and Morocco in particular, according to the availability of good quality pozzolans. In Latin America, some pozzolanic cements have been marketed, particularly in Chile and Mexico.

The blending of fly ashes is more recent. In 1953, 110000 tonnes of fly ash were used during the construction of the Hungry Horse dam on the Flathead river in Montana. The fly ash was introduced at the concrete plant. In France, in the early 1950s, Fouilloux developed the use of a fly-ash-blended cement (Papadakis et al. 1970; Jarrige 1971).

In Europe, the great need for cement that followed the Second World War resulted in an increased use in fly ash, as it was an easy way to increase cement production without investing in clinker production. This move was also beneficial to power companies who found an economical way of disposing of their fly ash.

In France, fly ash has never been commercialized separately; this was not the case in England, Germany and the USA. In Germany, cement and slag producers strongly opposed the use of fly ash, as a result, about 4 millions tonnes a year of fly ash are presently sold directly to concrete producers.

The use of limestone filler started to develop in France in the 1980s, before being used in Europe and in Canada. Americans are still fiercely opposed to the use of limestone filler.

During the late 1970s, silica fume started to be used, first in Scandinavia (Norway, Iceland and Sweden) then, at the beginning of the 1980s, in Canada and the USA. Silica fumes were either sold as a separate mineral component introduced at the concrete plant (Norway, Sweden and the USA) or as a blended cement (Iceland and Canada).

It is only very recently that American cement producers decided to adopt the new ecological policy and started to promote the use of blended cement in order to reduce their CO_2 emissions. It is certain that the binders of the twenty-first century will contain cementious materials other than Portland cement.

Presently, ternary cements are appearing on the market (Nehdi 2001), particularly in Europe, where blended cements containing both slag and fly ash are already marketed, and in Quebec, where cement producers have developed a ternary-blended cement containing slag and silica fume or fly ash and silica fume. Even quaternary cements are being proposed by an Australian and a Canadian company; these blended cements contain slag, fly ash and silica fume. On top of their ecological virtues, these blended cements are as efficient as low alkali cements in controlling alkali/aggregate reaction.

Many people believe that cement companies are introducing slag or fly ash to reduce their production costs. This is not necessarily true, because handling, transportation, stocking and processing costs have to be added to the initial cost of the cementitious material. Only the addition of limestone filler is for certain a beneficial activity for any cement producer. Limestone is obtained directly from the quarry; it is easy to grind and it does not need to pass through the kiln. It is only necessary to grind the blended cement somewhat finer that an ordinary Portland cement to increase its initial strength and match that of pure Portland cement.

On some occasions, cement plants that are licensed to eliminate all kinds of dangerous and toxic products could produce clinker at a negative cost when they are eliminating BPC (biphenol chloride), mad cows, animal flours, contaminated Coca-Cola, etc. In fact, the very high temperature reached in the flame (2500°C) destroys all organic molecules that otherwise are very difficult to eliminate.

3.3 The structure of the cement industry during the second half of the twentieth century

The cement industry is still a fragmented industry, in spite of the concentration that has occurred quite recently. The three biggest cement producers, which will be named in alphabetical order, the only order that has a chance of remaining the same in the near future are: Cemex, from Mexico; Holcim, from Switzerland; and Lafarge, from France. They produced in 2000 between 250 and 300 millions tonnes of cement, which is at once a lot and a little. This amount represents around 20 per cent of the worldwide cement production.

Acquisitions and concentrations are developing at a quite rapid rate presently, so that it is difficult to predict what the cement industry will look like in twenty-five years, except that it will be more concentrated than it is presently. For example, Cemex recently bought Southdown, a cement and concrete company operating in the south of the USA, essentially along the Mexican border. More recently Cemex bought two large ready-mix producers, the RMC company, operating in the USA and Europe, and the Rinker Group, operating in the USA and Australia. Similarly, Lafarge recently bought Blue Circle, which was covering the cement needs of many countries formerly part of the British Empire where Lafarge was almost absent.

It is surprising to see that none of the five major cement groups are based in the leading industrial countries of the world: the USA, Japan and Germany. In Japan and Germany in the year 2000, the cement industry was still fragmented. There were more than twenty cement accompanies operating in these countries where cement production is essentially a family business. In Japan and Germany, the small concentration seen recently occurred at the top. Only a few major companies were merging, for example, in Japan, Onoda Cement, which was the biggest cement producer, bought Chichibu cement which was number three and then merged with Nihon cement to create Taiheiyo Cement (Pacific Cement).

In the USA, 70 per cent of cement production is under the control of European, Japanese and Mexican groups (Cemex, Heidelberg, Holcim, Lafarge, Taiheiyo, Vicat, etc., named in alphabetical order). There are still a few cement companies controlled by US interests, but in general, they are not very large and not very up to date from a processing point of view (wet process). In order to explain such a situation, S. Berger (Hervé and Jeunemaître 2000) suggests three reasons.

Until 1950, the US cement industry operated in a well-protected market that favoured a weak competition between companies. During the 1960s and 1970s, many executives from these companies took wrong decisions because they were used to operating in a very protected market and were not ready to take aggressive decisions. The cement plants in operation had been designed to satisfy local markets and were not easy to modernize. Moreover, these plants were not well located to be transformed into large cement plants able to cover a vast territory. Consequently, in order to satisfy the increase in cement consumption and the yearly peak of consumption, US cement companies started importing cement. Finally, the economic period of stagnation between 1970 and 1980 was fatal to the US cement industry, as well as to the US steel industry, because it coincided with the end of the major highway-construction programme.

A series of political decisions on the quality of air, added to the petroleum crisis, accelerated the sale of many cement companies. Making cement was no longer a very profitable business, and there were many other opportunities to invest in more profitable and not so cyclic businesses.

Faced with a depressed market, it is normal that US cement companies were seduced by offers to buy from the major European cement companies which had known very good years until the first petroleum shock.

Finally, why was the European cement industry so healthy during the same period? After the Second World War, European governments decided to fix the price of cement in order to protect their cement industry and help them to rebuild modern plants. Therefore, the European cement industry was able to recover rapidly with up-to-date facilities. In France, a great number of family-owned small cement plants were bought by the bigger players. Consequently, the European cement industry was modern and profitable when the American cement industry was operating old outdated plants which were not so profitable. Moreover, it is not so easy to get loans from banks when you are facing such a depressed business situation. It is at that time that the Swiss company Holcim decided to develop a cement distribution network using essentially the Mississippi waterway.

In the past twenty years, some major European players started to slowly buy small US companies in a more or less friendly way. In the year 2000, the concentration of the cement industry is still not finished and, as previously mentioned, it is difficult to predict what it will look like in twenty-five years.

Market globalization favours a further concentration of the cement industry, because, in spite of the fact that cement has a low commercial value (it is sold at around US$60 per tonne), it can be transported easily by boat from a place where it is being overproduced, or in economic crisis, to a place where there is a strong economy and a great demand for cement or simply a peak in consumption (Dumez and Jeunemaître 2000). Globalization offers flexibility to the management of large multinational companies: for example, it is easy to enter a new market by simply operating or building an inexpensive terminal. As a matter of fact, the cement tankers presently crossing the seas full of cement can be used as floating silos.

When a cement company has succeeded in opening a market with imported cement or when the market is not so large, a grinding operation can be built. This is, for example, the case of Martinique and Guadeloupe, two French islands of the West Indies. It is only when a local market is large enough (around 500 000 tonnes), and politically stable, that a cement company will be willing to invest US$200 to $300 million to build a modern cement facility.

This concentration of the cement industry is only Phase II of its concentration. Phase I started in each country, before the Second World War, transforming the cement industry from an essentially highly fragmented, family-owned local business into a multinational business. For example, in 1930 in France, there were forty-nine cement companies, in 1974 only fifteen (Gourdin 1984) and only four in the year 2000, in alphabetical order: Ciment-Français/Italcimenti, Holcim, Lafarge and Vicat.

In Switzerland, the Holcim group started to buy foreign cement companies located close to the Swiss border to protect its local market before becoming one of the biggest cement companies in the world. Cemex did the same in Mexico, and the Votorantin group in Brazil is emerging as a major player. Recently, Votarantin bought the Blue Circle operations in Canada and around the Great Lakes from Lafarge, because Lafarge could not keep them due to the anti-trust policies of the Canadian and US Governments.

3.4 The law of the market

It is not the intention of the author to review all the economic aspects of cement marketing; it is not his specialty, and a recent book deals with this subject (Dumez and Jeunemaître 2000), but it is important to know the specificity of the cement market to understand the past and the future of the cement and concrete industries. The brief analysis presented is essentially based on the synthesis on the subject written by F.M. Scherer in the postscript of Dumez and Jeunemaître's (2000: 207–12) book.

On a worldwide basis, the cement industry is a somewhat modest industry when compared to other heavy industries such as steel, automobiles or energy. However, it is an industry that requires large capital investments, and the invested capital can only be used to produce cement. It is quite difficult to recycle a cement plant to produce something other than cement. Moreover, the cement industry is highly automated and, consequently, does not create many jobs so its political weight is negligible.

The cement market is cyclic and, up to recently, before the construction of cement tankers, it was not possible to transport cement economically over long distances. Transporting by railway cars or trucks is expensive for a material having such a low economic value as cement. According to an economic study done by Murphy in 1973 and cited by Berger (Dumez and Jeunemaître 2000), transporting cement over 564 km represented a little more than 50 per cent of the sale value of the cement. Only the transportation of industrial gas represents a higher relative value than its trade value (70 per cent). Average transportation costs for 101 industries were only 5.45 per cent, and the median value was 2.61 per cent of the sale value.

However, this analysis does not take into account the fact that, since 1973, a great volume of cement is now transported by boats or barges. Presently, whenever possible, cement plants are built in a harbour or very close to a harbour or near a major river network, in order to lower expensive long-distance transportation costs. We already mentioned the fact that the construction of cement tankers is another way to extend the market of a cement plant.

The cement industry is a heavy industry that is fighting to minimize its transportation costs. Moreover, its only customer, the concrete industry, is dealing with a highly segmened market which is essentially a function of the GDP and the density of the population.

As can be seen in Figures 3.5 and 3.6, cement marketing varies greatly according to the state of development of each country. Cement is sold in sacks when the GDP is low and essentially in bulk in developed countries. For example, in the year 2000 in Canada and the USA, only 5 per cent of the cement was sold in sacks, while in Sri Lanka, it stood at 98 per cent, in Morocco more than 90 per cent and almost 85 per cent in Mexico and Brazil. It can be said that in a developing country, the cement market is a 'sack' market while in developed countries it is a 'bulk' market.

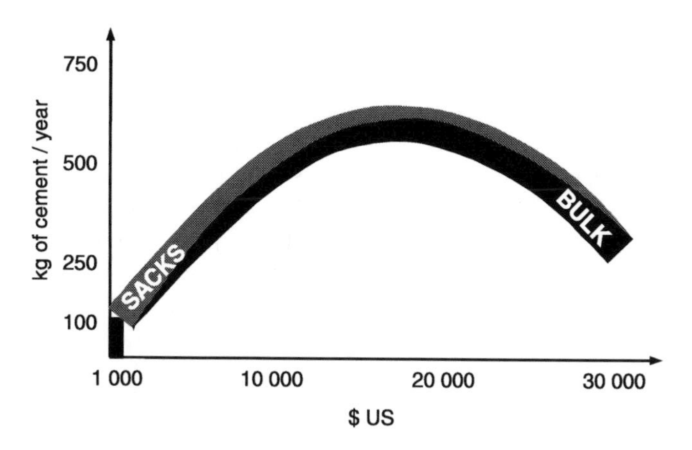

Figure 3.5 How cement is commercialized.

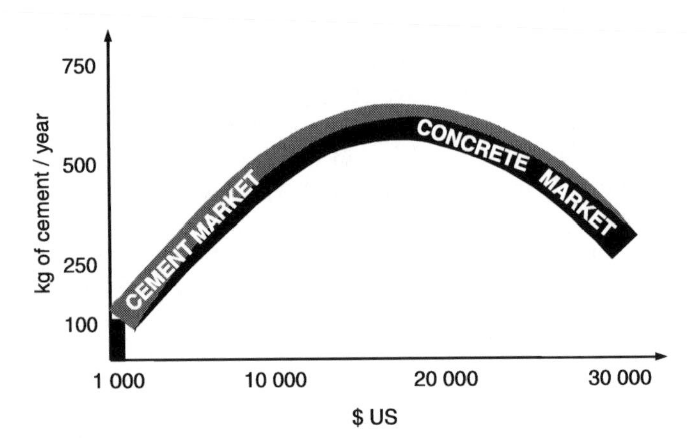

Figure 3.6 Cement and concrete markets.

3.5 Technological regulation of the cement industry

As will be seen in the next chapters, Portland cement is a quite complex multiphase material produced through a simple process. Making Portland cement consists in firing, at 1450°C, a well-proportioned mixture of two common natural materials, limestone and clay, or an impure limestone and some corrective additives. Experience shows that, in spite of the implementation of good-quality control programmes of the raw materials and of the process, it is almost impossible to produce two identical cements. Moreover, a well-controlled cement plant is unable to produce exactly the same clinker all year round, and, consequently, cement properties and characteristics vary within a more or less wide range. A cement plant operating two cement kilns using the same raw feed will simultaneously produce two clinkers that are not absolutely identical.

In order to provide customers with a product having a low variability, the cement industry is regrouped in professional associations that practise a kind of self-discipline. These associations have developed a series of acceptance criteria with their customers in order to limit the variability of the cement produced and provide a minimal performance.

It was in 1877 that the first professional association was created. This association resulted later on in the formation of the Verein Deutscher Zementwerke (VDZ). One of the first benefits of the creation of such an association was that it was in Germany that cement technology progressed more rapidly and the German Portland cement industry beat the English Portland cement industry on the export market. German Portland cement had such a reputation of quality that in 1884, 8000 barrels of cement (1360 tonnes) produced by Dyckerhoff of Germany were shipped to the USA to build the foundation of the Statue of Liberty. German cement continued to be imported in the USA for many years so that the Metropolitan Opera House and the Waldorf-Astoria

Hotel were built with German Portland cement. The New York Stock Exchange on Wall Street was built with German Portland cement provided by Lafarge (Dumez and Jeunemaître 2000) because American architects and engineers did not trust the quality of the cement produced in the USA at that time.

In France, the Syndicat des Producteurs de Ciment was created around 1900; the American Portland Cement Association was created in 1902 (Draffin 1943) and the British Cement Association in the UK in 1935. These associations operate through an annual charge imposed on their members which is proportional to the volume of cement produced.

Along with these national associations, national standard societies were created: in England, the Engineering Standards Committee was created and later became the British Standards Institution. The American Standard and Testing Material (ASTM) published the first standards on Portland cement in 1904 (Kett 2000). In France, the first cement standard was published in 1919 (Lafuma 1951).

As the performance of the cement industry was crucial to rebuild Europe after the last world war, European governments took a series of measures to protect their national cement industry and help it invest in research and development. For example, the French Government imposed a special tax on each tonne of cement or m^3 of concrete, and this was used to create associative research centres on cement and concrete. The idea was to favour non-competitive research and development, in order to progress more rapidly. The Centre d'Étude et de Recherche de l'Industrie des Liants Hydrauliques (CERILH) and the Centre d'Étude et de Recherche de l'Industrie du Béton (CERIB) were created in this way. In spite of the fact that the cement and concrete industries had their word to say in the management of these centres, they had some independence in their research programme. Moreover, these centres could act as arbitrators when a disagreement occurred between a cement or a concrete producer and some of their clients or a governmental agency. The rapid development of the French cement industry and its success after the Second World War can be partly linked to this political decision to create such research and development centres.

The same attitude, with local variations, produced similar effects in the USA, England, Germany and Japan. In the USA, theoretical and fundamental studies were conducted, particularly under the direction of Powers, Backstrom, Helmuth and Brownyard at the American Portland Cement Association. In France, theoretical and fundamental research was conducted at the CERILH under the direction of Guinier, Lafuma and Micheline Moranville-Regourd.

However, when the cement industry matured, the need for these cooperative research centres was not so crucial for the development of the industry. They were no longer supported by the cement industry, so that their name and vocation changed. In France, for example, the CERILH became l'Association Technique de l'Industrie des Liants Hydrauliques (ATILH).

3.6 International associations

There exists a second level of professional associations to defend the interests of the cement industry, for example, in Europe, CEMBUREAU bring together all the European and Turkish cement companies, and, up to recently, FICEM represented all the cement producers operating in Latin America, Spain and Portugal. These organizations keep track of all kinds of statistics concerning cement production and can lobby when general matters, such as environment issues, have to be debated with national and international governments.

3.7 Technical and educational societies

Along with these professional associations, there are national or supranational associations whose objectives are to promote the good use of cement and concrete. The American Concrete Institute (ACI), the Japanese Cement Association, the Japanese Concrete Institute, Le Rassemblement International des Laboratoires d'Essais sur les Matériaux (RILEM) are non-profit and unfunded organizations that mainly work on a voluntary basis to promote the good use of cement and concrete, or, more generally, construction materials in the case of RILEM. In these associations, engineers, researchers, university professors, contractors, technicians, cement, concrete, admixture and reinforcing-steel-fibres producers establish codes of good practice.

The most important of these professional associations is the American Concrete Institute, which has between 17000 and 20000 members (Jirsa 2000). ACI members work on a voluntary basis on 280 specific committees whose recommendations can be found in the *Manual of Concrete Practice*, which contains more than 5000 pages of very valuable technical documentation. These technical recommendations are continuously revised. A technical recommendation cannot be published if it is not approved by the members of the committees and voted upon. These committees meet in the USA, Canada or Mexico twice a year.

3.8 Research in the area of cement and concrete

The first research effort on cement and concrete was developed within cement companies to get an edge over their competitors. But, quite early, some governmental associations which were large users of cement and concrete also developed research activities. In France, for example, following the work of Vicat, Le Chatelier started doing research on cement hydration, as early as 1892, and later on Féret started doing research in concrete. A hundred years later, the work of these pioneers is still of great value and is regularly cited, in spite of the fact that they used very few of the powerful technical instruments available today, but these people were very clever and had time to think.

This scientific interest for cement and concrete was widespread in Europe, as well as in the USA. Michaëlis was working on cement hydration around 1893 in Germany; in 1897, Tornebon identified the four major phases present in Portland cement and gave them the names alite, belite, celite and felite. In 1905, the first ternary phase diagram $CaO–SiO_2–Al_2O_3$ was published by Rankin and Wright, and in 1906 the first comprehensive work on lime and silica phase diagrams was published in the USA by Day, Shephard and Wright.

3.9 Scientific literature

The results of the research on cement and concrete are presented in a certain number of journals published by self-sustaining organizations such as the American Concrete Institute (ACI). ACI publishes three journals: *Concrete International*, the *ACI Materials Journal* and the *ACI Structural Journal*; RILEM publishes *Materials and Structures*; ASTM publishes *Cement Concrete and Aggregates*; publishers such as Pergamon Press publish *Cement and Concrete Research*; Elsevier publishes *Cement and Concrete Composites*; the American Society of Civil Engineers (ASCE) publishes the *Journal of Materials in Civil Engineering*; the Canadian Society of Civil Engineers publishes the *Canadian Journal of Civil Engineering*, the French administration of Les Ponts et Chaussées publishes *Le Bulletin des Laboratoires des Ponts et Chaussées*, and so on.

In these journals, the proposed papers have to pass a peer review (Neville 2001) in order to verify their scientific quality and eliminate any commercial aspects.

Along with these scientific journals, there are a certain number of technical or scientific journals publishing papers and reports of general interest such as the *Civil Engineering Journal* in the USA, the *Moniteur and Constructions at Routes* in France, *Concrete*, the journal of the British Concrete Society and Betonwork plus Fertigteil-Tecknik (BFT) in Germany. These journals are read by people working in the construction industry. There are also technical journals, such as *Concrete Construction*, written for general users of concrete. It is now very easy to have access to this scientific literature, at least the more recent ones, through databanks available in universities or directly on the Internet.

Of course, with the years, English has became the lingua franca of the scientific community. However, some countries succeeded in keeping a national scientific literature in the area of cement and concrete through the use of bilingual journals. This is the case of Zement Kalk Gyps and BFT of Germany. In Italy, the journal *Il Cimento*, which was published with the same editing philosophy is, unfortunately, no longer published, but it has been replaced by *L'industrie italiana del Cemento*. Spain and Poland publish also bilingual journals. Recently, the *Bulletin des Laboratoires des Ponts et Chaussées* in France began providing electronic translations of the French papers it publishes. In Japan, the dissemination of Japanese research and technology in the area of cement and concrete is done differently. Every year, the Japanese Concrete

Institute (JCI) and the Cement Association of Japan (CAJ) publish two books in which the English version of the best Japanese papers of the year in cement and concrete technology are published.

3.10 Scientific meetings

Along with this scientific and technical literature, which is a not such a rapid process of knowledge transfer (it can take one or two years to have a paper published in certain journals), there are many (even too many) workshops, symposia, conventions, conferences and congresses organized every year. Some of these activities are of general interest and can bring together more than 1 000 people, such as the congresses on Cement Chemistry or the biannual ACI meetings; others bring together a few dozen attendees on a very specific topic. These activities are very important from a scientific and technological point of view since they favour a more rapid transfer of scientific and technological knowledge, because all the specialists in one field have an opportunity to meet and discuss their research in progress. In these meetings, breakfast and coffee-break meetings are almost as important as the organized sessions.

The location and dates of these meetings are fixed long in advance so that scientists and people working in the field have enough time to prepare their papers and plan these meetings in their busy schedules. These meetings can be organized by international organizations such as the ACI/CANMET and RILEM or by national or local organizations or even by individuals.

It is somewhat unfortunate that recently too many overlapping meetings have been organized, because they have contributed to a dilution of scientific knowledge rather than to the much needed transfer of scientific knowledge. Scientists have to make choices due to the high cost of participating in all these meetings and, sometimes, conflicting schedules. Presently, we are rather lacking papers presenting a synthesis of present knowledge because the promotion of university researchers still depends more on the number of papers they have published rather than on their quality.

3.11 Conclusion

This is the author's vision of the profile of the cement and concrete industries at the dawn of the twenty-first century. Others would have written a completely different profile. But I am convinced that these historical, socio-cultural and economical considerations are as important as scientific and technological knowledge; cement and concrete science and technology are not evolving in a closed system far from human and business activities. Cement and concrete are part of modern life; they have a history that must be considered in parallel to the laws of physics, chemistry, and thermodynamics, as well as the law of the market, of course.

References

CEMBUREAU (1968) *Statistiques sur la production cimentière*, Brussels: Cembureau.

Draffin, J.A. (1943) *A Brief History of Lime, Cement, Concrete and Reinforced concrete*, Urbana, Ill.: University of Illinois, Engineering Experiment Station.

Dumez, H. and Jeunemaître, A. (2000) *Understanding and Regulating the Market at a Time of Globalization: The Case of the Cement Industry*, Basingstoke: Macmillan.

Gourdin, P. (1984) 'La Cimenterie', *Revue Usine Nouvelle*, (May): 146–55.

Jarrige, A. (1971) *Les Cendres volantes: Propriétés, Applications industrielles*, Paris: Eyrolles.

Jirsa, J.O. (2000) 'ACI's Extended Family: 17,000 Strong', *Concrete International*, 22 (5, May): 5.

Kett, I. (2000) *Engineered Concrete*, Boca Raton, Fla.: CRC Press.

Lafuma, H. (1951) 'Le Contrôle NF-VP des ciments en usine dans le cadre de la normalisation française, Liants hydrauliques 5', *Annales de l'ITBTP*, 185 (March–April): 1–19.

Nehdi, M. (2001) 'Ternary and Quaternary Cements for Sustainable Development', *Concrete International*, 23 (4, April): 25–42.

Neville, A.M. (2001) 'Reviewing Publications: An Insider's View', *Concrete International*, 23 (9, September): 56–61.

Papadakis, M. and Venuat, M. (1966) *Fabrication et utilisation des liants hydrauliques*, n.p.

Papadakis, M., Venuat, M. and Vandamme, J. (1970) *Industrie de la chaux, du ciment et du plâtre*, Paris: Dunod.

Scheubel, B. and Nachtwey, W. (1997) 'Development of Cement Technology and its Influence on the Refractory Kiln Lining', Refra Kolloquium Berlin 1997, Göttingen: Refratecknik GmbH, pp. 25–43.

Weiss, J.C. (2000) Personal communication.

Wilson, E.O. (2000) *L'Unicité du savoir*, Paris: Robert Laffont.

Chapter 4

The chemical and phase composition of Portland cement

4.1 The chemical composition of Portland cement: its importance, its limitations

The chemical composition of Portland cement occupies an important place among the characteristics given in the data sheets provided by cement companies. As shown in Figure 4.1, this chemical composition is given in terms of oxide contents. Usually, these values represent average monthly values, sometimes given with an illusory precision of two digits after the decimal point. Figures 4.2 and 4.3 present the principal chemical elements usually found in Portland cements, with some important characteristics of these elements. In Table 4.1, it is seen that the CaO and SiO_2 contents are much greater than other oxide contents. The average CaO content is about 65 per cent, which is about three times the SiO_2 content. CaO and SiO_2 contents, when added, represent more than 85 per cent of the chemical composition of Portland cement. The next highest oxide contents are Al_2O_3 and Fe_2O_3 which are usually lower than 8 per cent. The four oxides CaO, SiO_2, Al_2O_3 and Fe_2O_3 represent what are usually called the main oxides (Lafuma 1965).

Chemical analysis (%)	
Alcalis (Na_2O equiv.)	0.84
Loss on ignition	2.9
Insolubles	0.4
Free lime	0.5
SiO_2	20.0
Al_2O_3	4.2
Fe_2O_3	3.1
CaO	62.4
MgO	1.9
SO_3	3.7
Bogue composition	
C_3S	58.6
C_2S	13.2
C_3A	5.9
C_4AF	9.5

Physical properties	
Fineness : Blaine	349 m^2/kg
Passing 45 μm	89 %
Autoclave expansion	-0.010 %
Expansion in water	+0.010 %
Setting time : Initial	150 min
Final	235 min
Air content	7.4 %
False set	84 %
Compressive strength	
at 3 days	22.2 MPa
at 7 days	27.8 MPa
at 28 days	36.4 MPa

Figure 4.1 Mill certificate of a Portland cement (production of one month).

1	2	3	4	5	6	7
1 **H** 1 Hydrogen $1s^1$ 0.32 - 1 2.08						
			6 **C** 12 Carbon $1s^2 2s^2 2p^2$ 0.77 0.91 6614 0.20 (+2)	7 **N** 14 Nitrogen $1s^2 2s^2 2p^3$ 0.75 0.2 6613 1.71 (-3)	8 **O** 16 Oxygen 0.73 1.40 (-2)	9 **F** 19 Fluor 0,72 - -1 1.36 (-1)
11 **Na** 23 Sodium $[Ne] 3s^1$ 1.54 1.90 1 0.95 (+1)	12 **Mg** 24,3 Magnesium $[Ne] 3s^2$ 1.36 1.60 2 0.65 (+2)	13 **Al** 27 Aluminium $[Ne] 3s^2 3p^1$ 1.18 1.43 3 0.50 (+3)	14 **Si** 28 Silicium $[Ne] 3s^3 3p^2$ 1.11 1.32 4 0.41 (+4)		16 **S** 32 Sulfur $[Ne] 3s^2 3p^4$ 1.02 1.26 6 1.84 (-2)	17 **Cl** 35,5 Chlorine $[Ne] 3s^2 3p^5$ 0.99 - -1 1.81 (-1)
19 **K** 39 Potassium $[Ar] 4s^1$ 2.03 2.35 1 1.33 (+1)	20 **Ca** 40 Calcium $[Ar] 4s^2$ 1.74 1.97 2 0.99 (+2)	26 **Fe** 56 Iron $[Ar] 3d^6 4s^2$ 1.17 1.26 2,3 0.76 (+2) 0.64 (+3) 0.,64 (+3)				

Figure 4.2 Simplified version of the periodic table of the elements (adapted to the needs of Portland cement chemistry).

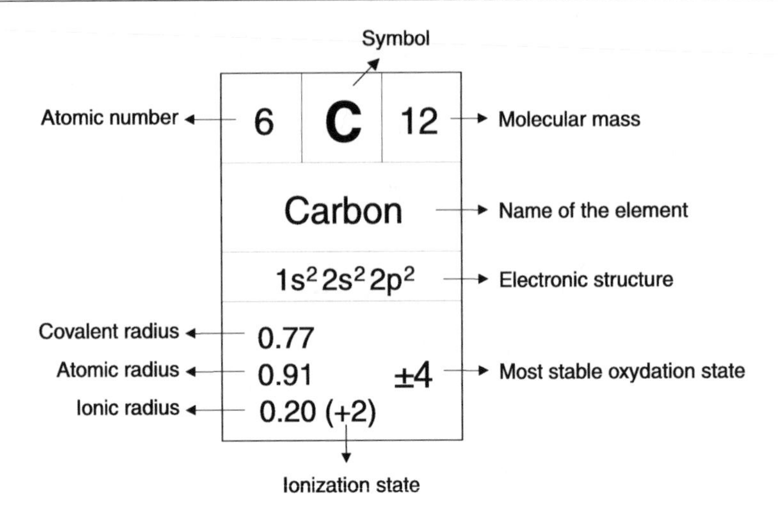

Figure 4.3 Meaning of the different symbols and numbers appearing in Figure 4.2.

Table 4.1 Oxyde composition in a Portland cement

Oxydes	*Maximum content (%)*	*Average content (%)*
CaO	60–69	65
SiO_2	18–24	21
Al_2O_3	4–8	6
Fe_2O_3	1–8	3
MgO	< 5	2
K_2O, Na_2O	< 2	1
SO_3	< 3	1

Table 4.2 shows that the Al_2O_3 and Fe_2O_3 contents vary from one cement to the other. For example, the Fe_2O_3 content of a white cement is lower than 1 per cent, while that of a sulphate resisting cement is much higher, consequently, sulphate resisting cements are much darker than usual Portland cements.

In Figure 4.1 and Table 4.2, it is seen that the content in oxides other than the main four, is much smaller: magnesia content is usually smaller than 5 per cent, SO_3 content is smaller than 3 per cent, and alkali oxides Na_2O and K_2O contents are lower than 1 per cent.

The content of these two alkali oxides is usually expressed in terms of Na_2O equivalent content (Na_2O equivalent). This Na_2O equivalent is calculated as Na_2O equivalent = $Na_2O + 0.658\ K_2O$. It is the Na_2O equivalent content that is calculated, in spite of the fact that the K_2O content is almost always much higher than the Na_2O content. As will be seen later, the Na_2O equivalent content is very important when concrete is made with potentially reactive aggregates.

Table 4.2 Average chemical and Bogue compositions of some Portland cements

	Cement type							
Oxyde	*CPA 32.5*	*CPA 52.5*	*US Type I without limestone filler*	*Canadian Type 10 with limestone filler*	*Type 20M low heat of hydration*	*US Type V sulphate resisting*	*Low alkali cemen*	*White*
CaO	64.40	66.28	63.92	63.21	63.42	61.29	65.44	69.53
SiO_2	20.55	20.66	20.57	20.52	24.13	21.34	21.13	23.84
Al_2O_3	5.21	5.55	4.28	4.63	3.21	2.92	4.53	4.65
Fe_2O_3	2.93	3.54	1.84	2.85	5.15	4.13	3.67	0.33
MgO	2.09	0.90	2.79	2.38	1.80	4.15	0.95	0.49
K_2O	0.90	0.69	0.52	0.82	0.68	0.68	0.21	0.06
Na_2O	0.20	0.30	0.34	0.28	0.17	0.17	0.10	0.03
Na_2O equiv.	0.79	0.75	0.63	0.74	0.30	0.56	0.22	0.07
SO_3	1.60	2.40	3.44	3.20	0.84	4.29	2.65	1.06
L.O.I.	–	–	1.51	1.69	0.30	1.20	1.12	1.60
Free lime	1.50	–	0.77	0.87	0.40	–	0.92	–
Insolubles	–	–	0.18	0.64	–	–	0.16	–
	Bogue composition							
C_3S	61	70	63	54	43	50	63	70
C_2S	13	6.1	12	18	37	24	13	20
C_3A	8.9	8.7	8.2	7.4	0	0.8	5.8	11.8
C_4AF	8.9	10.8	5.6	8.7	15	12.6	11.2	1.0
$C_3S + C_2S$	74	76	75	72	80	74	76	90
$C_3A + C_4AF$	17.8	19.5	13.8	16.1	15.0	13.4	17.0	12.8
Specific surface area m^2/kg	350	570	480	360	340	390	400	460

The oxides apart from the four principal ones, CaO, SiO_2, Al_2O_3 and Fe_2O_3, are called the minor oxides. Though minor from a content point of view, they can play a far from minor role on certain properties of fresh and hardened concrete. For example, when the MgO content is too high, a detrimental expansive action can occur in hardened concrete. The late hydration of periclase, the crystalline form of MgO in Portland cement clinkers, in the form of brucite $Mg(OH)_2$, results in a volumetric expansion that can lead to a severe cracking of concrete. Concrete can also crack when 'reactive' aggregates react with the alkalis in the cement. Workability losses with polysulphonate superplasticizers can also be caused by an inadequate solubility rate of SO_4^{2-} ions. In such a case, as it will be seen in Chapter 7, the SO_3^- terminations of the polysulphonate will react with the C_3A and result in a more or less rapid slump loss according to the reactivity of the C_3A and the amount of SO_4^{2-} ions available in the interstitial solution of the fresh concrete.

With the chemical composition the loss on ignition (LOI) is given. It represents the mass loss when a sample of cement is heated to 1050°C. Usually, the LOI value is a function of the amount of gypsum present in the Portland cement, the degree of prehydration of the cement, whether it has been stored in a humid environment and its limestone filler content, if present.

In Table 4.2, the chemical composition of eight different Portland cements is presented: two French cements, two North American ASTM Type I cements and four others that are particular Portland cements. These last four cements are: a low heat of hydration cement used by Hydro-Québec to build its dams, a sulphate-resistant cement, a low alkali cement and a white cement. As can be seen in Table 4.2, the greater variations in the chemical composition are observed at the level of the Al_2O_3, Fe_2O_3 and K_2O contents. This table emphasizes the importance of these three oxides, since by varying their respective proportions, a cement producer can produce such different cements as a white Portland cement, a low-heat-of-hydration Portland cement and a sulphate-resistant Portland cement in which the CaO and SiO_2 do not vary much.

Moreover, it is obvious from Table 4.2 that a Portland cement that does not have a constant chemical composition cannot have constant characteristics and properties. However, we will see later that even if a Portland cement has a constant chemical composition, it does not necessarily mean that it will have constant properties. Practical properties of Portland cement depend on other characteristics as important as its chemical composition. Of course, it is impossible to produce a cement having constant practical properties without controlling very carefully the chemical composition of the raw meal introduced in the kiln to produce Portland cement clinker. The chemical composition of the raw meal must remain within tight limits, as far as the values of the four main oxides are concerned, so that they may be almost entirely combined when exposed to the very high temperature found in the clinkering zone of the kiln.

Cement chemists calculate a certain number of factors or moduli (Pliskin 1993).

- the lime saturation factor (LSF). The most commonly used LSF is the one proposed by Küll.

$$LSF = \frac{CaO}{2.80\ SiO_2 + 1.10\ Al_2O_3 + 0.70\ Fe_2O_3}$$

- the silica modulus (SM)

$$SM = \frac{SiO_2}{Al_2O_3 + Fe_2O_3}$$

- the alumino-ferritic modulus (AF)

$$AF = \frac{Al_2O_3}{Fe_2O_3}$$

- and the degree of sulphatization (DS)

$$DS = \frac{SO_3}{1.292\ Na_2O + 0.85\ K_2O} \times 100$$

In the next chapter, we will see the usual values of these factors and moduli.

Of course, as previously said, the chemical composition of the raw meal is very important for the chemist who has to produce and control the quality of a given Portland cement, but for users it is the phasic composition of the cement that is more important, as well as the morphology of these phases after their cooling, especially when low W/B ratio concretes have to be made.

As will be seen, Portland cement is obtained by grinding simultaneously a small amount of calcium sulphate with Portland cement clinker. For a long time, gypsum was the calcium sulphate mineral used in this operation, but presently, different types of calcium sulphate are often used.

As long as most of the concrete used by the construction industry had a W/B ratio greater than 0.50, the usual W/B ratio used to verify the characteristics of Portland cement according to the standards, the nature of the calcium sulphate introduced during the grinding was not a real problem, but when using poly-sulphonate superplasticizers to produce concretes having W/B ratios lower than 0.40, it has been found that the solubility rate of the calcium sulphate added during the grinding of Portland cement clinker becomes critical (Kim 2000; Nkinamubazi et al. 2000). Therefore, the knowledge of the chemical composition, and even the phase composition, of some Portland cements is of no help to predict the rheological behavior of low W/B concretes made with this cement.

The amount of alkalis, the SO_3 content, and the *reactivity* of the interstitial phase play a key role in the rheology of low W/B ratio concretes. From a practical point of view, concrete rheology is as important, or even more, as the cube strength that is checked to evaluate the conformity of a cement to standards. As will be seen in Chapter 7, the soluble alkali content plays a key role on these rheological problems, known as compatibility problems or as the robustness of cement/superplasticizer combinations, when the superplasticizer used is a polysulphonate (polynaphthalene and polymelamine sulphonates, and lignosulphonates).

The chemical composition of a Portland cement is used to calculate the Bogue composition of this cement, also called its potential composition (Bogue 1952). Bogue composition is based on a certain number of hypotheses and assumptions that are detailed in Appendix V (p. 448). Formulas have been developed by Bogue to calculate the amount of di- and tricalcium silicates, tricalcium aluminate and tetracalcium ferroaluminate phases that can be formed when all the potential chemical reactions are fully developed. The predictive value of Bogue calculation will be discussed later on.

4.2 The phase composition of Portland cement

A simple way to identify the different mineral phases present in a crystallized solid is to conduct an X-ray diffractogram. The X-ray diffractogram of a Portland cement is shown in Figure 4.4(a). In this figure, it is seen that Portland cement is a complex multiphase material composed essentially of tricalcium and dicalcium silicates and of other mineral phases that exist in much smaller proportions, which makes them rather difficult to identify.

In order to identify these others phases, several techniques have been proposed. They consist essentially in selective chemical attacks. For example, when Portland cement is attacked by salicylic acid (Appendix IV, p. 447 the two silicate phases are dissolved. The resulting X-ray diffractogram, presented in Figure 4.4(b), shows that the remaining phases are tricalcium aluminate, tetracalcium ferroaluminate, calcium sulphate and periclase (MgO). When the cement is submitted to a KOSH treatment (Appendix IV, p. 447), only the silicate phases remain and can be identified in the X-ray diagram (Figure 4.4(c)).

The three diffractograms presented in Figure 4.4 show how the different oxides identified by the chemical analysis are combined. Therefore, Portland cement is essentially a mixture of

- tricalcium silicate
- dicalcium silicate
- tricalcium aluminate
- tetracalcium ferroaluminate
- calcium sulphates.

In order to have a much better idea, and even a quantitative idea, of the proportions of the different phases, a polished section of a piece of clinker can be made. Figure 4.5 represents such a surface after its etching with an aqueous solution that contains 1 per cent succinic acid. Large hexagonal crystals, as well as round and striated can be seen. These two types of crystals are surrounded by a more or less well crystallized interstitial phase. Using other

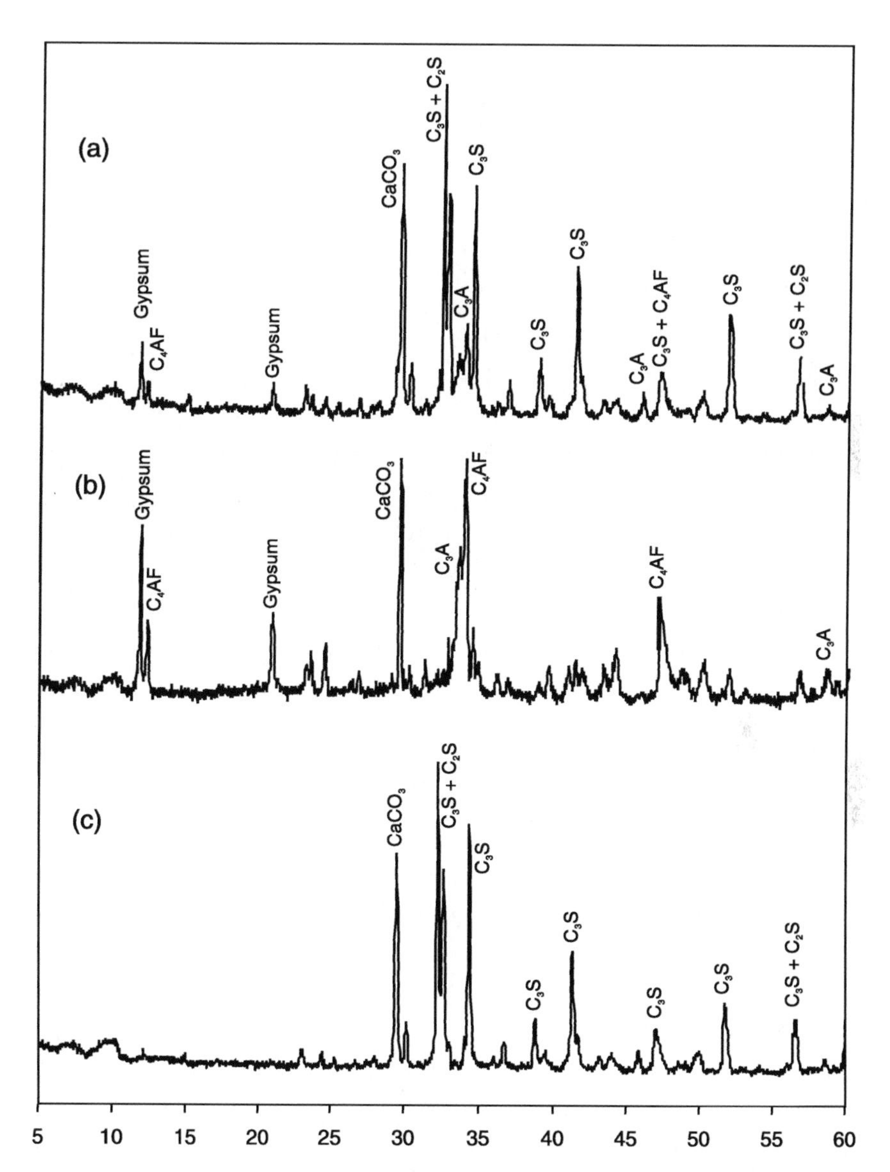

Figure 4.4 (a) X-ray diffractogram of a Portland cement; (b) X-ray diffractogram of the same cement, after a treatment with salicylic acid; (c) X-ray diffractogram of the same cement, after a KOSH treatment (courtesy of Mladenka Saric Coric).

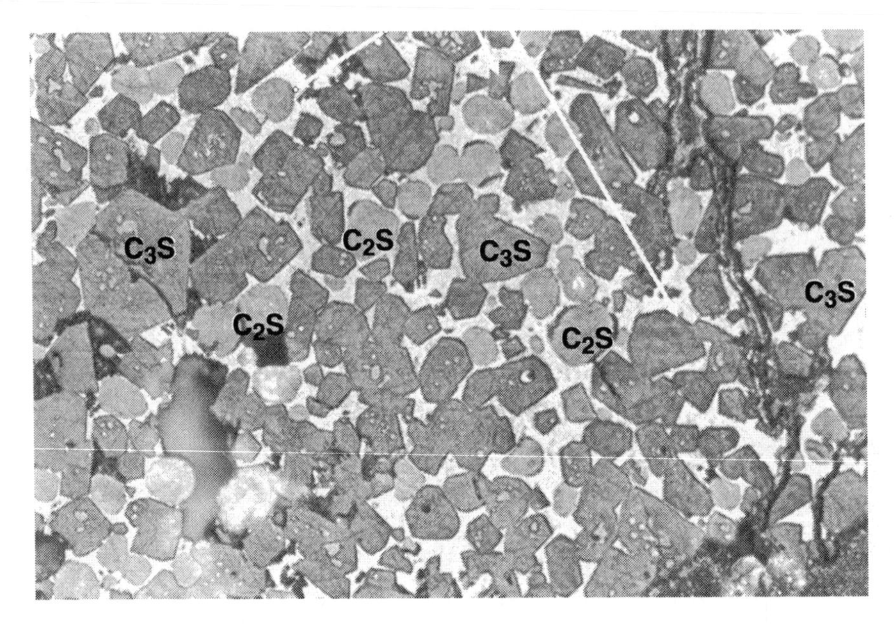

Figure 4.5 Polished surface of a clinker particle after its attack with succinic acid (courtesy of Arezki Tagnit-Hamou).

mineralogical techniques, it is easy to identify these three different phases. The large hexagonal crystals are more or less pure tricalcium silicate crystals, the rounded striated crystals are more or less pure dicalcium silicate, and the interstitial phase can be composed of a mixture of more or less well crystallized tricalcium aluminate, tetraferrocalcium aluminate, or it can be essentially vitreous depending on the quenching conditions that prevailed after the passage of the clinker in the clinkering zone.

The morphology of clinker grains can also be observed with a scanning electron microscope (SEM). Figure 4.6 represents the fractured surfaces of two clinkers where large polygonal crystals can be observerd (tricalcium silicate), as well as striated round ones (dicalcium silicate). In between these crystals, the interstitial phase can be observed. It is composed of tricalcium aluminate and tetracalcium ferroaluminate. The insterstitial phase acts as a cement which binds the silicate phases within the clinker particles.

A chemical spot analysis using a microprobe can be used to identify the qualitative chemical composition of the four main phases found in Portland cement clinker. This chemical analysis shows that pure phases do not exist in Portland cement clinker, but rather impure phases. This is why Thorborn, in 1897 (Bogue 1952), gave them different names: alite for the tricalcium silicate, belite for the bicalcium silicate, celite for the tricalcium aluminate and felite for the tetracalcium ferroaluminate. The terms 'celite' and 'felite' are no longer used.

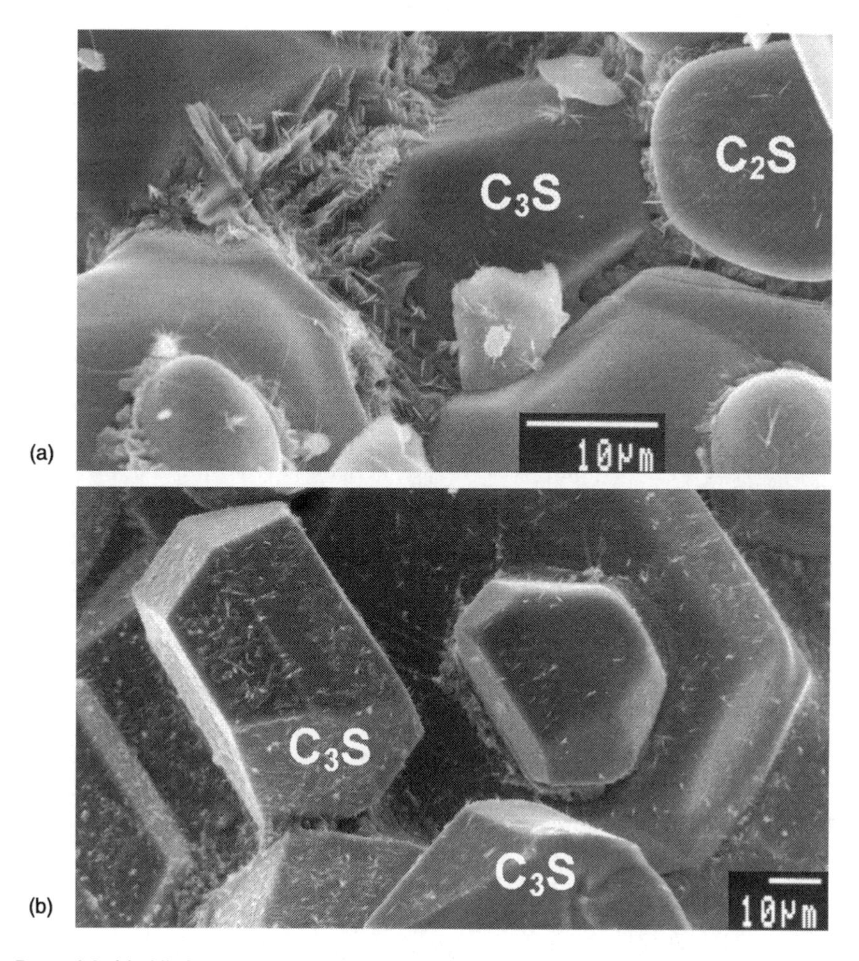

Figure 4.6 (a) Alkalis sulphates deposited on C_3S and C_2S and on the interstitial phase; (b) alkalis sulphates on C_3S crystals (courtesy of Arezki Tagnit-Hamou).

Optical and electronic microscopical examinations definitely show that Portland cement is composed of two coarsely grained silicate phases composed of impure crystals of di- and tricalcium silicate cemented by an interstitial phase composed of more or less well crystallized tricalcium aluminate and tetracalcium ferroaluminate phases. The examination of Portland cement clinker produced in different cement plants confirms such a crystallographical pattern. The tricalcium and dicalcium silicates are quite coarse and well defined, while the morphology of the interstitial phase is more variable from one clinker to the other.

4.3 Simplified chemical notations

As ceramic materials in general, and Portland cement in particular, are essentially composed of a limited number of oxides: SiO_2, Al_2O_3, CaO, Fe_2O_3, MgO, SO_3, etc., simplified chemical notations are used to identify these oxides in order to simplify the writing of the different minerals found in these materials. For example, S is used for SiO_2, C for CaO, A for Al_2O_3, F for Fe_2O_3, and $\bar{S}$ for SO_3, etc. In Tables 4.3 and 4.4, the symbols used and the simplified form of the most important crystalline phases found in Portland cement clinker can be found.

Consequently, in the following tricalcium silicate will no longer be identified as $3\ CaO \cdot SiO_2$ but as C_3S, dicalcium silicate will be written as C_2S, tricalcium aluminate as C_3A and tetracalcium ferroaluminate as C_4AF.

The hydrated form of calcium silicate that is formed when water reacts with C_3S as well as C_2S will be written as C-S-H (H representing H_2O). C, S and H are separated by a hyphen because the mineral composition of hydrated calcium silicate is not well defined. In fact, chemical analyses show that the C/S ratio can vary over an extended range (1.2 to 1.9) and even in some cases, be outside this range. Usually, the C/S ratio is taken as 1.5 so that some researchers instead of writing C-S-H, write rather $C_3S_2H_3$ which is the abbreviated form for $3\ CaO \cdot 2\ SiO_2 \cdot 3\ H_2O$.

4.4 Bogue potential composition

As it is clear that Portland cement clinker oxides are combined, after firing, in the form of four main crystalline phases, it is quite easy to calculate the *potential* phase composition of a Portland cement clinker from its chemical

Table 4.3 Simplified symbols used to write Portland cement chemical composition

Oxyde	SiO_2	CaO	Al_2O_3	Fe_2O_3	MgO	K_2O	Na_2O	SO_3	H_2O
Symbol	S	C	A	F	M	K	N	$\bar{S}$	H

Table 4.4 Simplified writing of the major cristalline phases found in a Portland cement

Pure phase	*Name*	*Simplified writing*	*Impure phase*
$3CaO.SiO_2$	Tricalcium silicate	C_3S	Alite
$2CaO.SiO_2$	Dicalcium silicate	C_2S	Belite
$3CaO.Al_2O_3$	Tricalcium aluminate	C_3A	Sometimes celite
$4CaO.Al_2O_3.Fe_2O_3$	Tetrtacalcium ferroaluminate	C_4AF	Sometimes felite
$CaSO_4$	Calcium sulphate	$C\bar{S}$	
$CaSO_4.2H_2O$	Gypsum	$C\bar{S}2H$	
	Hydrated calcium silicate	C-S-H	

composition. It is only necessary to assume that all the chemicals reactions necessary to form the four principal phases are completed in the clinkering zone, and that these four phases are absolutely pure.

As cement producers usually select a raw meal that is somewhat richer in lime than strictly necessary in order to form the maximum amount of C_3S possible, cement chemical analyses always contain the total lime content and the free lime content. Therefore, the Bogue calculation starts by substracting the free lime from the total lime content to calculate the amount of lime that could actually be combined with the other oxides.

In order to calculate the potential phase composition of a Portland cement, Bogue (1952) proposed the following hypotheses:

1 All the SO_3 is combined with lime as calcium sulphate $C\bar{S}$.
2 All the Fe_2O_3 has been combined to form C_4AF.
3 The rest of the Al_2O_3 is combined as C_3A.
4 The rest of the CaO and SiO_2 are combined into C_3S and C_2S.

As will be seen in the next chapter, not all the chemical reactions are fully developed during the short time the raw materials spend in the clinkering zone, which is why the Bogue composition is only a potential composition (Deloye 1991). Moreover, nearly everyone today forgets the ten pages of calculations proposed by Bogue (1952) to take into account the different conditions of clinker formation. However, as Bogue calculations are so easy to perform, the Bogue composition is still very popular within the cement industry. This method is presented in Appendix V (p. 448) to this volume.

Quantitative phase calculations through X-ray diffractogram analysis are quite long and complex and, of course, not much used. The quantitative determination of the different phases of a Portland cement is sometimes performed by optical microscopy because it gives a good idea of the actual phase composition of the Portland cement and also because it permits a direct observation of the size of the C_3S and C_2S crystals, of the porosity of the interstitial phase and of its degree of crystallinity. In Table 4.5, the Bogue potential composition and the optical phase composition of a low alkali clinker are compared. This is a case where Bogue hypotheses have the greater chances of being justified: very little SO_3 is combined into alkali sulphate and very little Na_2O can be trapped in the C_3A. In spite of these favourable conditions, some discrepancies between the phase compositions are found. It is also seen that a direct microscopical observation brings very useful information about the porosity of the clinker, its amount of free lime and MgO content. Moreover, the average size of C_3S and C_2S crystals gives an indication of the firing conditions that prevailed during the clinkerization process.

In conclusion, in spite of its limitations, it can be said that it is not only easy to calculate the potential Bogue composition of a Portland cement clinker but it is also useful. Calculating Bogue composition is also an easy way to control

Table 4.5 Comparison of the phase composition of a clinker calculated according to Bogue formulae and that found by optical microscopy

	Bogue formula	*Quantitative analysis Optical microscopy*
C_3S	70	75
C_2S	9	3
C_3A	6	6
C_4AF	12	14
CaO	nd	2
Periclase MgO	0	0
Porosity (% vol.)	–	12.5

the consistency of a particular cement, however, all the hypotheses on which the Bogue calculation depends must be kept in mind. Consequently, it is not necessary to give the results of Bogue calculations with two digits after the decimal point as is too often seen in mill data sheets. In the following, Bogue composition will always be rounded out to the nearest 1 per cent value, for C_2S and C_3S, and 0.1 per cent for C_3A and C_4AF, which is still very optimistic.

4.5 The ionic nature of the mineral phases found in Portland cement clinker

As has been seen, it is very important to have full control of the chemical composition of the raw meal introduced in a cement kiln, because the four main mineralogical phases found in Portland cement can be formed simultaneously in a relatively limited domain of chemical compositions. The Bogue composition of a Portland cement is also very useful, to have an idea of the relative proportions of its four main mineral phases, but it must be remembered that the actual phases found in a Portland cement clinker are not pure because of the inclusion or substitution of certain ions. Therefore, it is very important to see how the different chemical elements that constitute Portland cement are bound together.

The bonds combining the different elements are ionic in nature. It is not our intention to enter very deeply into crystallographic considerations about crystallographic structures, but it is important to realize that the four main crystallographic phases that constitute Portland cement – C_3S, C_2S, C_3A and C_4AF – do not exist as molecules, they exist as crystals. A crystal is composed of positive and negative ions. A positive ion can be considered as an atom which has lost one or several of its outer electrons while a negative ion can be considered as an atom that has gained some extra electrons in its outer layer. In a crystal, positive and negative ions are located in organized positions which are essentially governed by relatively simple rules:

- The overall crystalline system must be neutral from an electrical point of view.
- All the electrostatic forces must be balanced.
- The volumetric arrangement of the positive and negative ions is such that it minimizes the energy of the system in the particular thermodynamic conditions prevailing.

Therefore, the spatial organization of ions in a crystal depends on their size and electrovalence, that is, their number of extra or missing electrons.

It is easy to guess that positive ions (cations) that have lost one, two or three electrons from their outer layers will be smaller than the corresponding atom because the unchanged number of positive charges of the nucleus more strongly attract the remaining electrons. The opposite is true for negative ions (anions): the positive charges of the nucleus have to attract more electrons, therefore, anions are larger than the corresponding atoms. However, this very simplistic explanation does not explain why Al^{3+} and Fe^{3+} ions, which have quite different atomic masses, have very close ionic diameters and can easily be substituted one for the other, to a certain extent, in many crystalline structures and, in particular, in the interstitial phase of Portland cement.

Tables 4.6 and 4.7 give the diameter of the principal ions found in Portland cement. This table shows that the Si^{4+} ion is the smallest ion found among the four main minerals of Portland cement. The Si^{4+} ion can occasionally be replaced by Al^{3+}, which has an ionic diameter varying between 0.49 and 0.55 Å. Because of the difference in the size of these two positive ions, significant distortions will be caused by such a substitution, therefore, the number of substituted Si^{+4} and Al^{+3} ions will be limited. Moreover, such a substitution between two ions that do not have the same electrical valence implies the involvement of an accompanying monovalent ion to respect the principle of electrical neutrality.

A close observation of natural and industrial minerals shows that such impurities during formation create distortions in the crystallographic network and can alter the 'reactivity' of these crystals. The C_3S, C_2S, C_3A and C_4AF

Table 4.6 Ionic radius in angstrom of the most important ions found in hydraulic binders (for coordinance number of 6)

Cations	*Ionic radius*	*Anions*	*Ionic radius*
K^+	1.33	Cl^-	1.81
Ca^{2+}	1.06	OH^-	1.32
Al^{3+}			
Na^+	0.98	O^-	1.32
Mg^{2+}	0.78		
Fe^{3+}	0.67		
Si^{4+}	0.39		

Table 4.7 Ionic radius in angstrom of some other ions found in hydraulic binders (for a coordinance number of 6)

S^{2-}	F^{-}	Cr^{3+}	Ti^{3+}	Cr^{6+}	P^{5+}	C^{4+}
1.85	1.36	0.65	0.64	0.35	0.34	0.15

crystals found in Portland cement are far from being pure crystals; they always contains impurities. This is why cement chemists and mineralogists prefer to use the terms 'alite' and 'belite' to designate the impure C_3S and C_2S crystals found in Portland cement (Bogue 1952).

Moreover, the presence of certain foreign ions in a crystalline network can result in a change of the crystalline network when they are numerous enough to create significant distortions. This is the case of C_3A which, when it is pure, crystallizes in a cubic network. This cubic network is maintained as long as the Na_2O content of the C_3A remains lower than 2.4 per cent, after which Na_2O starts to crystallize in an orthorhombic network up to 5.3 per cent of Na_2O content. Above 5.3 per cent, C_3A crystallizes in a monoclinic network. As is observed, the 'rheological reactivity' of C_3A depends on its crystalline network and the morphology of the sulphoaluminate crystals that are formed when C_3A reacts with the calcium sulphate added during the grinding of Portland cement to control its setting.

4.6 Amorphous solids: glasses

Not all of the materials used to manufacture hydraulic binders are crystalline. Some of them are amorphous, that is, they exist on a glassy state. We will focus on the main structural difference between crystalline minerals and glasses. In a crystalline solid, ions are organized in an repetitive manner so that a crystal can be represented as a spatial repetition of a unit cell. On the contrary, in amorphous solids (glasses), ions or group of ions present a disorganized structure. At the most, at the level of a few elemental cells, a kind of spatial organization can be observed, but it never extends very far. Figure 4.7 represent a two-dimensional schematic representation of crystalline and amorphous materials.

When an amorphous solid is submitted to X-ray radiation, instead of observing a certain number of well-defined peaks as in a crystalline material, the diffractogram shows a diffuse hump. The position of this hump varies from one amorphous solid to another. In fact, this hump is usually located where the main peak of the crystalline solid having the same chemical composition as the amorphous solid under study would have been if the amorphous solid had had a chance to crystallize (Figure 4.8).

We have reached a point where most of the fundamental notions necessary to understand the binding properties of Portland cement have been reviewed.

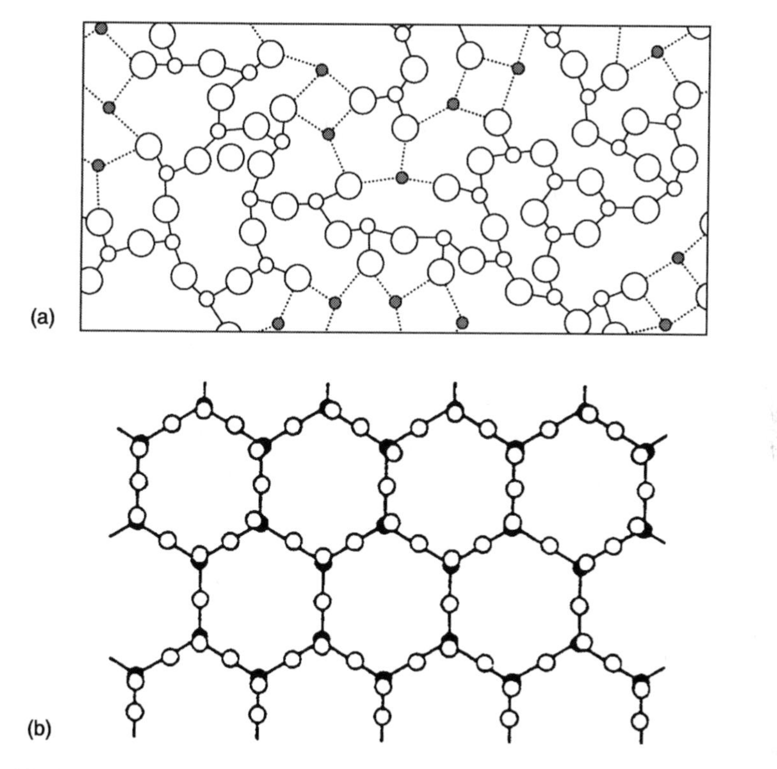

Figure 4.7 Comparison of the ionic structure of a glass and of a mineral (a) disorganized structure of a glass (2D); (b) structure of a layered silicate (phyllosilicate $(Si_2O_3)^{2-}$) (2D).

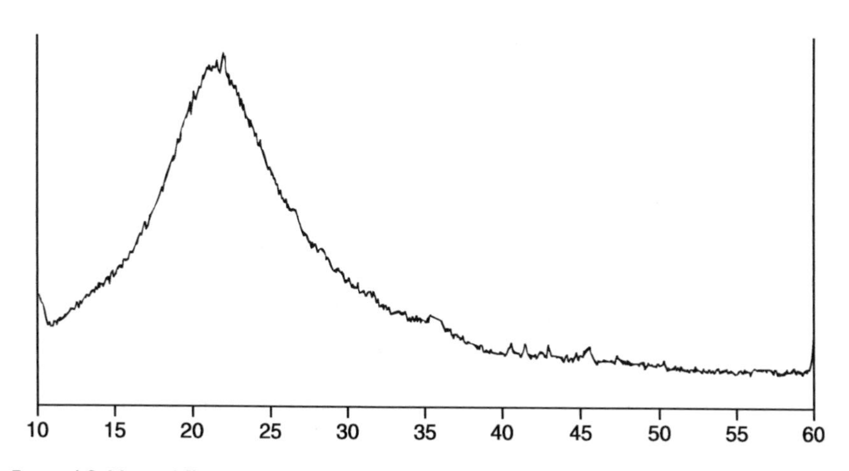

Figure 4.8 X-ray diffractogram of a vitreous slag.

Of course, there are fundamental textbooks where the influence of various ions on Portland cement reactivity can be found (Taylor 1997), but this matter is beyond the objective of this practical book for civil engineers. It is time now to look at how the different ions are combined, through the study of different phase diagrams, to understand how the raw meal is transformed into C_3S, C_2S, C_3A and C_4AF.

4.7 The $CaO–SiO_2$ and $SiO_2–CaO–Al_2O_3$ phase diagrams

4.7.1 Introduction

Microscopic observation of Portland cement clinker shows that it is composed of two parts: the silicate phases, which represent about 80 per cent of the clinker and the interstitial phase which represents 15 per cent. Therefore, we will look first at the $SiO_2–CaO$ phase diagram.

Before looking at this diagram in more detail, it must be kept in mind how these phase diagrams are obtained, in order to better understand not only what can be learned from them but also their limitations when they are used to explain the formation of Portland cement clinker from the raw meal.

The information presented in a phase diagram is obtained by heating to fusion a series of combinations of pure oxides which are cooled down at different temperatures before being quenched to fix the mineral phases present at that temperature. Afterwards, the mineral phases are identified either by X-ray analysis or by direct observation under an optical microscope.

Establishing a phase diagram is actually a long and painful process. For example, to establish the first $SiO_2–CaO–Al_2O_3$ phase diagram, Rankine and Wright studied 7000 compositions corresponding to different combinations of SiO_2, CaO and Al_2O_3 (Papadakis et al. 1970). The results are presented as fusibility diagrams or phase diagrams. This last expression is the one we will use. In a binary diagram, the contents of the two phases are reported in the x axis and the fusion and solidification temperatures on the y axis.

When the experimental work has been completed, the different mineral phases that can be obtained by mixing the two oxides are clearly identified, as well as the melting temperature of the mixture of these new phases (the liquidus) and their solidification temperature (the solidus). In between the liquidus and the solidus, some crystals are already formed in the liquid phase; the closer to the temperature of the solidus, the higher the amount of crystals. On these diagrams, the phase transformations that occur within the solid when it is cooled down very slowly are also reported. Moreover, in certain phase diagrams it can be observed that certain specific compositions melt at a constant temperature, like a pure phase. These points are called eutectics (see Appendix III, p. 446).

The $SiO_2–CaO$ phase diagram presented in Figure 4.9 appears to be quite complex because several minerals can be formed when combining CaO and SiO_2 in different proportions. In the context of this book, only the part of the diagram comprised between the C_3S and C_2S will be looked at more closely because this is the only part that is of interest when making Portland cement clinker (Figure 4.10).

4.7.2 The CaO–SiO_2 phase diagram

Figure 4.10 represents the part of the SiO_2–CaO phase diagram comprised between C_3S and C_2S. In this part of the phase diagram, it is seen that the lowest temperature at which a mixture of CaO and SiO_2 can be entirely melted is

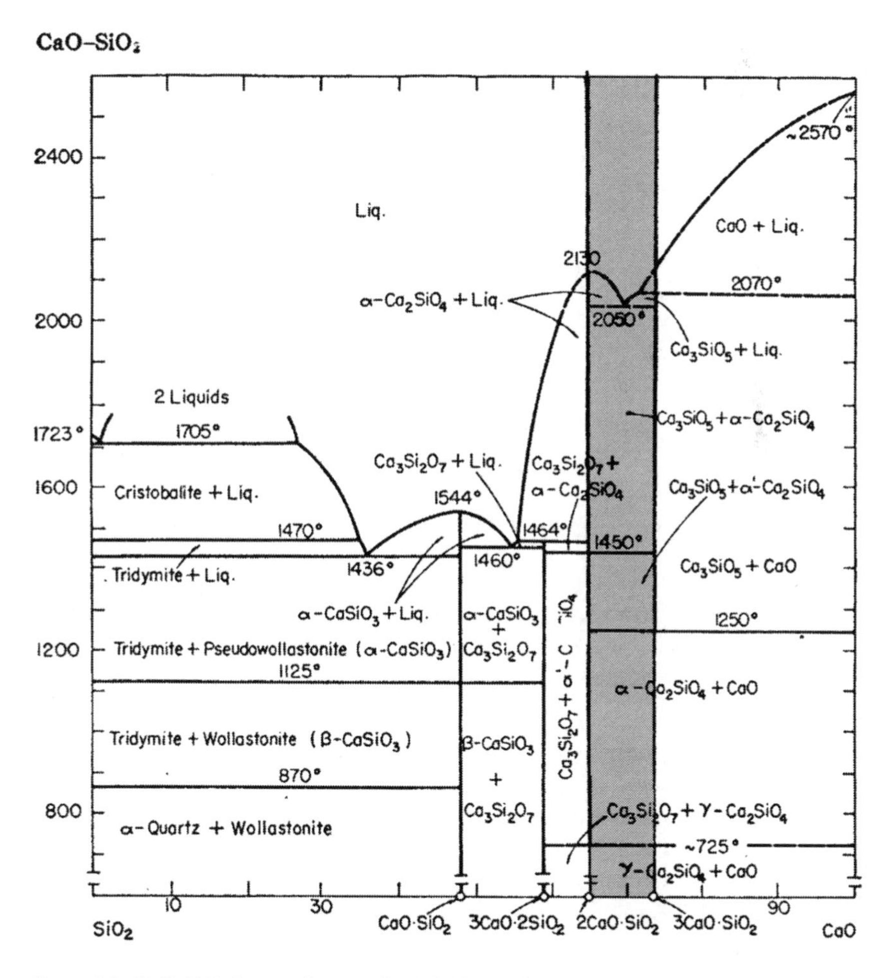

Figure 4.9 CaO–SiO_2 binary diagram from Phillips and Muan (1959) (Reprinted with permission of the American Ceramic Society, www.ceramics.org. All rights reserved).

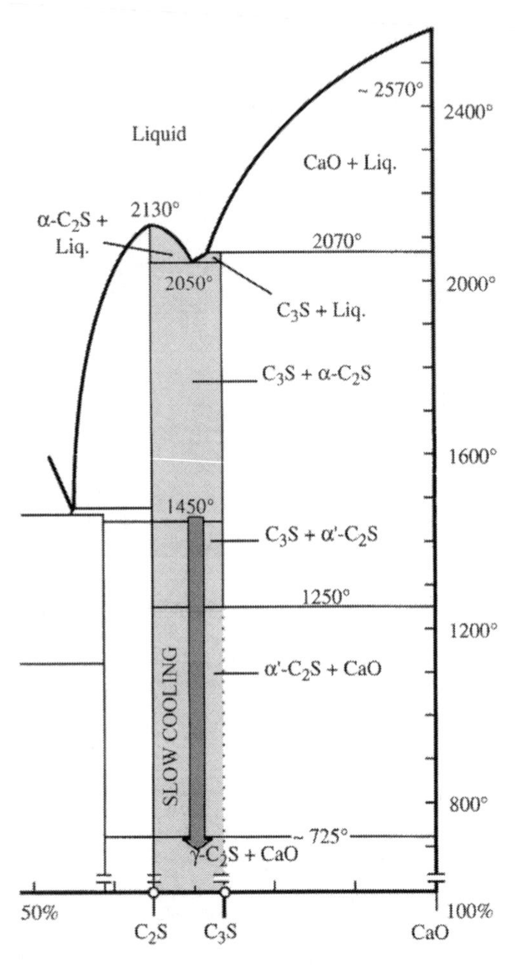

Figure 4.10 $CaO–SiO_2$ binary phase diagram in the C_2S and C_3S areas (Reprinted with permission of the American Ceramic Society, www.ceramics.org. All rights reserved).

2050°C. It is also seen that a mixture of C_3S and α C_2S (α C_2S is the hydraulic form of C_2S) can coexist if the temperature of the mix is greater than 1450°C. Later on, we will see the importance of this temperature from a processing point of view.

What is also observed in this phase diagram is that if a mixture having a chemical composition lying between that of C_3S and C_2S is cooled down very slowly, several mineralogical transformations occur in the solid phase during cooling. The mixture of C_3S and α–C_2S is transformed, first, into a mixture of C_3S and α′–C_2S as soon as the temperature decreases below 1450°C, then the mixture is transformed into α'–C_2S and CaO when a temperature of 1250°C is reached, and finally into γ–C_2S, which does not present any hydraulic properties and CaO when a temperature of 725°C is reached.

Therefore, this brief examination of the CaO–SiO_2 phase diagram shows that:

- it is out of the question to produce C_3S and C_2S through an industrial process consisting of the melting of an adequate mixture of CaO and SiO_2, as the melting temperature required is too high;
- the minimum temperature at which C_3S and α–C_2S can coexist is 1450°C;
- it is absolutely necessary to quench the mixture of C_3S and α–C_2S in order that they keep the structure they have at high temperature and to avoid them ending up as a mixture of lime and γ–C_2S at the end of a slow cooling process.

Consequently, in order to be able to produce Portland cement clinker at the lowest temperature possible, it is necessary to add fluxing elements. These fluxing elements are Al and Fe. As Al^{3+} and Fe^{3+} ions can be substituted quite easily one with the other (they are trivalent ions and have ionic diameters which are very close), we will study the phase diagram CaO–SiO_2–Al_2O_3 rather than the more complicated CaO–SiO_2–Al_2O_3–Fe_2O_3 quaternary phase diagram (Sorrentino 1974).

4.7.3 The CaO–SiO_2–Al_2O_3 phase diagram

Figure 4.11 represents the CaO–SiO_2–Al_2O_3 phase diagram in all its complexity. It is a very important phase diagram from a technological point of view because it is not only used by the Portland cement industry but also by the aluminous cement industry, the refractory industry, the glass industry and even metallurgists when they want to adjust the melting temperature of their slag (Figure 4.12).

This phase diagram is based on a triangular representation that uses the properties of the Rooseboon triangle presented in Appendix VII (p. 453). General definitions on ternary phase diagrams can also be found in this appendix.

Figure 4.13 represents the useful part of Figure 4.11 when studying Portland cement, which is the composition zone where the three phases C_3S, C_2S and C_3A coexist. This is quite a small domain within the whole phase diagram. This domain is situated close to the summit representing pure CaO. It is seen that alumina play the role of a fluxing oxide, which is quite unusual. Usually Al_2O_3 is more associated with a refractory behavior. As tricalcium aluminate has quite a high melting temperature, some Fe_2O_3 must be added during the manufacturing process of Portland cement.

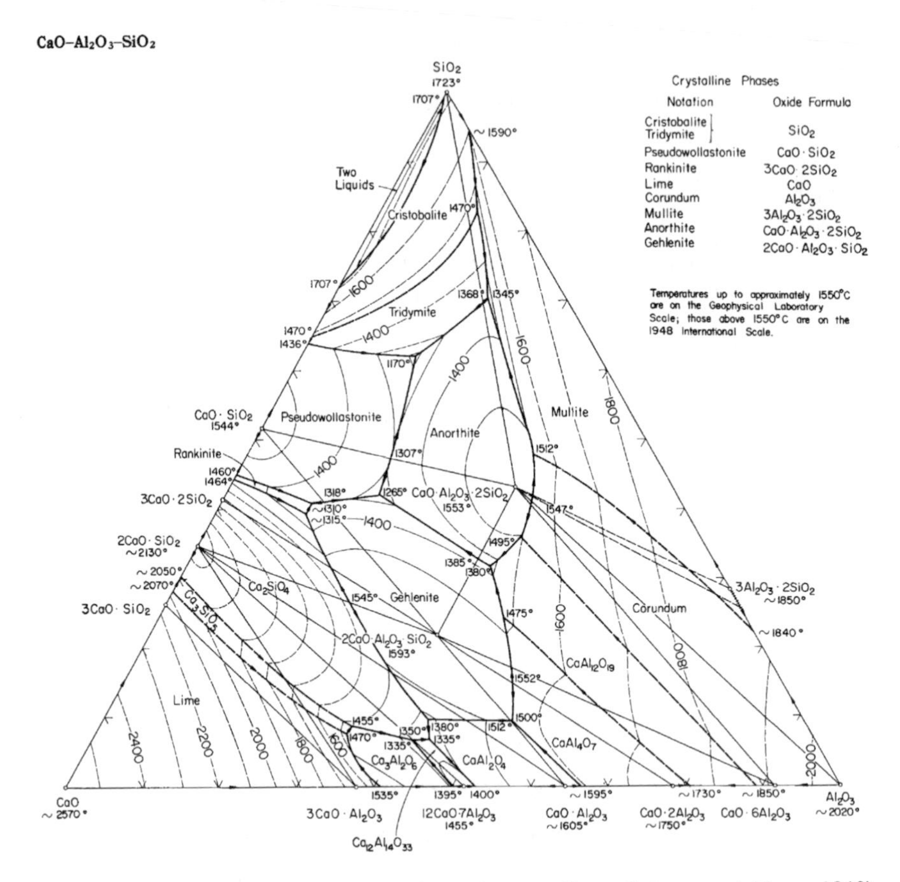

Figure 4.11 $CaO–SiO_2–Al_2O_3$ ternary phase diagram (from Osborn and Muan, 1960) (Reprinted with permission of the American Ceramic Society, www.ceramics.org. All rights reserved).

It has already been said that the manufacturing of Portland cement consists in developing two different mineralogical systems: a ferroaluminous calcium phase that is liquefied at a quite low temperature and a silicocalcium phase that is more 'refractory'. In the silicocalcium phase, mineralogical transformations can only be developed in solid phases through a diffusion process. Maintaining good balance and full control of these mineralogical processes constitutes the challenge of making Portland cement clinker, as will be seen in the next chapter.

In Appendix VII (pp. 453–9), a certain number of theoretical exercices show how to use phase diagrams.

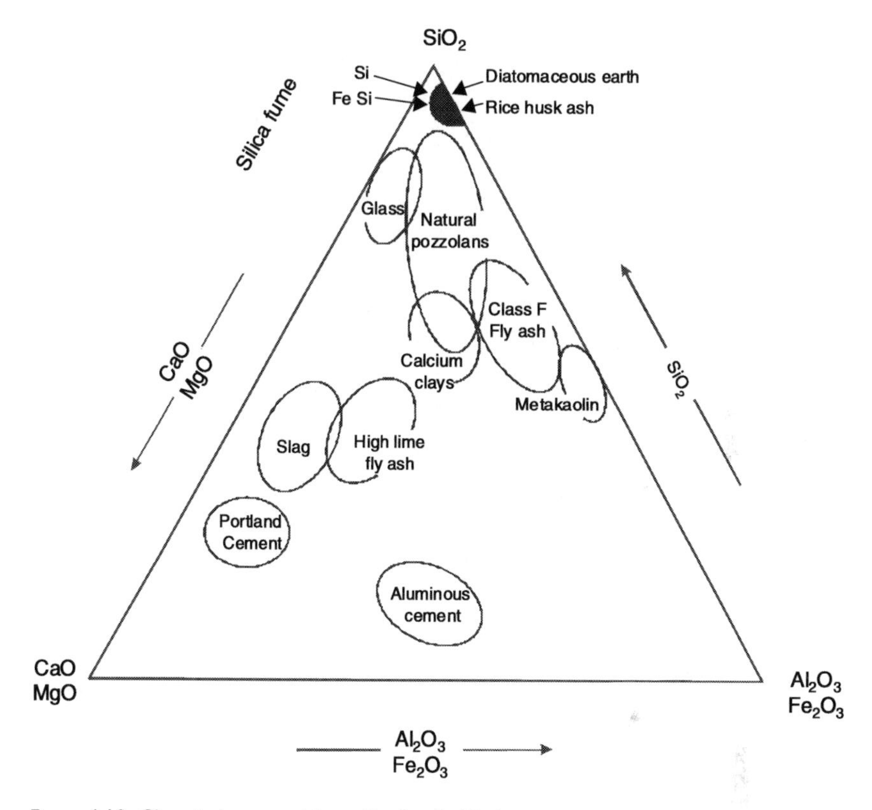

Figure 4.12 Chemical composition of hydraulic binder components.

4.8 Making of Portland cement clinker

It has been seen that it is out of the question to form C_3S and C_2S through a melting process; rather, a diffusion process in the solid phases must be used. In order to favour such a process, it is very important to introduce the raw meal in the kiln in the form of finely divided particles. It is therefore necessary to grind the raw material before introducing it into the kiln.

As the raw meal progresses within the kiln, it is exposed to increasing temperature and a certain number of chemical and mineralogical transformations begin occurring. Finally, when all the lime and silica have been transformed into C_3S and C_2S in the clinkering zone, it has been seen that it is necessary to quench the clinker to fix the C_3S and C_2S in their highly hydraulic forms.

We will now see how this is done from a theoretical point of view using the graphic representation developed by KHD Humboldt Wedag (in 1986) (Figure 4.14). In this representation, the raw meal is a mixture of pure

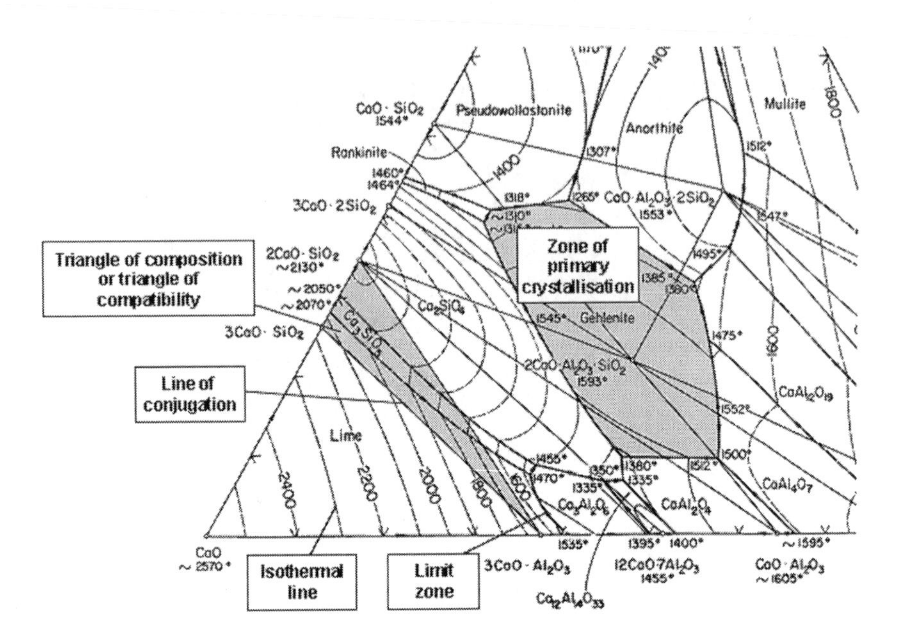

Figure 4.13 SiO_2–CaO–Al_2O_3 phase diagram in the C_3S–C_2S–C_3A (from Osborn and Muan, 1960) (Reprinted with permission of the American Ceramic Society, www.ceramics.org. All rights reserved).

limestone, clay, β quartz and Fe_2O_3. Figure 4.14 represents all the chemical transformations of the raw material until the final quenching of the Portland cement clinker. Temperature is recorded on the x axis and the y axis represents the mass of the different phases found at each temperature.

First, it is obvious that the clinkering of the raw meal occurs through a loss of the solid mass. This mass loss is equal to the mass of water and carbon dioxide injected into the atmosphere during processing. If L_0 and L_f are measured on the graph, it is found that to produce 1 tonne of clinker it is necessary to introduce about 1.5 tonnes of raw material in the kiln.

The first transformation that occurs in the kiln takes place between 500 and 600°C, which corresponds to a loss of water from the clay. The second transformation has very little consequence and corresponds to the transformation of β quartz into α quartz at 565°C. The next transformation during heating occurs at 700°C and is completed at 900°C; it corresponds to the decarbonation of the limestone. The first CaO groups that are liberated are quite active and start to combine with Al_2O_3 to form $C_{12}A_7$ and with some of the Fe^{3+} and Al^{3+} ions to form $C_2(A,F)$. At 700°C some belite (C_2S) is already formed.

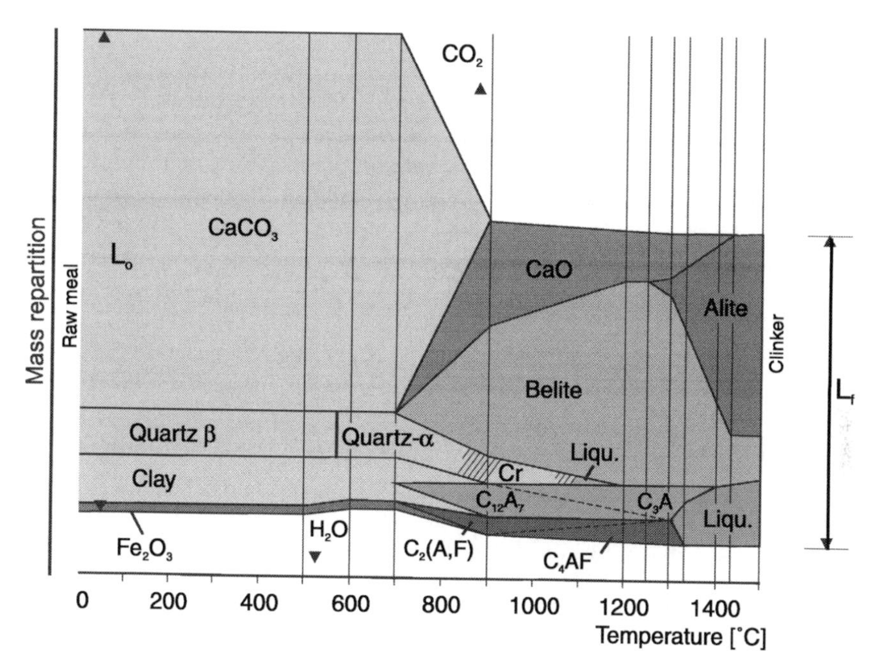

Figure 4.14 Transformation of the raw meal into clinker (with the permission of KHD Humboldt Wedag, 1986).

At 900°C:

- all the limestone has been decarbonated;
- some alumina has already reacted with the lime to form $C_{12}A_7$;
- some lime has reacted with Al^{3+} and Fe^{3+} to form $C_2(A,F)$;
- some lime has already been transformed into belite (C_2S).

However, there is still plenty of lime that has not been combined and some SiO_2 in the form of α quartz.

Beyond 900°C, the first C_3A and C_4AF crystals appear and α quartz is transformed into cristobalite (a mineral variety of SiO_2, stable at high temperature).

At 1100°C, cristobalite melts which accelerates its combination with lime. At 1200°C there is no longer any form of free silica.

The first C_3S crystals, or rather alite crystals, begin to form at 1250°C. These crystals are formed through the reaction of the excess free lime with the already-formed belite crystals. At 1300°C, the mixture of C_3A and C_4Af starts to melt; C_4AF disappears at 1325°C and C_3A around 1400°C. During this time, the ferroaluminous calcium melt acts as a catalyst for the transformation of belite into alite by favouring the diffusion of Ca^{2+} ions into the belite crystals. Above 1425°C, there is practically no more excess lime. In fact, as it will be seen in the next chapter, in practice, a small amount of uncombined lime is always found in Portland cement clinkers. In the chemical analysis of Portland cement, this excess lime is identified as free lime.

All the chemical and mineralogical reactions necessary to transform the raw meal into Portland cement have now been accomplished. In the clinkering zone, three different phases have been formed: alite (impure C_3S), belite (impure C_2S) and a liquid interstitial phase. If the length of the segments representing the proportion of each phase in the clinker is measured in Figure 4.14, it is seen that in the particular case corresponding to that of the figure, the clinker that has been produced contains 64 per cent of alite, 14 per cent of belite and 21 per cent interstitial phase.

When the clinker has been exposed to the clinkering temperature, it is only necessary to quench it to fix C_3S and C_2S in their active form. According to the quenching rate, the interstitial phase will have time or not to start to crystallize or to simply crystallize partially. Therefore, microscopical observation of the interstitial phases indicates the quenching process, which is important from rheological and strength points of view.

4.9 Conclusion

Portland cement is a complex multiphase material; its full characterization involves several analysis techniques that have to be cross-checked. Presently, all of the required techniques are available, but most of them are time-consuming and necessitate the use of expensive apparatus as well as the training of skilled technicians. This is why most cement producers and users have to rely on the data that can be deduced from the chemical analysis, which is easy and fast to perform in a repetitive manner and from some empirical tests, as will be seen in the next chapter. However, each time a better knowledge of Portland cement characteristics is needed, it is always a good and profitable investment to perform a more detailed characterization. Those interested by these techniques will find many specialized books of the matter (Regourd 1982; Taylor 1997; Hewlett 1998; Ramachandran and Beaudoin 2001).

In the next chapter, we will see how the theoretical principles that have been presented in this chapter are put into application to produce millions of tonnes of Portland cement every day.

References

Aïtcin, P.-C. (1983) *Technologie des granulats*, Quebec: Éditions du Griffon d'Argile.

Bogue, R.H. (1952) *La Chimie du ciment Portland*, Paris: Eyrolles.

Callister, W.D. Jr (2000) *Material Science and Engineering: An Introduction*, New York: John Wiley and Sons.

Campbell, D.H. (1986) *Microscopical Examination and Interpretation of Portland Cement and Clinker*, Skoke, Ill.: Construction Technology Laboratories.

Deloye, F.-X. (1991) 'Du bon usage de la formule de Bogue', *Bulletin de Liaison des Laboratoires des Ponts et Chaussées*, 176 (November–December): 81–5.

Dorlot, J.M., Baïlon, J.-P., and Masounave, J. (1986) *Des matériaux*, Montreal: Éditions de l'École Polytechnique de Montréal.

Hewlett, P.C. (1998) *Lea's Chemistry of Cement and Concrete*, London: Arnold.

Humboldt Wedag (1986) 'L'Évolution technologique du four rotatif avec préchauffeur à cyclones vers le four court PYRORAPIDfi' technical document 11.0/f by KHD Humboldt Wedag AG.

Jouenne, C.A. (1990) *Traité de céramiques et matériaux minéraux*, Paris: Édition Septima.

Kim, B.-G. (2000) 'Compatibility Between Cements and Superplasticizers in High-Performance Concrete: Influence of Alkali Content in Cement and of the Molecular Weight of PNS on the Properties of Cement Pastes and Concretes', Ph.D. thesis, Université de Sherbrooke.

Lafuma, H. (1965) *Liants hydrauliques*, Paris: Dunod.

Levin, E.M., Robbins, C.R. and McMurdie, H.F. (1964) *Phase Diagrams for Ceramists*, Vol. 1, Columbus, Ohio: American Ceramic Society.

Liévin, A. (1935a) 'Propriétés du diagramme ternaire appliquées à l'industrie des ciments' Part I, *Revue des Matériaux de Construction et de Travaux Publics*, 309 (June): 137–41.

Liévin, A. (1935b) 'Propriétés du diagramme ternaire appliquées à l'industrie des ciments', *Revue des Matériaux de Construction et de Travaux Publics*, 310 (July): 166–73.

Liebau, F. (1985) *Structural Chemistry of Silicates*, Berlin: Springer-Verlag.

Nkinamubanzi, P.-C., Kim, B.-G., Saric Coric, M. and Aïtcin, P.-C. (2000) 'Key Cement Factors Controlling the Compatibility Between Naphthalene-Based Superplasticizers and Ordinary Portland Cements', *Sixth CANMET/ACI International Conference on Superplasticizers and other Chemical Admixtures in Concrete, Supplementary Papers*, Nice, October, pp. 33–54.

Osborn, E.F. and Muan, A. (1960) 'Phase Equilibrium Diagrams of Oxyde systems', Columbus, Ohio: American Ceramic Society and Edward Orton Jr Foundation.

Papadakis, M., Venuat, M. and Vandamme, J. (1970) 'Industrie de la chaux, du ciment et du plâtre', Paris: Dunod.

Phillips, B. and Muan, A. (1959) 'System $CaO–SiO_2$', *Journal of the American Ceramic Society*, 42 (9): 414.

Pliskin, L. (1993) *La Fabrication du ciment Portland*, Paris: Eyrolles.

Ramachandran, V.S. and Beaudoin, J.J. (2001) Handbook of Analytical Techniques in Concrete Science and Technology: Principles, Techniques and Applications, Norwich, NJ: Noyes Publications.

Regourd, M. (1982) 'L'Hydratation du ciment Portland', in J. Baron and R. Sauterey, Paris: *Le Béton hydraulique*, Presses de l'École Nationale des Ponts et Chaussées, pp. 143–221.

Sorrentino, F. (1974) 'Studies in the system $CaO–Al_2O_3–SiO_2–Fe_2O_3$', doctoral thesis, University of Aberdeen.

Taylor, H.F.W. (1997) *Cement Chemistry*, London: Thomas Telford.

Chapter 5

Production of Portland cement

(This chapter was written in collaboration with J.-C. Weiss)

The objective of this chapter is not to describe in minute detail all the equipment used to transform limestone and clay into clinker and then clinker into Portland cement, it is instead to show how the basic notions seen in the previous chapter are applied to produce the most widely used artificial construction material: Portland cement. In this context, artificial is used as opposed to natural.

These who would like to have a better knowledge of the technological aspects of the production of Portland cement should consult the following books: Bogue (1952), Papadakis and Venuat (1966), Pliskin (1993) or *The World Cement Plant and Equipment Handbook* by Hardy et al. (1998).

5.1 Historical background of Portland cement production

The processing of the raw material and the production of Portland cement have evolved over the years. For example, in order to prepare his 'artificial lime', Vicat used a two-step method: he first produced hydrated lime and then mixed it with clay until he obtained a homogenous mix that he shaped into bowls (Mary 1862). When these bowls dried, they were piled in vertical kilns (Figures 5.1 and 5.2) similar to those used to produce lime (Candlot 1906; Leduc 1925; Bogue 1952; Gourdin 1984). In order to avoid the workers getting their hands burned with the small quick lime particles that always remain unhydrated in hydrated lime, Vicat provided them with pails of tar with which to protect their hands (Gourdin 1984).

Vicat thought of grinding dry chalk and mixing it with clay, but this method could be used only when chalk was available, whereas the previous two-step method could be used with any type of limestone (Bogue 1952). Moreover, because he wanted to be sure to use the exact proportions of raw materials to manufacture his artificial lime, he preferred the two-step method, whereas Aspdin and Johnson favoured a one-step firing process.

The first industrial processes used to make Portland cement were particularly long and involved a great number of workers, so that Portland cement could not

Figure 5.1 Old kilns in Genevray-de-Vif in France (courtesy of Groupe Vicat).

Figure 5.2 Old cement kilns (courtesy of Portland Cement Association).

compete on a price basis with the best hydraulic limes and later with the best natural cements. In England, the price of Portland cement was much higher than the price of the best natural cement, and in France 3 times the price of Le Teil lime that was the reference binder at that time (Gourdin 1984). This mark-up was accepted by some engineers who preferred the use of Portland cement rather than that of hydraulic lime. For example, as long ago as 1838, the English engineer Isambar K. Brunel preferred to use Portland cement rather than natural cement to build the first tunnel under the Thames river in London (Davis 1924). Later, as already mentioned, the foundations of the Statue of Liberty in New York were built with a Portland cement imported from the Dyckerhoff Company of Germany. This imported Portland cement was selected rather than an American natural cement because it had more regular and predictable properties. At that time, the transportation of Portland cement barrels was not very expensive because these barrels were used to ballast ships (Figure 5.3) (Dumez and Jeunemaître 2000).

In order to compete on a price basis with natural cements, it was necessary to develop an efficient homogenization process that could be used to deal with large amounts of raw materials. It was therefore necessary to begin by grinding and homogenizing the raw materials.

Initially, raw-meal homogenization used the so-called wet process (Figure 5.4a). The pulverized raw meal was mixed with water in large agitated basins to form a homogenous slurry. The first series of basins used to properly dose

Figure 5.3 Transportation of Portland cement in barrels (courtesy of Portland Cement Association).

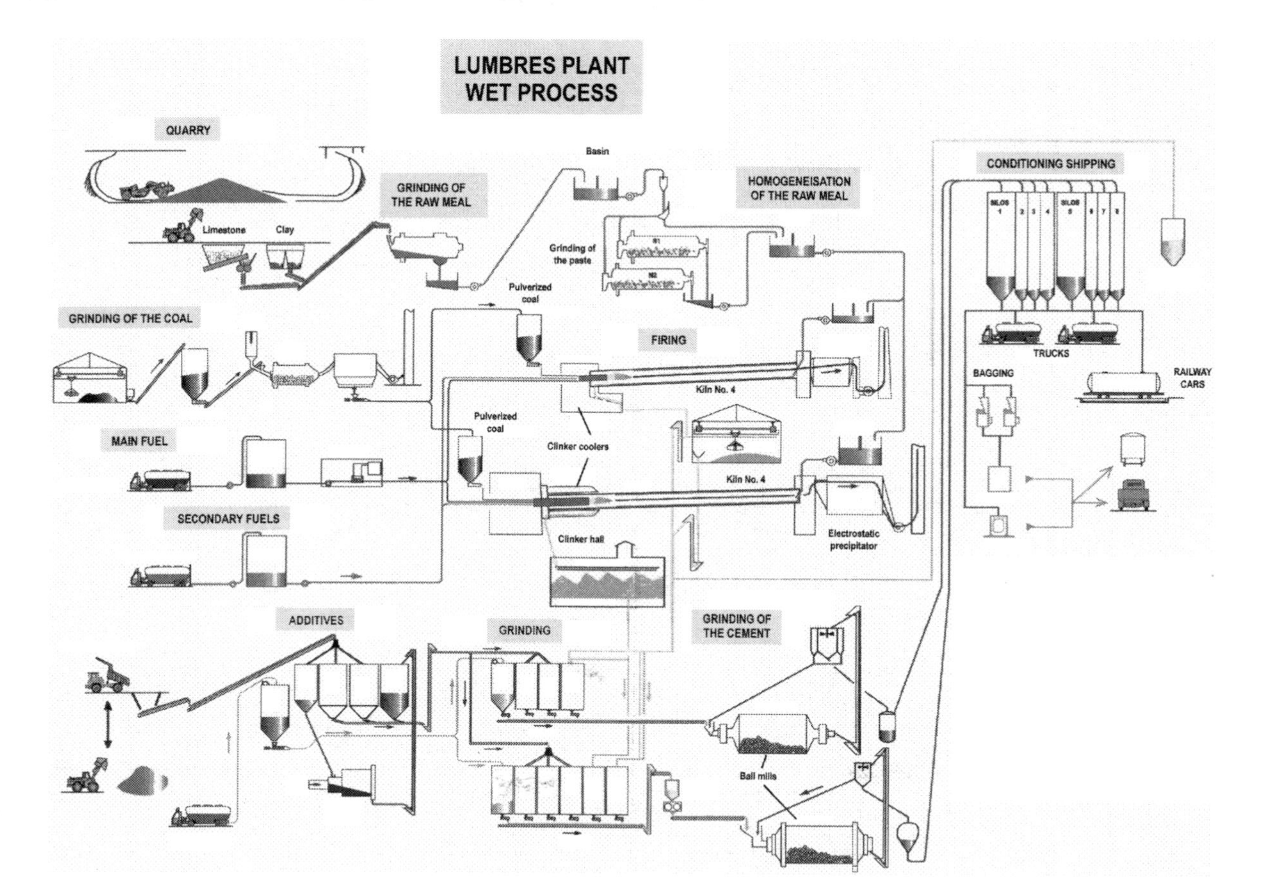

Figure 5.4(a) Wet process (courtesy of J.-C. Weiss from Ciments d'Obourg).

the raw materials were installed in Boulogne-sur-Mer in 1869 by the Dupont and Demare company (Papadakis and Venuat 1966). The slurry was then transferred into a large basin to be homogenized. In this basin, the slurry was continuously agitated by mechanical arms and by compressed air injected from the bottom of the basin. A few cement plants still use this wet process. In order to limit the amount of water necessary to form the slurry as much as possible, dispersants are used, essentially lignosulphonates, because they decrease the amount of water necessary to obtain a slurry of a given fluidity. Therefore, dispersants limit the amount of water to be evaporated. The water content of the slurry can be decreased to about 28–30 per cent. This amount of water is decreased in the semi-wet process (Figure 5.4b).

The wet process is a batch process. It presents the advantage of adjusting very precisely the chemical composition of the raw meal before its introduction in the kiln. As long as energy costs were not too high, it was not very important to devote a significant amount of energy to evaporate the water in the slurry, but the situation changed suddenly with the first petroleum crisis. It is at that time that the dry process started to be used (Figure 5.5a) and that it finally supplanted the wet process or the semi-wet or semi-dry processes. Semi-wet and semi-dry processes (Figure 5.5b) essentially consisted in granulating or extruding a paste containing 35 per cent of water before drying it first mechanically in filter presses or a pelletizer and finally with the heat recovered from the cooling of the clinker or with the gas leaving the kiln (Gourdin 1984). Presently, the dry process is used more and more, but it necessitates an efficient homogenization of the raw materials before their grinding. This homogenization is realized in large halls (Figures 5.6a, b and c). This pre-homogenization is followed by a second step of homogenization taking place after the grinding. The raw meal is homogenized in large silos where it is kept in movement with pressurized dry air.

Whatever the homogenization process used, the grinding of the raw meal is absolutely necessary in order to favour the chemical transformations that during firing will transform the raw meal into the two silicate phases and the interstitial liquid.

The firing of the raw meal evolved slowly but with spectacular results. In the beginning, the first clinkers were fired in vertical kilns looking very much like lime furnaces (Figures 5.1 and 5.2). In 1924, Lafarge was operating forty-two vertical kilns in one of its plants and eighty-two in its Beffes plant (Gourdin 1984).

The first rotating kilns, like the ones presently used in cement plants, appeared in 1895 in the USA. The use of rotating kilns resulted, in only five years, in a tenfold increase in cement output (Candlot 1906; Bogue 1952; Gourdin 1984). This kiln technology rapidly crossed the Atlantic to be applied in Europe. Fifty years later, the firing technology remains basically the same in spite of the fact that it is more sophisticated (Figure 5.7). Presently, it depends heavily on electronic controls and computer technology.

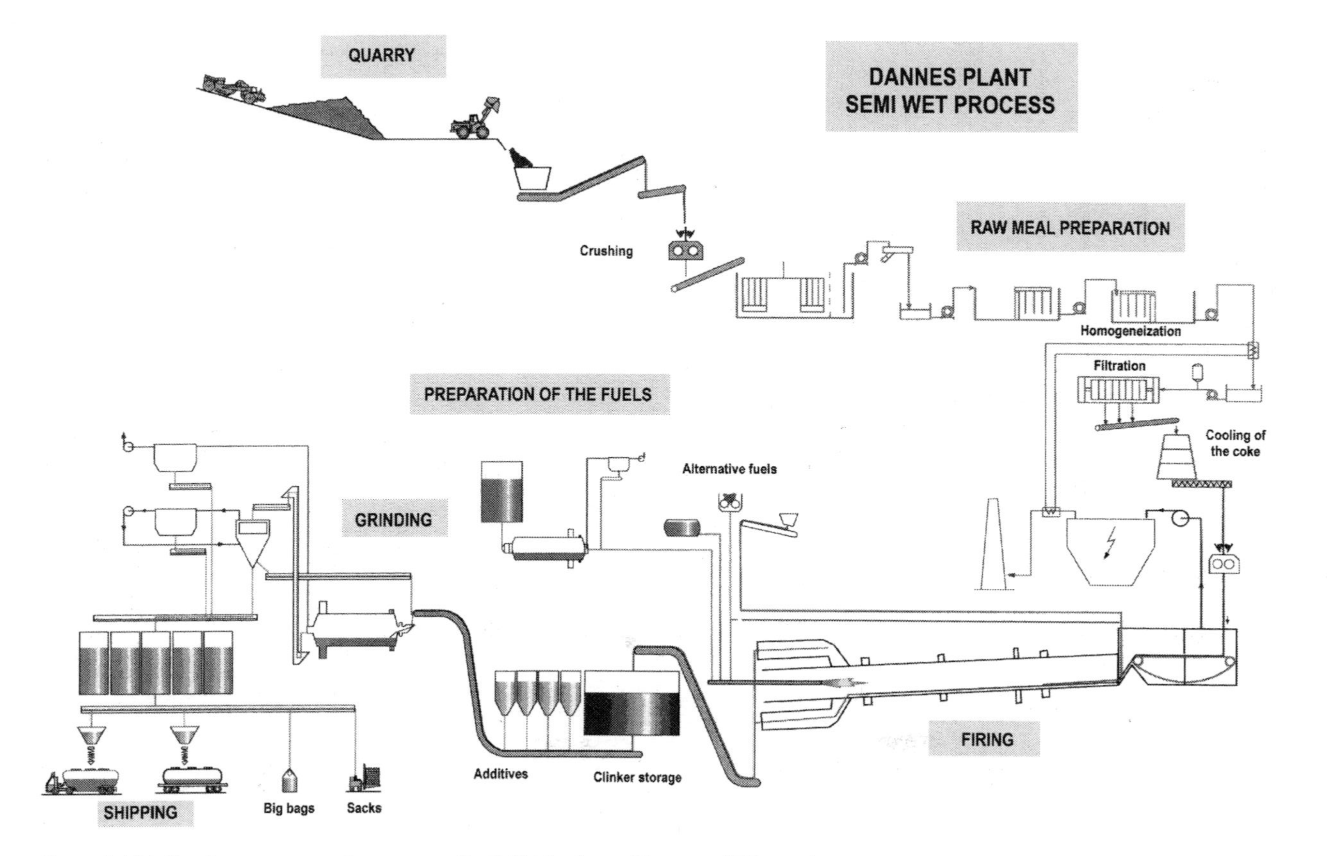

Figure 5.4(b) Semi-wet process (courtesy of J.-C. Weiss from Ciments d'Obourg).

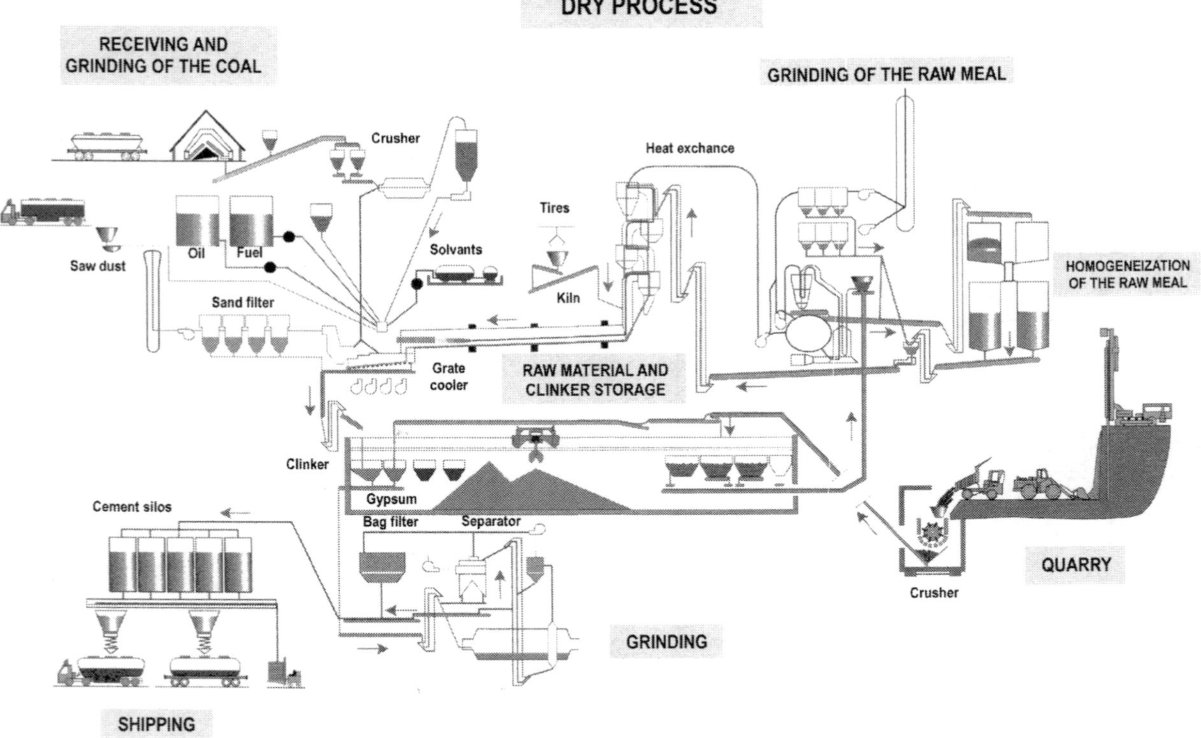

Figure 5.5(a) Dry process (courtesy of J.-C. Weiss from Ciments d'Obourg).

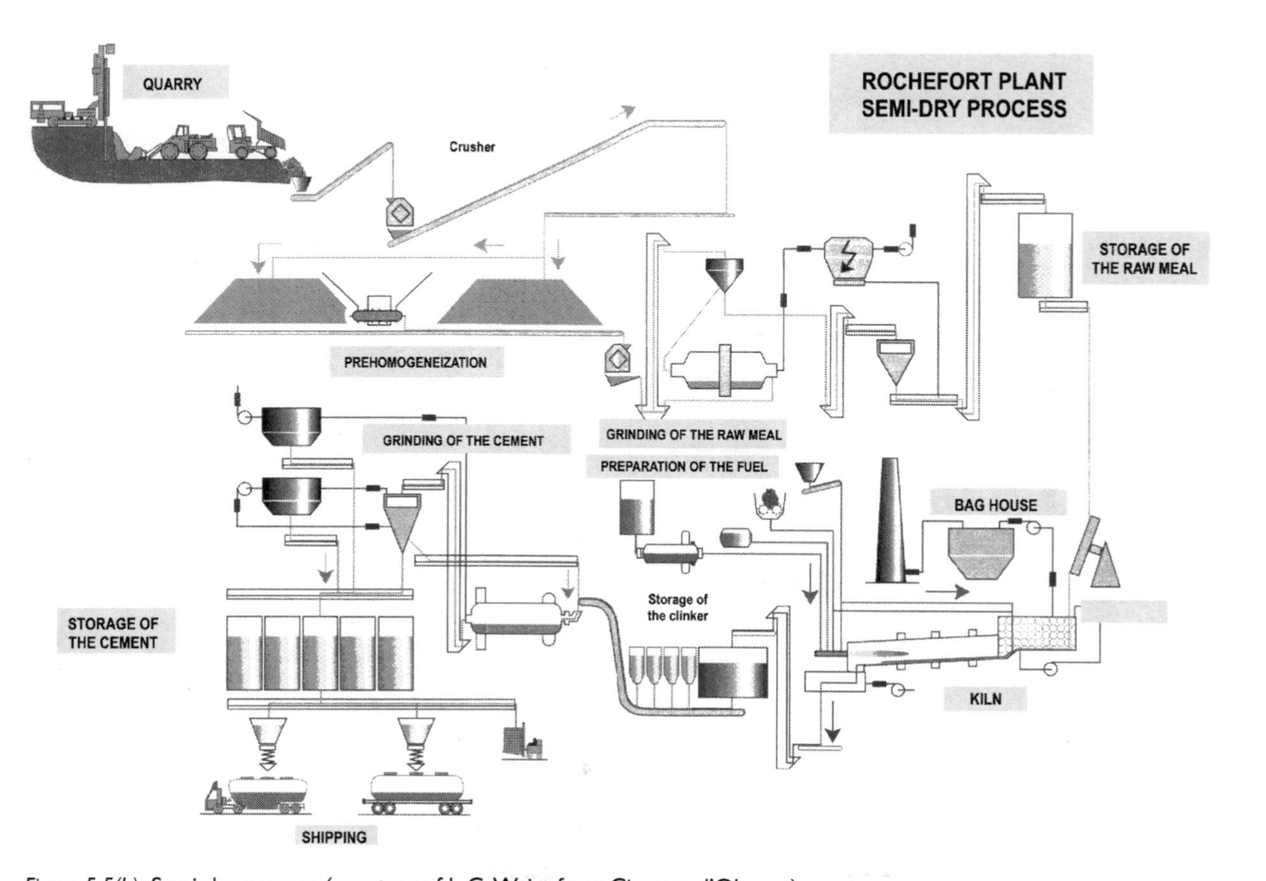

Figure 5.5(b) Semi-dry process (courtesy of J.-C. Weiss from Ciments d'Obourg).

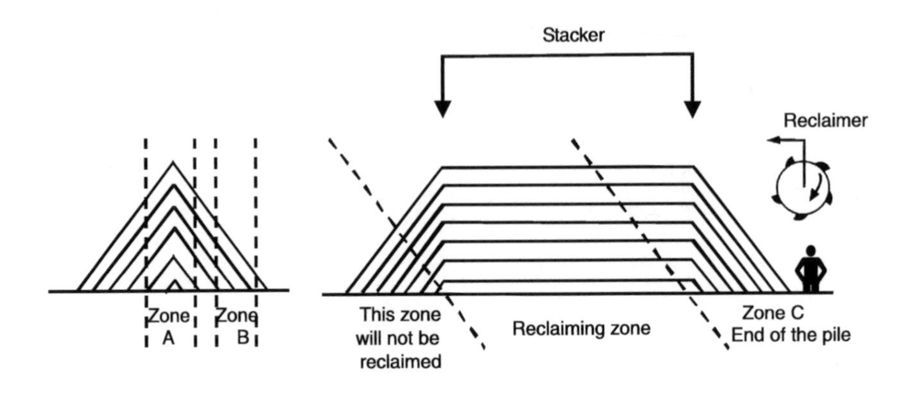

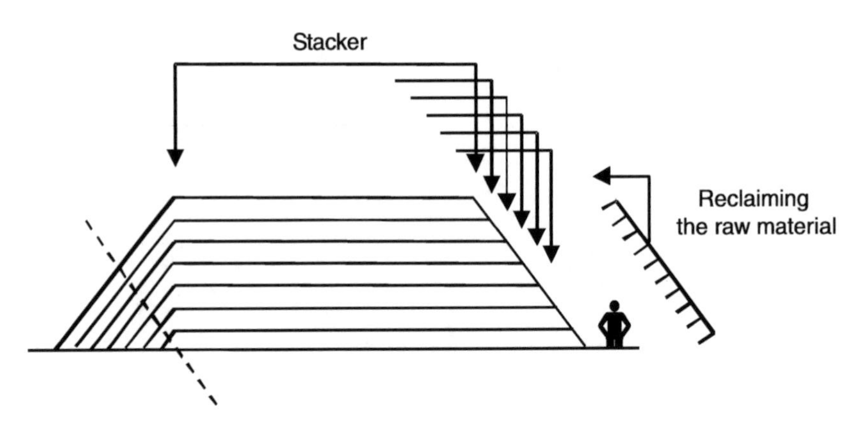

Figure 5.6(a) Structure of pre-homogenization stack piles (Pliskin 1993) (with permission of Ciments Français).

During the 1950s, the first pre-heaters were developed. In a pre-heater, the raw meal is partially decarbonated in a series of cyclones by the hot gases coming from the kiln. After its passage in the last cyclone, the raw meal presents a degree of decarbonation equal to 40 per cent.

Finally, during the 1970s, the first precalciners were developed (Figure 5.8). Now they are installed in almost all modern kilns. With an additional power input in the pre-heaters, it is possible to reach a 90 per cent degree of decarbonation. Consequently, the cement kiln can be shortened drastically, its diameter increased, and the heat losses reduced. Presently, some kilns can produce up to 10000 tonnes of clinker per day which is far greater than the 10 to 50 tonnes per day produced in the first vertical kilns in 1870 and 1880.

Figure 5.6(b) Reclaiming the raw materials in a pre-homogenization stock yard (courtesy of Lafarge).

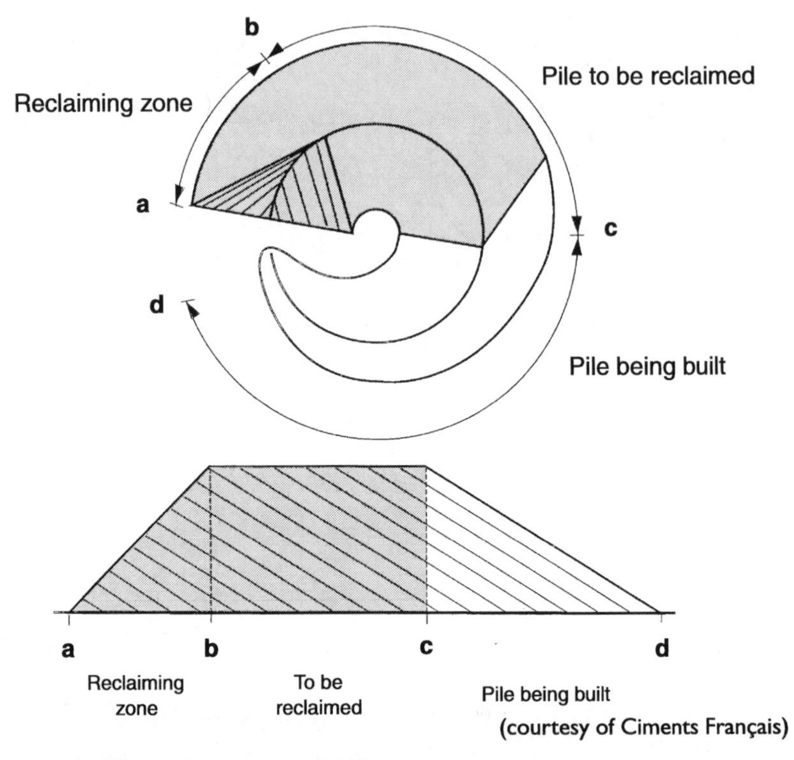

(courtesy of Ciments Français)

(courtesy of Lafarge)

Figure 5.6(c) Circular pre-homogenization system.

Figure 5.7 Long cement kiln without preheating and precalcination tower (St Lawrence Cement, Joliette Plant) (courtesy of Philippe Pinsonneault, Ciments St-Laurent).

Figure 5.8 Short cement kiln with a precalciner (courtesy of Groupe Vicat).

In order to transform clinker into Portland cement, it is only necessary to grind it with an appropriate amount of calcium sulphate. At first, the grinding was done using millstones made of hard rocks. Horizontal millstones were usually used. Later on, pendular millstones were used before ball mills such as those used in the mining industry were introduced in the cement industry (Figures 5.9a and b).

The first grinding mill was designed by Davidsen in France in 1892; its lining was composed of pieces of quartz and the grinding load was composed of beach pebbles. It was not until 1920 that the first steel balls were used to grind Portland cement. This technological development resulted in a drastic increase in the grinding capacity of cement plants (Papadakis and Venuat 1966). However, the energetic efficiency of a ball mill is quite low. Only about 5 to 10 per

Figure 5.9(a) Ball mills (courtesy of Ciment Québec).

Figure 5.9(b) Ball mill (courtesy of Ciment Québec).

Figure 5.10 Roller mill (courtesy of Polysius).

cent of the energy is actually used in comminution, the rest of the energy is dissipated as heat or noise. In order to improve the energy efficiency of the grinding system, particles are extracted from the ball mill by an air separator as soon as they have reached a satisfactory diameter. Recently, very efficient grinding units based on press rolls have been developed (Figure 5.10). After the first grinding stage with press rolls has been done, the final grinding is still carried out in a ball mill (Figure 5.9b).

As previously stated, air separator systems are becoming very sophisticated and efficient so that no cement particle passes more than three to four times within a ball mill, consequently the daily output of a grinding unit has increased spectacularly (Figure 5.11a and b).

In a ball mill, grinding is done in two compartments separated by a diaphragm (Figure 5.9a). The first compartment is filled with large balls and the second with smaller ones. In order to improve the efficiency of the ball mill, very small amounts of very specific chemicals called grinding aids are added within the grinding mill. In the best cases, the increase in output of the mill can

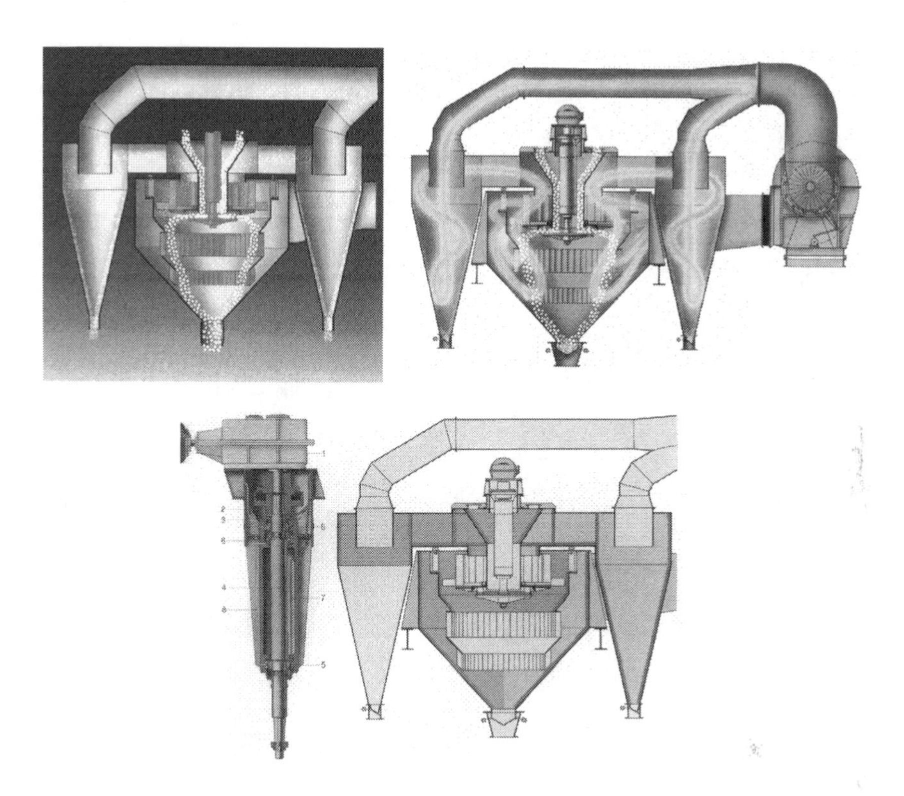

Figure 5.11a Schematic of cyclone separators (courtesy of KHD Humboldt Wedag).

reach 15 per cent and pack set problems can be eliminated during storage of the cement in the silos or during transportation. The ground particles are separated in an open air circuit and pneumatically transported to storage silos.

Cement conditioning evolved with time. At first, cement was sold by the barrel. A cement barrel contained 170 kg of cement (Figure 5.3). Cement was also conditioned in burlap sacks that had to be returned to the cement plant (Gourdin 1984) (Figure 5.12). It was a shortage of burlap that forced German producers to develop paper bags (Gourdin 1984). At this time, paper bags are used all over the world (Figures 5.13 and 5.14). The most common weight of a cement bag is 40 or 50 kg, but presently cement can also be sold in big bags of 1 or 2 tonnes.

When delivered in bulk, hydraulic binders are transported by railway car (Figure 5.15) or by truck (Figure 5.16) or, finally, by boat (Figure 5.17). In any case, loading and unloading operations are done pneumatically. Recently, large boats equipped with bagging and pneumatic facilities have appeared and act as floating silos. They are totally autonomous when they are moored in a wharf (Dumez and Jeunemaître 2000).

Figure 5.11b Cyclone separator (courtesy of Philippe Pinsonneault, Ciment St-Laurent).

In Figure 5.18, the production cost of Portland cement has been split into different components for a cement plant producing 1 million tonnes of cement. In this figure, it is seen that the two most important costs are financial costs and energy cost, followed by salaries and electricity. The cost of the paper can be significant in Third World countries, where most of the production is sold in paper bags. On the contrary, it is quite negligible, in the USA and Canada in particular, where less than 5 per cent of the cement production is bagged.

Figure 5.19 represents very schematically the economic challenge that all cement companies have to face in order to increase their productivity: production and marketing costs have to be squeezed to increase profits.

5.2 Manufacturing processes

After this historical introduction, we will come back in more detail on each principal step of the manufacturing of Portland cement. In a modern plant, Portland cement is manufactured in three major stages: the first one consists of the preparation of the raw meal, the second of the production of clinker and the third of the final grinding of the mixture of clinker and calcium sulphate, more and more often with the help of a grinding aid.

Figure 5.12 Fabrication of fabric sacks (courtesy of the Portland Cement Association).

Figure 5.13 Semi-automatic bagger (courtesy of Philippe Pinsonneault, Ciment St-Laurent).

Figure 5.14 Automatic roto-packer (courtesy of Philippe Pinsonneault, Ciment St-Laurent).

Figure 5.15 Railway car used to transport cement (courtesy of Philippe Pinsonneault, Ciment St-Laurent).

The preparation of the raw meal can be separated into different operations. The extraction of the raw materials, their preparation, their homogenization, their analysis, their dosage and the grinding of the raw meal before its introduction in the kiln.

The raw meal is then heated up to about 1450°C to form clinker. Clinker is quenched down to 1350°C in order to stabilize alite (C_3S) and belite (C_2S) crystals in their most reactive form. The clinker is then cooled down to 200°C. During this cooling operation, most of the heat is recovered to heat the secondary air that is injected in the burner of the kiln.

The clinker is then stored in special silos or halls; in some extreme cases it can be stored outside (which is not recommended). Afterwards, the clinker is ground with an appropriate amount of calcium sulphate or the clinker is mixed with an appropriate amount of cementitious material to produce Portland cement or a blended cement. Portland cement or blended cement are stored in silos before shipping (Kosmatka et al. 2002).

The hot gases used in the pre-heater to start the decarbonation of the limestone in the precalciner correspond to the tertiary air coming out of the cooling system. To increase the rate of decarbonation, an auxillary source of heat is necessary.

Figure 5.16 Cement tanker (courtesy of Groupe Vicat).

Figure 5.17 Loading of a ship used for cement transportation (courtesy of Philippe Pinsonneault, Ciment St-Laurent).

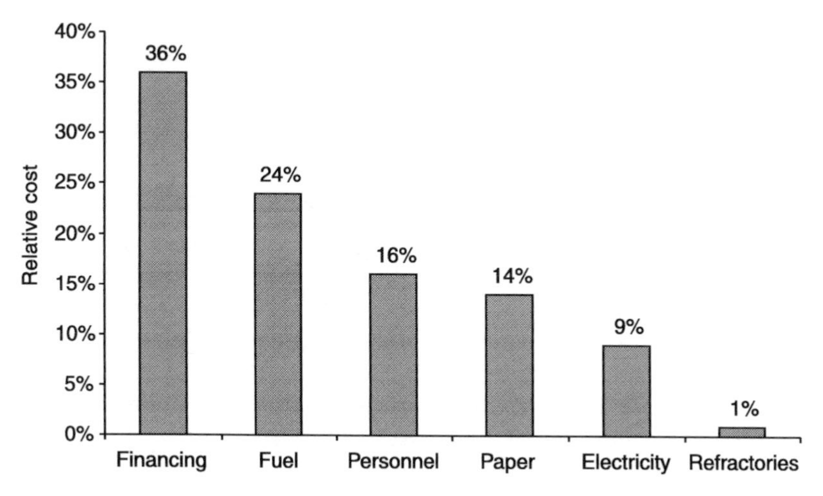

Figure 5.18 Relative cost in a 1-million-tonnes-per-year cement plant (from Scheubel and Nachtwey 1997).

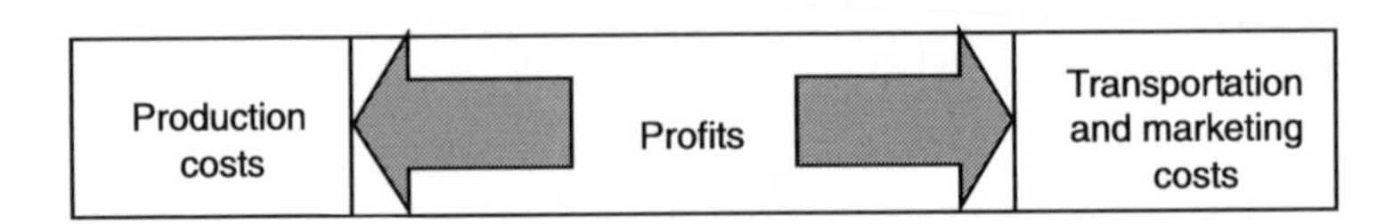

Figure 5.19 Commercialization of Portland cement.

Before their release into the atmosphere, the gases are cooled down to 200°C and passed through a dedusting system (very often electrostatic precipitators). The particles collected by the dedusting system are usually called cement kiln dust (CKD). The clean gases rejected in the atmosphere by the chimney are therefore essentially composed of CO_2, some water vapour and some NO_x and SO_2, as will be seen in Section 5.14.

The manufacturing of Portland cement requires a double grinding in a grinding mill and/or in a pressure roller. The first grinding is absolutely necessary to favour the solid/solid reactions that will occur at 1450°C and result in the formation of alite, belite and the interstitial phase. The second grinding is necessary to activate the reactivity of Portland cement clinker.

The mechanical improvements previously described have been very closely accompanied by remote control monitoring and automation. The control of a modern cement plant is done by computer using an expert system (Figure 5.20). In spite of this great effort to improve the efficiency of modern cement plants through automation, it must be mentioned that there are still unsophisticated but efficient cement plants operating 'manually' in some Third World countries, where electronic and computing support is limited. These plants also operate with a large workforce to create jobs.

In the following, we will describe in further detail two types of cement kilns, one operating with a pre-heater and another operating with a precalciner.

Presently, some cement plants still use the wet process through they are becoming less and less numerous due to their lack of thermal efficiency and very high CO_2 emission level. They remain in operation when there is no other way to produce clinker, especially where chalk quarries are below the water table. The energetic efficiency of the wet process, from a cost point of view, can be improved by burning alternative fuels (domestic waste, scrap tyres, used oils, varnishes, resins, etc.), as will be seen in Chapter 11.

5.3 Selection and preparation of the raw materials

The construction of a modern cement plant capable of producing 1 million tonnes of cement per year represents an investment of more than US$200 million. It is an important investment for a plant that cannot produce anything other than cement. It is essential to secure enough raw materials to operate the cement plant over about fifty years and it is essential to be sure that the local

Figure 5.20 Control room in recent cement plants.

market and/or the export market will be strong enough to absorb this production.

As the market price of Portland cement is not very high, it rapidly becomes uneconomical to transport the raw materials over long distances. As the lime content of Portland cement is about 65 per cent, a cement plant is always built as close as possible to a limestone quarry (Pliskin 1993). The selected limestone or lime source should have a constant chemical composition, as well as shale or clay deposits. The raw material used to produce Portland cement can also be a clayey marl or an impure limestone having approximately the chemical composition of Portland cement. The chemical composition of the raw material is often corrected by small additions that compensate for the low content or the lack of one or several of the four major oxides in the raw materials available. Of course, depending on the process used (wet or dry), homogenization requirements are different but always essential.

When it is certain that a fifty-year supply of raw materials having a satisfactory chemical composition is available, the first problem to address it how to homogenize the raw materials in order to produce a clinker having characteristics and properties as constant as possible. Of course, as the raw materials are unprocessed natural materials, it is necessary to be sure to feed the kiln with a constant raw meal, because when this raw meal enters the kiln, cement manufacturing becomes a very rigid process, lacking flexibility or having a long response time of several hours for any modification of the composition of the raw meal. At present, pre-homogenization halls are widely used. It is still only a dream to be able to control the raw meal 'in line' by appropriate additions in the first grinding unit.

The raw materials extracted from the quarry are transferred to large crushers that reduce their size to a maximum diameter of 80 to 50 mm. Crushed aggregates are then piled in a 150-mm thick layer. When the piling is finished, the aggregates are retrieved vertically by a bucket wheel operating in a direction perpendicular to the one used to build the pile, as seen in Figures 5.6a, b and c. In fact, a pre-homogenization hall contains two piles of raw materials, one which is being built, the other which provides raw materials having a constant chemical composition to the grinding unit. The different corrective materials (iron oxide, alumina, silica) necessary to adjust the composition of the raw mill are introduced in the primary grinder.

The fineness of the raw meal is crucial because it conditions, to a certain extent, the characteristics of the clinker. Particularly, quartz particles must be crushed finely enough (Figure 5.21) to avoid the occurrence of too many belite nests in the final clinker. In the same way, limestone particles should be fine enough to avoid clusters of free lime.

After grinding, the raw meal is stored in a silo where it is homogenized by air before being sent to the kiln or in a pre-heater or a precalciner. The use of the dry process imposes, after the pre-homogenization hall, the use of special silos where the raw meal is continually mixed to complete the homogenization of the raw meal.

In the case of a wet process, the raw materials are ground under water in a ball mill. The raw meal is then sent into large basins where it is agitated by rotating arms and by compressed air that homogenize the slurry. We will return later to the different types of grinding systems used when we examine the final grinding of the clinker.

The dosage of the different chemical oxides is adjusted to produce the kind of clinker sought. It is, of course, essential to combine the maximum amount of lime while avoiding too great an excess. Usually, clinkers contain about 1 per cent excess lime after firing. CaO dosage is controlled using the lime saturation factor LSF:

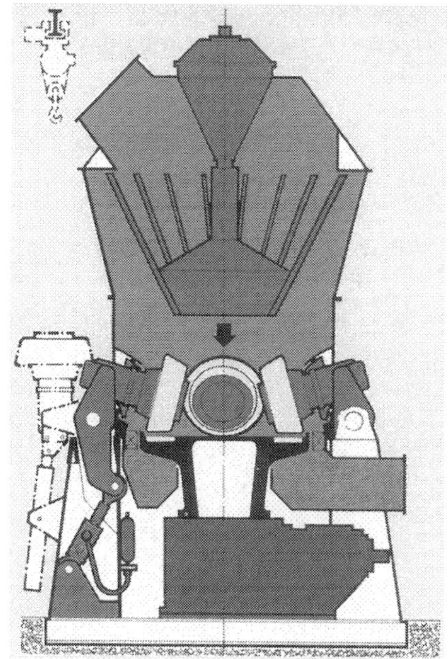

Figure 5.21 Grinding machine (courtesy of Loesche GmbH).

$$LSF_{Kühl} = \frac{CaO}{2.80\ SiO_2 + 1.10\ Al_2O_3 + 0.70\ Fe_2O_3}$$

From a practical point of view, the optimal value of the LSF is between 0.95 and 0.97 in order to leave 1 per cent free lime in the clinker. An excess lime value that is too low (0.5 per cent) corresponds to a hard-burned clinker, while one that is too high (2 to 3 per cent) corresponds to an underfiring, which means that the clinker will not be as reactive as it could have been. According to Pliskin (1993), Type I normal Portland cement usually have a LSF of 0.90 to 0.95 and of 0.95 to 0.98 for high early strength cements.

The silica modulus $SM = \frac{SiO_2}{Al_2O_3 + Fe_2O_3}$ usually lies between 2 and 3 for grey Portland cements and between 4 to 7 for white Portland cements (Pliskin 1993) in order to maintain a sufficient amount of liquid phase during the clinkering to favor the formation of C_3S.

The alumino-ferrite modulus $AF = \frac{Al_2O_3}{Fe_2O_3}$ varies according to the type of cement being produced (heat of hydration, resistance to sulphates). This ratio is very important because it controls the viscosity of the interstitial phase. If the AF modulus is lower than 0.64, C_3A cannot be formed, and the corresponding clinker has a very low heat of hydration and a very good resistance to sulphates. In Type 10 clinkers, the AF modulus lies between 1.8 and 2.8 but it can be as low as 1.3 or as high as 4.0 in some cases (Pliskin 1993).

5.4 The fuels

The selection of the type of fuel to be used is crucial from an economic point of view, considering the effect its cost has on the cost of production of Portland cement clinker, as already seen in Figure 5.18. This cost represents 24 per cent of the total cost of manufacturing Portland cement. The selection of the fuel is therefore essentially guided by local economic considerations. Pulverized coal, heavy fuels, gas, petroleum coke (Petcoke), organic waste or a mixture of several of these fuels can be used. In some cases, the fuel used in a cement plant can vary during the year to take advantage of seasonal opportunities. Pulverized coal and heavy fuels are, however, the most frequently used fuels.

During the past ten to twenty years, cement plants were offered coals and fuels rich in sulphur at a very good price, due to the fact that the cement industry is one of the few industries that can burn sulphur-rich fuels without emitting SO_3 because the volatile alkalis contained in the raw meal fix this sulphur in the form of alkali sulphates. But the use of such high sulphur fuels has a strong influence on the characteristics of the clinker, as will be seen later in Section 5.13.

High-ash-content coals can be also used when producing clinker; it is only necessary to take into account the amount and the chemical composition of the ashes generated during the burning of the coal.

5.5 Schematic representation of a cement kiln

The kiln is the key equipment where the pyroprocess by which the raw meal is transformed into clinker occurs (see Figure 4.14). In this section, we will describe three types of kilns found in modern cement plants that use the dry process: a long kiln equipped with a pre-heater, a short kiln equipped with a precalcinator and a long kiln equipped with a precalcinator. Their description is based on schematic representations developed by KHD Humboldt Wedag (1986). The three schematic representations are in fact the stretching of Figure 4.14 all along the different temperature zones in which a cement kiln can be divided.

In these figures, the horizontal scale represents the residence time in each different zone, from the introduction of the raw meal to the exit of the clinker. As the kiln turns at a constant speed, this time corresponds to a particular position of the raw meal within the kiln. On the y axis, two scales are used, on one of which the evolution of the mass of the materials can be followed and a second one on which the temperature profile is reported.

In Figures 5.22, 5.23 and 5.24 it is seen that the kiln has been divided into four zones:

- the *calcination* zone
- the *transition* zone
- the *firing* zone
- the *cooling* zone.

The three kilns presented have an average daily production rate of about 2500 tonnes of clinker which corresponds to an annual production of 700000 to 800000 tonnes of clinker per year.

5.5.1 Kiln with a pre-heater (Figure 5.22)

In order to produce 2500 tonnes of clinker with such a kiln, the kiln should have a diameter of 4.8 to 5.0 metres and its length should vary between 67 and 74 metres. The L/D ratio is therefore of the order of 14 and its rotational speed is of about 2 revolutions/minute.

The raw meal is transferred from the homogenization silo to the top of the pre-heater tower. As it goes down the pre-heater tower the raw meal comes into contact with hot gases that are moving in the opposite direction. While in contact with these hot gases, about 40 per cent of the limestone is transformed into lime. Therefore, when the raw meal enters the kiln, the first 28 minutes of

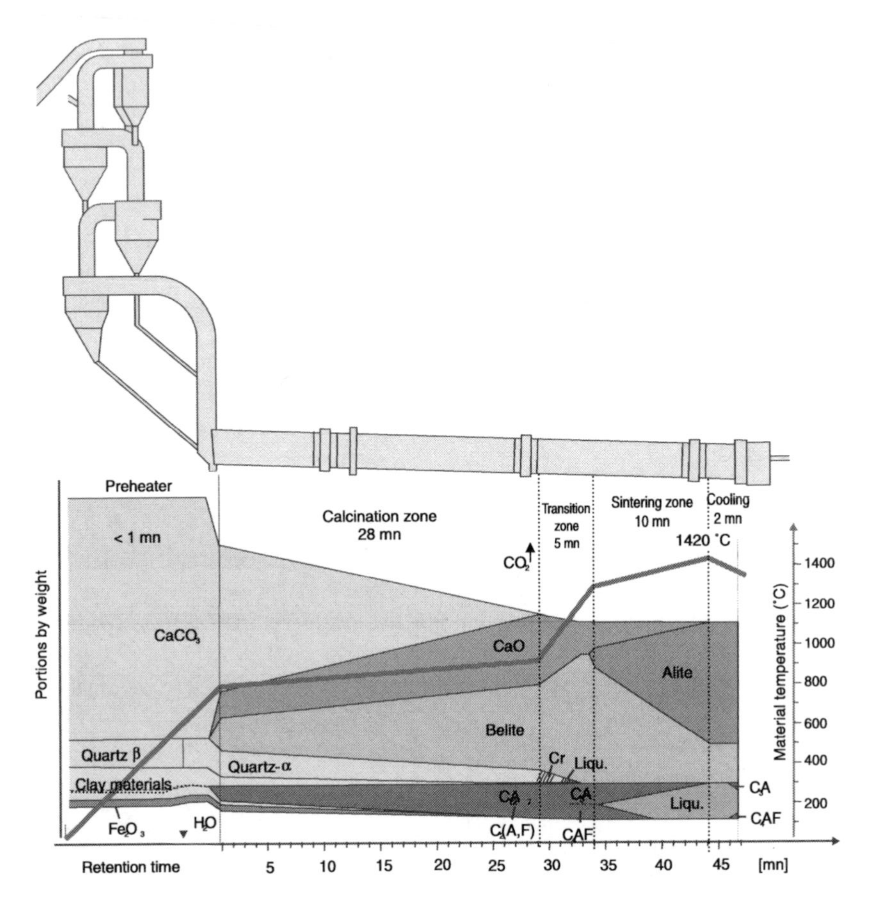

Figure 5.22 Formation of clinker in a kiln with a pre-heater (courtesy of KHD Humboldt Wedag).

the residence time in the kiln are spent calcining the 60 per cent of remaining limestone. This decarbonation process monopolizes 60 per cent of the length of the kiln. In this zone, the temperature does not rise very much because the decarbonation process is strongly endothermic (222 kJ/kg).

It is in this zone that the first belite is formed through the reaction of newly formed CaO with silica. Also in this zone, it is seen that $C_{12}A_7$ and $C_2(A_1F)$ are formed as part of the interstitial phase.

When all the limestone has been calcined, the temperature rises quite rapidly within the zone called the *transition zone*, just ahead of the firing zone. In 5 minutes, the temperature of the raw meal goes from 900°C to 1250°C, in a length corresponding to less than 10 per cent of the length of the kiln. The raw meal activated by the decarbonation process arrives quite rapidly in the firing zone. During this rapid progression, the amount of belite increases, all the silica disappears and a small amount of alite is formed. All the while, the interstitial phase is already composed of C_3A and C_4AF. However, there still exists an excess of CaO that will react in the *firing zone* with part of the already formed belite to produce alite crystals.

As it leaves the transition zone, the material arrives at the hottest zone of the kiln where the temperature reaches about 1450°C. At the end of this zone, theoretically, all the residual lime has been combined with belite. Consequently, the clinker contains only alite, belite and a liquid interstitial phase. In fact, there is always a small amount of uncombined lime remaining in the clinker, in the order of 0.5 to 1.0 per cent. This amount of lime is indicated as free lime in the chemical analysis of the clinker. Minor components, such as periclase (MgO) and alkali sulphates, are also found in the clinker, as will be seen in Section 5.6.

After the firing zone comes the *cooling zone* just behind the burner. In this zone the clinker is quenched so that its C_3S and C_2S keep their highly reactive forms, which are stable only at a very high temperature.

In this kind of kiln, the clinker leaves the kiln after 45 minutes while its temperature is still of the order of 1350°C. It is then cooled down to 200°C. During this cooling phase, a maximum amount of heat is recovered. The clinker emerging from such a kiln is finely crystallized, slightly porous and, therefore, relatively easy to grind. When ground with an adequate amount of calcium sulphate, it should develop good short- and long-term strengths.

5.5.2 Two-pier short kiln equipped with a precalcinator (Figure 5.23)

In order to produce 2500 tonnes of clinker per day in such a kiln, it must have a diameter of only 4 to 4.4 metres and its length should be between 40 and 50 metres. Such a short kiln can be installed on only two supporting piers instead of three in the previous case. The L/D ratio is of about 10 and the kiln makes 3.5 revolutions per minute. The residence time of the raw materials is reduced to 20 minutes instead of 45 as in the previous case.

Instead of describing the progression of the raw mill downwards, we will look at it in the reverse direction. In Figure 5.23, it is seen that:

- the retention time in the *cooling zone* is about 2 minutes as in the previous case;
- the retention time in the *firing zone* is about 10 minutes as in the previous case;
- the residence in the *transition zone* lasts 6 minutes instead of 5 minutes in the previous case, which is not so different;
- however, calcination last only 2 minutes in a kiln equipped with a pre-calcinator.

This last point is the only major difference in the processing of the clinker.

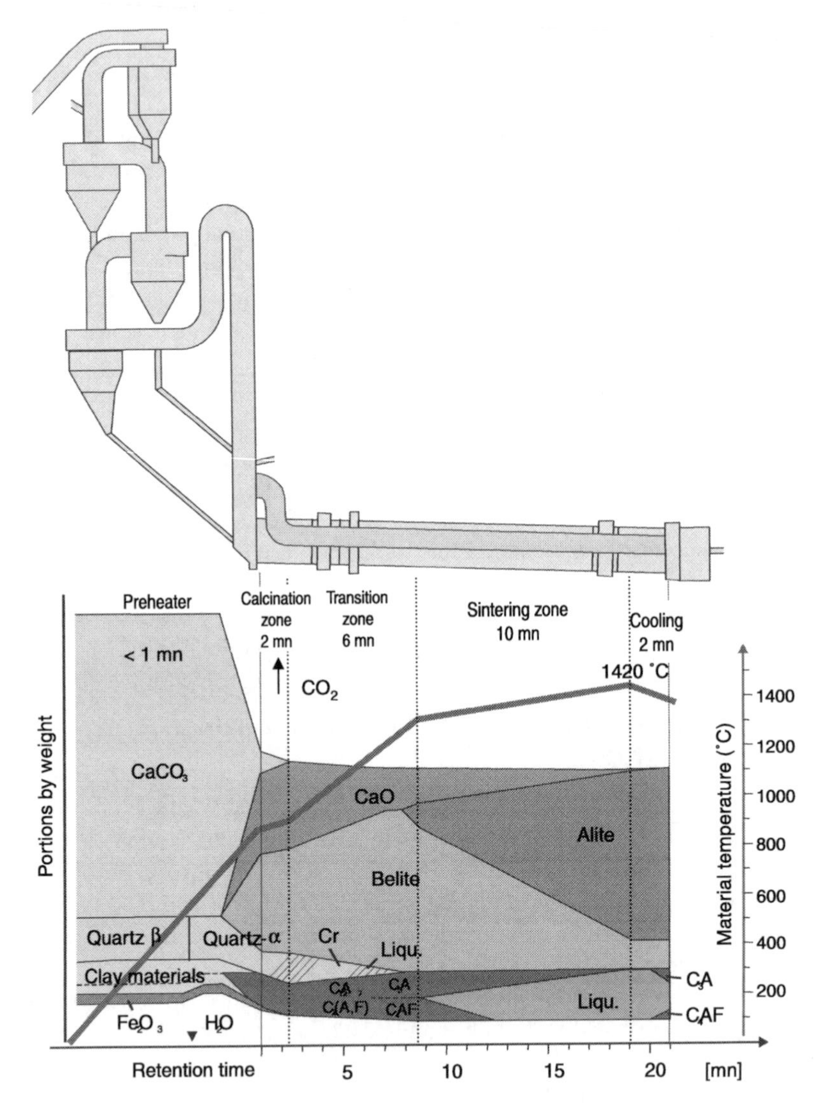

Figure 5.23 Formation of clinker in a *short* kiln equipped with a *precalcinator* (courtesy of KHD Humboldt Wedag).

From an energetic, financial and mechanical point of view, a short kiln equipped with a precalciner is particularly interesting, however, we will see later that the operation of a short kiln can create some technical problems when the raw materials are rich in alkalis.

To conclude, the use of a precalcinator results in:

- a shorter kiln;
- a reduced residence time;
- a rapid temperature increase in the transition zone which reduces the crystalline growth of the belite;
- the production of finely crsytallized alite and a porous interstitial phase that produces a clinker that is easy to grind.

As the kilns are shorter, their diameter can be increased. Presently, the biggest short kilns have a diameter of 10 metres and they can produce up to 10 000 tonnes of clinker per day. In 1984, Gourdin predicted the technical feasibility of kilns producing 10 000 tonnes per day. Such large kilns can only be built in markets that can consume such a huge amount of cement.

5.5.3 Long kiln with three piers equipped with a precalcinator (Figure 5.24)

In order to produce 2 500 tonnes per day, such a kiln must have a diameter of 4 to 4.4 metres and a length of 56 to 64 metres. The L/D ratio should be in the order of 14 and the kiln should make 3 revolutions per minute. In such a case the decarbonation obtained before the entry of the kiln is also about 95 per cent.

If the retention times in the four zones of the kiln are compared, it is seen that the decarbonation and cooling times are the same, the retention time in the *firing zone* is slightly longer (12 minutes instead of 10), but the retention time in the *transition zone* is 28 minutes which much longer than the 5 to 6 minutes in the two previous cases. In such a kiln, the raw meal stays in the transition zone for too long and the temperature increase is too low, which has the following adverse consequences:

- The reactivity of CaO is greatly decreased because of the growth of belite crystals.
- The formation of clinker particles becomes more difficult and it is necessary to increase the temperature of the clinker from 1 420°C to 1 450°C.
- The structure of the clinker is coarser, the clinker is less porous and harder to grind.

The clinker produced is much less interesting and less reactive than in the two previous cases. Consequently, it is not interesting to install a precalcinator when the cement plant already has a long kiln.

5.6 Circulation of volatiles

During firing, the chemical elements K, Na, S, Cl are partially or totally volatilized in the hotter part of the kiln (Taylor 1997; Jackson 1998). As they are entrained by combustion gases, they condense when they reach the coldest part of the kiln where they can build up, which impairs the normal operation of the kiln.

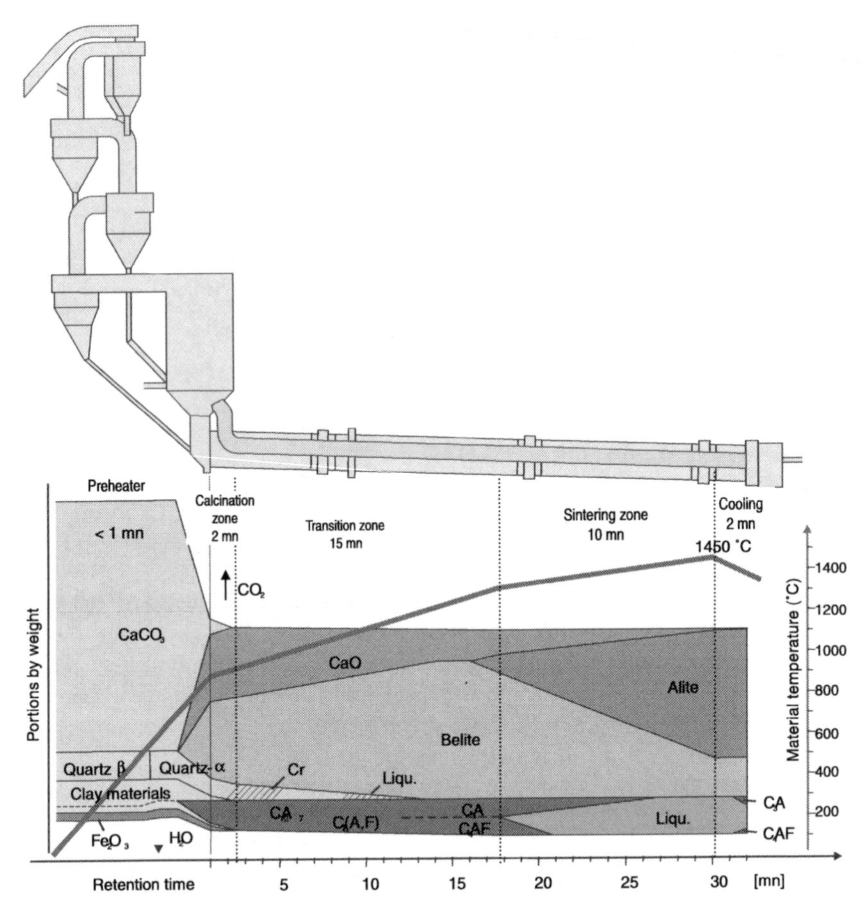

Figure 5.24 Formation of clinker in a *long* kiln equipped with a *precalcinator* (courtesy of KHD Humboldt Wedag).

The sublimation of the volatiles starts at a temperature as low as 700°C in the order Na < S < K < Cl. The circulation of volatiles is presented schematically in Figure 5.25. The volatiles condense at the entrance of the kiln, which is of course the coldest part, the firing zone being the hottest part. When these volatiles condense on the raw material, they progress again towards the hottest part of the kiln. When the volatiles condense on the kiln dust, they make it sticky and start to create concretions or rings. Moreover, chlorine and sulphur attack certain refractories in the kiln.

5.6.1 Chlorine cycle

Because they have a high vapour pressure, chlorides are more volatile than sulphates. Potassium chloride is the most easily vaporized chloride (Taylor

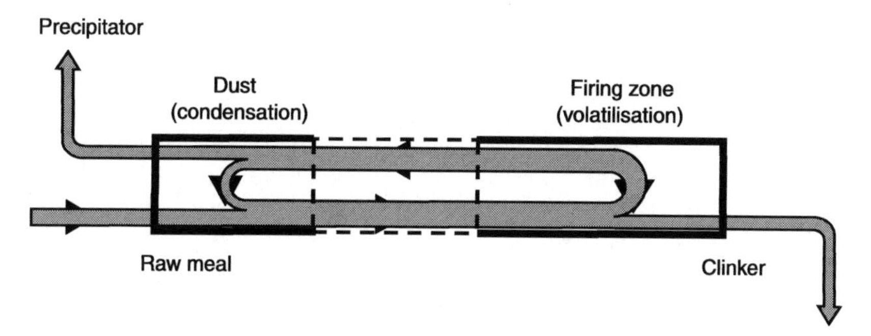

Figure 5.25 Cycle of volatiles in a kiln (after Taylor 1997; Jackson 1998).

1997; Jackson 1998). Due to cycling, the amount of chlorine in the kiln can be much greater than its original content in the raw materials and the fuel.

5.6.2 Sulphur and alkali cycles

A long residence time and an increase in temperature favour the evaporation of SO_3, K_2O and Na_2O. In the clinker, sulphur is essentially combined in the form of alkali sulphates $M_e\ SO_4$, where M_e represents an alkali metal. These sulphates are decomposed at high temperature:

$$M_e\ SO_4 \rightarrow SO_3 + M_eO$$

Alkali oxides K_2O, NaO_2 and SO_3 have a great influence on the viscosity of the interstitial phase and, consequently, on the formation of the clinker. When they are volatilized, SO_3 and the alkalis are carried into the coldest part of the kiln where they condense as alkali sulphates. SO_3 combines with alkalis to form preferentially K_2SO_4 (arcanite) or double sulphates as $3\ K_2\ SO_4 \cdot Na_2SO_4$ (aphtitalite) or $2\ Ca\ SO_4 \cdot K_2\ SO_4$ (Ca–langbeinite) (Grzeszczyk 1994; Taylor 1997; Jackson 1998). In a very few cases, when there is an excess of SO_3, some anhydrite is formed.

Because of their low vapour pressure, part of the alkali sulphates are trapped within the clinker nodules. In SEM micrographs, they are seen as tiny crystals or small concretions deposited on C_3S and C_2S, as will be seen in the next paragraph.

The major problem from a manufacturing point of view is the formation of build-ups that must be broken down with air guns (Palmer 1990). In order to avoid these build-ups, a bypass is installed which allows the removal of the alkali sulphate, with, of course, some heat and kiln-dust losses. In some cement plants, this problem is so crucial that kiln dusts cannot be reintroduced in the process.

5.7 Microscopical examination of some industrial clinkers

It is out of question to attempt a comprehensive study of the different morphologies of the clinker produced in cement kilns. Campbell's book (1986) deals with this subject in great detail. The objective of this paragraph is rather to point out the great variety of clinker morphology that can be found, even when these clinkers have a very close chemical composition. This helps to understand why each Portland cement is quite unique (like wines), in spite of the fact that they can be classified in broad categories. In the following, we will describe SEM micrographs obtained on fresh fractures of a few industrial clinkers.

5.7.1 *Examination of Figure 5.26*

The top of the figure presents a clinker composed of small crystals of alite having an average size of 10 to 20 μm. In this picture, only one alite crystal is coarser than 40 μm. Alite crystals are polygonal and well individualized and not very well bonded to each other. This clinker is relatively porous and is easy to grind.

The bottom picture shows much coarser alite crystals which have a diameter larger than 40 μm. The alite crystals are strongly bonded to each other and they form a solid having a low porosity. This clinker is difficult and costly to grind. Most probably, it is a clinker nodule that stayed quite a long time in the hottest part of the kiln. It will be less reactive than the one pictured above.

5.7.2 *Examination of Figure 5.27*

The top of the picture represents rounded belite crystals having a characteristic lamellar texture. This clinker nodule is porous, and it will be easy to grind. The bottom of the picture represents a belite grain at a larger scale. The lamellar structure of belite indicates the presence of various polymorphic forms of C_2S. This clinker is also porous.

5.7.3 *Examination of Figure 5.28*

The top picture represents, at a large scale, a part of the interstitial phase around a coarse belite grain. The porous nature of the interstitial phase is quite visible, it is essentially composed of hollow crystals of C_4AF that bind alite and belite crystals together.

The bottom picture represents, on a smaller scale, some of the interstitial phase that binds alite and belite crystals. One part of the interstitial phase is well crystallized while the other part is rather vitreous.

Figure 5.26 Alite crystals in two clinkers: (a) small alite crystals (C_3S); (b) coarse alite crystals (C_3S) (courtesy of Arezki Tagnit-Hamou).

5.7.4 Examination of Figure 5.29

The two pictures in this figure represent coarse polygonal alite crystals on which alkali sulphate deposits can be seen. The bottom picture shows secondary belite formed through the decomposition of C_3S.

5.7.5 Examination of Figure 5.30

On the two pictures, it is possible to observe secondary belite which appears as a protuberance on alite crystals. This kind of morphology is characteristic of a slow cooling.

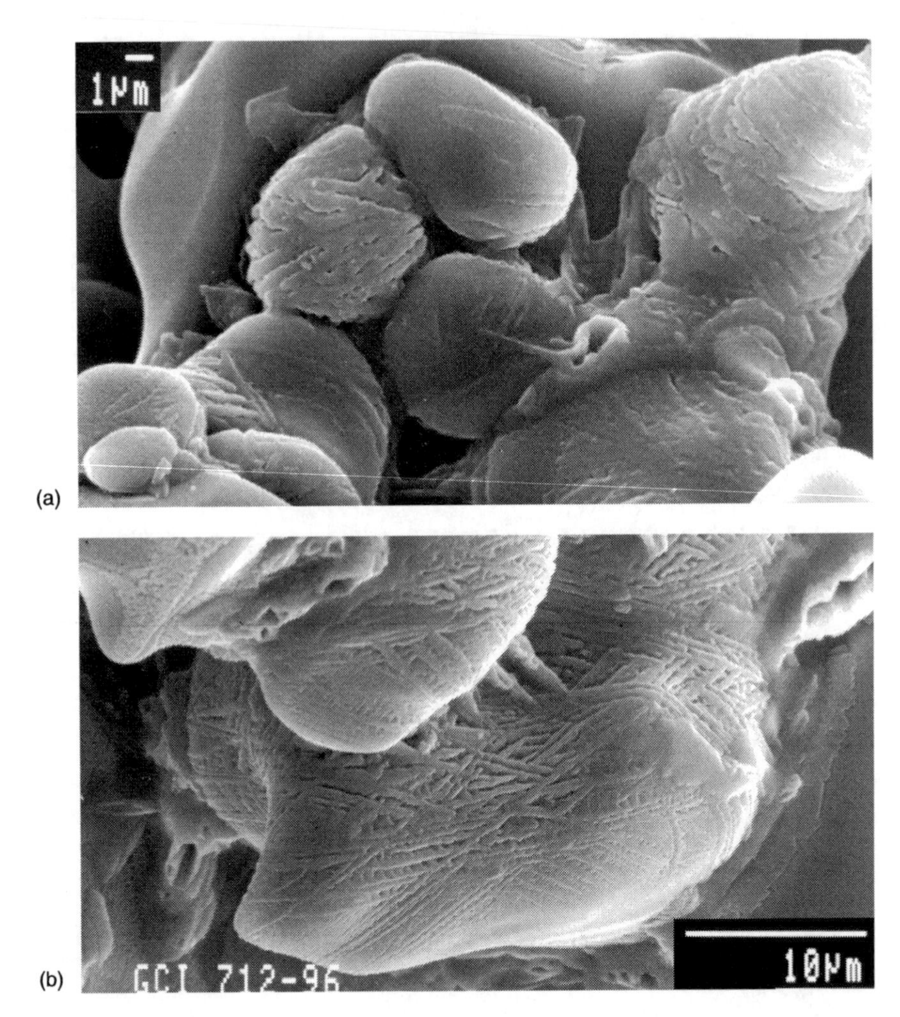

Figure 5.27 Belite crystals in two clinkers: (a) belite crystal (C_2S); (b) detail of a belite crystal (C_2S) (courtesy of Arezki Tagnit-Hamou).

5.7.6 Examination of Figure 5.31

This figure presents a belite nest formed by a cluster of small belite crystals. The presence of belite nests in a clinker is usually associated with the presence of excessively coarse quartz particles in the raw meal. This SiO_2 rich zone did not allow the transformation of all belite crystals into alite due to a lack of lime in this area.

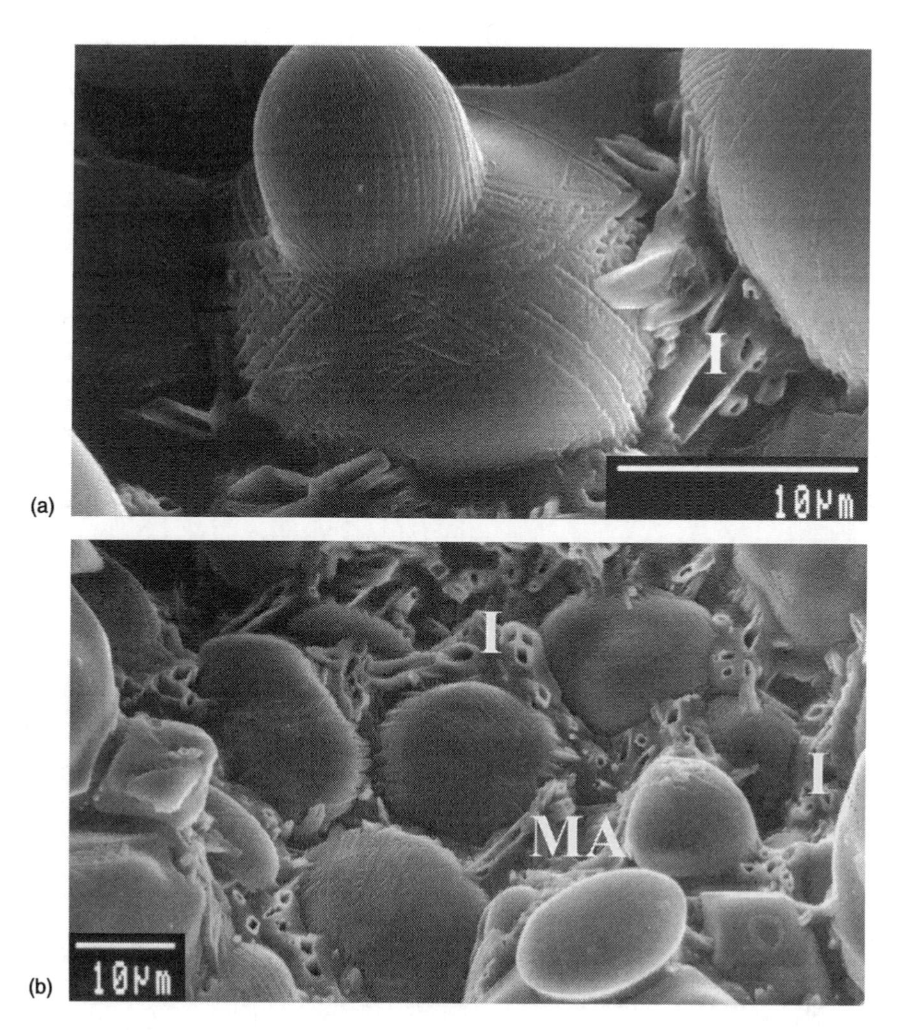

Figure 5.28 Interstitial phase (I) in two clinkers: (a) detailed view of a clinker showing the interstitial phase (I) between two belite crystals and the tubular nature of C_4AF crystals; (b) detailed view of a clinker showing the interstititalphase (I) composed of well crystallized C_3A and C_4AF among an amorphous phase (MA) (courtesy of Arezki Tagnit-Hamou).

5.7.7 Examination of Figure 5.32

This picture represents a cluster of free lime which indicates an excess of limestone in this particular area with regards to the available SiO_2. This excessive concentration of lime resulted in the formation of this cluster of free lime. Clinkers usually contain between 0.5 and 2.0 per cent of free lime.

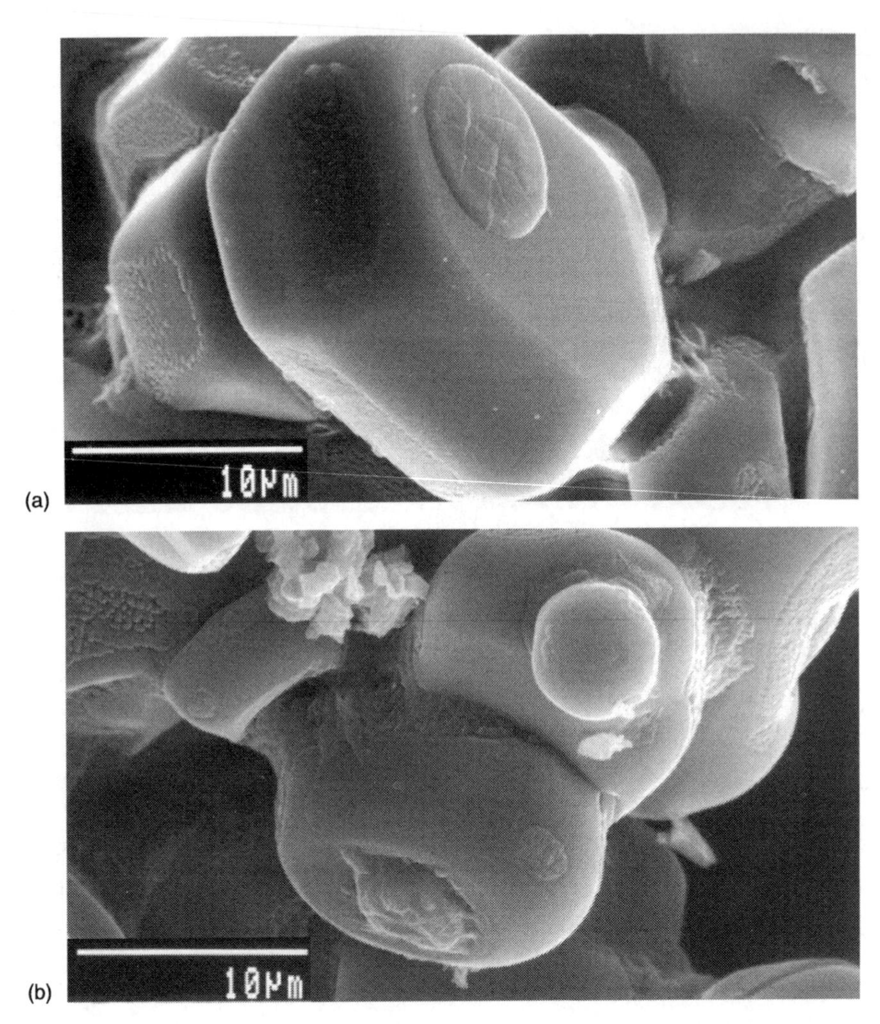

Figure 5.29 Deposits on two clinker grains: (a) deposits on alite crystals (C_3S); (b) deposits on belite crystals (C_2S) (courtesy of Arezki Tagnit-Hamou).

5.8 Clinker storage

Clinker is usually stored in halls or in special silos (Figure 5.33) to maintain their reactivity. When a clinker is stored outside, it is necessary to grind it finer to restore its potential strength and rheology so that the cement can comply with acceptance standards.

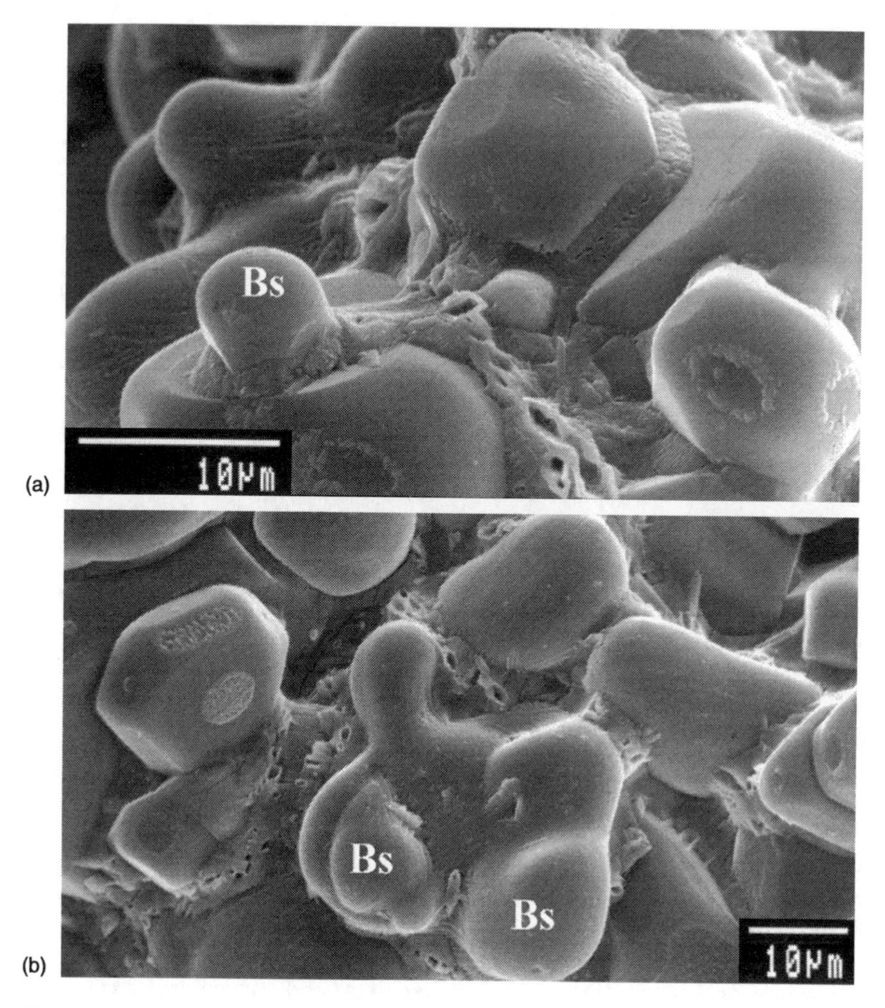

Figure 5.30 Secondary belite (Bs) formed during a slow-cooling (a) and (b) (courtesy of Arezki Tagnit-Hamou).

5.9 The addition of calcium sulphate

As we will see in the next chapter, the hydration of C_3A is so rapid that it is not possible to use ground clinker as a cement. We will also see how calcium sulphate reacts with C_3A to form initially a mineral called ettringite and later a second mineral called calcium monosulphoaluminate when there are no longer any SO_4^{2-} ions available in the interstitial solution of the hardening concrete.

It was around 1900 that the French chemist Giron (Candlot 1906; Draffin 1943), who was trying to improve the technology of rotary kilns in the USA,

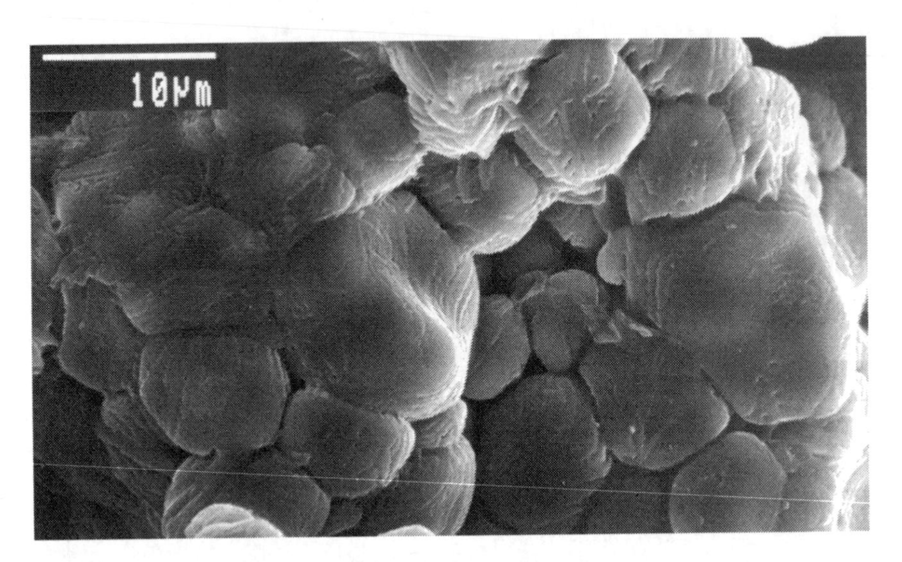

Figure 5.31 Belite nest (courtesy of Arezki Tagnit-Hamou).

had the idea of incorporating a small amount of gypsum to clinker during its final grinding to control its initial set and hardening. Thereafter, le Chatelier and Candlot scientifically explained the chemical reactions involved in that process.

For a long time, cement producers used pure gypsum as a source of calcium sulphate. However, in order to reduce production costs, low-grade gypsum started to be used (Hansen and Offutt 1962). As the process of gypsum-board manufacturing has become more and more automated, the gypsum industry is using more pure gypsum. Consequently, gypsum quarries are looking for a market for their low-grade gypsum which is unacceptable for the gypsum industry. The major 'contaminants' of these low-grade gypsums are anhydrite (a crystallized form of anhydrous calcium sulphate) and calcium carbonate which have no adverse effect on the quality of Portland cement. However, in the presence of water, anhydrite releases its SO_4^{2-} ions at a slower rate than gypsum. For a cement manufacturer, it is always possible to compensate for the slower reactivity of anhydrite by dehydrating some gypsum during the final grinding to form hemihydrate which is more rapidly soluble than gypsum. Hemihydrate is the most rapidly soluble form of calcium sulphate. We will see in the next chapter that it is necessary to control closely the amount of hemihydrate, in order to avoid a phenomenon called *false set* that corresponds to the precipitation of hemihydrate in the form of gypsum in the fresh concrete. This phenomenon is accompanied by a rapid stiffening of the fresh concrete (Hansen and Offutt 1962).

Controlling the calcium sulphate content became more complicated when the cement industry started using fuels rich in sulphur (Miller and Tang 1996).

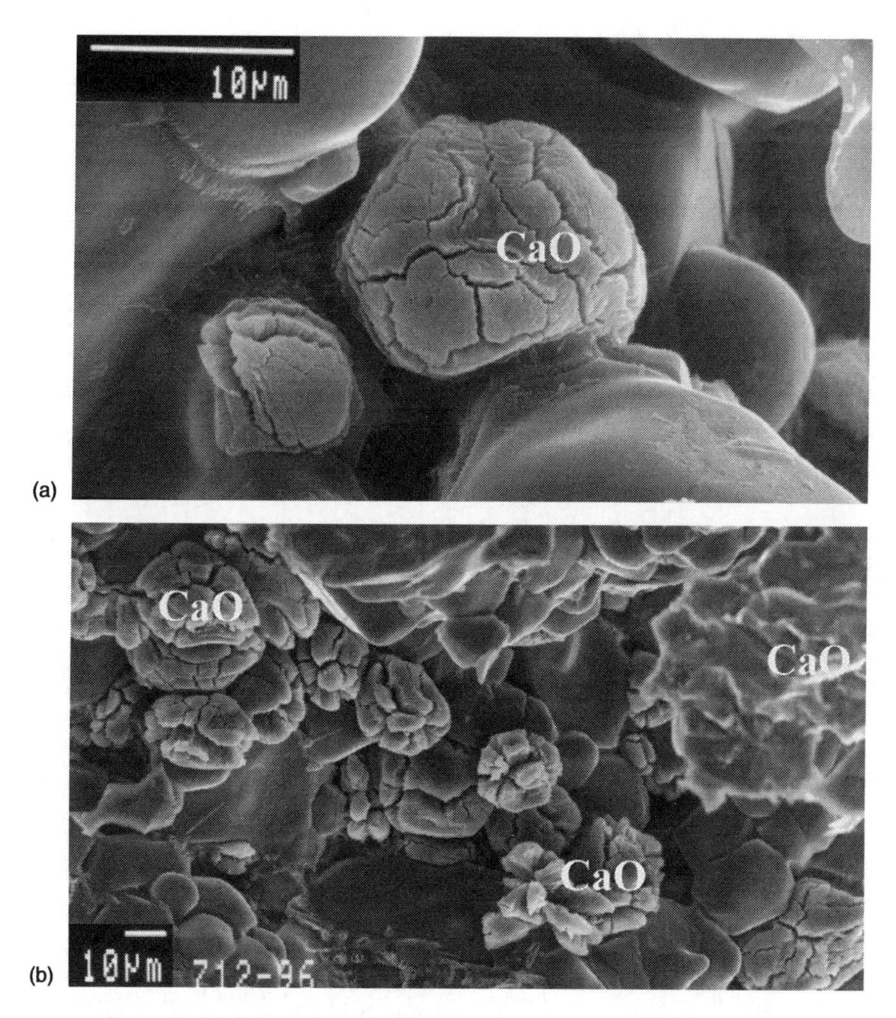

Figure 5.32 Free lime in two clinkers (courtesy of Arezki Tagnit-Hamou).

In a few years, the average SO_3 content of Portland cement clinker rose from 0.5 per cent to 1.5 per cent and even in some cases up to 2.5 per cent (Zhang and Odler 1996a and 1996b). As cement standards usually limit to 2.5 to 3.5 per cent the maximum amount of SO_3 in Portland cement, there is sometimes very little room left for an addition of calcium sulphate. Consequently, the solubility of the SO_4^{2-} ions liberated by Portland cement, when it comes into contact with water, is becoming a critical factor, as will be seen in Section 5.13, particularly in the case of low water/binder concretes when polysulphonate superplasticizers are used (see Chapter 7).

Figure 5.33 Silos for clinker (courtesy of Lafarge).

The problem became even more complicated when, in some countries, a quantity of iron sulphate was added to complex hexavalent chromium which is responsible for mason scabies (essentially in masonry cements or in Portland cements that could be used by masons). When iron sulphate or pyrite is introduced in the raw meal for the same reason, it decreases the amount of calcium sulphate that can be added during the final grinding of Portland cement.

Alongside this evolution of the SO_3 content of Portland cement clinker it must be pointed out that cement companies are 'helping' different industries eliminate their calcium sulphate waste (plaster mould from the ceramic industry, gypsum-board waste, synthetic calcium sulphate produced during the neutralization of SO_3 gases, etc.). In some cases, very interesting synergies have been developed between a cement plant and an industry generating SO_3. For example, in Japan, a cement company is providing its CKD to an adjacent oil refinery in order to help it lower its SO_3 emissions. The cement plant gets these sulphated CKD back to use as a source of calcium sulphate. This kind of synergy makes this oil refinery, the cement company, the transportation company, and the Japanese Department of the Environment very happy.

5.10 Final grinding

The final grinding of the clinker is a very important step in the manufacturing of Portland cement. In fact, Portland cement fineness significantly influences the reactivity and rheological characteristics of Portland cement. Moreover, final grinding is the last chance a cement producer has to adjust the characteristics of the cement he or she will put on the market to the requirements of standards.

The grinding of clinker and calcium sulphate is not in itself a very complicated operation, but it must be well controlled because cement fineness directly influences the technological properties of concrete, in the fresh state as much as in the hardened state. The finer a cement, the more reactive it is, the more difficult it is to control its rheology; the higher its initial heat of hydration, the higher the amount of alkalis released and the closer to one the ratios between its long-term compressive strength and its twenty-eight-day strength. This means that after twenty-eight days, this cement will not gain very much compressive strength.

Portland cement fineness is usually controlled in three different manners: By measuring

- the number of particles retained on a 45 μm sieve;
- the Blaine specific surface area, which is in fact a measure of the permeability of a layer of Portland cement that has been compacted in a standardized manner;
- the slope of the grain-size distribution which is expressed as a function of the Rosin-Rammler number (Figure 5.34) (Pliskin 1993).

Another important factor to be verified during final grinding is the temperature within the ball mill. We have already seen that the energetic efficiency of a ball mill is very low, almost 95 per cent of the energy is dissipated as heat and noise during the grinding process. It is important to be sure that the

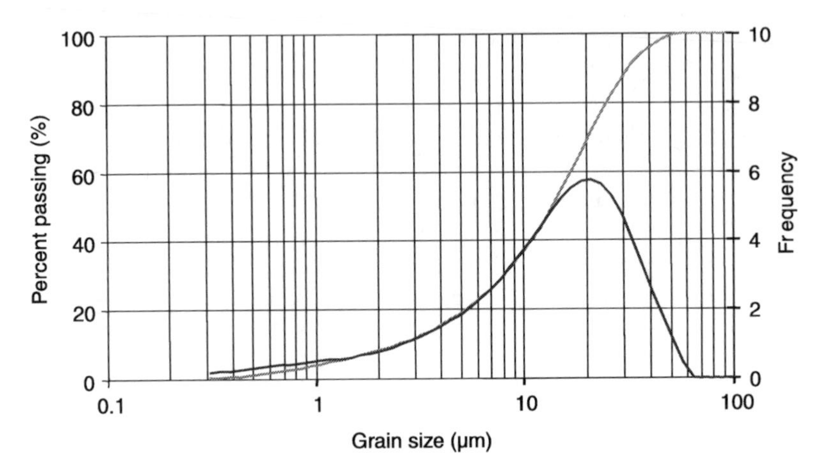

Figure 5.34 Grain-size distribution of a cement (courtesy of Mladenka Saric Coric).

temperature within the ball mill does not increase too much, because too much gypsum could be dehydrated as hemihydrate, and this excessive dehydration of gypsum could result in a false set situation as will be seen in the next chapter (Hansen and Offutt 1962).

When the temperature inside a ball mill has a tendency to become too high, the exterior part of the ball mill can be cooled with water or with fresh air, or water can even be pulverized within the ball mill. Of course, a small amount of C_3A and C_3S will hydrate, but experience shows that it is the hemihydrate that preferentially fixes this pulverized water. It must be remembered that gypsum is a very soft material that is ground much finer than clinker in a ball mill.

As some clinkers have a higher SO_3 content that does not always go very fast into solution, it is therefore necessary to compensate by dehydrating a part of the gypsum because hemihydrate is releasing its SO_4^{2-} ions much faster than gypsum. In order to reach a good balance of the SO_4^{2-} ions solubility, some cement producers are using a mixture of anhydrite and gypsum and let it heat up so that at the end of the grinding almost all the gypsum is dehydrated without causing any risk of false set (Hansen and Offutt 1962). This concept is not too bad when dealing with high water/binder concretes, but it can result in serious rheological problems when using low water/binder concretes and polysulphonate water reducers or superplasticizers, as will be seen in Chapter 7.

One of the latest developments to have taken place in cement grinding is the use of press-rolls to perform a preliminary grinding before the finishing ball mill (Figure 5.8). This development increases the efficiency of the grinding process very much, it shortens the retention time of cement particles in the ball mill and increases the daily output of the grinding operation. When a press-roll is used with a very efficient air separator, cement particles do not pass more than two or three times in the ball mill. These Portland cements are 'cold ground' cements having a more uniform grain-size distribution than cements that are totally ground in ball mills. In such a cold process, gypsum has very little chance of being dehydrated.

As clinker is much harder than calcium sulphate, after grinding it is essentially found in the finest part of the cement, while the harder C_2S is rather found in the coarser fraction of the cement (Hansen and Offutt 1962). There is not very much to do in that domain except carefully check the clinker formation process so that its porosity is high and grindability easy. The specific surface area of a cement can be modified to correct the reactivity of the clinker. Finally, grinding aids are used to increase the output of the grinding operation and to avoid pack-set problems. The technology of grinding aids will not be developed in this chapter; it is beyond the objectives of this book.

5.11 Quality control

In spite of all the care taken when preparing the raw meal, during the firing and the grinding of the clinker cement producers must constantly control the characteristics and the quality of the cement they are storing in their silos before shipment. This constant control is absolutely necessary in order to guarantee the delivery of Portland cements or of blended cements which can be used to make concretes having a predictable behaviour in the fresh state as well as in the hardened state, and also, of course, to comply with the minimum quality requirements of acceptance standards (see Appendix XI).

In this section, it is not our intention to go into details about all the acceptance standards for the different types of Portland cements or of blended cements but, rather, to see what are the main tests done on a routine basis in a cement plant, why they are done, why they are specified, how their results can be interpreted by a concrete producer, their practical value and, sometimes, their inadequacy. In Appendix XI, the principal ASTM standards relative to Portland cement and hydraulic binders are listed. Kett (2000) has written a book describing in detail all these tests, the material and equipment to be used and how to organize the calculations.

5.11.1 Chemical analysis

As it has been seen, the $CaO–SiO_2–Al_2O_3–Fe_2O_3$ domain of composition where C_3S, C_2S, C_3A and C_4AF can coexist is quite limited. It is, therefore, very important to control the chemical composition of the raw meal, of the clinker and of the cement produced. It is very important to be sure that these values do not deviate too far from the average chemical composition that has been selected. Of course, it is certain that any Portland cement or blended cement that does not have a constant chemical composition does not have constant technological properties, however, as already written a constant chemical composition does not guarantee constant technological properties. Chemical composition consistency is a necessary, but not sufficient, condition for making a cement having predictable characteristics.

From the chemical analysis results, the potential phase composition can be calculated using Bogue's formulas. In a modern cement plant, where the process is well under control, an average C_3S content of 60 to 65 per cent can be maintained within a range of 2 to 3 per cent and the C_3A content within a range of 1 to 2 per cent.

Acceptance standards sometimes set a limit on C_3A content for some cements. For example, for certain types of applications, a European cement of PM or ES type must contain less than 5 per cent C_3A. An ASTM Type V cement must also contain also less than 5 per cent C_3A.

Acceptance standards also limit the SO_3 content of Portland cements within a 2.5 to 3.5 per cent bracket, depending on the type of Portland cement. In some cases, when C_3A is lower than 5 per cent, the maximum amount of SO_3 is 2.3 per cent. We have seen that these limits do not take into account the origin of this SO_3, which, from a practical point of view, is very important, because it is not the total amount of SO_3 that counts but rather the balance between the solubility rate of the SO_4^{2-} and the 'reactivity' of the C_3A. To qualify as a low-alkali cement, the Na_2O equivalent of Portland cement must be lower than 0.6 per cent.

The autoclave test has been developed to check the volumetric stability of a cement. In fact, during the 1930s, it was observed that some cements produced in the US Midwest had a tendency to swell on a long-term basis. A detailed study showed that this swelling was caused by the presence of an excess of periclase in the clinker. This hard-burned periclase (MgO) was transformed into brucite, $Mg(OH)_2$, which induced the swelling. This brucite is developed late in the hardened cement paste due to its very slow hydration process, whereas hard-burnt lime, CaO, is transformed very rapidly during mixing into slacked lime $Ca(OH)_2$. The resulting volume expansion of lime is not harmful at all, because it happens when the concrete is still in the fresh state. When heating a small cube of cement paste or mortar in an autoclave (1.5 atm and 150°C), the hydration of periclase is accelerated, so that the measurement of the swelling developed after 3 hours of 'cooking' in the autoclave can be used to discriminate safe cements from unacceptable cements with regards to this type of volumetric expansion. This test is efficient because, since it has been introduced as routine test in cement plants, no more periclase swelling problems have been reported.

5.11.2 Physical properties

Cement producers are controlling the value of the specific surface area through Blaine fineness measurements. This specific surface area is expressed in m^2/kg (cm^2/g is not a proper SI unit to express the specific surface area).

Unfortunately, standard requirements do not usually specify bracket values for Blaine fineness but rather minimum values (personally, I would prefer that standards specify maximum values). It is very important to control cement fineness because it essentially controls the 'physical reactivity' of a cement and its properties in the fresh state (Uchikawa et al. 1990).

This test is, in fact, a permeability test. In North America, it is performed according to ASTM C204 (Kett 2000). Using a certain number of hypotheses it is possible to link the air permeability of the tested cement to its specific surface area (Papadakis and Venuat 1966; Jouenne 1990; Pliskin 1993). Acceptance standards can also require a maximum amount of particles retained on a 45 μm sieve to prevent a cement from containing too many coarse particles.

All cements must also comply to setting requirements measured using Vicat needles. Such a test is quite empirical, but very convenient. On a routine basis, cement producers measure the initial and final setting time. This test is also very useful to identify false and flash set problems, as will be explained in the next chapter.

5.11.3 Mechanical properties

Cement producers measure the 'strength' of their cement on small mortar cubes having a water/binder ratio of about 0.50; more precisely 0.50 for ENV 196-1 European standard and 0.485 in ASTM C-150. According to the type of cement produced, different strength requirements are to be fulfilled either at twenty-four hours, three days, seven days, and twenty-eight days. Unfortunately, at each age, only a minimum value is specified (the author would have preferred bracket values). These routine tests are unfortunately perceived by cement producers as the key tests to determine the quality of their cement; for them, the higher the compressive strength, the better the quality of their cement.

The initial strength of standardized mortar having a constant water/binder ratio is a function of the C_3S and C_3A content of a cement as well as of its fineness and its calcium-sulphate content. Usually, the calcium-sulphate content of a cement is optimized to increase initial compressive strength through the formation of ettringite, as will be seen in the following chapter (Hansen and Offutt 1962).

The author, and he is not the only one, would like to see the importance of the cube strength of cements be reduced, because it is more and more important to control the rheology of specialized concretes. It is always easy to increase concrete compressive strength: it is only necessary to decrease its water/binder ratio (any decrease of the water/binder ratio favours concrete durability as it will be seen in Chapter 8) which is good for concrete. On the contrary the increase of cement fineness and the increase of C_3A content can result very rapidly in a decrease is the durability of concrete, because to adjust the rheology of the concretes, it is necessary to increase its water/binder ratio, as will be seen in Chapter 10.

5.12 Conditioning and shipping

Portland cement was initially shipped in barrels weighing 170, 230 or 350 kg. In England and in the USA a cement barrel weighed 170 kg (Figure 5.3). Then Portland cement was conditioned in 50 kg burlap sacks, as previously written. These sacks had to be returned to the cement plant where women cleaned and repaired them. The first paper bags appeared in 1920. When Bates had the idea of incorporating a valve to the paper sack, it resulted in an increase in the rate of bagging to up to 1200 bags per hour (Papadakis and Venuat 1966). These days, the bagging rate is much higher.

Presently, Portland cement can be transported in different forms. First, it can be shipped as clinker, which is later ground with an appropriate addition of calcium sulphate at the delivery site. This is an easy and safe way of transporting Portland cement to overseas markets where the local consumption of Portland cement cannot support the construction of a cement plant. The construction of a local grinding facility does not imply a large investment and the implementation of a very sophisticated and costly quality-control department. Clinker transportation can be useful to overcome a strong Portland consumption or to export to a temporary market during a period of overproduction. It is also an easy way for a cement group to make its first approach of a new market without making a large investment. It is also easy, in this way, to adjust the production of Portland cement to fulfil local Portland cement requirements.

Portland cement can be shipped in sacks, in 40 or 50 kg paper bags, or in paper bags having an additional plastic ply when waterproof bags are required. At this moment in time, the three-ply paper bags can be used for general field applications, but must be protected from humidity and rain in the field.

Portland cement bags are piled on wood pallets weighing 1 or 2 tonnes; most of the time the pallet is protected with shrink-wrap plastic. Bagging and palleting are entirely automated operations. A modern bagging system can bundle thousands of bags per hour (Figure 5.35).

Portland cement can be shipped in large bags weighing approximately 1.5 tonnes (Figure 5.35). In highly industrialized countries, Portland cement and hydraulic binders are most often transported in bulk in railway cars of 40 to 50 tonnes (Figure 5.15) or by road in 20 to 30 tonnes tanks (Figure 5.16). The loading and unloading of a cement tank is done pneumatically and takes one to two hours. The proportion of bagged cement versus bulk cement varies greatly from one country to another, depending on the nature of concrete use and the evolution of the local concrete market.

Portland cement can be shipped in bulk in barges or in boats equipped with pneumatic delivery facilities. Usually, these special boats can transport from 5000 to 20000 tonnes of Portland cement and deliver Portland cement to the local market in bulk, in bags or in large bags (Figure 5.17).

For example, for a while, certain Norwegian and Spanish cement plants were

(a)

(b)

Figure 5.35 Bagging: (a) big bags (courtesy of Groupe Vicat); (b) paletized loads (courtesy of Lafarge).

able to sell their cement to the Houston, Boston and New York markets at a very competitive price (Dumez and Jeunemaître 2000). In New York City, in particular, it is as expensive to transport Portland cement by road from a cement plant located in the Hudson river valley, 200 km away from downtown New York as it is to ship it by boat from a European cement plant operating in a harbour.

More recently, larger ships have been built that can considered as cement silos able to deliver cement in several locations. Dumez and Jeunemaître (2000) state that in the year 2000, approximately 90 millions tonnes of cement were transported by boat. It is in extreme cases that Portland cement is transported by plane or by helicopter.

5.13 Evolution of the characteristics of Portland cement

The Portland cement and hydraulic binders presently produced are not exactly the same as those produced fifty or 100 years ago. Portland cement's chemical and phase compositions have evolved over the past 100 years. From a chemical point of view, Portland cements are presently richer in lime and alumina and, consequently, richer in C_3A and C_3S (Bogue 1952). In the next chapter, it will be seen than C_3S and C_3A are the two phases responsible for early strength. Moreover, the SO_3 content of modern clinker increased drastically when cement producers started to burn fuels rich in sulphur after the first petroleum crisis and when they were forced to meet tighter regulations on SO_3 emissions which cause acid rains.

Petroleum and coal companies have been proposing fuels rich in sulphur at a lower price, because very few industries can use them without incorporating into their production facilities a costly desulphurization unit. The cement industry is one of the rare industries that can accommodate a fuel with a high sulphur content, because the SO_3 produced during combustion combines with the alkalis and ends up trapped in alkali sulphates. Within the past fifteen years, the SO_3 content of the clinker went from an average of 0.5 per cent to 1.5 per cent, in some cases, and to an even higher content when the cement industry started using high sulphur pet cokes or coals.

As the maximum SO_3 content of Portland cement has not been modified in the standards to limit the amount of sulphoalumiantes produced during cement hydration, some modern Portland cements contain less and less calcium sulphate. In some cases, the amount of rapidly soluble SO_4^{2-} ions is not high enough and can cause compatibility problems with polysulphonate superplasticizers in low water/binder ratio concretes. This problem will be dealt with in more detail in Chapter 7.

Consequently, it can be said that the principal characteristics of Portland cement evolved over the past fifteen to twenty years and not always to facilitate the production of a constant and durable concrete.

It is certain that, overall, the consistency of Portland cement characteristics has improved; modern cement plants are more and more efficient; processes are under tighter control; and globalization has resulted in stronger competition between large cement companies. However, the recent increases in Portland cement fineness and in C_3S and C_3A content to make them more 'nervous' is a step in the wrong direction for concrete durability. It is supposedly under the pressure of concrete producers and precasters that cement companies have been 'forced' to increase fineness and the C_3A and C_3S contents of their cements. The author does not share this view, because concrete producers and precasters do not gain any competitive advantage when all cement producers offer all contractors the same 'nervous' cement.

When making concrete, the use of a fine cement, rich in C_3A and C_3S, presents major technological disadvantages, because these cements are very prone to cracking due to their higher shrinkage and higher heat of hydration. Moreover, as the design of concrete structures is still based on the twenty-eight-day compressive strength and not on the value of the water/binder ratio, with 'nervous' cements, the final strength is nearly reached at twenty-eight days, whereas with less nervous cements, the compressive strength of concrete continues to increase significantly after twenty-eight days. It was this late strength increase that improved the durability of the good old concretes made with non-'nervous' cement.

Another hidden modification of the characteristic of modern Portland cement is related to the grain-size distribution of cement particles. Of course, Blaine fineness still remains the most practical way of evaluating cement fineness, but it does not take into account the changes in the shape of the grain-size distribution. A given Blaine fineness can be obtained from cement housing different particle-size distributions due to technological changes in the grinding of modern Portland cement. While until recently Portland cement was ground in an open system until the adequate fineness was achieved, Portland cement is now round in closed circuits, which means that as soon as a cement particle has reached an adequate size, it is extracted from the grinding system. Only the coarser particles return into the grinding system. This new technology represents a significant increase in productivity in one of the most inefficient parts of the process of cement manufacturing. A closed circuit grinding system produces 20 per cent more cement than an open system. Consequently, modern cements have a more uniform grain-size distribution than the cements ground some years ago in open circuits. This difference in the grain-size distribution can make a significant difference in the 'reactivity' of the cement and its rheological behavior (Uchikawa et al. 1990). When examining the grain-size distribution of modern Portland cement, it is often found that their Rosin–Ramler number, which roughly represents the slope of the middle part of the grain-size distribution curve and which is almost linear, has increased. As a result, modern cements are more sensitive to bleeding and segregation than the cements of yesterday.

Finally, another change in the way modern cements are ground that has modified the grain-size distribution and the morphology of cement particles is the use of press-rolls before final grinding in the ball mill. Presently, the residence time of a cement particle in the ball mill can be very short, because the fine particles are extracted very rapidly from the grinding system by the highly efficient air separators. Unfortunately, the author does not have enough experience in this matter to evaluate the actual technological impact of this situation on the properties of fresh and hardened concrete. It is certain that there are no concerns about the setting time and cube strength of these cements which have been adjusted to fulfil standard requirements, but it is not certain that these modifications will not have any impact on bleeding, segregation, slump retention and air entrainment.

Concrete producers will have to learn to live with these new cements, because cement producers will never go back to open-circuit mills in their quest to lower production costs. Considering the economic advantages that this new technology represents for cement producers, concrete producers will simply have to quickly learn not only how to take advantage of it, but also how to correct its disadvantages.

5.14 The ecological impact of Portland cement manufacturing

As has been seen in Chapter 4, Portland cement clinker has an average lime content of between 60 and 70 per cent. In order to simplify calculations, let us suppose that the lime content is 65 per cent. As the principal source of this lime on a worldwide basis is still limestone, it very easy to calculate how much carbon dioxide is liberated during decarbonation of the limestone to produce a cement containing 65 per cent CaO.

$$CaCO_3 \rightarrow CaO + \overline{CO_2}^{\uparrow}$$

$$100 \rightarrow 56 + 44$$

In order to make 1 tonne of clinker having or 65 per cent CaO content, it is necessary to start with $\frac{65}{56}$ tonnes of limestone, that is 1.16 tonnes, and emit $\frac{65}{56} \times \frac{44}{100}$ tonne of CO_2, that is 0.51 tonne of CO_2.

In modern cement plants depending on the energy efficiency of the process, it is necessary to burn between 100 and 200 kg of 'coal equivalent' to produce 1 tonne of clinker.

$$C + O_2 \rightarrow \overline{CO_2}^{\uparrow}$$

$$12 + 32 \rightarrow 44$$

Therefore, the amount of CO_2 generated during the process varies between $\frac{44}{12} \times 0.1 = 0.37$ tonne of CO_2 and $\frac{44}{12} \times 0.2 = 0.74$ tonne of CO_2.

Overall, the amount of CO_2 generated in a cement plant can vary from 0.9 tonne to 1.25 tonnes of CO_2 for each tonne of clinker. In this calculation, the amount of CO_2 generated to produce the electrical energy used during Portland cements grinding is not taken into account.

As a rough figure, it can be said that the production of 1 tonne of clinker generates 1 tonne of CO_2. In fact, when making more precise calculations on actual energy consumption during the processing of Portland cement, it is found that the amount of CO_2 varies between 0.8 and 1.2 tonnes of CO_2 per tonne of clinker.

As in some countries a non-negligible amount of blended cements is produced, less CO_2 is actually emitted into the atmosphere. Let us suppose, optimistically, that in the year 2002, the production of blended cement reduces the production of clinker by 10 per cent and that 1 tonne of Portland cement contains 5 per cent of gypsum. It can then be estimated that the 85 per cent of clinker contained in the 1.5 billion tonnes of cement actually produced would liberate 1 275 billion tonnes of CO_2 into the atmosphere. This represents a huge amount of CO_2, but it must be remembered that this amount constitutes only 6 to 8 per cent of the total amount of CO_2 presently being emitted worldwide. As it will be seen in Chapters 11 and 12, the reduction of CO_2 emissions will be a great challenge for the cement industry in the years to come.

Aside from the CO_2 emissions produced during its passage in the burning zone, some of the nitrogen in the air is transformed into NO_x and some of the sulphur into SO_3. The amount of these two pollutants can be decreased only by lowering the temperature at which clinker is produced. The only other gas emitted through the chimney of a cement plant is water vapour.

As far as the emission of solid particles is concerned, cement plants have complied with legal requirements in this matter for many years. Almost all the solid particles coming out of the kiln are collected in very efficient electrostatic precipitators. Most of this dust is reintroduced into the kiln. However, in some cases, when the alkali content of the CKD is very high, the CKD cannot be reinjected into the process, and the cement producer has to stockpile it. This stockpiling can cause environmental problems by contaminating ground-water through a leaching process. As will be seen in Chapter 6, the chemical stabilization of CKDs is becoming a priority in some cement plants.

5.15 Does an ideal Portland cement exist?

The answer is yes. Because the author believes, as do the President of MAPEI in Italy, the Executive Vice-President of DRAGADOS in Spain, the Scientific Director of BOUYGUES in France, and many others, that an ideal cement is simply a CONSTANT cement in terms of its *rheological* and *strength* properties. As far as the cement of this author's dreams is concerned, it will be described in Chapter 13.

5.16 Conclusion

The cement industry has made great progress over the past 100 years, going from a process adapted from the manufacturing of lime to an entirely automated process in modern cement kilns that can produce up to 10000 tonnes of Portland cement clinker per day.

The first petroleum crisis forced cement producers to start looking for more energy-efficient ways of producing Portland cement as early as the 1970s. Therefore, the cement industry immediately decreased its CO_2 emissions significantly. In order to decrease its CO_2 emissions further, the industry has no other choice than to promote the use of blended cement containing fly ash, slag, metakaolin, silica fume, limestone from silica filler, before being able to produce mineralized clinkers (Jackson 1998) on an industrial scale or a new process based on fluidized beds (Yuko et al. 2000) for which the firing temperature will be lower.

It must be pointed out that the cement industry has already made significant progress in reducing its energy needs. Its success in the twenty-first century will depend on how rapidly it will be able to adapt to new environmental regulations brought about by the more stringent application of a sustainable development approach.

This new adaptation of cement processing is well under way in the leading multinational cement groups, and these leading companies will force the rest of the industry to follow them.

References

Bogue, R.H. (1952) *La Chimie du ciment Portland*, Paris: Dunod.

Campbell, D.H. (1986) *Microscopic Examination and Interpretation of Portland Cement and Clinker*, Skokre, Ill.: Construction Technology Laboratories.

Candlot, E. (1906) *Ciments et chaux hydrauliques: fabrication, propriétés, emploi*, 3rd edn, Paris: Polytechnique Library.

Davis, A.C. (1924) *A Hundred Years of Portland Cement*, London: Concrete Publications Limited.

Draffin, J.O. (1976) *A Brief History of Lime, Cement, Concrete and Reinforced Concrete*, Farmington Hills, Mich.: American Concrete Institute.

Dumez, H. and Jeunemaître, A. (2000) *Understanding and Regulating the Market at a Time of Globalization: The Case of the Cement Industry*, Basingstoke: Macmillan.

Gourdin, P. (1984) 'La Cimenterie', *Usine Nouvelle*, (May): 146–55.

Grzeszczyk, S. (1994) 'Distribution of Potassium in Clinker Phases', *Silicates Industriels*, 61 (7–8): 241–6.

Hansen, W.C. and Offutt, J.S. (1962) *Gypsum and Anhydrite in Portland Cement*, Chicago, Ill.: United States Gypsum.

Hardy, N., Thomson, R. and Hardy, R. (1998) *The World Cement Plant and Equipment Handbook* 2nd edn, Farnham: Palladium Publications.

Jackson, P.J. (1998) 'Portland Cement: Classification and Manufacture' in P.C. Hewlett (ed.) *Lea's Chemistry of Cement and Concrete*, 4th edn, London: Arnold, pp. 25–94.

Jouenne, C.A. (1990) *Traité de céramiques et matériaux minéraux*, Paris: Éditions Septima.

Kett, J. (2000) *Engineered Concrete: Mix Design and Test Methods*, Boca Raton, Fla.: CRC Press.

KHD Humboldt Wedag (1986) *L'Évolution technologique du four rotatif avec préchauffeur à cyclones vers le four court PYRORAPIDfi*, Cologne: KHD Humboldt Wedag.

Kosmatka, S.H., Kerkhoff, B., Panarese, W.C., Macleod, N.F. and McGrath, R.J. (2002) *Design and Control of Concrete Mixtures*, Ottawa: Cement Association of Canada.

Leduc, E. (1925) *Chaux et ciments*, Paris: Librairie J.-G. Baillère et Fils.

Maki, I., Taniska, T., Imoto, Y. and Ohsato, H. (1990) 'Influence of Firing Modes on the Fine Texture of Alite in Portland Clinker', *Il Cemento*, (2): 71–8.

Mary, M. (1862) *Louis Vicat invente le ciment artificiel*, Publish en 2000 by Ciment Vicat Company.

Miller, F.M. and Tang, F.J. (1996) 'The Distribution of Sulphur in Present-Day Clinkers of Variable Sulphur Content', *Cement and Concrete Research*, 26 (12): 1821–9.

Palmer, G. (1990) 'Ring Formation in Cement Kilns', *World Cement*, 21 (12, December): 533–43.

Papadakis, M. and Venuat, M. (1966) *Fabrication et utilisation des liants hydrauliques*, 2nd edn, n.p.

Pliskin, L. (1993) *La Fabrication du ciment*, Paris: Eyrolles.

Scheubel, B. and Nachtwey, W. (1998) 'Development of Cement Technology and its Influence on the Refractory Kiln Lining', *REFRA Kolloquium Berlin 1997*, Göttingen: Refratechnik GmbH, pp. 25–43.

Taylor, H.F.W. (1997) *Cement Chemistry*, 2nd edn, Oxford: Thomas Telford.

Uchikawa, H., Uchida, S., Okamura, T. (1990) 'Influence of Fineness and Particle Size Distribution of Cement on Fluidity of Fresh Cement Paste, Mortar and Concrete', *Journal of Research of the Onoda Cement Company*, 42 (122): 75–84.

Weiss, J.-C. (1999) Personnal communication.

Yuko, T., Ikabata, T., Akiyama, T., Yamamoto, T. and Kurunada, N. (2000) 'New Clinker Formation Process by the Fluidized Bed Kiln System', *Cement and Concrete Research*, 30: 1113–20.

Zhang, M.-H. and Odler, I. (1996a) 'Investigations on High SO_3 Portland Clinkers and Cements: Part I. Clinker Synthesis and Cement Preparation', *Cement and Concrete Research*, 26 (9): 1307–13.

Zhang, M.-H. and Odler, I. (1996b) 'Investigations on High SO_3 Portland Clinkers and Cements: Part II. Properties of Cements', *Cement and Concrete Research*, 26 (9): 1315–24.

Chapter 6

Portland cement hydration

6.1 Introduction

The expressions 'Portland cement hydration' and 'hydration reaction' are used to cover a certain number of physico-chemical and thermodynamic processes that develop, simultaneously or sequentially, when Portland cement comes into contact with water.

Considering that:

- Portland cement is essentially a multiphase material of which the composition can vary over wide limits;
- the phases present in Portland cement are not pure, because they have trapped some impurities during the firing process of the clinker;
- some of these impurities can act as retarders, stabilizers or accelerators of some specific phases and even modify the crystallographic network of some phases and consequently their reactivity;
- the different phases found in clinkers do not always have the same morphology;
- in some cases the firing conditions favor the growth of large crystals, and in other cases very fine crystals or a mixture of large and fine crystals;
- the interstitial phase is more or less crystallized after the quenching that follows the passage of the clinker through the clinkering zone;
- commercial Portland cements always contain some so-called minor components in variable amounts, such as uncombined lime, periclase, alkali sulphates, free lime, etc.;
- many other factors of the process can alter the characteristics of the clinker (oxidizing or reducing conditions, etc.).

the complexity and variability of the physico-chemical and thermodynamic processes that are hidden under the very simplistic expression 'hydration reaction' must be realized.

This complexity explains the reasons for so much research work and why so many publications and books have been written on the subject. Among my readings I can cite Le Chatelier (1904); Powers and Brownyard (1948); Lafuma (1965); Papadakis and Venuat (1966); Mindess and Young (1981); Folliot and Buil (1982); Regourd (1982a and 1982b); Vernet and Cadoret (1992); Nonat (1994); Eckart et al. (1995); Taylor (1997); Damidot et al. (1997); Odler (1998); Gauffinet et al. (2000); Jensen and Hansen (2001); and Bensted (2001). This doesn't even take into account the numerous papers found in the proceedings of the two RILEM meetings on Portland cement hydration and setting organized at Dijon in 1991 and 1997 (edited by A. Nonat and J.C. Mutin) and the Kurdowski symposium held in Crakov (Kurdowski and Gawlicki 2001).

It is surprising that, 100 years after the work done by Le Chatelier and Michaëlis, there are so many researchers passionately studying Portland cement hydration, such a simple technological reaction that occurs many times every day in concrete mixers and trucks. In fact, as pointed out earlier, the chemical reactions occurring during Portland cement hydration are complex. Moreover, currently, they are made even more complex by the use of admixtures or mineral components, as will be shown in the following two chapters.

It should be emphasized that currently we still do not understand the detailed features of cement hydration; we know only the broad outlines within which Portland cement hydration evolves. In spite of all its complexity, Portland cement hydration results in a series of fundamental phenomena that characterize the action of water on Portland cement. Portland cement hydration corresponds to:

- a succession of chemical reactions that transform the anhydrous phases of Portland cement into hydrated phases;
- the appearance of a progressive development of structure in the hydrated cement paste that in the case of concrete, is manifested initially by a more or less progressive slump loss and then a strength increase;
- a temperature rise, more or less rapid and intense;
- changes of the absolute and apparent volume of the paste;
- a decrease in the electrical conductivity of the interstitial solution in the fresh concrete.

It has been said that Portland cement hydration occurs within the eternal triangle: strength, heat and volumetric variations, which is presented in Figure 6.1. The users of Portland cement have no other choice than to learn how to deal with these three concomitant phenomena.

From a technological point of view, the action of water on Portland cement is even more complex because water plays a key role in the rheological behaviour of the fresh concrete. In fact, it is the initial reaction of water and Portland cement particles that gives to concrete its initial slump and workability and its resistance to bleeding and segregation. From a practical point of view, the initial slump and workability are two very important characteristics, because they are responsible for the easy use and transportion of concrete. Finally, the effect of

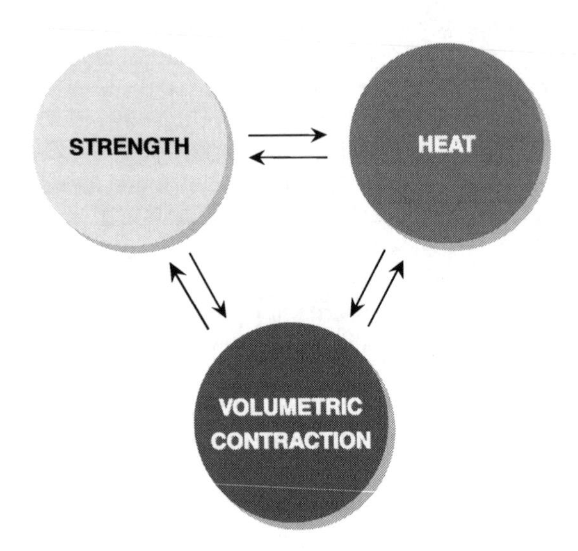

Figure 6.1 Portland cement hydration.

temperature on the kinetics of hydration, and its microstructural and mechanical consequences, are very important from a technological point of view.

6.2 The scientific complexity of a simple technological phenomenon

Portland cement hydration is at the same time a very simple technological phenomenon that results in the solidification of fresh concrete and its mechanical strengthening and, also, a very complex chemical process (Bensted 2001).

Concrete is, therefore, the fruit of a very simple technology and a very complex science: this duality is at the same time the cause of concrete's success but also of its weakness when this complex science is not mastered.

When presenting the simplified chemical notations used in the cement and concrete field, we have seen in Chapter 4 in particular, that in order to translate the chemical reactions that occur during the hydration of the silicate phase (C_3S and C_2S), the hydrated phase is written as C-S-H. Such a vague notation is used because the structure of calcium silicate hydrate is not well known and defined. In particular, the chemical microanalysis of C-S-H shows that C-S-H presents a wide range of C/S ratios within the different types of C-S-H that are observed under a scanning electron microscope (Diamond 1976).

As has already been seen in Chapter 2, although it is easy to write a very simple chemical reaction to describe the hydration of calcium sulphate hemihydrate (plaster of Paris) when it hardens as gypsum, or for the chemical reactions that transform quicklime (CaO) into calcium carbonate ($CaCO_3$), it is not possible to associate any simple chemical reaction to translate the formation of calcium silicate hydrate from the two anhydrous forms C_3S and C_2S.

All the authors that write that the hydration of C_3S and C_2S is as follows:

$$2\ C_3S + 6\ H \rightarrow C_3S_2H_3 + 3\ CH$$

$$2\ C_2S + 4\ H \rightarrow C_3S_2H_3 + CH$$

also state that writing $C_3S_2H_3$ is only approximate, because the actual composition of calcium silicate hydrate can vary widely, and that it is still difficult to describe its actual morphology (Mindess and Young 1981; Bensted 2001).

Karen Scrivener (2001) prefers to write the hydration reaction of C_3S and C_2S as follows:

$$C_3S + 5{,}3\ H \rightarrow C_{1,7}\,SH_4 + 1{,}3\ CH$$

and

$$C_2S + 4{,}3\ H \rightarrow C_{1,7}\,SH_4 + 0{,}3\ CH$$

The first chemical theories on Portland cement hydration began to be studied more than 100 years ago by great chemists like Le Chatelier and Michaëlis, who tried to translate into chemical terms the phenomena they were observing with the limited experimental tools that were available to them. Le Chatelier, for example, refers in his writings to calcium silicate hydrate in the form $SiO_2 \cdot CaO \cdot A_q$ (A_q for the Latin term *aqua* for water). This looks very much like the modern writing C-S-H. It must be admitted that we have not made much progress in the intimate knowledge of calcium silicate hydrate in the past 100 years.

In his remarkable work, Le Chatelier (1904) has pointed out the importance and the consequences of the physico-chemical and thermodynamic phenomena that are induced by the hydration reaction, such that the absolute volumetric contraction resulting from the hydration of the anhydrous phases that is still known in the French literature as Le Chatelier contraction. However, Le Chatelier himself admitted that he was unable to explain why this absolute volumetric contraction could be transformed into an expansion of the apparent volume when the cement paste was cured under water, as seen in Figure 6.17 on p. 172.

In fact, Portland cement hydration is more than simply a series of chemical reactions; it is also a spatial reorganization of the ionic species present in the clinker into hydrated phases where water can be found in different forms or even be transformed into hydroxyl groups, OH^- (Bensted 2001). For a mineralogist preoccupied by the spatial organization of structural elements, calcium silicate hydrate is built around an O_2^- tetrahedral having a Si^{4+} ion at the centre and an octahedral cavity having Ca^{2+} ions at its centre. However, although the structure of aluminum silicate hydrate, kaolinite ($2\ SiO_2 \cdot Al_2O_3 \cdot 2\ H_2O$), is well known and easy to describe, as is the structure of the magnesium silicate hydrate, the serpentine or asbestos or chrysotile ($2\ SiO_2 \cdot 3\ MgO \cdot 2\ H_2O$) as shown in Figure 6.2, the crystallographic structure of C-S-H is still debated among mineralogists.

For a mineralogist, kaolinite can be written also as $Al_2Si_2O_5\ (OH)_4$, and chrysotile as $Mg_3\ (Si_2O_5)\ (OH)_4$. The spatial organization of the silicon tetrahedral and 'alumina' octahedral fits perfectly to form a layered flat structure in kaolinite but a curved one in chrysotile because Mg^{2+} ions are bigger than Al^{3+} ions in the octahedral layer (Yada 1967) (Figure 6.2). However, it is observed that it is impossible to obtain such a simple spatial organization when Al^{3+} and Mg^{2+} ions are replaced by a larger Ca^{2+} ion (Liebau 1985; Jouenne 1990) therefore, it is not possible to write C-S-H in the form of $Ca_3\ (Si_2O_5)\ (OH)_4$.

The size of the ion at the centre of the octahedral position has a great technological importance because it is possible to get materials as different as a clay (aluminum silicate hydrate), an asbestos (magnesium silicate hydrate) and mineral glue (calcium silicate hydrate). This glue is so easy to use and to develop its mechanical properties at room temperature that within less than 100 years, concrete has become the most widely used artificial construction material in the world (Aïtcin 1995).

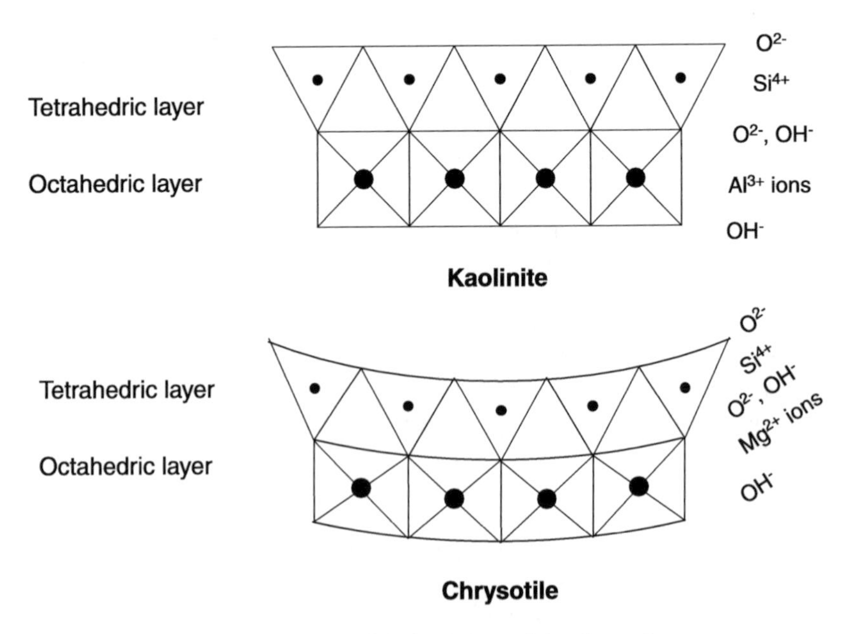

Figure 6.2 Schematic representation of the kaolinite and the chrysotile.

6.3 How to approach the hydration reaction

In order to overcome its complexity, the hydration reaction can be approached in several different ways. One approach consists of studying separately the hydration of each of the four main hydraulic phases found in Portland cement and hoping that the hydration reaction is globally equivalent to the sum of each of these individual reactions. Such an approach has been used extensively, although it is recognized that during hydration some interaction and synergies develop amongst the four individual components. For example, it is known that C_3A hydration accelerates the hydration of C_3S (Bensted 2001; Stark et al. 2001b). Some observations and results from this approach are presented in Section 6.6.3.

Another approach consists of examining under a scanning electron microscope (SEM) the surface of freshly fractured samples of a cement paste, mortar or concrete, in which the hydration reaction has been stopped with acetone, and observing the crystalline species that develop at a particular time. This method presents two major drawbacks; first, the hydration reaction has to be stopped brutally by acetone and, second, the samples must be dried before being observed under the SEM. However, this method of observation of the hydration reaction has the advantage of permitting direct observation of the crystals formed and offering the possibility of making a microchemical analysis on these crystals or even mapping some key elements. The use of an environmental SEM eliminates the need to dry the sample, but the reduced vacuum that has to be established within the microscope somewhat dries the sample under observation, leading some researchers to question the validity of the observations made with an environmental microscope. With the most up-to-date environmental SEM, it seems that this is no longer a major problem and that the observations are quite realistic (Stark et al. 2001a and 2001b) (Appendix IX).

Another approach consists of monitoring simultaneous phenomena that are consequences of the hydration reaction, for example: the strength increase, the increase of the elastic modulus, the variation of the electrical conductivity of the interstitial solution, or the volumetric variations under different curing conditions (Figure 6.3).

Different chemical equations can be written and discussed with all the limitations that have to be taken into account with such an approach (Bensted 2001). Mathematical models can be established when making some hypotheses. Nuclear magnetic resonance can be used to study the nanostructure of the hydrates.

The author has finally opted for a 'civil-engineering approach' resulting from his experience in observing different types of concrete, of all ages and composition under a SEM. In this engineering approach, the rheology of concrete, the volumetric variations and the heat development within the hardened concrete, are characteristics as important, or even more important from a durability and increased strength point of view.

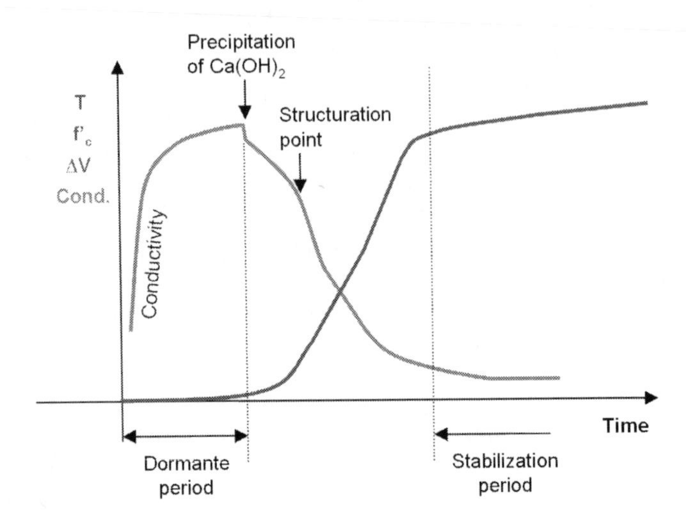

Figure 6.3 The consequences of hydration reaction.

The role played by the 'gypsum' ion in the development of the hydration reaction will be analysed first because it is a very important parameter of the hydration reaction. Two not-so-frequent accidents will be discussed: flash set and false set. Thereafter, some of the consequences of the hydration reaction will be analysed: variations of the electrical conductivity of the interstitial solution, setting, volumetric variations, heat development and strength increase. Some SEM observations carried out on hydrating cement paste will complete the picture and finally, some hydration models will also be briefly presented as well as some of the most recent C-S-H structural models suggested by certain authors.

At the end of this chapter, it will be seen that, in spite of all the efforts that have been devoted to trying to better understand hydration reaction during the past 100 years, it must be admitted that we do not yet have a clear and simple theory that explains why every year 6 billion cubic metres of concrete are hardening. The author does not agree with some engineers and cement manufacturers who maintain that comprehensive studies on cement hydration are only of limited value and that the only thing that matters is that Portland cement hardens. Some of these people even argue that is not necessary to know how an engine runs in order to drive a car, or to have a notion of ballistics to shoot a rabbit.

Of course, up to now it must be admitted that quite often technological progress has preceded scientific knowledge in the field of Portland cement and concrete, but recently giant steps forward have been made in concrete technology due to a scientific approach of the material rather than by chance or by trial and error. The author is definitely convinced that the future of the cement and concrete industries depends more on a deeper understanding of hydration reaction than on luck. In spite of the fact that Portland cement hydration is a very complex phenomenon and not easy to understand, it obeys the laws of physics, chemistry and thermodynamics. Only a better understanding of the hydration reaction will significantly expand the limits of utilization of Portland cement.

6.4 Portland cement and gypsum

6.4.1 Controlling the hydration of C_3A

When C_3A comes into contact with water, it reacts very rapidly and releases a great amount of heat as it is transformed into calcium aluminate hydrates of the hydrogamet type (Regourd 1978; Bensted 2001). In some books, this reaction is said to be *vigorous* or even *violent*.

C_3A reacts with water according to the following reaction:

$$C_3A + 21\,H \rightarrow C_4AH_{13} + C_2AH_8 + \text{heat}$$

As these two hydrates are not very stable, they are transformed more or less rapidly into hydrogamet according to the following reaction

$$C_4AH_{13} + C_2AH_8 \rightarrow 2\,C_3AH_6 + 9H$$

In a preceding chapter, it was mentioned that it was the French chemist Giron who, while working in the USA, found that the addition of a small amount of gypsum could be used to control C_3A hydration, and that later on Le Chatelier (1904) and Candlot (1906) were able to explain how calcium sulphate was able to control C_3A hydration. In the presence of gypsum and water, C_3A hydrates as tricalcium sulphoaluminate hydrate, known as ettringite according to the following chemical reaction.

$$C_3A + 3C\bar{S}H_2 + 26H \rightarrow C_6A\bar{S}H_{32}$$

Some purists call this compound trisulphate calcium aluminate (32) hydrate (Mindess and Young 1981), and some French researchers call it Candlot salt (D'Aloia 1998).

The developed chemical composition of ettringite is

$$3\,CaO \cdot Al_2O_3 \cdot 3\,CaSO_4 \cdot 32\,H_2O$$

but in some cases it will be useful to use a semi-developed formula and to write it as

$$C_3A \cdot 3\,CaSO_4 \cdot 32H_2O$$

This notation shows clearly the three chemical groups that are at the origin of ettringite during Portland cement hydration.

In some books the ettringite composition is written as

$$C_3A \cdot 3\,CaSO_4 \cdot 30H_2O \quad \text{as well as} \quad C_3A \cdot 3\,CaSO_4 \cdot 31H_2O$$

For a long time I was amazed to see that such a well-known crystal, which crystallizes so easily and 'beautifully' into a characteristic hexagonal shape had a chemical composition about which researchers did not agree. I have even read that ettringite could not have thirty-one water molecules because 31 is a prime number, and it would be impossible for a crystal having an hexagonal symmetry to end up with a prime number of water molecules.

The answer to the actual number of water molecules contained in ettringite was given to me by Micheline Moranville-Regourd who explained that on the hexagonal ettringite there were two water molecules that were loosely attached to the main frame of the ettringite crystal, as is shown in the schematic representation of the ettringite crystal by Vershaeve (1994) (Figure 6.4a), and that these two water molecules can be easily detached (30 H_2O) or partially detached (31 H_2O).

Therefore, some calcium sulphate is added to Portland cement clinker to transform C_3A into ettringite, and as will be shown, to stabilize Portland cement hydration (Collepardi et al. 1979; Clark and Brown 1999 and 2000; Bensted 2001).

However, when making a rapid calculation, taking into account the amount of C_3A as given by the Bogue calculation, it is seen that not enough gypsum is added to Portland cement clinker during its final grinding to transform all the C_3A into ettringite. This is done purposely for two reasons: the first reason is that not all the C_3A contained in Portland cement hydrates instantaneously, because it is not necessarily situated at the surface of the cement particles, and, second, because the C_3A that is at the surface of the cement grains is not disintegrated instantaneously by water. Therefore, from a rheological point of view, it is not necessary to control the hydration of all the C_3A present in Portland cement. Finally, it is important to point out that **ettringite is not a stable compound and that the less there is in a concrete the better it is.** In fact, 100 years ago, Michaëlis called ettringite the 'concrete bacillus'.

When all the calcium sulphate added during grinding has been combined with C_3A the remaining C_3A decomposes ettringite to form another calcium sulphoaluminate less rich in calcium sulphate which is more stable. This is called hydrated tetracalcium monosulphoaluminate, or simply, monosulphoaluminate and is formed according to the following chemical reaction.

$$2C_3A + C_6A\bar{S}H_{32} + 4H \rightarrow 3C_4A\bar{S}H_{12}$$

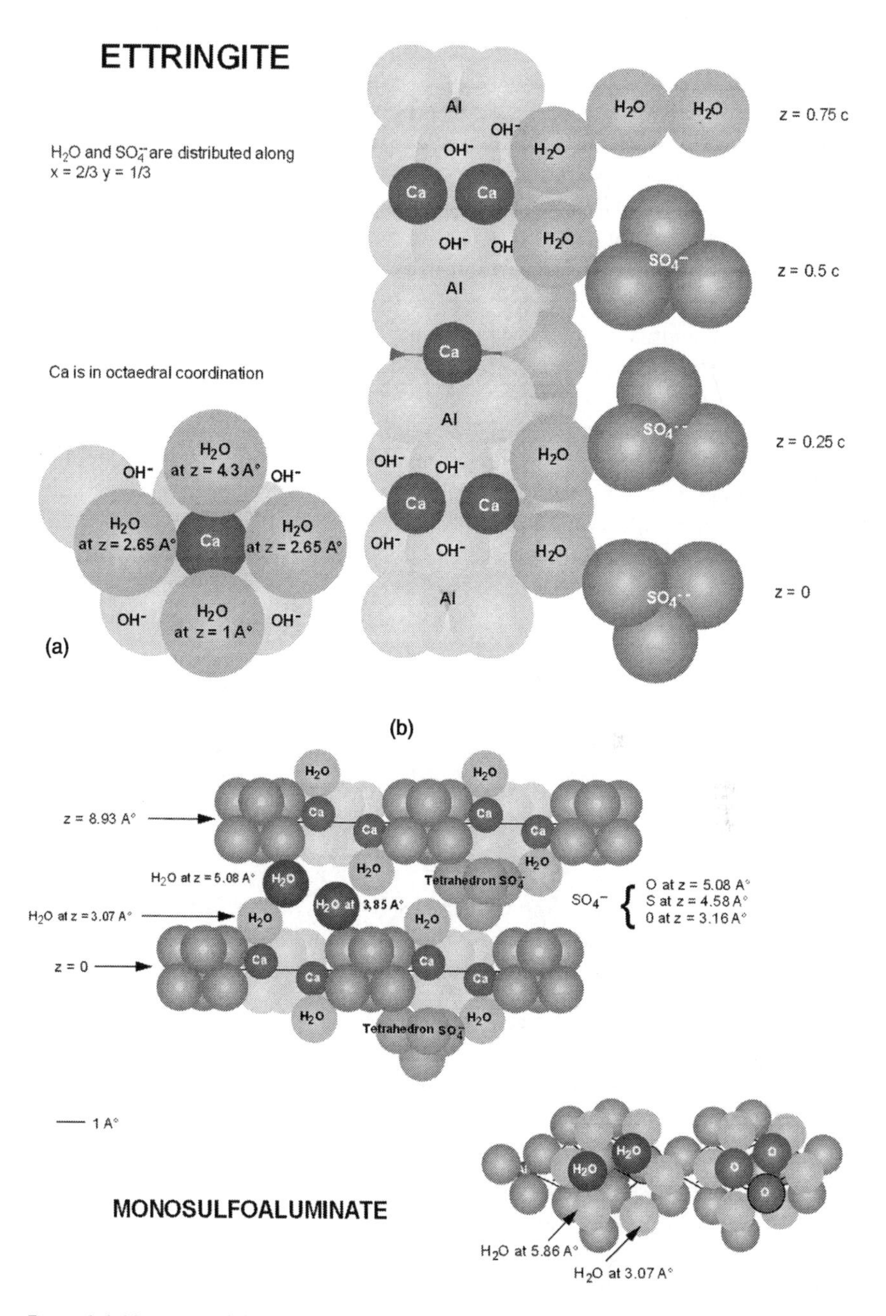

Figure 6.4 Unit cell of (a) ettringite and (b) calcium monosulphoaluminate (courtesy of M. Vershaeve).

Monosulphoaluminate can also be written as $C_3A \cdot C\bar{S} \cdot H_{12}$ or in a semi-developed form as $C_3A \cdot CaSO_4 \cdot 12\ H_2O$. The structure of monosulphoaluminate is presented in Figure 6.4b.

Consequently, when standards purposely limit the amount of calcium sulphate to control the setting time and increase the 24-hour strength of Portland cement, they favour the partial or total decomposition of the ettringite that has been initially formed. However, this reaction of decomposition occurs within the hardened cement and has no interest from a setting and rheological point of view. It is possible in a Portland cement paste that is hydrating, that some C_3A hydrates directly as monosulphoaluminate or as hydrogarnet due to a partial or total lack of sulphate ions close by (Bensted 2001).

In the literature, ettringite is also referred as A_{ft} and monosulphoaluminate as A_{fm}.

When trying to match the calcium sulphate available to make ettringite and monosulphoaluminate, it is almost always found that there is not enough calcium sulphate added during the grinding of Portland cement clinker, and, consequently, that some C_3A finally hydrates directly as hydrogarnet. But, this hydrogarnet formation occurs very late in the hardening of concrete. Moreover, practical experience shows that some sulphate ions end up trapped in the C-S-H and, therefore, do not react to form ettringite or calcium sulphoaluminate (Bensted 2001).

However, when later, for any reason, sulphate ions penetrate in the interstitial solution (acid rain, underground water rich in sulphate, thiobacillus ferroxydans attack), the monosulphoaluminate and the hydrogarnet present in the hardened concrete can be transformed into so-called secondary 'swelling' ettringite, with all the pernicious effects that this 'swelling' ettringite produces. In northern countries where concrete is exposed to freezing and thawing cycles, ettringite clusters are almost always found inside the air-entrained bubbles as seen in Figure 6.5. Ettringite is also found in cracks and in the transition zone between aggregates and the cement paste in concrete samples destroyed by freezing and thawing cycles, although this ettringite is not directly responsible for the destruction of concrete (Baalbaki et al. 1998).

6.4.2 *The true nature of the calcium sulphate found in Portland cement*

For a long time, gypsum has been used to control Portland cement-setting but currently it is better to speak about calcium sulphate addition. In fact, even when gypsum was introduced in the grinding mill, it was not certain that the same amount of gypsum would be found in Portland cement because, following the increase of the temperature in the grinding mill, some gypsum could have been partially dehydrated as calcium sulphate hemihydrate $CaSO_4 \cdot \frac{1}{2}H_2O$, or even totally dehydrated as $CaSO_4$, sometimes improperly called soluble anhydrite. (Anhydrite is a natural mineral having the same chemical

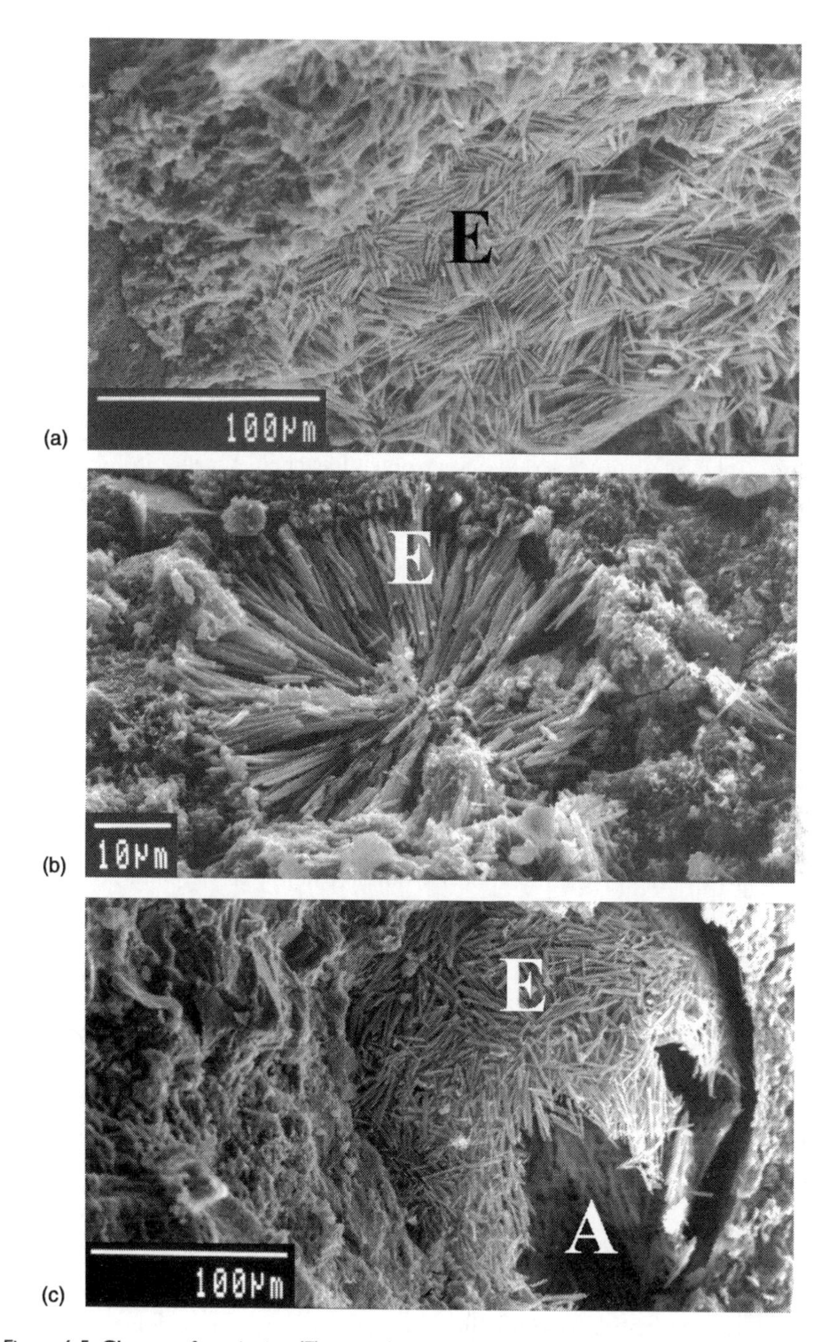

Figure 6.5 Cluster of ettringite (E) crystals in some concretes: (a) ettringite found in old concretes exposed to freezing and thawing cycles in Quebec at the paste aggregate interface or in fissures; (b) ettringite needles have almost filled an air-entrained bubble; (c) ettringite needles lining an air-entrained bubble (A) (photos taken by Guanshu Li).

composition but a mineralogical structure and *solubility rate* that has nothing to do with that of a totally dehydrated gypsum, in spite of the fact that they have the same chemical composition.) This kind of calcium sulphate should rather be called dehydrated gypsum to remember its origin and its high solubility rate. Gypsum dehydration could occur at a low temperature in storage silos (Hansen and Offutt 1962).

Partial dehydration of gypsum is currently used by some cement producers to control the setting of their 'nervous' cement that is finely ground and rich in C_3A; 30 to 40 per cent of the gypsum can be partially dehydrated in certain Portland cements.

However, when the temperature increases excessively during grinding, too much gypsum can be dehydrated, resulting in a premature stiffening of concrete. This type of accident is known in the industry as 'false set'. The premature stiffening of concrete looks very much like that observed in the case of a flash set, but, if the mixing is continued, the gypsum that has been formed during the hydration of the hemihydrate within the fresh concrete can dissolve and the concrete can thus recover its plasticity (Hansen and Offutt 1962; Bensted 2001).

The problem when facing a rapid stiffening of concrete in a mixer is that it is impossible to discriminate between false set or flash set; therefore, in either case, it is better to add as much water as possible and to empty the mixer. Following this, it is suggested to mix a small batch in a mortar mixer or in a small concrete mixer to determine if it is a flash set or a false set.

If it is a false set, it will only be necessary to lengthen the mixing time and to add some more water, or even better, some more superplasticizer, but if it is a flash set situation, it is necessary to stop concrete production, to empty the cement silo and refill it with a good Portland cement. This type of accident is very costly in term of loss of production; fortunately, it does not happen too often.

False set sometimes happens in the summer when too hot a clinker is introduced directly within the grinding mill (Bensted 2001).

Another source of rheological problems resulting from an improper addition of calcium sulphate is due to the fact that cement producers are adding less and less pure gypsum but rather a cocktail of calcium sulphates of different origin with different solubility rates (Bensted 2001). Of course, when making the chemical analysis of a Portland cement, the SO_3 content of the cement stays the same, but the release of SO_4^{2-} ions in the interstitial solution, which is the most important factor from a rheological point of view, can vary, and it can be difficult to control concrete rheology.

Sometimes natural anhydrite is used to partially replace gypsum (Hansen and Offutt 1962). Natural anhydrite is a crystalline form of calcium sulphate that is found in large natural deposits, or which is mixed with gypsum in gypsum deposits, because some gypsum deposits result from the hydrothermal alteration of an anhydrite deposit (Hansen and Offutt 1962; Bensted 2001). As the manufacturing of gypsum boards has become more and more automated, cement manufacturers can buy for a few dollars less more or less 'contaminated' gypsum containing some natural anhydrite. As long as cement manufacturers are only preoccupied to comply with the SO_3 content prescribed in cement standards, they don't care about the nature of the SO_3-bearing mineral. In fact, for water/cement ratios greater than 0.50, the nature of calcium sulphate does not make much difference on the setting properties and rheological characteristics of the cement. The rheology of concrete with such a high water/cement ratio is still essentially controlled by the amount of water. However, when dealing with concretes having a lower water/cement ratio, the very slow solubility rate of natural anhydrite creates problems because as high-performance concretes contain a high amount of cement, there is more C_3A to control and less water to dissolve SO_4^{2-} ions so that superplasticizer molecules are consumed to control the hydration of C_3A.

As will be seen in the next chapter, the SO_3^- end groups of the polysulphonate superplasticizer can react with the C_3A (Baussant 1990; Flatt and Houst 2001; Prince et al. 2002). This consumption of superplasticizer molecules results in a slump loss.

In their objective to lower their production costs, as much as possible cement producers have started to use all kinds of calcium sulphate to 'help' some industries get rid of their calcium sulphate wastes. This is definitely the right attitude in a society concerned with the beneficial recycling of industrial wastes, because it fits perfectly into the implementation of a policy of sustainable development. The only problem for the cement producer is to be sure that the grinding mill is fed with a calcium sulphate cocktail that always has the same composition. Some cements plants are already homogenizing their calcium sulphate sources.

Therefore, the calcium sulphate that is currently found in a Portland cement can be: natural anhydrite, gypsum, hemihydrate, totally dehydrated gypsum or synthetic calcium sulphate (Bensted 2001). As previously pointed out when looking only at the SO_3 content of these different forms of calcium sulphate, it is easy to make the necessary adjustments to comply with the SO_3 requirements of current cement standards, but, when dealing with the rheology of low water/binder concretes, serious rheological problems can be encountered.

6.4.3 The SO_3 content of clinker

For a long time, the SO_3 content of a cement and its loss on ignition (LOI) could be used to provide an idea of the amount of gypsum contained in a Portland cement, but this is less and less true, because the SO_3 content of some modern clinkers has changed drastically. This change in the SO_3 content of the clinker is hidden by the fact that cement producers always give only the SO_3 content of their cement (Skalny et al. 1997). The SO_3 content of present cements comes from the SO_3 content of the clinker and of the various sources of calcium sulphate introduced in the grinding mill.

Some years ago, when oil prices were very low and refineries and coal users did not care about the amount of SO_3 they were releasing into the atmosphere, cement producers could buy fuels with a low sulphur content or clean coal from an SO_3 point of view. Other industries were also releasing large amounts of SO_3, which resulted in an increase of devastating acid rains. When governments decided to put limits on SO_3 emissions, the price of clean fuel and coal increased while the price of fuels and coal rich in sulphur decreased. As the cement industry is one of the few industries that can support a high amount of sulphur in its fuels because this sulphur is easily trapped by the alkalis contained in the raw material or by the C_2S and C_3S, it became a great business opportunity for the cement industry to burn fuels rich in sulphur. We have already seen the importance of the cost of fuel in the cement manufacturing process.

Consequently, the SO_3 content of clinker that for a long time was around 0.5 per cent jumped in a certain number of clinkers to 1.5 to 2 per cent and even to 2.5 per cent in extreme cases. As cement standards always require a maximum amount of SO_3 lower than 2.3 to 3.5 per cent according to the Portland cement type, some cement producers have very little room for introducing calcium sulphate. Presently, it is not sure that all the SO_3 contained in the clinker has the same solubility rate as the SO_3 provided by calcium sulphate. Such a situation has resulted, in some cases, in rheological problems, particularly in the case of low water/binder ratio concretes.

Another SO_3-related problem concerns the use of $Fe_2(SO_4)_3$ to complex hexavalent chromium, responsible for masons' scabies, which limit the amount of SO_3 that can be incorporated in Portland cement. Usually $Fe_2(SO_4)_3$ is included in Portland cement sold in bags and not in Portland cement sold in bulk to the ready mix or precast industry.

This dissertation on the role of calcium sulphate could be perceived as too long or as giving too much emphasis to a minor component of Portland cement, since problems in low water/binder concrete represent a very small percentage of the Portland cement market. But, if cement producers are obsessed by the cube strength of the cement they are producing, contractors and concrete users are much more concerned by the control of the rheology of the concrete they are using, since currently it is increasingly common for owners and governmental agencies to strictly forbid the addition of water in the field to control concrete rheology. As will be seen later, the cost of Portland cement represents only 1 or 2 per cent of the total cost of a civil-engineering project, but when things go wrong with the cement and/or the admixture, the direct costs of a rheological problem rapidly become dramatic for a contractor, because the clock runs at a very expensive rate.

6.5 Direct observation of the structural modifications occurring during cement hydration

X-ray diffractometry and scanning electron microscopy can be used to follow the structural transformations occurring during Portland cement hydration.

Historically, X-ray diffractometry has been used to identify the different components of Portland cement clinker and the hydrates formed during hydration. Thereafter, scanning electron microscopy has been used extensively to visualize what had been observed by X-ray diffractograms. More recently, environmental SEMs have been used to follow the growth of the products of hydration (see Appendix IX).

The combination of scanning electron microscopy and the EDAX microprobe has been very useful in identifying the mineral species involved in cement hydration and in seeing where the different impurities were trapped in clinker, as well as in the products resulting from the hydration of Portland cement.

Some SEM pictures will be presented here. In Figure 6.6, beautiful hexagonal crystals of portlandite, $Ca(OH)_2$, are presented. Hexagonal needles of ettringite are very often observed (Figure 6.7) as well as monosulphoaluminate platelets (Figure 6.8). The more 'beautiful' these pictures are from an aesthetic point of view, the higher the water/binder ratio of the concrete, that is, the poorer the quality of the concrete. Ill-defined C-S-H crystals can also be observed in SEM pictures; these small crystals grow like very short needles, and consequently, partially hydrated, C_3S and C_2S crystals look like sea urchins in the first hours that follow the beginning of hydration. C-S-H crystals are much smaller than portlandite, ettringite and monosulphoaluminate crystals.

Hydration can be stopped at any time by acetone immersion to calculate the degree of hydration of Portland cement paste. However, some researchers question the 'brutality' of such a treatment as it interferes so drastically with the process of hydration.

Scanning electron microscopy can also be used to observe the influence of the water/binder ratio on the morphology of the hydrates that are formed. It is observed that as Portland cement particles become closer to each other, that is when the water/binder ratio decreases, it is more and more difficult to observe beautiful crystals; rather, a continuous mass looking 'amorphous' is observed (Figure 6.9). In very low water/binder ratio pastes, portlandite and ettringite crystal are no longer visible when observed in an SEM. However, when the same pastes are submitted to a thermogravimetric analysis, it is possible to see that they contain some portlandite because at 450°C there is a mass loss observed.

The C-S-H observed in low water/binder paste (Figure 6.10) seems to be 'amorphous' as the hydration of C_3S and C_2S results from a diffusion process (inner product) (Figure 6.11) as opposed to the rather well crystallized outer products when hydration occurs through a dissolution and recrystallization

Figure 6.6 Crystals of portlandite $Ca(OH)_2$ (P): (a) hexagonal platelets of portlandite (P); (b) hexagonal platelets of portlandite (P) and ettringite needles (E); (c) clusters of portlandite crystals (photos taken by Irène Kelsey-Lévesque).

Figure 6.7 Ettringite crystals in a cement paste hydrated for three days: (a) general aspect; (b) scale-up of the preceding photo showing bundles of ettringite needles (E) as well as isolated needles (photos taken by Irène Kelsey-Lévesque).

process. Figure 6.12 represents C_2S at different stages of hydration when observing the development of hydration with X-ray diffractometry. The decrease of the intensity of the C_3S and C_2S peaks is observed as well as the increase of the peaks of portlandite and ettringite.

When measuring the C/S ratio of C-S-H with a microprobe, it is observed that this ratio varies between 1.2 and 2, which is quite a large range. However, some researchers point out that due to the very small size of C-S-H crystallites, this result can be influenced by C_2S crystals hidden behind the thin layer of C-S-H.

Figure 6.8 Monosulphoaluminate crystals (Afm) (photo taken by Irène Kelsey-Lévesque).

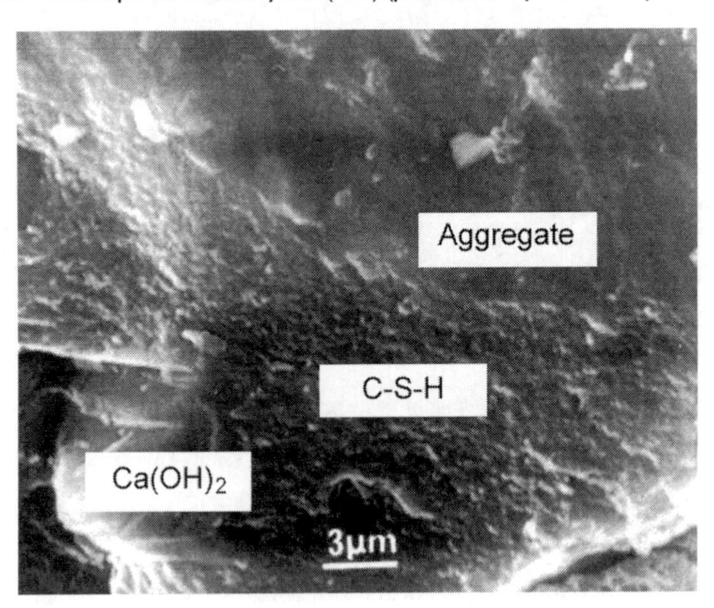

Figure 6.9 Vitreous aspect of C-S-H in low W/C non-air-entrained concrete (photo taken by Irène Kelsey-Lévesque).

It is difficult to make quantitative measurements from EDAX microprobe analysis. The only quantitative analysis that can be done is a thermogravimetric one for portlandite, because from the measured mass loss it is easy to calculate the actual amount of portlandite that has been dehydrated.

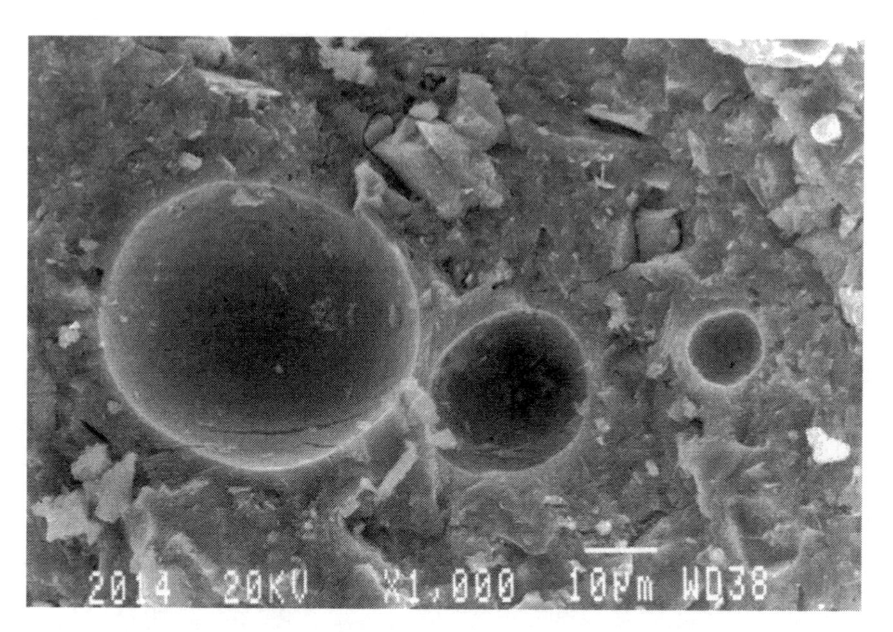

Figure 6.10 Vitreous aspect of C-S-H in a low air-entrained W/C concrete (photo taken by Irène Kelsey-Lévesque).

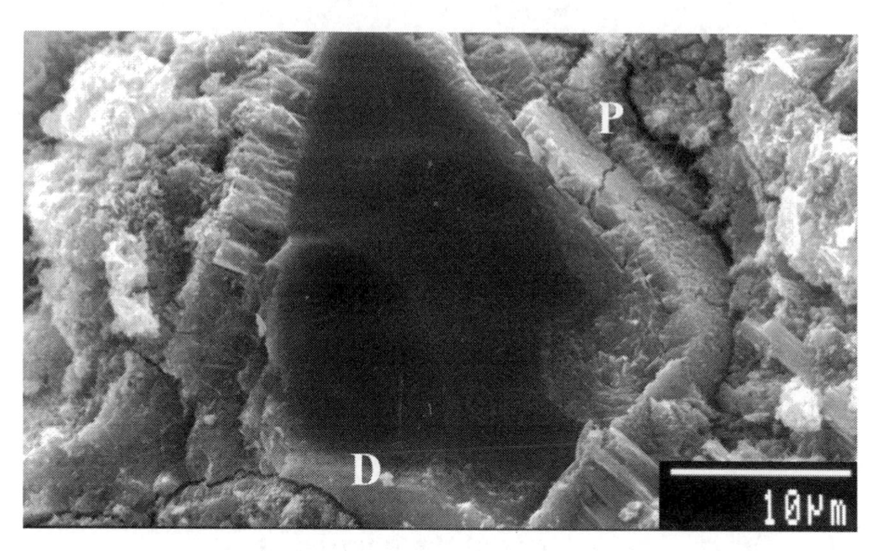

Figure 6.11 C_2S crystal hydrating. It is possible to see some C-S-H hydrated by precipitation (P) and by diffusion (D) (photo taken by Irène Kelsey-Lévesque).

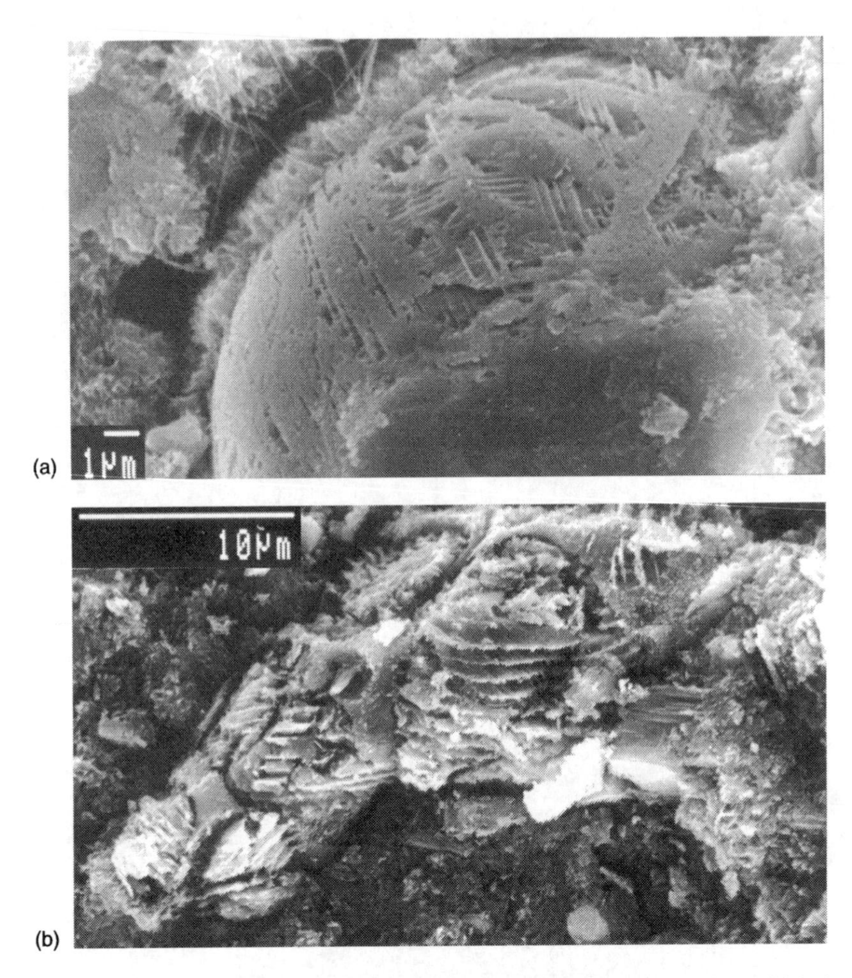

Figure 6.12 C_2S crystal hydrating: (a) belite particle (C_2S) beginning to hydrate; (b) belite particle (C_2S) in a state of advanced hydration (photos taken by Irène Kelsey-Lévesque).

In Appendix IX, some beautiful pictures of hydrating cement pastes taken by Stark et al. (2001a and 2001b) are presented.

6.6 Indirect observations of the physico-chemical and thermodynamic changes occurring during Portland cement hydration

The effects of the hydration reaction can be followed by monitoring some simultaneously occurring phenomena in the hydrating cement paste, such as:

- the variation of the electrical conductivity of the interstitial solution;
- the setting of the paste;
- the heat development;
- the volumetric changes;
- the increase of compressive strength and elastic modulus.

6.6.1 Variation of the electrical conductivity

This is a very simple and easy measurement that traces the rate at which the ions are released or combined during the hydration process. Figure 6.13 represents a characteristic curve showing the variation of electrical conductivity with time. This curve can be decomposed into different parts. The first part, A, corresponds to the ascending part that follows the contact of Portland cement particles with water. It is due to an intense chemical activity that corresponds to the dissolution of the various chemical species found in Portland cement. This chemical activity is superficial and does not lead to the formation of large crystals that could rigidify the matrix; in parallel, a rapid and strong increase of the pH is observed. It ends up as a peak P followed by a small drop, C. This small drop corresponds to the quasi-instantaneous precipitation of the first portlandite crystals when the interstitial solution has reached an over-sursaturation level of Ca^{2+} ions.

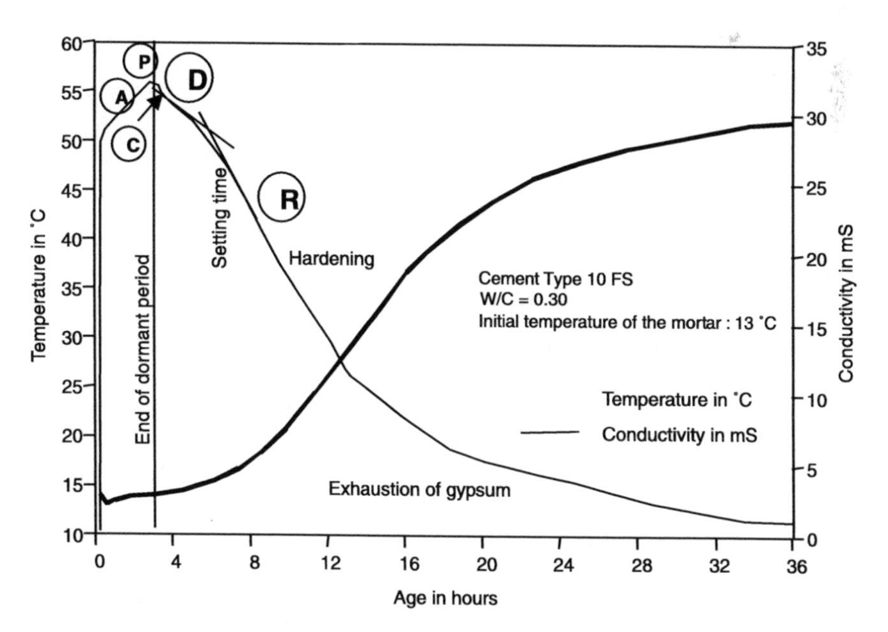

Figure 6.13 Electrical conductivity and heat release of a blended cement containing 8 per cent silica fume.

This small drop, C, is followed by a more progressive decrease of the conductivity of short duration, D, followed by a more rapid decrease in part R. This rather rapid decrease in the R part of the curve corresponds to the 'structuration' of the hydrated cement paste; some French authors prefer the word 'percolation' (Vernet 1995; Acker 2001; Barcelo et al. 2001).

At this stage of the hydration process, the first hydrates are bridging cement particles so that the motion of the ions in the interstitial solution becomes more difficult due to the tortuosity of their path; moreover, less water is available in the interstitial solution, because some water has been combined in hydrated products.

When measuring the electrical conductivity of different cement pastes, it is easy to follow the influence of one particular parameter on Portland cement hydration, such as the effect of an admixture, the effect of a mineral component, the effect of a particular component of a clinker, the effect of the type of sulphate, etc.

6.6.2 Portland cement setting

Portland cement setting can be followed using a test as simple as the one developed by Vicat. It consists of measuring the penetration of a standard needle on which a standard load is applied. More or less arbitrary times, in relation to cement hydration, are determined and identified as initial and final setting times. These two setting times follow the precipitation of portlandite and the acceleration of C_3S hydration, corresponding to two different degrees of structure formation reached by the hydrated cement paste.

The determination of these two setting times is useful when controlling the production of Portland cement, because it is a good indicator of the consistency of the production. Moreover, it gives a good idea of the degree of control of the rheology of the cement paste (initial set) and of the initial rate of hardening (final set).

6.6.3 Development of the heat of hydration

The development of heat during the hydration process, or rather the rate of heat evolution (the change of the heat rate with time) can be monitored in adiabatic conditions. A typical curve is presented in Figure 6.14.

Before trying to interpret such a curve directly it is best to look at the evolution of the heat rate of C_3S and C_3A pastes. These curves are presented in Figures 6.15 and 6.16.

The curve representing the rate of evolution of heat of C_3S (Figure 6.15) can be decomposed into five different stages which are numbered 1 to 5. Stage 1 corresponds to a short heat release that follows the first contact of C_3S and water. This heat release is associated with the dissolution of the most active ionic species located at the surface of the C_3S particles. It results in the

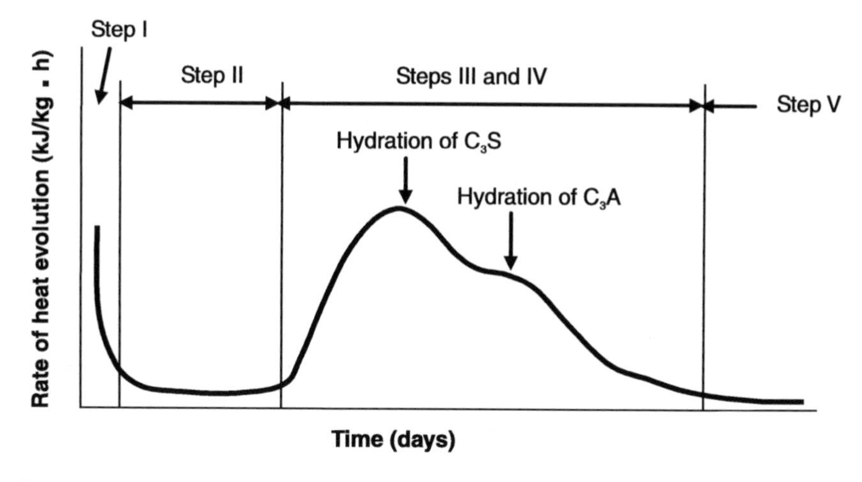

Figure 6.14 Rate of the heat evolution during the hydration of Portland cement (Mindess and Young, *Concrete*, © 1981, reprinted with permission of Pearson Education, Inc., Upper Saddle River, NJ).

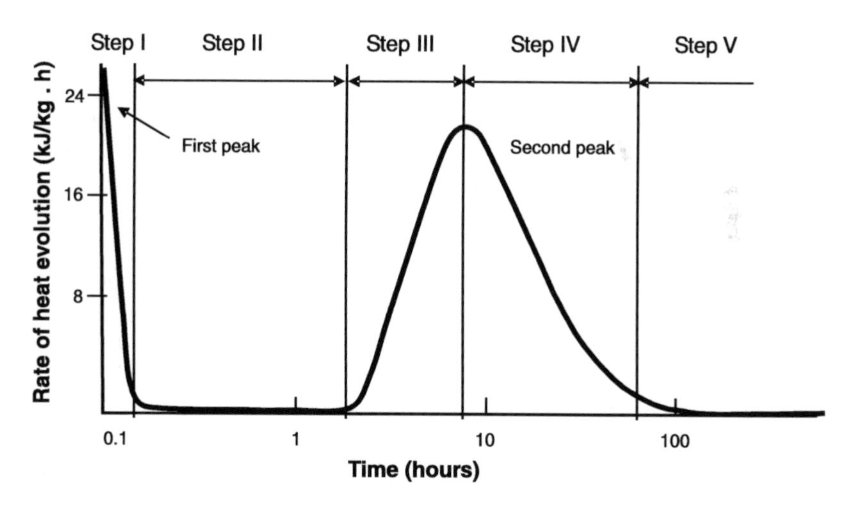

Figure 6.15 Rate of the heat evolution during the hydration of C_3S (Mindess and Young 1981).

formation of a thin layer of C-S-H over the C_3S particles. This thin layer of C-S-H blocks, or significantly slows down at least, further hydration of C_3S. This first stage has a duration of about 15 minutes.

The second stage corresponds to what is usually called the dormant period. It is not a period during which 'nothing' happens, because during this time, it is observed that the pH as well as the electrical conductivity increase and

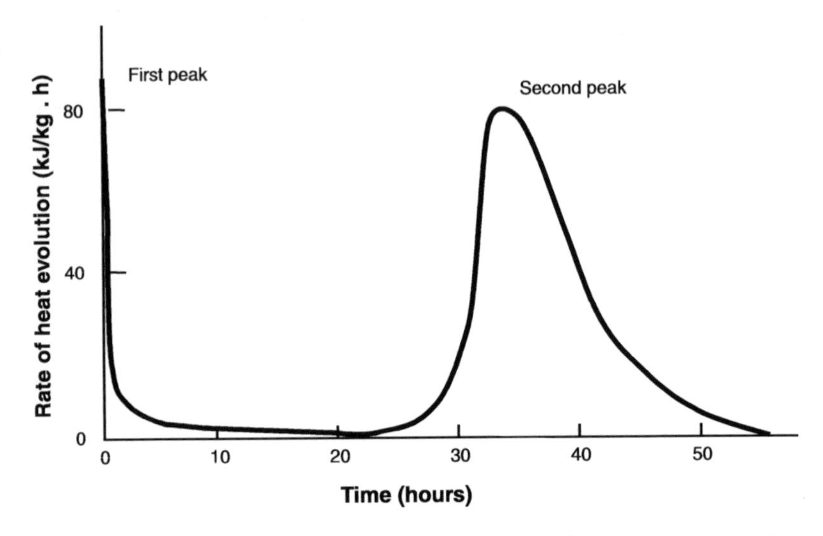

Figure 6.16 Rate of the heat evolution during C_3A hydration (Mindess and Young 1981).

that the C_3S paste stiffens slightly. This stage has a duration of about 1 hour or 1.5 hours.

The third stage corresponds to a rapid release of heat due to the rapid hydration of C_3S that follows the precipitation of the first crystals of portlandite. The rate of evolution of heat presents a peak after which it starts to decrease as rapidly as it had increased. This decrease corresponds to a slowdown of the hydration reaction: the first hydrates formed at the surface of C_3S particles by a dissolution/precipitation process interfere with the migration of water towards the inner part of C_3S particles. Consequently, the hydration reaction proceeds thereafter as a diffusion process, rather than a dissolution/precipitation process as in the ascending part of the peak.

The last stage of the rate of evolution of heat, Part 5, shows a small release of heat that decreases with time. The hydration reaction does not stop but continues at a very slow rate.

In Figure 6.16 it is seen that the rate of evolution of heat of C_3A has a shape that looks like that of C_3S; however, on the y axis, it is seen that the rate of evolution of heat is much larger (for an equal mass). Moreover, on the x axis it is seen that the second peak occurs later than in the case of C_3S.

The first peak observed in the C_3A curve corresponds to the initial formation of ettringite on C_3A crystals. This first layer of ettringite temporarily blocks the hydration of C_3A in the same way as the first C-S-H layer of small crystals blocked the hydration of C_3S. This first peak corresponds to a duration of about 15 minutes. Later on, the protection of this first layer of ettringite stops when ettringite starts to transform into monosulphoaluminate and, consequently,

C_3A hydration resumes. This rapid heat release is represented by a second peak. When a cement contains a high amount of gypsum, the second peak is somewhat delayed. This second peak usually occurs between 10 and 15 hours, at a moment when the hydrated cement paste has hardened and when the concrete has been in the forms for several hours. The rate of evolution of heat for C_2S and C_4AF shows that these two compounds are much less reactive than C_3A and C_4AF and that their effect on the rate of evolution of Portland cement is negligible.

When observing Figure 6.12, representing the change of the heat rate with time, it is seen that this curve can be obtained by combining the curves obtained for C_3S and C_3A, taking into account that there is five to ten times more C_3S than C_3A in Portland cement. Consequently, in spite of the fact that C_3A develops three times more heat than the same mass of C_3S (in the curve representing the rate of evolution of heat of Portland cement), it is the heat developed by C_3S that is predominant.

However, upon a closer look at Portland cement hydration, it is seen that the curve representing the rate of evolution of heat of Portland cement is not exactly the sum of the corresponding curves of C_3S, C_3A, C_2S and C_4AF, as there are some synergies that are developed during hydration. For example C_3A and C_4AF are in direct competition vis-à-vis sulphate ions, but as C_3A is more reactive than C_4AF, it consumes SO_4^{2-} ions more rapidly. Thus, calcium sulphate dissolves more rapidly, which indirectly results in the activation of C_4AF hydration.

Moreover, gypsum is known to accelerate the hydration of C_3S and C_2S and some SO_4^{2-} ions can be trapped in the C-S-H formed during the hydration of the C_3S and C_2S. According to Mindess and Young (1981) and Bensted (2001), the amount of monosulphoaluminate found in Portland cement is usually half the theoretical amount that should be found.

Cement producers always optimize the amount of gypsum they add in Portland cement from a cube strength point of view. In fact, too high an amount of ettringite results in a swelling of the paste and its microcracking, while too low an amount results in the early formation of monosulphoaluminate. This rapid formation of monosulphoaluminate consumes some lime and results in a slow-down of the hydration of C_3S. Some researchers maintain that the existence of an optimal gypsum dosage results from the acceleration of C_3S hydration but, at the same time, in a weakening of the C-S-H formed, due to the presence of sulphate ions trapped within the C-S-H.

The heat of hydration of each phase can be measured separately. C_3A is definitely the compound that releases the highest amount of heat per unit mass, at 900 kJ/kg, followed by C_3S, at 500 kJ/kg, then by C_4AF, at 400 kJ/kg and finally by C_2S at 250 kJ/kg. As a typical Portland cement contains between 50 and 60 per cent of C_3S, 20 to 30 per cent of C_2S and 5 to 10 per cent of C_3A and C_4AF, it is definitely the heat liberated by the hydration of C_3S that influences most significantly the heat of hydration of Portland cement. The contributions of C_3A, C_2S and C_4AF are definitely much smaller. When making a Portland cement for mass concrete, it is necessary to lower the amounts of C_3S and C_3A and, of course, to increase the C_2S and C_4AF content.

6.6.4 Volumetric variations associated with hydration reaction

It is on purpose that we use the plural in this subtitle, because, as already pointed out by Le Chatelier 100 years ago, the apparent volume of a cement paste increases or decreases depending upon the conditions under which hydration reaction takes place, as shown in Figure 6.17.

The volumetric variations of the cement paste during its hydration had negligible technological consequences when most industrial concretes had a water/binder ratio of greater than 0.50. In particular, these volumetric variations usually represented less than 10 per cent of the volumetric variation associated with drying shrinkage (Aïtcin et al. 1998). However, since low water/cement ratio concretes have been used to produce high-strength or high-performance concretes, it has been found that in the *absence* of adequate water

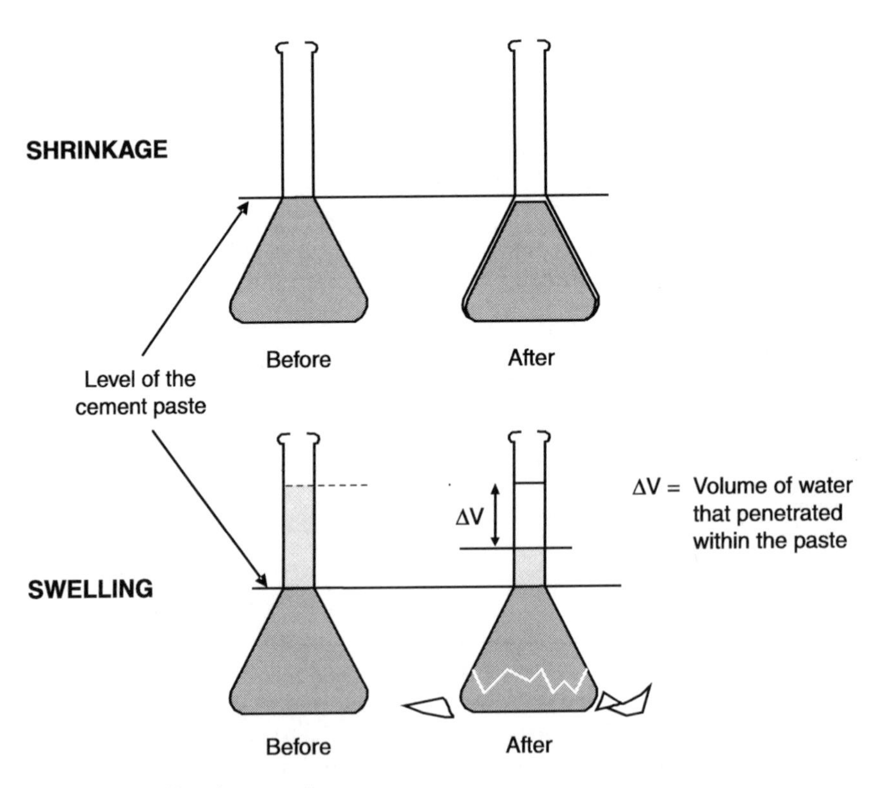

Figure 6.17 Le Chatelier experiment.

curing, such concretes could develop an initial *autogenous shrinkage*, equal to or higher than the drying shrinkage of regular concrete. This means that, when these concretes are drying, they could develop a total shrinkage twice as great as the drying shrinkage of regular concretes. Of course, the lower the water/binder ratio, the greater the autogenous shrinkage. Ignoring the importance of autogenous shrinkage results in very poor behaviour of some uncured high-performance concretes in the field.

In the following section, this subject will be treated in great detail in order to clarify what is meant by chemical contraction, self-desiccation and autogenous shrinkage. When the theoretical principles involved in the development of autogenous shrinkage are well understood, it is easier to apply appropriate solutions in the field to prevent or at least to control its development, so that it does not harm the concrete cast in a structure.

6.6.4.1 Le Chatelier's observations

In his experimental work on Portland cement hydration, Le Chatelier (1904) conducted the two very simple experiments presented in Figure 6.17. He filled two glass flasks with long vertical tubes with a cement paste. The top of the tube was closed by a glass stopper in which there was a small hole so that the hydration proceeded at atmospheric pressure. In one of the two flasks the upper tube was filled with water so that hydration reaction proceeded under water while in the second flask the hydration reaction proceeded in air. As a scrupulous experimenter, Le Chatelier checked that the masses of the two flasks did not change during his experiment.

After a while, Le Chatelier observed that the cement paste contained in flask A that had hydrated in air had shrunk and did not occupy the whole volume of the base of the flask. It was easy to check it by agitating the flask. The cement paste had definitely shrunk and was now loose in the base of the flask. Therefore, in the absence of any drying and of any external source of water, the apparent volume of the cement paste had decreased.

In the case of flask B, where the hydrated cement paste was under a column of water, he first observed a decrease in the height of the water in the upper tube, as some water had been absorbed by the hydrating cement paste. After a while, the rate of this absorption slowed down until the glass flask broke. Consequently, when a cement paste hydrates under water, it absorbs some water at the beginning of its hydration and it swells. Therefore, when a cement paste hydrates with an external source of water, its absolute volume decreases (absorption of the water) but its apparent volume increases (disintegration of the flask).

Consequently, depending on curing conditions, a cement paste can shrink or swell.

In order to explain his observations, Le Chatelier concluded that cement hydration resulted in an absolute volumetric contraction and that, according to curing conditions, this absolute volumetric contraction resulted in an increase or decrease of its apparent volume.

It was easy for Le Chatelier to explain the chemical contraction of Portland cement: the absolute volume of a hydrated cement paste is less than the sum of the absolute volume of the anhydrous cement particles and of the water that has reacted with them. However, he was unable to find an explanation for the swelling, and it must be admitted that 100 years later we are no closer to an explanation that is unaminously accepted (Nonat 2005). Vernet and Cadoret (1992) believe that it is the pressure developed by the growth of portlandite crystals that is the prime cause of this swelling. Barcelo et al. (2001) believe rather that this swelling is created by the growth of C-S-H due to the diffusion process in confined conditions created by the aggregate skeleton and by the self-structuring of the paste.

6.6.4.2 Autogenous shrinkage

It was not until the 1930s and 1940s that some researchers became interested in the volumetric changes of concrete and rediscovered this type of shrinkage linked to the chemical contraction of Portland cement (Lynam 1934; Davis 1940). Lynam defined autogenous shrinkage as the shrinkage that is observed in the absence of drying at a constant temperature in an unloaded concrete, and, of course, in the absence of any external source of water. He called this type of shrinkage autogenous shrinkage.

Autogenous shrinkage develops in a concrete hydrating as a closed system without any exchange with the exterior and is a consequence of chemical shrinkage. Whatever the water/binder ratio, the chemical shrinkage occurring during cement hydration is restrained by the structuring of the paste and the aggregate skeleton so that it generates very fine porosity. This very fine porosity drains the water contained in the coarse capillaries, which can be considered at that stage as an internal source of water. Consequently, the coarse capillaries dry while the concrete keeps the same mass. It is this internal movement of water in the closed system that is responsible for the drying of the coarse capillaries and for the appearance of menisci within the capillary network. As the hydration reaction develops, more water is drained from the capillary network so that the menisci are formed in capillaries that are finer and finer and the tensile stresses developed at the surface of these menisci therefore get larger and larger. It is these stresses developed within the capillaries that are responsible for the apparent volumetric contraction of flask A in Le Chatelier experiment. When the hydration reaction stops, autogenous shrinkage also stops because no more menisci are formed in the closed system.

In the case of flask B, where there is an external source of water, when the absolute volumetric contraction of the paste occurs, the very fine porosity that is created sucks the water from the larger capillaries, as in the previous case, but this time, the larger capillaries pump an equivalent volume of water from the external source. Consequently, in this case, no menisci are formed within the capillary network, and, thus the apparent volume of the cement paste does not shrink; on the contrary, it swells. The chemical contraction is the same in each of the flasks A and B, but in the case of flask B, it does not result in the formation of menisci (Figure 6.4).

Of course, when there is an external source of water, but, in a part of the concrete where the nanoporosity created by the chemical contraction is disconnected from the capillary network that links it to the external source of water, this part of the concrete continues to hydrate as a closed system, as in the previous case, which results in the development of some autogenous shrinkage in this part of the concrete. This disconnection between the capillary system and the nanoporosity can result from some fine capillaries being filled by hydration products for which the water came from the external source, or also, as proposed by Barcelo et al. (2001), due to the swelling of the restrained hydrated paste that is hydrating through a diffusion process. In such cases, menisci are developed within the volume of concrete disconnected from the external source of water and are responsible for the local volumetric contraction.

Autogenous shrinkage has not been, and is not, harmful as long as the water/binder ratio of concrete is high (over 0.50); such concretes have a large capillary network in terms of volume, and the coarser capillaries have large diameters; therefore, capillary water is very easily drained by the nanoporosity resulting from the chemical contraction. The corresponding menisci that are created are developed in large capillaries, and this generates very weak tensile forces. In such cases, autogenous shrinkage is negligible. Davis (1940: 1103) wrote:

> However, it would ordinarily be neither practical nor desirable to attempt to differentiate between autogenous movements, which are believed usually to be of relatively small magnitude, and the direct effects of drying and temperature changes.

But when it recently became possible to make low water/binder concretes with the help of superplasticizers, the volume of the initial capillary network was drastically reduced as well as the diameter of the capillaries. In low water/binder mixtures more cement particles hydrate per unit of volume. Consequently, great tensile forces can be developed in the early stage of hydration when concrete has not yet developed a very strong tensile strength. The lower the water/binder ratio, the earlier and greater is the autogenous shrinkage in a closed system. In some cases, autogenous shrinkage can be greater than the drying shrinkage that is developed later (Tazawa and Miyiazawa 1993).

The development and the control of the development of autogenous shrinkage is becoming a major subject of interest, and, to date, at least four major international symposia have been held on this subject, without taking into account the numerous workshops held on the subject (in Hiroshima 1998; in Paris 2000; in Haïfa 2001; and in Phoenix, Arizona 2002).

The lack of fundamental knowledge about autogenous shrinkage and the different ways to control it has sometimes produced catastrophic results in the field. Too many concrete structures made of high-performance concrete are currently presenting a catastrophic cracking network through which aggressive agents can easily reach the first rank of reinforcing steel. Instead of getting the best high-performance concrete, owners are getting the worst.

As engineers, it is absolutely mandatory that we take all the necessary measures to prevent or as least control as much as possible the harmful development of autogenous shrinkage. We must also learn how to take advantage of some of the positive aspects of the development of late autogenous shrinkage, because from a durability point of view, it is very important to let some menisci develop within the hydrated cement paste. These menisci develop 'air plugs' within the capillary network and these 'air plugs' decrease considerably the absorptivity and permeability of the concrete, as will be seen later.

6.7 Portland cement hydration

Fifty years after Le Chatelier's work, Powers and Brownyard (1948) conducted a series of fundamental quantitative experiments on Portland cement hydration in the laboratories of the Portland Cement Association in Skokie, Illinois. This work resulted in a quantitative physico-chemical model of hydration reaction that has not yet been contested. Based on this experimental work, Powers was able to differentiate three types of water within a hydrated cement paste, although it is rather difficult to define a clear-cut frontier between them.

The first type of water is composed of water molecules *chemically linked* within the C-S-H. This type of water can only evaporate at a temperature much higher than 105°C. Another type of water that has been called *free water* by Powers is the water contained in the capillary network. Between these two types of water, Powers showed that there exists yet a third type which is physically linked to the C-S-H, but which can be evaporated at 105°C, and he called this water the *gel water*.

6.7.1 Some experimental data on hydration reaction

Powers found that the amount of chemically bound water was equal to 0.23 g for each gram of cement, and that the amount of gel water physically bound to the C-S-H was equal to 0.19 g per gram of cement; consequently, to fully hydrate 1 g of cement it is necessary to have at least $0.23 + 0.19 = 0.42$ grams of water. Therefore, the minimum water/cement ratio necessary to reach full hydration in a closed system is 0.42. When the water/binder ratio is greater than this value, the hydrated cement paste will still contain some free water after its full hydration. If the water/binder ratio is smaller, in a closed system, hydration will stop due to the lack of water and some cement particles will not hydrate.

Powers was able to measure the chemical contraction associated with Portland cement hydration, and he found that it was equal to 6.4 ml for 100 g of cement. Aïtcin et al. (1998) have shown that this value corresponds exactly to the one expressed in a different way by Le Chatelier fifty years earlier. Le Chatelier wrote that the volumetric contraction was equal to 8 per cent.

6.7.2 *Jensen and Hansen's graphical representation of hydration reaction*

Recently O. Jensen and F. Hansen (2001) proposed a very clever schematic representation of the relations and equations developed by Powers. This graphical representation has been reproduced in Figures 6.18, 6.19 and 6.20. When considering these figures it is easy to understand the origin of the development of self-dessication in any cement paste, whatever its water/binder ratio, when it is hydrating in a closed system. (A closed system is a system that has no water exchange with the exterior.)

On the contrary, Figures 6.21, 6.22 and 6.23 represent cement hydration in an open system where there is an external source of water.

It is easy to understand why, in a closed system, in isothermal conditions, and in the absence of any external load, chemical contraction results in self-desiccation, which, in turn, results in the contraction of the apparent volume of concrete. It is this apparent volumetric contraction that has been called autogenous shrinkage by Lynam (1934).

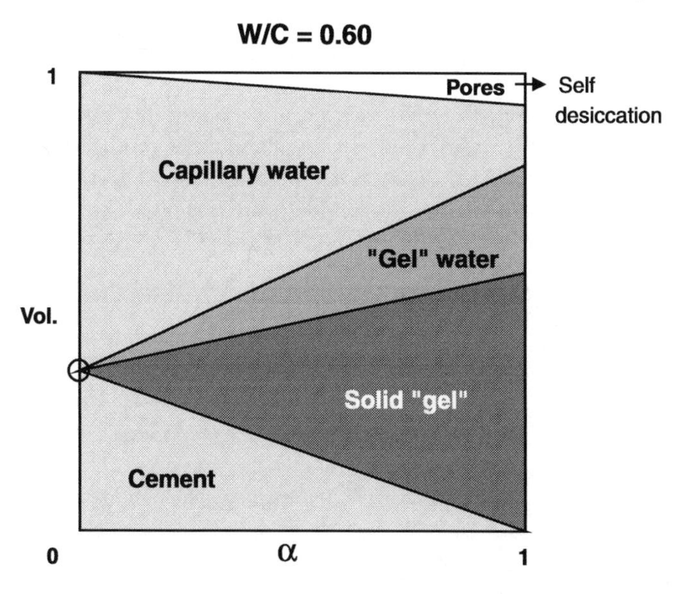

Figure 6.18 Schematic representation of the hydration of a Portland cement paste having a W/C equal to 0.60 (Jensen and Hansen 2001).

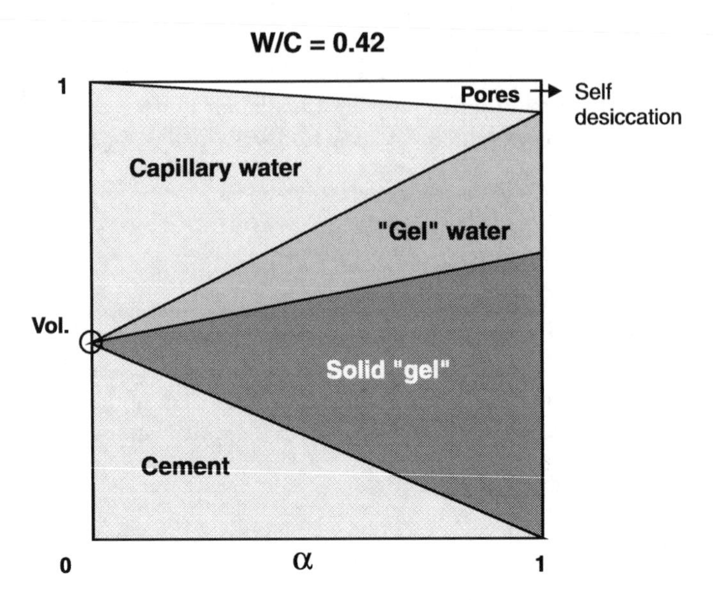

Figure 6.19 Schematic representation of the hydration of a Portland cement having a W/C equal to 0.42 (Jensen and Hansen 2001).

As long as the hydrating cement paste behaves like a soft deformable material, chemical shrinkage and autogenous shrinkage are equal. But, after hardening, final autogenous shrinkage is, of course, smaller than chemical contraction, because, when the hydrated cement paste develop a structure, it opposes the chemical contraction.

When hydration happens in non-isothermal conditions, and the thermal dilatation has been substracted, theoretically it should be more correct to speak about isothermal shrinkage, because temperature modifies the kinetics of hydration reaction. Usually, when the temperature increase linked to hydration is not too high, researchers do not differentiate between autogenous shrinkage and isothermal shrinkage.

First, we will consider the volumetric variations of a cement paste during its hydration and then we will look at the effect of the aggregate skeleton on the development of shrinkage in a concrete. The schematical representation proposed by Jensen and Hansen (2001) will be used in three particular situations:

1 a cement paste having a water/binder ratio greater than 0.42, that is, a cement paste in which there is initially more water than necessary to fully hydrate, in a closed system, all the cement particles (Figure 6.18);
2 a cement paste having a water/binder ratio equal to 0.42 (Figure 6.19) containing just enough water to fully hydrate the cement particles.

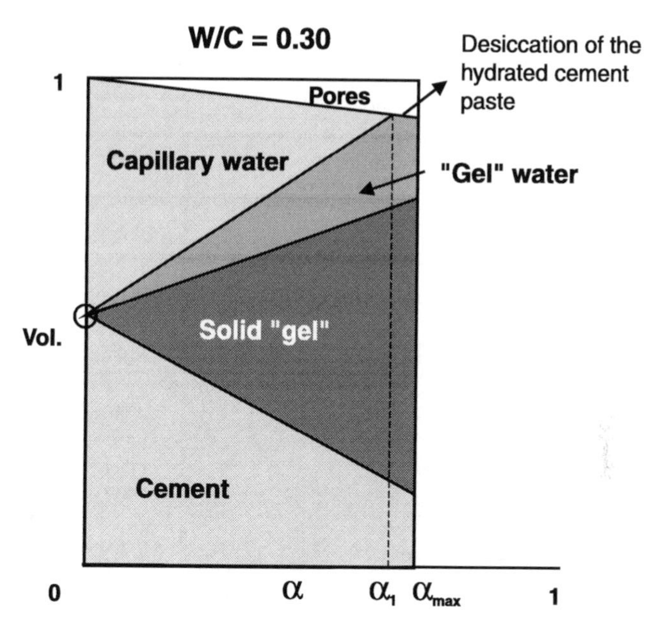

Figure 6.20 Schematic representation of the hydration of a Portland cement having a W/C equal to 0.30 (Jensen and Hansen 2001).

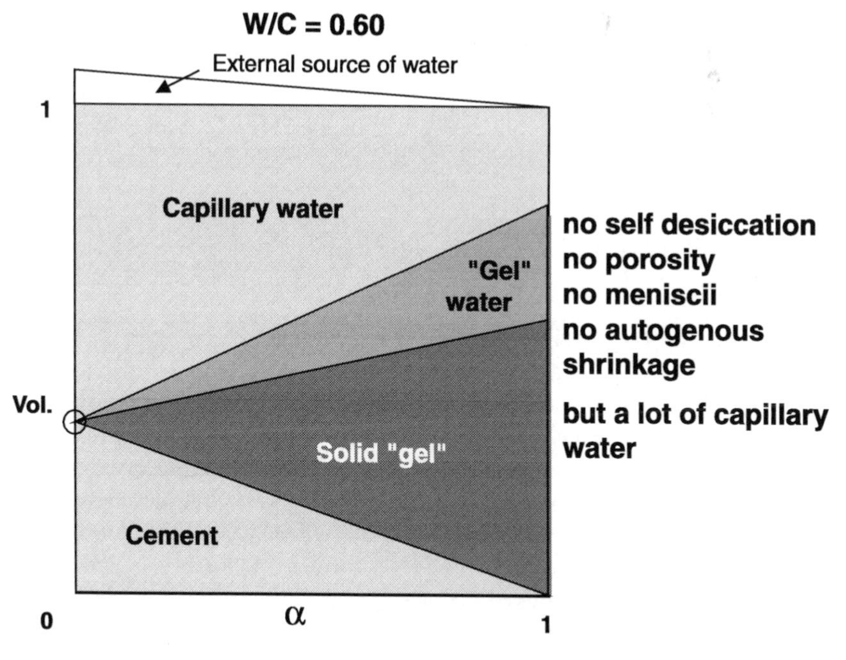

Figure 6.21 Schematic representation of the hydration of a Portland cement paste having a W/C equal to 0.60 in the presence of an external water source (adapted from Jensen and Hansen 2001).

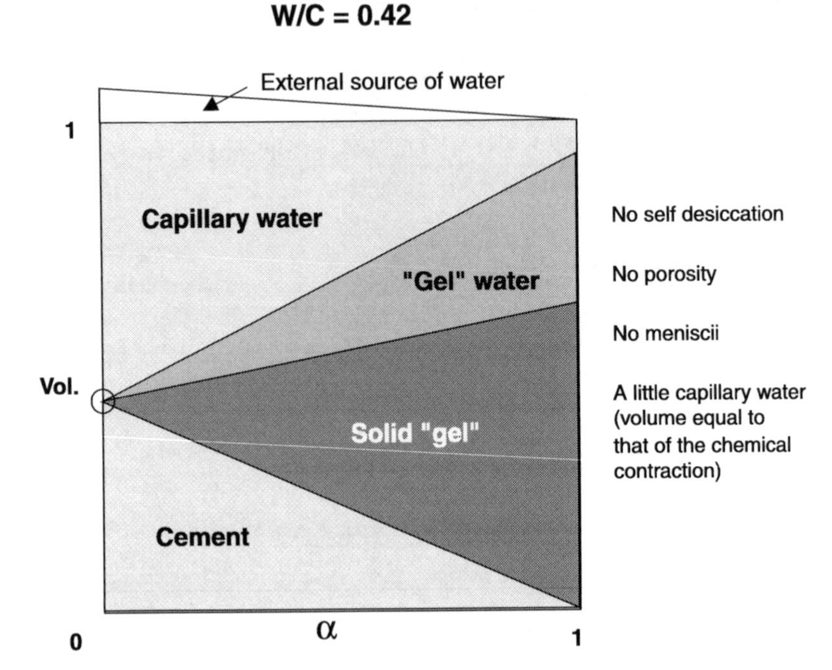

Figure 6.22 Schematic representation of the hydration of a Portland cement paste having a W/C equal to 0.42 in the presence of an external water source (adapted from Jensen and Hansen 2001).

3 a cement paste having a water/binder ratio smaller than 0.42 that initially does not contain enough water to fully hydrate all the cement particles (Figure 6.20);

Finally we will consider the case of the three previous cement pastes when they hydrate in the presence of an external source of water (Figures 6.21, 6.22 and 6.23).

In their graphical representation Jensen and Hansen have plotted on the x-axis the degree of hydration, α, which represents the mass of cement that has been hydrated with respect to the initial mass of anhydrous cement, and on the y-axis, the absolute volume occupied by each phase. Entrapped air has not been included in their representation.

At time t = 0 when cement enters in contact with water, the system is composed of two phases; Portland cement and water. When hydration reaction starts to develop, a certain volume of *gel* (C-S-H) is formed and a certain quantity of water is strongly physically linked to this gel, which Powers called the *gel water*. In parallel, there is a decrease in the amount of *capillary water* and a *gaseous volume* appears within the closed system due to the *chemical contraction*.

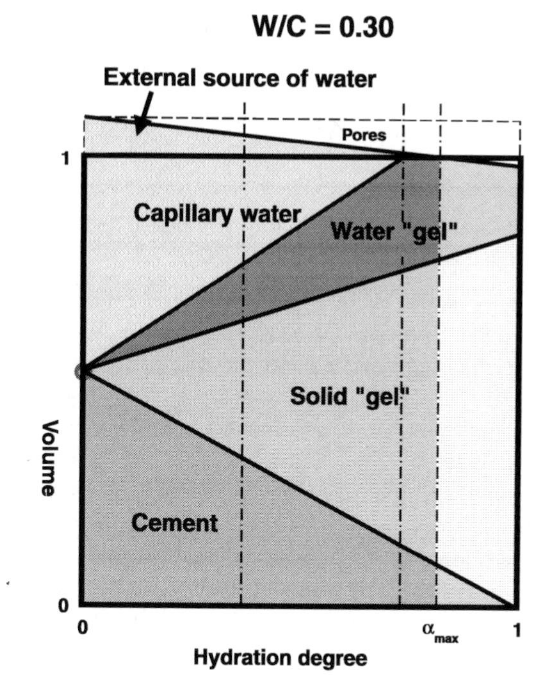

Figure 6.23 Schematic representation of the hydration of a Portland cement paste having a W/C equal to 0.30 in the presence of an external water source (adapted from Jensen and Hansen 2001).

6.7.2.1 *Case of cement paste having a water/binder ratio equal to 0.60 (Figure 6.18)*

In this case, the initial system contains more water than necessary to fully hydrate the cement particles and to be fixed on the gel. The hydration reaction stops when there is no more cement to hydrate. In such a case, theoretically, a degree of hydration equal to 1 can be reached. At that time there is still some water left in the capillary system. The higher the water/binder ratio, the larger the amount of water that remains within the capillary system. In such a case, a gaseous phase also develops within the cement paste, and its volume is proportional to the initial amount of cement.

Such a paste will present a poor durability because its capillary system is well developed and interconnected, so that it will allow easy penetration of any aggressive agent within the paste. The higher the water/binder ratio the poorer the durability.

6.7.2.2 *Case of cement paste having a water/binder ratio equal to 0.42 (Figure 6.19)*

This is almost the same as the previous case, except that, when a degree of hydration of 1 is attained, there is no capillary water left, and, as in the previous case, a gaseous phase develops during hydration.

6.7.2.3 *Case of cement paste having a water/binder ratio equal to 0.30 (Figure 6.20)*

The hydration reaction starts as in the previous cases, but it stops when there is no more water available to further hydrate the remaining cement particles. The hydration degree is then equal to α_1, and the RH within the pores created by hydration is equal to 100 per cent. Powers observed that the hydration reaction did not stop there, and that it could continue for a while, because the anhydrous cement particles were able to dry a part of the gel water until a RH of 80 per cent was reached in the capillaries. Therefore, the final degree of hydration α_{max} is slightly greater than α_1. This partial drying of the gel water was called by Powers *self-desiccation* of the paste. Unfortunately, there are currently many researchers who use the expression 'self-desiccation' to designate the drying of the capillary network in a closed system and this creates some confusion.

6.7.2.4 *Case of cement paste having a water/binder ratio equal to 0.60 that is hydrating in the presence of an external source of water (Figure 6.21)*

As in Figure 6.20, at the bottom of the graph it is seen that hydration develops in the same way as in a closed system, but that, at the top of the graph a major difference can be observed. The pores created by chemical contraction which were filled by a gaseous phase in Figure 6.20 are this time being filled with water coming from the external source. The level of water in the external source decreases as had been observed by Le Chatelier (1904) (Figure 6.18). When full hydration is reached, there is no gaseous phase in the hardened cement paste and a certain volume of water contained in the nanoporosity created by cement hydration has been added to the initial capillary water. There are no menisci in the hardened paste.

In Figure 6.23 we have considered that the amount of water provided by the external source was equal to the amount of water drained by chemical contraction. If initially there was more water than this amount, some external water will remain on top of the saturated cement paste.

It will be shown later that it is not advantageous to have a fully saturated cement paste at the end of the hydration process, because aggressive ions can penetrate within the concrete due to osmotic pressures. If the external water is contaminated with a particular aggressive ion, the concentration of this ion within the interstitial solution and the external source will tend to equilibrate.

6.7.2.5 *Case of cement paste having a water/cement ratio of 0.42 that is hydrating in the presence of an external source of water (Figure 6.24)*

Hydration proceeds as in the previous case, but when a degree of hydration equal to one is reached, there is no longer any capillary water, nor is there a gaseous phase in the hardened cement paste. The volume of water drained from the external source has filled the nanoporosity created by chemical contraction. There are no menisci within the fully hydrated cement paste and, consequently, no autogenous shrinkage.

6.7.2.6 *Case of cement paste having a water/cement ratio of 0.30 that is hydrating in the presence of an external source of water (Figure 6.25)*

As in Figure 6.22, the hydration reaction stops when there is no more capillary water available, however, in this case, the hydrated cement paste does not dessicate. The paste does not present any phenomenon of self-desiccation.

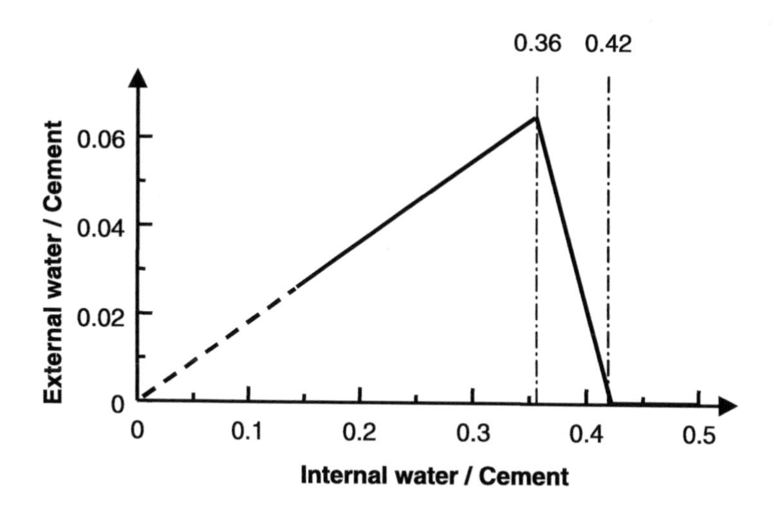

Figure 6.24 Minimal amount of external water to obtain a maximum hydrating degree α max and consequently eliminate self-desiccation according to Powers' equations.

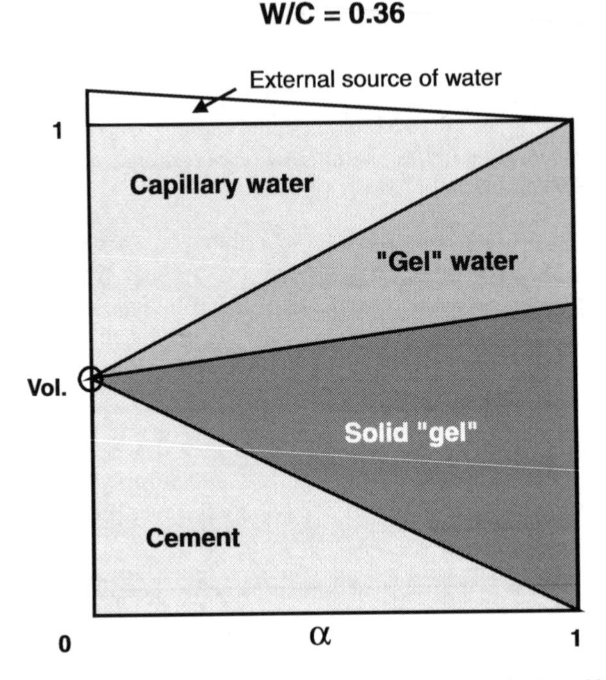

Figure 6.25 Schemical representation of a hydrated cement paste having a W/C of 0.36 in the presence of an external source of water (adapted from Jensen and Hansen 2001).

6.7.3 *Amount of external water necessary to avoid self-desiccation of the cement paste (using Power's definition of self-desiccation)*

Using Power's equations, Jensen and Hansen (2001) have calculated the minimum amount of water necessary to avoid the self-desiccation of cement paste as a function of the water/binder ratio. The results have been plotted in Figure 6.24. The water/binder ratio has been reported on the x-axis, and on the y-axis, the amount of water needed to avoid self-desiccation. This amount of water has been reported as the mass ratio of the external water consumed to the initial mass of cement. The graph is composed of three linear parts:

- For water/binder ratios greater than 0.42, there is no need to add any external water to eliminate the self-desiccation of the gel, because there is some capillary water left when a degree of hydration equal to one is reached.

- For water/binder ratio values between 0.36 and 0.42, it is possible to reach a degree of hydration of 1 when adding an appropriate amount of external water (or more).
- For water/binder ratio values lower than 0.36 it is impossible to reach a degree of hydration of 1 because the cement paste has no more porosity.

Some cement particles remain unhydrated in the paste. The lower the water/binder ratio the higher the amount of unhydrated cement particles.

This theoretical graph presents a great technological interest: it shows, in particular, that when there is an external source of water, all the cement particles can be hydrated if the water/binder ratio is greater than 0.36. As can be seen in Figure 6.25, in the case of a 0.36 water/binder ratio cement paste, full hydration in the presence of an external source of water ends up with a non-porous system that does not contain any capillary water or gaseous porosity. It is a non-porous solid.

Moreover, it is seen in Figure 6.24 that the lower the water/binder ratio, the lower the amount of external water that is necessary to fill the voids created by chemical contraction, because fewer and fewer cement particles are hydrated and less and less chemical contraction occurs.

These theoretical considerations demonstrate the importance of using concretes having a water/binder ratio equal to or close to 0.36 and of curing them with an external source of water as soon as possible, before the hydration of C_3S develops.

It has been found in the field that these concretes are more robust vis-à-vis early cracking, because initially they are permeable enough to let the external water fill the hydrated cement paste (at least the paste very close to the external source of water) and the continuity between the liquid phase within the paste and the external water is maintained.

NB: It is important to re-emphasize the distinction between the two types of self-desiccation that can occur within concrete. First, there is the phenomenon of self-desiccation which is observed in a concrete that is hydrating in a closed system and which results in a drying of the capillary network due to the drainage of some capillary water by the nanoporosity created by chemical contraction. Second, there is the phenomenon of self-desiccation of hydrated cement paste that occurs in concrete having a water/binder ratio lower than 0.42 and that is hydrating in a closed system. This phenomenon results in a decrease of the R.H. of the gaseous phase which finally stabilizes around 80 per cent. In a closed system when such a R.H. is reached, anhydrous cement particles can no larger attract any water from the gel water.

6.7.4 Volumetric changes during the hardening of concrete

Now we will leave hydrated cement paste and consider concrete. First, we will discuss the volumetric variations of concrete that is hardening in isothermal conditions without any external source of water and without any water loss due to external drying. In such conditions concrete develops only some autogenous shrinkage.

In concrete that is hydrating in a closed system, the hydrating cement paste undergoes a chemical contraction, but this volumetric contraction is restrained by the structure formation of the paste itself and by the presence of the more or less interlocked aggregate skeleton that can restrain to a certain extent the global deformation of concrete. Consequently, concrete always presents an autogenous shrinkage much lower than that of a cement paste having the same water/binder ratio. This restriction of the shrinkage of the hydrated cement paste will possibly result in the microcracking of the hydrated cement paste depending on the ratio between the tensile strength of the paste and the tensile stresses generated by the menisci developed by the chemical contraction.

Moreover, it is well known that a simple way to decrease concrete shrinkage (whatever its origin) is to increase the amount of aggregates within the mix. The chemical contraction of the paste remains the same but it is more restrained. Consequently, the global volumetric deformation of concrete is decreased, but the hydrated cement paste has more possibility of being cracked, though it will be a diffuse homogeneous microcracking.

When high tensile stresses are developed very early in a concrete having a low tensile strength, microcracks develop, and the microcracks coalesce to form macrocracks. According to the width of these macrocracks, aggressive ions may or may not easily penetrate the concrete. In the case of reinforced concrete, the restraint due to the reinforcing steel and any other inclusions has also to be taken into account.

Therefore, from a practical and durability point of view, it is very important to control the development of autogenous shrinkage that happens at the very beginning of hydration in low water/binder concretes, because it occurs when concrete has not yet developed a very strong tensile strength that can oppose shrinkage.

But, from a durability point of view, it is also essential to allow some late autogenous shrinkage to develop because it creates very small well-distributed menisci within the concrete which disconnect to some extent the capillary network from the nanoporosity. As in the case of cement pastes, these menisci create 'air plugs' which drastically decrease concrete absorptivity and permeability and prevent it from reaching full saturation. By opposing water percolation within the mass of concrete, late self-desiccation drastically increases concrete durability, of course, on the condition that early self-desiccation has been controlled so that the concrete does not present any micro or macrocracks.

There is one case in which early self-desiccation can be beneficial, that of massive elements just after their casting when the temperature of concrete increases due to the development of the heat of hydration. Until the temperature peak is reached, chemical contraction partially opposes thermal dilatation, but, after the peak, chemical contraction and thermal contraction have to be added. The addition of these two contractions occurs at a very critical moment in the life of concrete, because concrete has not yet developed its full tensile strength. Consequently, from a practical point of view, it is very important to delay as long as possible the peak temperature due to the heat developed by the hydration of the C_3S and to decrease it by spraying concrete with water that is not too cold in order to avoid thermal gradients (Figure 6.26).

Drying shrinkage of concrete which is not related to cement hydration will not be developed in this book.

6.7.5 *Influence of the temperature in the kinetics of hydration reaction*

Temperature modifies the kinetics of hydration: a low temperature slows down the hydration reaction, and a high temperature accelerates it. In this chapter, it is our quest to study the influence of thermodynamic conditions on the structure of hydrates, a subject that has not received too much attention until now.

The kinetics of hydration reaction varies as a function of the temperature according to an exponential law called the Arrhenius law.

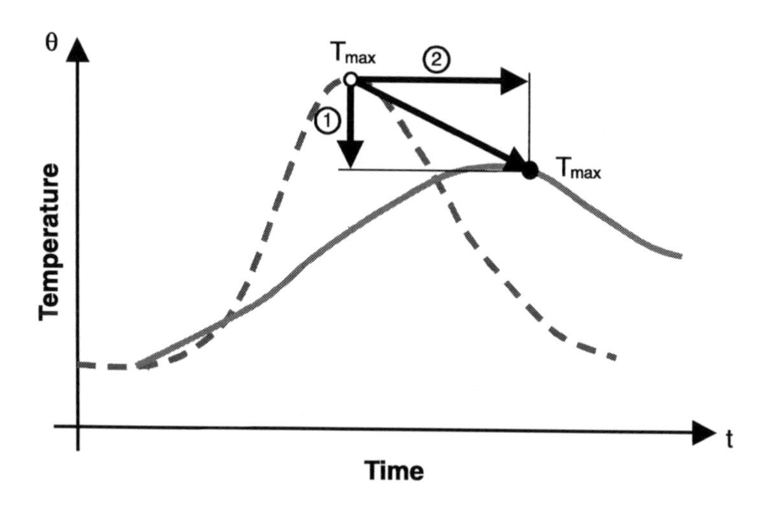

Figure 6.26 Shifting of the temperature peak in massive elements.

$$K = A\,e^{-\frac{E_a}{RT}}$$

where K is the specific hydrating rate
A is a constant
E_a is the activation energy
R the gas constant
T the temperature expressed in °K

Hydration being an exothermic reaction, Portland cement hydration is an autoactivated reaction until the temperature stabilizes when thermal losses at the boundaries are equal to the heat liberated by hydration. At ambient temperature in a concrete element, the temperature of the concrete starts to rise quite rapidly, to reach a peak and then to decrease more slowly (Figure 6.18). The peak temperature and the decrease of the temperature varies from one location to another within a concrete element cast with the same concrete, so that not all cement particles hydrate at the same rate. This non-uniformity of thermal conditions within a concrete element that is hydrating creates thermal gradients, and, when these thermal gradients are too high, they can generate cracks because concrete deformations are restrained.

Thermal activation or autogenous curing is frequently used in pre-cast concrete plants to accelerate the development of a high early strength in order to cut the pre-stressing cables earlier or to tension earlier the post-tensioning cables.

The monitoring of the variation of the temperature of concrete can be used to determine the degree of hydration and, consequently, concrete strength. Maturity meters translate into MPa the recorded temperature variations using the Arrhenius Law. This technique of strength evaluation uses the notion of equivalent time, that is, the time necessary, at a constant temperature of 20°C, to reach the same strength (Laplante 1993; Chanvillard and D'Aloia 1994 and 1997; D'Aloia 1998).

6.8 Sequential description of hydration reaction

Now that all the individual phenomena which are developed during cement hydration have been described, it is time to integrate all this knowledge and to present sequentially the various reactions that happen in a cement paste that is hydrating. We will use schematic graphical representations as proposed by Vernet and Cadoret (1992) which were later reviewed by Vernet (1995). This schematic representation is based on a few explicit graphics made previously before Vernet and Cadoret divided the hydration reaction into five stages.

6.8.1 First stage: mixing period

During this stage, different ions are dissolved from the different phases. This dissolution is very rapid and exothermic. The surface of the cement particles starts to be covered with C-S-H formed by Ca^{2+}, $H_2SiO_4^-$ and OH^- provided by the silicates of the clinker and water, and, by ettringite (hydrated trisulphoaluminate of calcium) formed by the combination of AlO_2, SO_4^{2-} and OH^- provided by the interstitial phase and water, as well as by the different forms of calcium and alkali sulphates present in Portland cement.

6.8.2 Second stage: dormant period (Figures 6.27 and 6.28)

This stage develops between one and two hours after mixing. The rapid increase of the pH and of the Ca^{2+} content of the aqueous solution slow down the dissolution of the clinker phases; the heat release slows down considerably but never stops. Small amounts of C-S-H and ettringite continue to be formed,

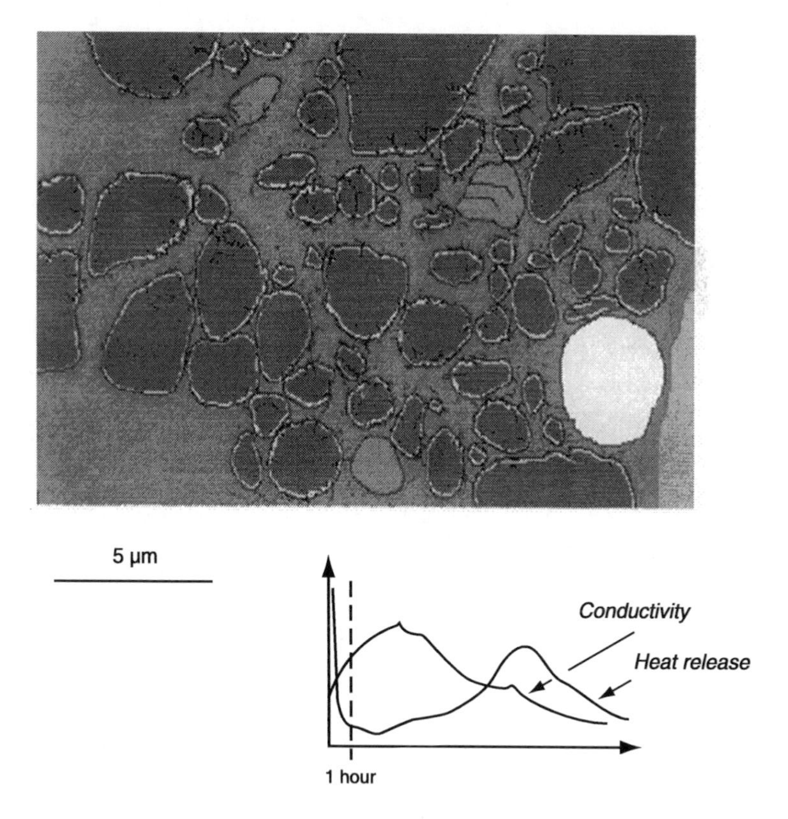

Figure 6.27 Schematical representation of a cement paste during its hydration one hour after its mixing (according to Vernet 1995).

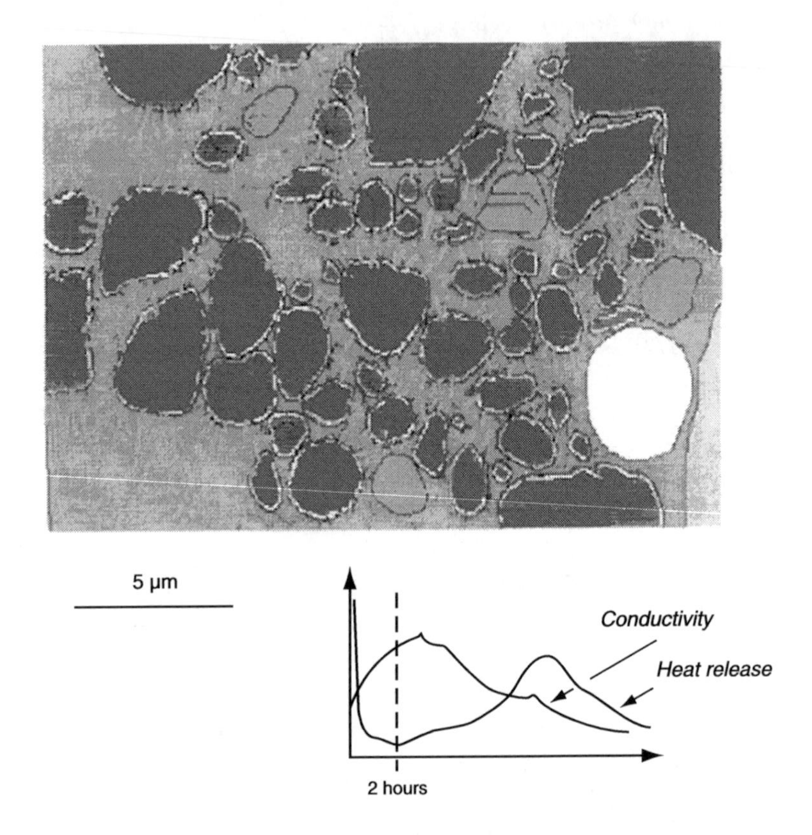

Figure 6.28 Schematical representation of a cement paste during its hydration two hours after its mixing (according to Vernet 1995).

and the aqueous solution becomes saturated in Ca^{2+} as evidenced by the continuous increase of the conductivity. Portlandite is not formed due to its very low rate of germination, which is much lower than that of C-S-H.

6.8.3 Third stage: setting and acceleration of hydration (Figure 6.29)

This stage occurs about five hours after mixing. Setting is 'triggered off', according to Vernet, by the sudden precipitation of portlandite crystals, the 'chemical trigger' of the hydration reaction. The precipitation of portlandite crystals occurs when the concentration of silicates in the aqueous phase is lower than a few micromoles per litre. The rapid consumption of Ca^{2+} and OH^- during the formation and growth of portlandite crystals accelerates the dissolution of the anhydrous phases of the clinker. The conductivity curve presents a sudden drop, then stabilizes for a while, and some heat starts to be

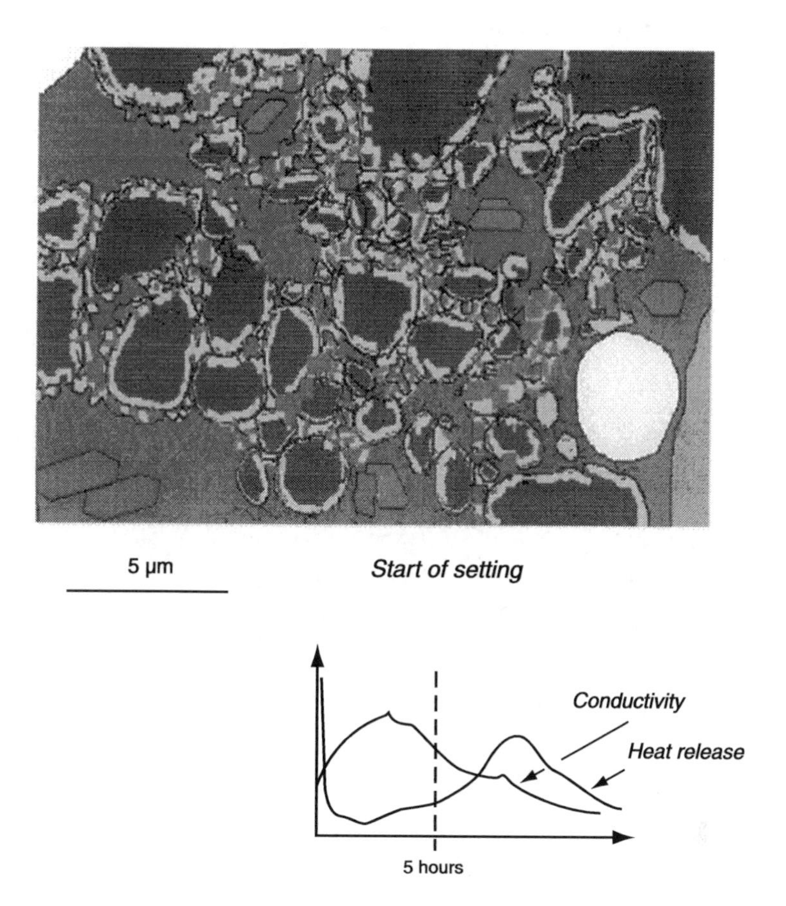

Figure 6.29 Schematical representation of a cement paste during its hydration five hours after its mixing (according to Vernet 1995).

released. The heat released is not very high at the beginning because portlandite precipitation is endothermic.

When monitoring the progression of an ultrasonic signal through a Portland cement paste that is hydrating, it is observed that the signal starts to be attenuated due to the first bridgings between the cement particles caused by the newly formed crystals. Initial setting time, as measured by a Vicat needle, occurs in the descending part of the conductivity curve where the first rigid structure of the paste due to the formation of ettringite crystals occurs. The end of setting corresponds to a very weak ultrasonic signal and to a rapid heat release. Conductivity becomes very low and the first compressive strength testing can be done.

6.8.4 *Fourth stage: hardening (Figure 6.30)*

This fourth stage occurs about nine hours after mixing. In most Portland cements, the molar proportion of calcium and alkali sulphate is smaller than that of aluminates, consequently the rapid formation of ettringite at the end of the dormant period results in the exhaustion of SO_4^{2-} ions which happens between nine and fifteen hours after mixing. At that time, ettringite starts to decompose and to release SO_4^{2-} that will form calcium monosulphate crystals. This reaction consumes 2 moles of aluminates for each mole of ettringite that is decomposed. Consequently, aluminates start to react quite rapidly and an exothermal peak develops. The intensity of the peak depends on the C_3A content in the cement. The heat released accelerates the rate of hydration of the silicate phase. A small peak is observed in the conductivity curve due to the passage into solution of the alkalis trapped in the C_3A network. Hydration

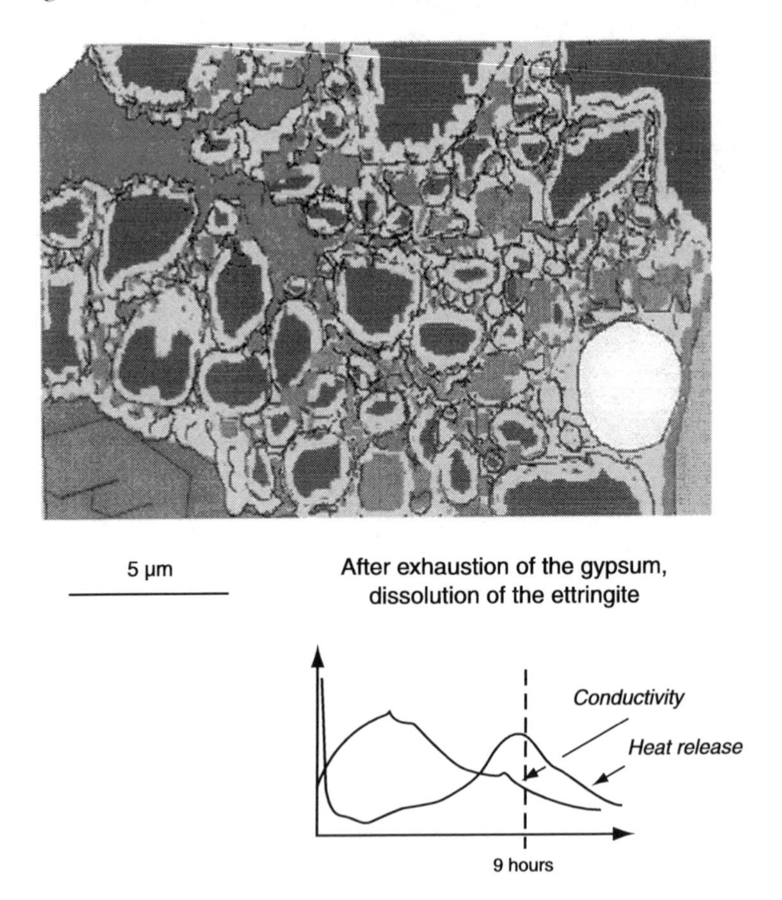

Figure 6.30 Schematical representation of a cement paste during its hydration nine hours after its mixing (according to Vernet 1995).

heat is released very quickly but without any significant improvement in compressive strength. In a mortar, the plateau observed in compressive strength lasts up to twenty hours, in spite of the fact that some heat continues to be developed. It is only after twenty hours that compressive strength starts to substantially increase.

6.8.5 Fifth stage: slow-down period

Cement particles get covered with more or less thick layers of hydrates which slow down the progression of water toward the interface between hydrates and anhydrous constituents. The hydration reaction slows down, but chemical reactions continue at a slow rate. At that moment, autogenous shrinkage can become a very important factor that strongly influences concrete volumetric variations in low water/binder ratio mixtures. Electrical conductivity is very low and is a good indicator of the water content of concrete.

6.8.6 Schematic representation of a hydrated cement paste at twenty-eight days (Figure 6.31)

The hydrated cement paste is compact, and C-S-H is the essential component of the concrete matrix. Some portlandite crystals have grown bigger than some initial cement particles. In some very impervious zones, hydration can proceed

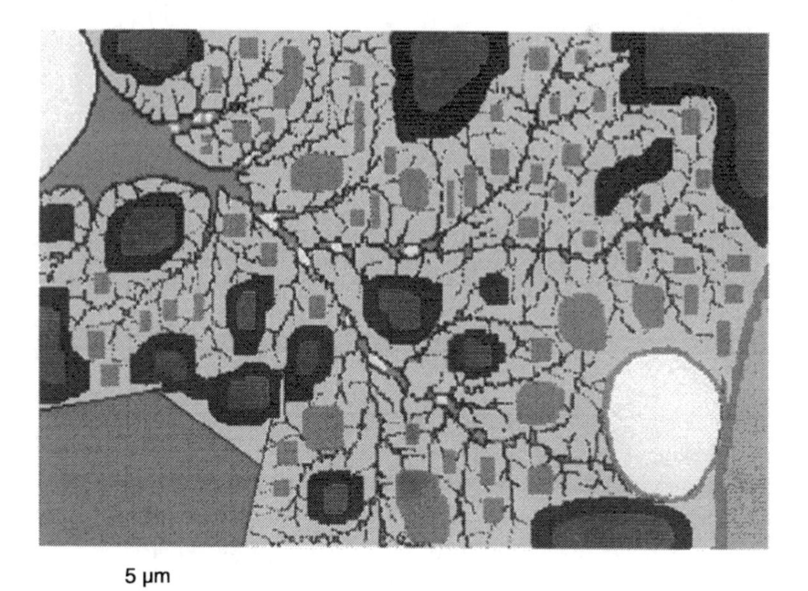

Figure 6.31 Schematic representation of a hardened cement paste at twenty-eight days (Vernet 1995).

for a while only because anhydrous components are able to drain some water from the gel water, which corresponds to the self-desiccation of the cement paste as found by Powers.

6.9 Linking hydration and mechanical strength

One of the most striking features of the different monitoring techniques carried out during hydration is the difference between the development of heat and compressive strength during the hardening process. Vernet explains this difference by what is observed at a microscopic level. During the hardening, there are phenomena that contribute to the development of the 'glue' of concrete, essentially C-S-H and, to a lesser extent, ettringite, but there are also phenomena that create internal stresses and even microcracks.

Heat of hydration provides information about the amount of 'glue' that is formed, but the chemical process at the origin of the formation of C-S-H results in a volumetric contraction. If these processes were acting individually, a simple relationship between compressive strength and the degree of hydration could be established, although it would be necessary to consider a lower gluing effect due to the volumetric contraction. Moreover, some of these chemical reactions produce crystals that during their growth develop a pressure of crystallization (Riecke law) linked to their rate of oversaturation μ

$$P = R.T.Ln\ \mu/\upsilon$$

where υ is the molar volume of the considered species, R the gas constant and T the absolute temperature in °K. During their growth, crystals behave like microcracks that create additional porosity. This also applies to ettringite and portlandite crystals (Figure 6.32) that are used in some expansive cements. In such cement, apparent volumetric expansion occurs in spite of the absolute volume reduction. The resulting volumetric variation is caused by the increase of the porosity due to crystal growth and chemical contraction.

Concrete shrinkage results from these phenomena but also from self-desiccation that drains the water from the capillary network towards the nanoporosity created by chemical contraction. In concrete structures, thermal contraction, or thermal shrinkage as it is most often called, contributes to the global shrinkage of concrete. Thermal shrinkage varies as a function of heat loss, the thickness of the concrete element and the heat conductivity of the hardening concrete.

Consequently, at a microscopic level, shear stresses can be developed at the level of the different interface present in a concrete: sand particles, coarse aggregate, but also near portlandite crystals. In a certain manner, C-S-H growth can partially compensate for the effects of these local volumetric variations. Some of the newly formed 'glue' can be used to heal these microcracks linked to the growth of crystals.

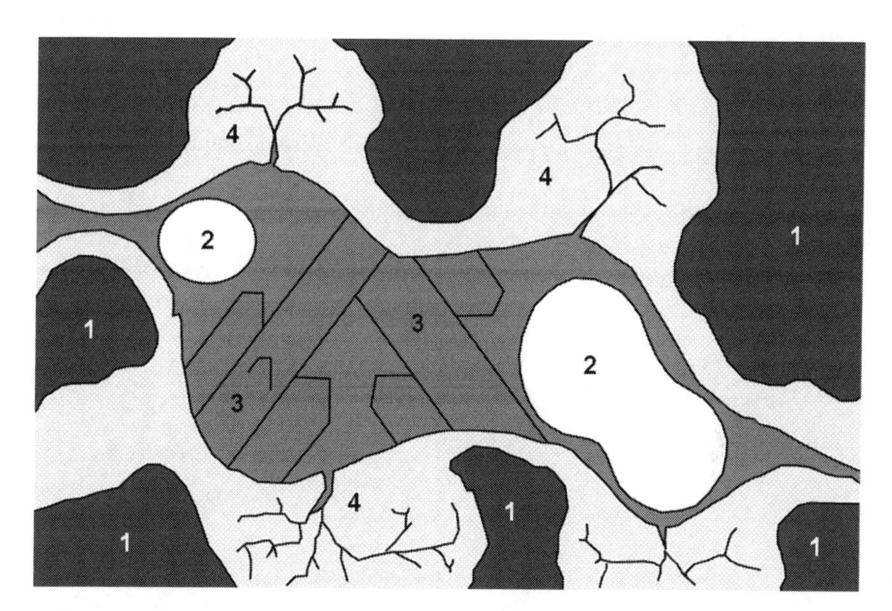

Figure 6.32 Increase of the porosity due to the growth of massive crystals (according to Vernet 1995): (1) cement particles hydrating; (2) new pores; (3) massive portlandite; (4) capillary pores and microfissures.

Mechanical strength is, therefore, the result of this complex situation. During the hardening process, when crystal growth is more rapid, the heating rate is sufficient to result in a strength increase. From experimental results, it has been found that during the hardening period the crystallization pressure stays high, while C-S-H production slows down, which can explain the change of slope in the strength curve.

6.10 Structural models of C-S-H

It has already been explained that because we don't know exactly the crystallographic form of the calcium silicate hydrate, it is still written as C-S-H. There is no doubt that C-S-H has a layered structure. Between the 1950s and 1970s some researchers have proposed theoretical schematic representations of C-S-H that could explain the general properties of C-S-H, but none of them had any mineralogical support. The usual tools used to define the crystallographic network of minerals were inoperative in the case of C-S-H. From among these models, those of Powers (1958), Feldman and Sereda (1970) and the so-called Munich model, are presented in Figure 6.33. As a comparison, Figure 6.34 represents schematically the structural model of a clay and of C-S-H.

Recently, observations with high-resolution electron microscopes on C-S-H prepared through a pozzolanic reaction have definitely shown the layered structure of C-S-H (Figure 6.35). This figure looks very much like the one taken by Yada (1967) on chrysotile asbestos.

All researchers agree that C-S-H crystallites cannot develop over long distances, such as the layered structure based on SiO^{4+}, Al^{3+} and Mg^{2+} can do in clays and chrysotile asbestos, and different models of ionic arrangements have been proposed.

The first results obtained through magnetic resonance indicate that the layered structure generated by Ca^{2+} ions can develop over four to five reticular distances before distortions that are too large stop the growth of crystallites

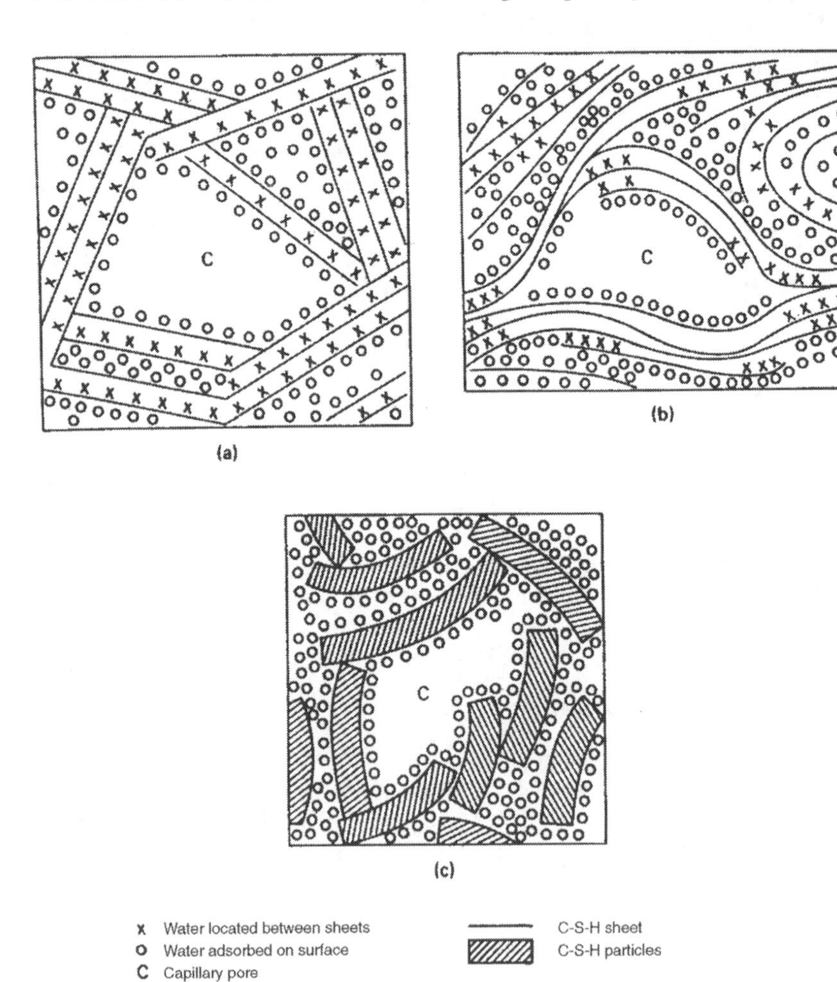

Figure 6.33 Schematic representation of different C-S-H models (Mindess and Young, Concrete, © 1981, reprinted with permission of Pearson Education, Inc., Upper Saddle River, NJ): (a) model of Powers and Brunauer adapted from Powers; (b) model of Feldman-Sereda; (c) model of Munich.

(Rodger 1991; Porteneuve 2001). The integration of all the results already obtained has resulted in the development of models composed of a double chain of silica tetrahedra linked by Ca^{2+} ions with a coordination number of 6 on which more or less ordered silica chains are grafted on both side of the double layer. In Figure 6.36, the structural model proposed by Zhang et al. (2000) is presented.

The structure of C-S-H is viewed as the structure of partially disorganized tobermorite. Tobermorite is a well-crystallized form of C-S-H obtained when autoclaving Portland cement at a pressure of 15 MPa and a temperature of 180°C. When the C/S ratio in the C-S-H increases, the size of the crystallites decreases, for a C-S-H ratio of 1. C-S-H crystallites are composed of pentamers. However, the C/S ratio of 1.5 is equal to crystallites that are composed of dimers. Speaking of C-S-H, Henri Van Damme (2001: 1090) has written: 'It is a hydrated lamellar material, ill defined, poorly organized and highly charged that is at the base of the strength of most of our concrete structures.'

Recently Nonat (2005: 12) has written a synthesis of our present knowledge on C-S-H structure. He discusses the models proposed by Taylor and Cony and Kirkpatrick and proposes a slightly different model. He concludes his presentation as follows:

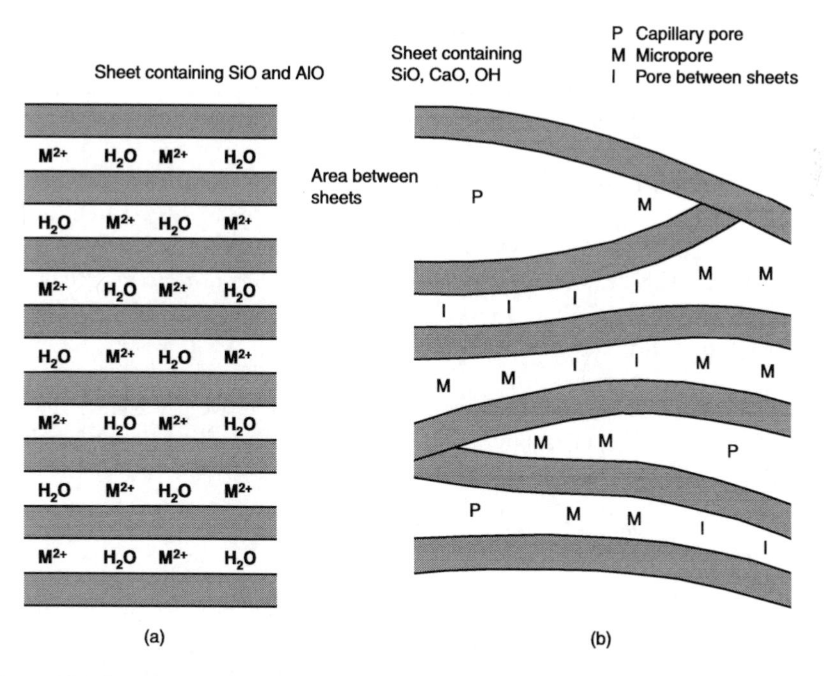

Figure 6.34 Schematic representation of a well-crystallized clay (a) and of a C-S-H more or less well crystallized (b) (Mindess and Young, *Concrete* © 1981, Reprinted with permission of Pearson Education, Inc., Upper Saddle River, NJ).

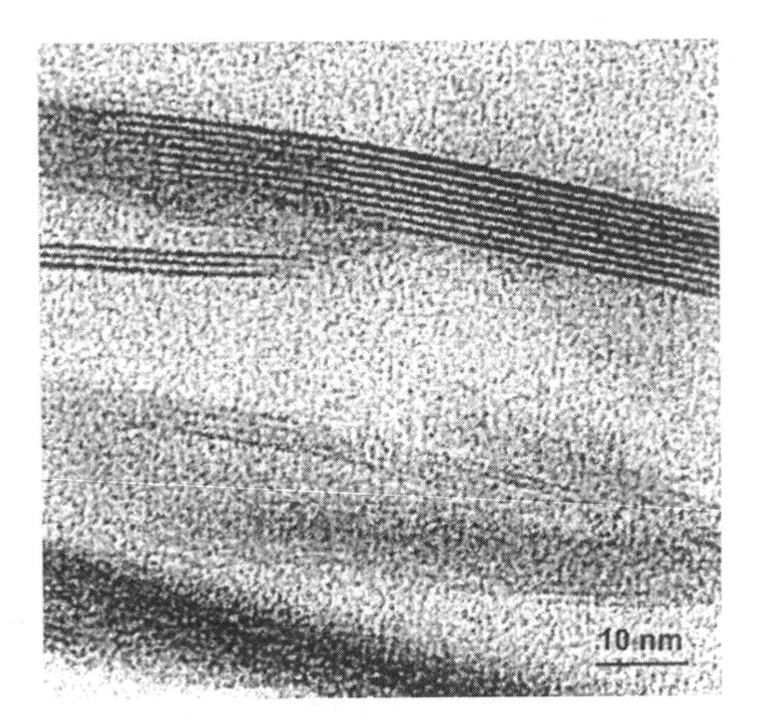

Figure 6.35 Transmission microscope photograph of a C-S-H prepared by a pozzolanic reaction between silica and lime (courtesy of N. Lequeux and C. Clinard).

> The lamellar character of C-S-H has long been recognized . . . The data presented in this paper indicate that C-S-H presents undoubtedly a structural order at relatively short range. C-S-H are nanocrystals in cement paste . . . The structure of C-S-H is close to the one of tobermorite whatever the Ca/Si ratio but is not the same . . . some work remains to do in order to precise it.

In this matter, much progress has to be made before we get a clear understanding of how C-S-H is built, and how C-S-H crystallites organize their structure.

6.11 The origin of the cohesive forces

This is another domain where progress is needed. Currently, as stated by H. Van Damme, we have no clear explanation for such a simple phenomenon as the hardening of cement paste. It is accepted that the origin of the mechanical strength of hydrated cement paste must be linked to the growth of C-S-H; in fact, it is a trivial observation that the richer a cement is in C_3S, the stronger

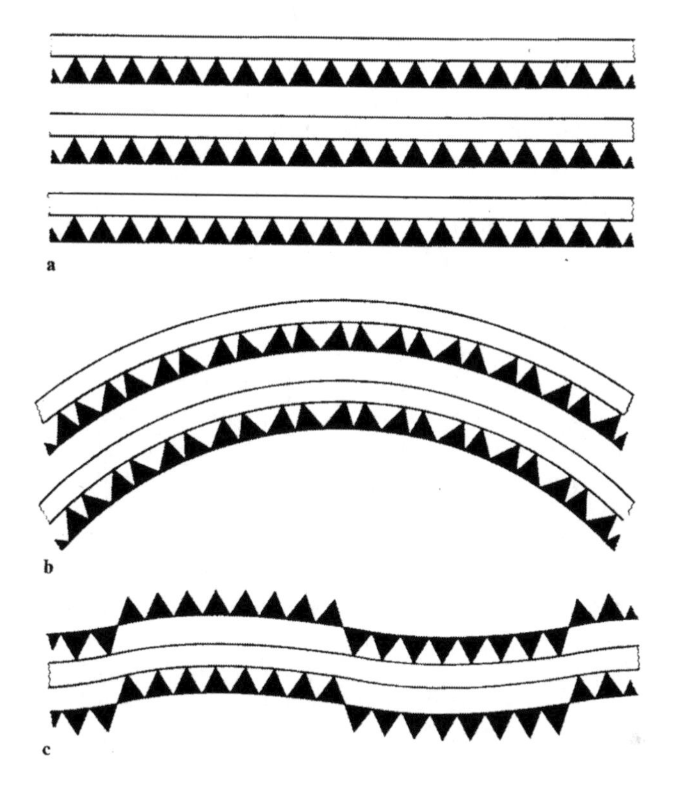

Figure 6.36 Structural model proposed by Zhang et al. (2000) (courtesy of American Ceramic Society): (a) kaolinite; (b) chrysotile; (c) C-S-H. Black triangles represent SiO_4 tetrahedra.

the hydrated cement paste, at a constant water/binder ratio, of course. In his paper 'Colloidal Chemo-Mechanics of Cement Hydrates and Smectite Clay: Cohesion Versus Swelling', H. Van Damme (2001) presents and discusses the most plausible hypothesis to explain concrete mechanical strength.

It is out of the question to try to explain the origin of the cohesive forces developed during cement hydration in terms of ionic or covalent forces as has been proposed in some models. All researchers currently agree that the forces that tighten C-S-H layers have a physical origin, and there are surface forces.

Moreover, it is agreed that neither the DLVO theory (Dejarguin, Landau, Vervey, Overbeck) nor Van der Wall's forces can be used to explain the cohesion of C-S-H. The role played by capillary forces as proposed by Barcello et al. (2000) is more subtle, and it can neither be totally accepted nor rejected to explain C-S-H cohesion.

Therefore, the only mechanism left to explain C-S-H cohesion is the existence of correlation forces due to ionic attractions that can be evidenced using the Monte Carlo method.

When measuring the forces developed between C-S-H particles with a force microscope, it is possible to observe the ionic and electrostatic nature of the cohesive forces developed within a hydrated cement paste. Therefore, the only thing that has yet to be understood is how C-S-H crystallites gets agglomerated to give cohesion to the hydrated cement paste.

It seems that the C-S-H crystallites start to agglomerate to form disc-shaped C-S-H arrangements 5 nm thick, having an average diameter of 60 nm (Van Damme 2001). These arrangements are sometimes called quasi-crystallites (QCs). These QCs can be developed laterally, in parallel with the surface of the substrate or axially perpendicular to the surface. This organization of QCs is used to explain the passage to a diffusion mode of hydration after the dissolution/precipitation mode of the hydration reaction.

Finally, the strength of the hydrated cement paste that is observed at a macroscopic level originates in the connectivity and in the compactress of the arrangements of C-S-H crystallites. Recently, it has been suggested that the bonds develop essentially at the edges of the lamellae. Acker (2001) is also considering the role played by water molecules in the cohesion of hydrated cement paste.

It is hoped that in the near future a clearer understanding of the mechanism of the cohesive forces that are developed within a cement paste will result in a better and more efficient use of Portland cement in order to build more durable and environmentally friendly concrete structures.

6.12 Modelling hydration reaction

Based on the present knowledge of the hydration process, some researchers have developed 2D and 3D mathematical models in order to deduce out mechanisms that can improve or impair the hydration reaction, or even interact with the durability of the matrix (leaching, chemical attack, alkali/aggregate reaction, etc.). These models are very important for complex parametric studies or for extrapolating the results of accelerated tests on a long-term basis. Several models have been developed, among them those of van Breugel (Figure 6.37), of Bentz (1997) and of Jennings (2000).

6.13 Conclusion

It is quite surprising that, in spite of all the efforts that have been devoted in the past 100 years, the hydration reaction is not better understood, because from a technological point of view it is such a simple and important reaction.

Of course, since the first studies of Le Chatelier, Powers and many others, the science of Portland cement and concrete has definitely progressed and will

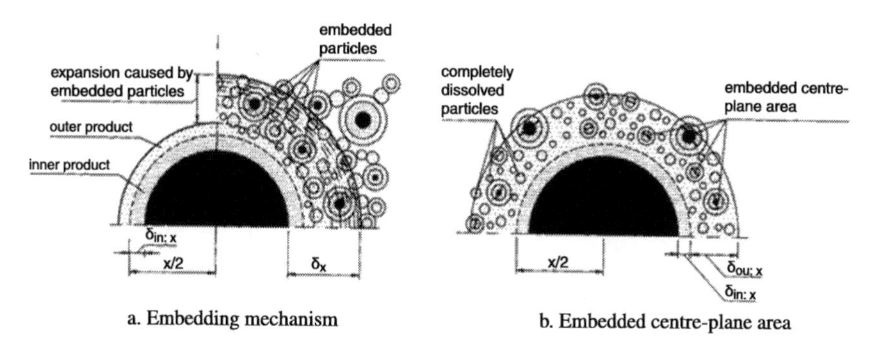

Figure 6.37 Van Breugel model (with the permission of K. van Breugel).

continue to progress. Hydration will be studied with more sophisticated scientific equipment that will allow attainment of a nanometer scale comprehension of hydration reaction, because this is the scale of ionic species involved in the hydration process.

The scientific complexity of hydration explains why there are still problems in the field, because we have not yet mastered the complexity of the hydration reaction, and, of course, because too many people have a poor knowledge of what has already been established. For those who think that years of practice are sufficient to master all the aspects of Portland cement hydration, it must be remembered that ten years of experience can be either ten years of doing it right or ten years of doing it wrong.

Certainly, Portland cement hydration is a complex phenomenon, but it is a phenomenon that obeys the law of physics, chemistry and thermodynamics. Only benefits can be generated by a better understanding of the scientific aspects involved in the hydration reaction.

References

Acker, P. (2001) Micromechanical Analysis of Creep and Shrinkage Mechanisms, Cambridge, Mass.: MIT Press.

Aïtcin, P.-C. (1995) *Concrete: The Most Widely Used Construction Material*, Ottawa: M. Malhotra, CANMET.

—— (1999) 'The Volumetric Changes of Concrete; Or Does Concrete Shrink or Does it Swell?' *Concrete International*, 21 (12): 77–80.

Aïtcin, P.-C., Neville, A.M. and Acker, P. (1998) 'Les Différents Types de retrait du béton', *Bulletin des Laboratoires des Ponts et Chaussées*, 215: 41–51.

Baalbaki, W., Aïtcin, P.-C., Li, G. and Moranville, M. (1998) 'Microstructure, Elastic Properties and Freeze-Thaw Durability of High-Performance Concrete', M. Cohen, S. Mindess and J. Skalny (eds) *Material Science of Concrete: The Sydney Diamond Symposium*, Westerville, Ohio: American Ceramic Society, pp. 339–56.

Bache, H.H. (1981) 'Densified Cement/Ultra-Fine Particle-Based Materials', *Second*

Conférence Internationale sur les Superplastifiants dans le Béton, Ottawa, Canada, 10–12 June.

Barcelo, L., Boivin, S., Acker, P., Toupin, J. and Clavaud, B. (2001) 'Early Age Shrinkage of Concrete: Back to Physical Mechanisms', *Concrete Science and Engineering*, 3 (10): 85–91.

Baussant, J.B. (1990) 'Nouvelle Méthode d'étude de la formation d'hydrates des ciments: applications à l'analyse de l'effet d'adjuvants organiques', doctoral thesis, Université de Franche-Comté.

Bensted, J. (2001) 'Cement Science: Is it Simple, Cement', *Wapno Beton*, 6/67 (1): 6–19.

Bentz, D.P. (1997) 'Three Dimensional Computer Simulation of Portland Cement Hydration and Microstructure Development', *Journal of the American Ceramic Society*, 80 (1): 3–2.

Candlot, E. (1906) *Ciments et chaux hydrauliques*, Paris: Ch. Béranger.

Chanvillard, G. and D'Aloia, L. (1994) 'Prévision de la résistance en compression au jeune âge du béton: application de la méthode du temps équivalent', *Bulletin de Liaison des Laboratoires des Ponts et Chaussées*, 193 (September–October): 39–51.

—— (1997) 'Concrete Strength Estimation at Early Ages: Modification of the Method of Equivalent Age', *ACI Materials Journal*, 94 (6, November–December): 520–30.

Clark, B.A. and Brown, P.W. (1999) 'The Formation of Calcium Sulphoaluminate Hydrate Compounds: Part I', *Cement and Concrete Research*, 29 (12): 1943–8.

—— (2000) 'The Formation of Calcium Sulphoaluminate Hydrate Compounds: Part II', *Cement and Concrete Research*, 30 (2): 233–240.

Collepardi, M., Baldini, G., Pauri, M. and Conradi, M. (1979) 'Retardation of Tricalcium Aluminate Hydratation by Calcium Sulphate', *Journal of the American Ceramic Society*, 62 (1–2, January–February): 33–5.

D'Aloia, L. (1998), 'Détermination de l'énergie d'activation apparente du béton dans le cadre de l'application de la méthode du temps équivalent à la prévision de la résistance au jeune âge: approches expérimentales mécanique et calorimétrique, simulation numérique', doctoral thesis, Institut National des Sciences.

Damidot, D., Sorrentino, D. and Guinot, D. (1997) 'Factors Influencing the Nucleation and Growth of the Hydrates in Cementitious Systems: An Experimental Approach', Second RILEM Workshop, Hydration and Setting, 11–13 June, Dijon.

Davis, H.E. (1940) 'Autogenous Volume Changes of Concrete', Forty-Third AGM of ASTM, 40: 1103–12.

Diamond, S. (1976) 'Cement Paste Microstructure: An Overview at Several Levels', paper delivered, at a conference at Tapton Hall, University of Sheffield.

Eckart, V.A., Ludwig, H.-M. and Stark, J. (1995) 'Hydration of the Four Main Portland Cement Clinker Phases', *Zement-Kalk-Gips International*, 48 (48): 443–52.

Feldman, R.F. and Sereda, P.J. (1970) 'A New Model for Hydrated Portland Cement and its Practical Implications', *Engineering Journal (Canada)*, 53 (8/9): 53–9.

Flatt, R.I. and Houst, Y.F. (2001) 'A Simplified View on Chemical Effects Perturbing the Action of Superplasticizers', *Cement and Concrete Research*, 31 (8): 1169–76.

Folliot, A. and Buil, M. (1982) 'La Structuration progressive de la pierre de ciment', in *Le Béton hydraulique*, Paris: Presses de l'École Nationale des Ponts et Chaussées, pp. 223–36.

Gauffinet, S., Finot, E. and Nonat, A. (2000) 'Experimental Study and Simulation of C-S-H Nucleation and Growth', in A. Nonat (ed.) *Hydration and Setting: Why Does*

Cement Set? An Interdisciplinary Approach, proceedings of Second, International RILEM Symposium.

Jennings, H.M. (2000) 'A Model for the Microstructure of Calcium Silicate Hydrate in Cement Paste', *Cement and Concrete Research*, 30 (1): 101–16.

Jensen, O. and Hansen, F. (2001) 'Water-Entrained Cement-Based Materials: Principle and Theoretical Background', *Cement and Concrete Research*, 31 (4): 647–54. Figures 6.18–6.25 reprinted with permission of Elsevier.

Jouenne, C.A. (1990) *Traité de céramique générale*, Paris: Éditions Septima.

Kurdowski, W. and Gawlicki, M. (eds) (2001) Proceedings of Kurdowski Symposium, Krákow.

Lafuma, H. (1995) *Liants hydrauliques*, Paris: Dunod.

Laplante, P. (1993) '*Propriétés mécaniques de bétons durcissants: analyse comparée des bétons classiques et à très hautes performances*', *Études et Recherches des Laboratoires des Ponts et Chaussées*, Série Ouvrage d'art OA13, ISSN 1161–028X, 299 p.

Le Chatelier, H. (1904) *Recherches expérimentales sur la constitution des mortiers hydrauliques*, Paris: Dunod.

Lequeux, N. and Clinard, C. (2001) Personal communication.

Liebau, F. (1985) *Structural Chemistry of Silicates: Structure Bonding and Classification*, Berlin: Springer-Verlag.

Locher, F.W., Richartz, W. and Sprung, S. (1980) 'Studies on the Behaviour of C_3A in the Early Stages of Cement Hydration', Seminar on the Reaction of Aluminates During the Setting of Cement, Eindhoven, The Netherlands, 13–14 April.

Lynam, C.G. (1934) *Growth and Movement in Portland Cement Concrete*, Oxford: Oxford University Press.

Michaud, V. and Suderman, R. (1998) Solubility of sulphates in high SO_3 clinkers, ACI SP-177, pp. 15–25.

Mindess, S. and Young, J.F. (1981) *Concrete*, Englewood Cliffs, NJ: Prentice Hall.

Neville, A.M. (2000) *Les Propriétés des bétons*, Paris: Eyrolles.

Nonat, A. (1994) 'Interactions Between Chemical Evolution (Hydration) and Physical Evolution Setting in the Case of Tricalcium Silicate', *Matériaux et constructions*, 27 (186): 187–95.

—— (1998) 'Du gâchage jusqu'à l'état durci, ce sont les mêmes liaisons qui sont à l'heure', *Journée technique de l'industrie cimentière*, 21 January.

—— (ed.) (2000) *Hydration and Setting: Why Does Cement Set? An Interdisciplinary Approach*, Bagneux: RILEM Publications.

—— (2005) 'The Structure of C-S-H', *Cement Wapno Beton*, 2 (March): 65–73.

Nonat, A. and Mutin, J.C. (eds) (1992) *Hydration and Setting of Cements*, London: E & FN Spon.

Odler, I. (1998) 'Hydration, Setting and Hardening of Portland Cement', in P.C. Hewlett (ed.) *Lea's Chemistry of Cement and Concrete*, 4th edn, London: Arnold, pp. 241–97.

—— (1958) 'Structure and Physical Properties of Hardened Portland Cement Paste', *Journal of the American Ceramic Society*, 41 (January): 1–6.

—— (1968) *The Properties of Fresh Concrete*, New York: John Wiley & Sons.

Papadakis, M. and Venuat, M. (1966) *Fabrication et utilisation des liants hydrauliques* (self-published).

Porteneuve, C. (2001) 'Study of the Water Leaching of Concrete: The Nuclear Magnetic

Resonance Approach', RILEM Workshop, Improving Concrete Resistance to Leaching, Paris, 19 February.
Powers, T.C. (1947) 'A Discussion of Cement Hydration in Relation to the Curing of Concrete', *Proceedings of the Highway Research Board*, 27: 178–88.
Powers, T.C. and Brownyard, T.L. (1948) *Studies of the Physical Properties of Hardened Portland Cement Paste, Bulletin* 22, Skokie, Ill.: Research Laboratories of the Portland Cement Association.
Prince, W., Edwards-Lajnef, M. and Aïtcin, P.C. (2002) Interaction between Ettringite and a Polynaphtalene sulphonate superplasticizer in a Cementitious Paste, *Cement and Concrete Research*, 32 (1): 79–85.
Regourd, M. (1978) 'Cristallisation et réactivité de l'aluminate tricalcique dans les ciments Portland', *Il Cemento*, 3: 323–35.
Regourd, M. (1982a) 'L'hydratation du ciment Portland' in J. Baron et R. Sauterey (eds) *Le Béton hydraulique*, Paris: Presses de l'École Nationale des Ponts et Chaussées, pp. 193–221.
Regourd, M. (1982b) 'Microstructure et propriétés des ciments, mortiers et bétons', *Bétons, Plâtres, Chaux*, 734: 41–9.
Richard, P. and Cheyrezy, M. (1994) Reactive Powder Concrete with High Ductility and 200–800 MPa Compressive Strength, ACI SP–144, p. 507–518.
Richardson, I.G. (1999) 'The Nature of C-S-H in Hardened Cements', *Cement and Concrete Research*, 29 (8): 1131–47.
Rodger, S.A. (1991) 'The Use of Solid State Nuclear Magnetic Resonance in the Study of Cement Hydration', in A. Nonat and J.C. Mutin (eds), *Hydration and Setting of Cements*, London: E & FN Spon, pp. 65–76.
Scrivener, K. (2001) personal communication.
Skalny, J., Johansen, V. and Miller, F.M. (1997) Sulphates in Cement Clinker and Their Effect on Concrete Durability, ACI SP–171–30, pp. 625–631.
Sorrentino, D. and Damidot, D. (1998) 'La Mesure de la cinétique d'hydratation des liants hydrauliques: un outil essentiel pour la maîtrise des propriétés à court terme', Symposium RILEM en hommage à Micheline Moranville, Arles.
Stark, J. Möser, B. and Eckart, A. (2001a) 'New Approach to Cement Hydration', *ZKG International*, 54 (1): 52–60.
—— (2001b) 'New Approach to Cement Hydration', *ZKG International*, 54 (2): 114–19.
Taylor, H.F.W. (1993) 'A Discussion of the Papers "Models for the Composition and Structure of Calcium Silicate Hydrate (C-S-H) Gel in Hardened Tricalcium Silicate Pastes" and "THe Incorporation of Minor and Trace Elements into Calcium Silicate Hydrate C-S-H Gel in Hardened Cement Pastes", by I.G. Richardson and C.W. Groves', *Cement and Concrete Research*, 23 (4): 995–1000.
—— (1997) *Cement Chemistry*, 2nd edn, London: Thomas Telford.
—— (1999) 'Distribution of Sulphate between Phases in Portland Cement Clinkers', *Cement and Concrete Research*, 29: 1173–9.
Tazawa, E.I. and Miyazawa, S. (1993), *Autogenous shrinkage of concrete and its importance in concrete technology, Creep and Shrinkage of Concrete*, London: E & FN Spon.
van Breugel, K. (1991a) 'Hymostruc: A Computer-Based Simulation Model for Hydration and Formation of Structure in Cement-Based Materials', in A. Nonat and J.C. Mutin (eds), *Hydration and Setting of Cements*, London: E & FN Spon, pp. 361–8.

—— (1991b) 'Simulation of Hydration and Formation of Structure in Hardening Cement-Based Materials', doctoral thesis, Delft University.

Van Damme, H. (1994) 'Et si Le Chatelier s'était trompé: pour une physico-chimie mécanique des liants hydrauliques et des géomatériaux', *Annales des Ponts et Chaussées*, 71 (14): 30–41.

—— (1998) 'La Physique des liaisons entre les hydrates et les moyens d'agir au niveau moléculaire', *Journée Technique de l'Industrie Cimentière*, 21 January.

—— (2001) 'Colloidal Chemo-Mechanics of Cement Hydrates and Smeclite Clays: Cohesion vs. Swelling', in A. Hubbard (ed.) *Encyclopedia of Surface Colloid Science*, New York: Marcel Dekker.

Vernet, C. (1995) 'Mécanismes chimiques d'interactions ciment-adjuvants, GTG Spa, Guerville', *Service Physico-Chimie du Ciment* (January).

Vernet, C. et Cadoret, G. (1992) 'Suivi en continu de l'évolution chimique et mécanique des BHP pendant les premiers jours', in Y. Malier (ed.) *Les Bétons à hautes performances: Caractérisation, Durabilitém, Applications*, Paris: Presse de l'École Nationale des Ponts et Chaussées, pp. 115–56.

Verschaeve, M. (1994) Introduction to the structure of calcium aluminate hydrates, personal communication.

Yada, K. (1967) 'Study of Chrysotile Asbestos by a High Resolution Electron Microscope', *Acta Chrystallographica*, 2(5): 704–7.

Zhang, X., Chang, W., Zhang, T. and Ony, C.K. (2000) 'Nanostructure of Calcium Silicate Hydrate Gels in Cement Paste', *Journal of the American Ceramic Society*, 83(10): 2600–4.

Chapter 7

Admixtures

Usually the action and utilization of admixtures is treated in books related to concrete rather than in books dealing with Portland cement and hydraulic binders. This is unfortunate because the action of admixtures is most often intimately linked to that of Portland cement, hydraulic binders and their hydration. Moreover, under the action of some admixtures, the hydration reaction can be altered significantly. Of course, there are some admixtures that have only a 'physical' action on concrete or a chemical action that has nothing to do with the hydration of Portland cement and hydraulic binders, but they are not too numerous.

Due to the recent progress in the understanding of the mode of action and effects of admixtures and their scientific and technological importance, it has been decided to devote a whole chapter on admixtures in this book on hydraulic binders, because the author is deeply convinced that the future of concrete, Portland cement and hydraulic binders, is closely linked to a greater and better use of admixtures. When they are used appropriately, admixtures add more flexibility to the formulation of concrete and more durability to concrete structures.

Presently, for various reasons, not enough admixtures are used by the concrete industry worldwide. There are very few countries where almost all concretes contain at least one or several admixtures. Unfortunately, in too many developing countries, the existence and the necessity of using admixtures are ignored. In Chapter 11, the negative environmental impact of the quasi-absence of the use of admixtures will be seen. This is very unfortunate, because it is in these developing countries that cement consumption will increase. For example, 5 to 10 per cent of cement could be saved if a water reducer was systematically added to Portland cement and hydraulic binders. This 5- to 10-per cent saving would result in a reduction in fuel imports and a corresponding decrease in CO_2 emissions.

7.1 Historical background

As seen in Chapter 2, the idea of adding chemical compounds when making concrete is not new. Some Roman authors have reported that several hundred years before Christ, Roman masons were adding blood or eggs to the lime and the pozzolans they were using when making concrete (Mindess and Young 1981; Venuat 1984; Mielenz 1984). This idea of adding 'exotic' products when making concrete lasted many years, since, according to Garrison (1991), in around 1230 Villard de Honnecourt, a cathedral builder, recommended impregnating concrete cisterns with linseed oil to make them more impervious.

More recently, it was just before the Second World War that the beneficial effect of what we now call admixtures was fortuitously discovered in the USA. There are two accounts to explain this discovery. The first one says that it was a faulty bearing that had been releasing some heavy oil in a grinding mill that resulted in the discovery of the beneficial effects of air-entraining agents (Mindess and Young 1981). The second story (Dodson 1990) says that it is the tenacity and perspicacity of a DOT (Department of Transportation) engineer in New York state who discovered the beneficial effect of air-entraining agents. This engineer was building one of the first three-lane concrete highways in New York state and, in order to prevent accidents, he had the idea of colouring the concrete of the central lane in black so that the drivers in the shared central lane noticed that they were driving in the dangerous lane.

The contractor added some carbon black to its concrete, but the DOT engineer was not satisfied by this first trial, because the colour of the concrete was not very uniform. Therefore, he asked the contractor to find a way to improve the dispersion of the carbon black within the concrete to end up with a more aesthetic concrete lane. The concrete contractor asked his carbon-black supplier if he had a special chemical product that could help the dispersion of carbon black-particles in concrete. The carbon-black supplier suggested that he use a sodium salt of a polynaphthalene sulphonate based dispersant that was already being used by the painting industry to disperse carbon black in paints. After the addition of a small amount of this magic chemical, the uniformity of the coloured concrete was significantly improved and the New York DOT. engineer was satisfied.

Some years (winters) later, the same engineer observed that the black concrete was in much better condition than the concrete used in the other two pale lanes. He could have concluded that black carbon improved the freeze/thaw resistance of concrete. (We will see in the next chapter that Detwiler and Mehta [1989] have even found that carbon black could have the same effects as silica fume in improving concrete compressive strength.) However, this very conscientious engineer was not satisfied by such a direct and simplistic explanation, and he asked a friend of his to look at a sample of the pale concrete and a sample of the black concrete under a microscope to see if they were different.

The microscopist noticed that the cement paste of the dark concrete contained very small air bubbles uniformly distributed within the concrete. Our engineer could have concluded that carbon black entrained air bubbles within a concrete, but he remembered that a dispersing agent was also added. Doing some experimental work: concrete with and without the carbon black, with and without the sodium salt of polynaphthalene sulphonate (or lignosulphonate), he found that it was the dispersing agent that was responsible for the entrainment of the small air bubbles.

Later on, Powers demonstrated that it is the bubble network that improves so drastically the freezing and thawing durability of concrete. Sometimes, when this story is told, it is added that the contractor noticed that when he added the dispersing agent the concrete became creamy, that it could be batched with less mixing water, that it was bleeding less, that it was easier to finish and even that it was stronger. Moreover, in low-strength concrete, a better dispersion of cement particles compensates for the strength loss of the first few percent of entrained air.

This is how the beneficial dispersive properties of polysulphonate might have been discovered. Although this version could be perceived as a kind of fairy tale, the concrete industry could have its anonymous Alexander Fleming. The story does not tell if the initial objective of that tenacious DOT engineer was achieved, and, if there were less frontal collisions in the black lane.

Powers and the research team of the Portland Cement Association explained in the end of the 1940s and the beginning of the 1950s all the beneficial effects of these small air bubbles on the properties of fresh and hardened concrete (Powers 1956). In North America, air-entraining agents started being systematically introduced in concrete, even in concrete that was not exposed frequently to freezing/thawing cycles.

For a long time, admixture technology was closer to alchemy than to chemistry, the few companies involved in the admixture business being very secretive and jealous of their secrets. This is no longer the case, as there are some scientific books devoted to the science of admixtures (Joisel 1973; Venuat 1984; Dodson 1990; Ramachandran 1995; Rixom and Mailvaganam 1999; and Ramachandran et al. 1998). International conferences are held on the subject such as the six ACI/CANMET international conferences on superplasticizers; Ottawa (1978) ACI SP–72, Ottawa (1981) ACI SP–68; Ottawa (1989) ACI SP–119; Montreal (1994) ACI SP–148; Rome (1997) ACI SP–173; and Nice (2000) ACI SP–195. The seventh one was held in 2003 in Berlin and the eighth one in Sorrento in Italy in 2006.

In spite of recent knowledge and increasing research done on the mechanism of the action of superplasticizers, there are still too many engineers who question the beneficial aspects of concrete admixtures and who believe that admixture salesmen are pedlars. There are even some engineers and cement producers who question the existence of a science of admixtures. Some engineers still believe that admixtures should be banned to avoid the compatibility problems that are so costly for contractors when they occur in the field. Would it not be better to simply rather add some water to favour cement-particle dispersion?

We will see in Chapter 11 that admixtures are essential components of modern concrete and all the beneficial aspects of their use from a durability and sustainable development point of view.

The use of admixtures varies greatly from one country to another. For example, in Japan and Canada, almost 100 per cent of concrete contains at least a water reducer and an air-entraining agent. But in the USA, France and Switzerland, less than 50 per cent of concrete contains admixtures, though this percentage is increasing constantly.

So we will see that when the beneficial role of admixtures is understood, it becomes evident that they are key components of modern concretes. Indeed, the chemical system operating during cement hydration is becoming very complex due to the intrinsic complexity of Portland cement and hydraulic binders and due to the complexity of the organic molecules that are used as admixtures. However, it must be recognized that presently this complexity is more or less well understood and mastered.

In the past years, concrete technology has evolved rapidly, much faster than the standards of cement, of concrete and of admixtures, so that it is very important to close the gap between what is written in standards and what is known and done in practice. If such corrective action is not rapidly undertaken, artificial problems that will slow down the evolution of concrete technology could impair an efficient use of Portland cement and hydraulic binders.

Modern admixtures are made increasingly less frequently from by-products of other industries but rather from synthetic polymers especially produced for the concrete industry. Modern admixtures modify one or several properties of fresh and hardened concrete by correcting some of the technological weaknesses of Portland cement and hydraulic binders (Ramachandran et al. 1998). Their use is no longer a commercial issue, it is a matter of progress, of science and of sustainable development.

7.2 An artificially complicated terminology

A great source of confusion in the field of admixtures comes from the fact that an artificially complicated terminology has been developed over the years, not only in the technical and commercial literature but also in the standards. Some researchers maintain that this complex technology is a consequence of the ignorance of the actual action of admixtures on concrete, while others says admixture companies purposely hide their activity behind a complex terminology in order to create the message: 'admixtures are a very complicated field, don't waste your time, we are the specialists.'

For example, in the 2001 edition of *Design of Concrete Mixtures* edited by the Canadian Cement Association (in metric units), in Chapter 7, which deals with admixtures, the following list of admixtures (in alphabetic order) is found (see Table 7.1): accelerators, air detrainers, air-entraining admixtures, alkali-aggregate reactivity inhibitors, antiwashout admixtures, bonding admixtures, colouring admixtures, corrosion inhibitors, dampproofing admixtures, foaming agents, fungicides, germicides, insecticides, gas formers, grouting admixtures, hydration control admixtures, permeability reducers, pumping aids, retarders, shrinkage reducers superplasticizers, superplasticizer and retarder, water reducers, water reducers and accelerator, water reducers and retarder, water reducer – high range, water reducer – high range – and retarder, water reducer – mid range.

Table 7.1 Concrete admixtures by classification (reprinted with permission of the Portland Cement Association)

Desired effect	**Material**
Accelerate setting and early-strength development	Calcium chloride (ASTM D 98) Triethanolamine, sodium thiocyanate, calcium formate, calcium nitrite, calcium nitrate
Decrease air content	Tributyl phosphate, dibutyl phthalate, octyl alcohol, with insoluble esters of carbonic and boric acid, silicones
Improve durability in freeze/thaw, deicer, sulphate, and alkali-reactive environments improve workability	Salts of wood resins (Vinsol resin), some synthetic detergents, salts of sulphonated lighin, salts of petroleum acids, salts of proteinaceous material, fatty and resine acids and their salts, alkybenzene sulphonates, salts of sulphonated hydrocarbons
Reduce alkali-aggregate reactivity expansion	Barium salts, lithium nitrate, lithium carbonate, lithium hydroxide
Cohesive concrete for underwater placements	Cellulose, acrylic polymer
Increase bond strength	Polyvinyl chloride, polyvinyl acetate, acrylics, butadiene-styrene copolymers
Coloured concrete	Modified carbon black, iron oxide, phthalocyanine, chromium oxide, titanium oxide, cobalt blue
Reduce steel corrosion activity in a chloride-laden environment	Calcium nitrite, sodium nitrite, sodium benzoate, phosphates or fluosilicates, fluoaluminates, ester
Retard moisture penetration into dry concrete	Soaps of calcium or ammonium stearate or oleate Butyl stearate Petroleum products
Produce foamed concrete with low density	Cationic and anionic surfactants Hydrolized protein
Inhibit or control bacterial and fungal growth	Polyhalogenated phenols Dieldrin emulsions Copper compounds
Cause expansion before setting	Aluminum powder

Table 7.1—continued

Desired effect	Material
Adjust grout properties for specific applications	See Air-entraining admixtures, Accelerators, Retarders and Water reducers
Suspend and reactivate cement hydration with stabilizer and activator	Carboxylic acids Phosphorus-containing organic acid salts
Suspend and reactivate cement hydration with stabilizer and activator	Carboxylic acids Phosphorus-containing organic acid salts
Decrease permeability	Latex Calcium stearate
Improve pumpability	Organic and synthetic polymers Organic flocculents Organic emulsions of paraffin, coal tar, asphalt, acrylic Bentonite and pyrogenic silicas Hydrated lime (ASTM C 141)
Retard setting time	Lignin Borax Sugars Tartaric acid and salts
Reduce drying shrinkage	Polyoxyalkylene alkyl ether Propylene glycol
Increase flowability of concrete. Reduce water-cementing materials ratio	Sulphonated melamine formadelhyde condensates Sulphonated naphthalene formaldehyde condensated Lignosulphonates Polycarboxylates
Increase flowability with retarded set. Reduce water-cementing materials ratio	See superplasticizers and also water reducers
Reduce water content at least 5%	Lignosulphonates Hydroxylated carboxylic acids Carbohydrates (Also tend to retard set so accelerator is often added)
Reduce water content (minimum 5%) and accelerate set	See water reducer, Type A (accelerator is added)
Reduce water content (minimum 12%)	See superplasticizers
Reduce water content (minimum 12%) and retard set	See superplasticizers and also water reducers
Reduce water content (between 6 and 12%) without retarding	Lignosulphonates Polycarboxylates

The European terminology is as rich as the North American one. This rich terminology is also found in standards, for example the ASTM C494 standard recognizes different types of water reducers identified by the letter A to G. As far as it is concerned, the new European standard on admixtures recognizes nine different types of admixtures.

This complex terminology creates more confusion than clarification. It seems to be related to the strong marketing orientation of the admixture industry and to the fact that for a long time most concrete admixtures were made from industrial by-products having more than one specific action on the hydration reaction. As it was too costly to purify these by-products, it was easier to create a new subcategory in the standards.

When looking more closely at admixtures, it is easy to greatly simplify the terminology of admixtures by looking at their physico-chemical mechanism of action rather than at their effect on the properties of fresh and hardened concrete (Dodson 1990). Dodson classifies admixtures in the following four categories according to their mechanism of action.

1 Dispersion of the cement in the aqueous phase of concrete.
2 Alteration of the normal rate of hydration of cement, in particular the tricalcium silicate phase.
3 Reaction with the by-products of hydrating cement, such as alkalis and calcium hydroxide.
4 No reaction with either the cement or its by-products.

Mindess and Young (1981) recognize that admixture terminology is complex and propose classification of concrete admixtures into four different categories.

1 Air-entraining agents (ASTM C260) are added primarily to improve the frost resistance of concrete.
2 Chemical admixtures (ASTM C494 and BS 5075) are water-soluble compounds added primarily to control setting and early hardening of fresh concrete or to reduce its water requirements.
3 Mineral admixtures are finely divided solids added to concrete to improve its durability or to provide additional cementing properties. Slags and pozzolans are important categories of mineral admixtures.
4 Miscellaneous admixtures include all those materials that do not come under one of the foregoing categories, many of which have been developed for specialized applications.

On the following pages, we will use the Dodson classification. We will examine the principal mechanism of action of each of these four categories of admixtures that are presently used when making concrete (Hewlett and Young 1987; Dodson 1990). We will not develop the influence of the molecular structures and the action of each of these admixture on each individual phase of Portland cement because this can be found in specialized books and it is beyond the scope of this book (Venuat 1984; Dodson 1990; Ramachandran 1995; Ramachandran et al. 1998; Rixom and Mailvaganam 1999). Neither will we enter the field of practical uses of admixtures (dosage, sequence of introduction, duration of their effect, etc.) but will rather look at their principal effects on the properties of fresh and hardened concrete to show that, when they are used appropriately, they can improve concrete quality and durability.

7.3 Dispersing agents

7.3.1 The dispersion of cement particles

Experience shows that when the water/cement ratio is lower than 0.40, it is usually impossible to make a concrete having a slump greater than 100 mm when mixing only water and cement (Legrand 1982). But experiments also show that when using a superplasticizer it is possible to make concretes with a water/binder ratio lower than 0.30 having a slump greater than 200 mm. It is even possible to make fluid reactive powder concretes having a water/binder ratio lower than 0.20. These low water/binder systems are flowing and do not present any sign of bleeding or of segregation in spite of their quasi-fluidity.

How can such different behaviours be obtained using the same cement? How can small amounts of organic molecules have such a spectacular effect on the rheology of cement paste and concrete?

The grinding of clinker and gypsum results in the development of a great number of electrical charges on the surface of cement particles: essentially, negative charges on C_3S and C_2S particles and positive charges on C_3A and C_4AF particles. When these cement particles come into contact with a liquid as polar as water, they have a strong tendency to flocculate as can be seen schematically in Figure 7.1 (Kreijger 1980; Chatterji 1988; Dodson 1990; Rixom and Mailvaganam 1999). The cement flocs trap water that is no longer available to fluidify the cement suspension.

Cement-particle flocculation can easily be observed; it is only necessary to do the very simple experiment presented in Figure 7.2.

- 50 g of cement are placed in a 1 litre graduate full of water.
- A rubber top is placed on the graduate, and the graduate is agitated upside down twenty times in order to disperse as much as possible all the cement particles.
- The suspension is then left to settle.

In a matter of one or two minutes, cement particles start to flocculate and settle very rapidly at the bottom of the graduate leaving almost clear water in the upper part of the graduate after fifteen to thirty minutes. It can also be seen that if the top level of the cement flocs at the bottom of the graduate is

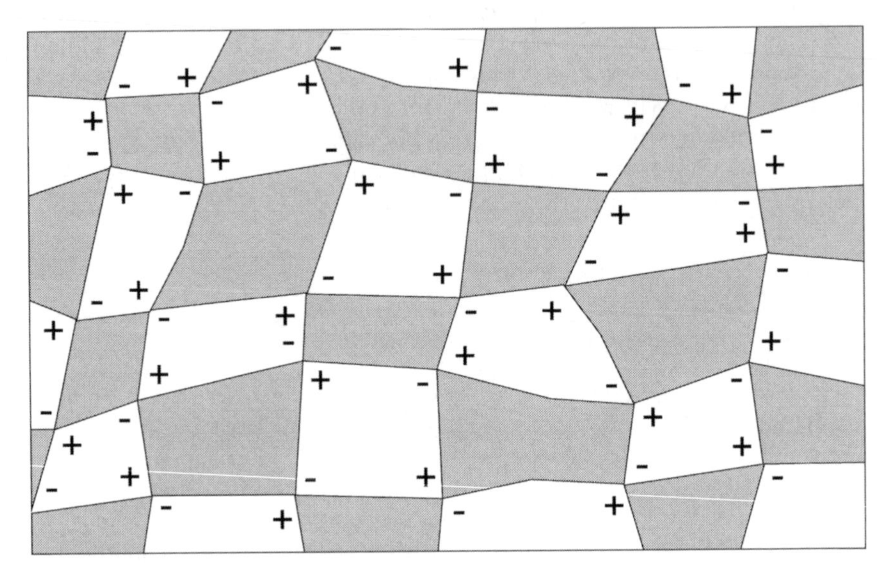

Figure 7.1 Flocks of cement particles (based on Kreijger 1980).

compared to the top level of 50 g of unpacked cement in an empty graduate, the flocculated suspension occupies a volume two or three times greater than the dry cement.

The same experiment can be repeated but this time with the addition of 5 to 10 cm^3 of a superplasticizer to the water. This time it is observed that the cement suspension is very stable and does not flocculate. Of course, it is observed that the coarser particles start to settle at the bottom of the graduate, but that it can take twenty-four or forty-eight hours before seeing the finest particles settle at the bottom of the graduate. This time, the final volume occupied by the layers of cement particles has about the same thickness as the dry cement when conducting this experiment with a water reducer and a superplasticizer. It can also be observed that superplasticizers deflocculate cement particles much better than water reducers.

From this very simple experiment, it can be concluded that some organic molecules strongly reduce or even destroy the flocculation of cement particles so that all the water introduced during the mixing of concrete can be efficiently used to give plasticity and workability to the concrete (Black et al. 1963; Dodson 1990).

Of course, the interaction between cement particles and these organic molecules depends on the physico-chemical characteristics of the cement particles and the architecture of the molecules of the dispersing agent used. In fact, the interaction mechanism can usually be decomposed into two parts: a physical effect that is similar to the one produced when these molecules interact

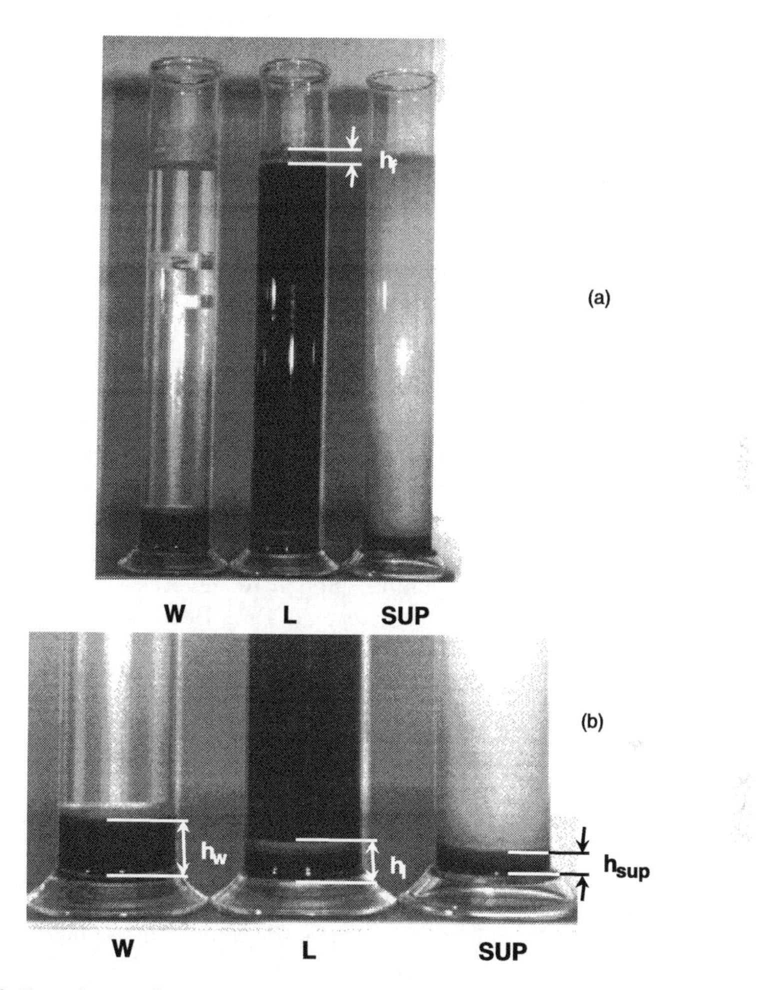

Figure 7.2 Flocculation of cement particles: (W) in water; (L) in water + water reducer; (SUP) in water + superplasticizer.

with a suspension of non-reactive (with water) particles like TiO_2, Al_2O_3, SiO_2, etc., and a chemical effect that can be linked to the chemical reaction of these molecules with the most active sites of the anhydrous cement particles or with cement particles during their hydration. As will be seen later, the interaction of the polysulphonates (the principal molecules that are used as dispersing agents) on cement particles is not simply physical. The sulphonate group of the superplasticizer can react with active sites of C_3A if the cement does not liberate SO_4^{2-} ions fast enough.

According to Jolicoeur et al. (1994), the physical effect of superplasticizer molecules can be decomposed into three parts:

- a physical adsorption under the effect of Van der Waals and electrostatic forces;
- long-range repulsive forces due to a reconfiguration of the superficial charges;
- steric repulsive forces between two adjacent superplasticizer molecules and two adjacent cement particles.

Based on numerous experimental observations, Jolicoeur et al. (1994) have proposed the schematic mechanism of action presented in Figure 7.3 to explain the action of a superplasticizer during cement hydration. The adsorption of the superplasticizer molecules is possible on negative sites due to the presence of Ca^{2+} ions that bridge the cement particle and the superplasticizer molecule.

Figure 7.4 represents a case of electrostatic repulsion between two cement particles, one positively charged and the other negatively charged. In the absence of a superplasticizer, these two particles would have been attracted and would have started to create a floc. When these two cement particles are covered with superplasticizer molecules, they repel each other. It is well accepted that polynaphthalene and polymelamine sulphonate molecules work essentially under this mode of repulsion (Uchikawa et al. 1992; Uchikawa 1994).

Figure 7.5 represents a case of steric repulsion between two molecules of superplasticizer that are adsorbed on two cement particles. Polycarboxylate and polyacrylate superplasticizers work essentially in this mode of dispersion, but also slightly in an electrostatic repulsive mode (Uchikawa 1994; Uchikawa et al. 1997).

Figure 7.6 represents the interaction between superplasticizer molecules and the reactive sites of a cement grain, essentially sites having a natural affinity for SO_3^{2-} terminations and SO_4^{2-} ions. In such a case, the superplasticizer is in competition with SO_4^{2-} ions liberated by calcium and alkali sulphates for the neutralization of these sites. It is this competition that explains why in some cases the initial improvement of the rheology due to the superplasticizer is of short duration, since a non-negligible portion of the superplasticizer molecules react with the interstitial phase and are no longer available for the dispersion of cement particles (Baussant 1990; Vichot 1990; Kim 2000).

This double physico-chemical action of the molecules of a superplasticizer can be very easily seen by adiabatic calorimetry: the use of a superplasticizer moves the maximum of the heat flux. When a small amount of superplasticizer is used the peak of heat appears sooner than in a water/cement system, while it is delayed when the superplasticizer dosage is higher. This initial accelerating effect is explained by a better deflocculation of the cement particles which results in the hydration of greater number of them and a more homogeneous dispersion of the cement particles within the suspension. But, as the superplasticizer dosage increases, this accelerating effect diappears, because the

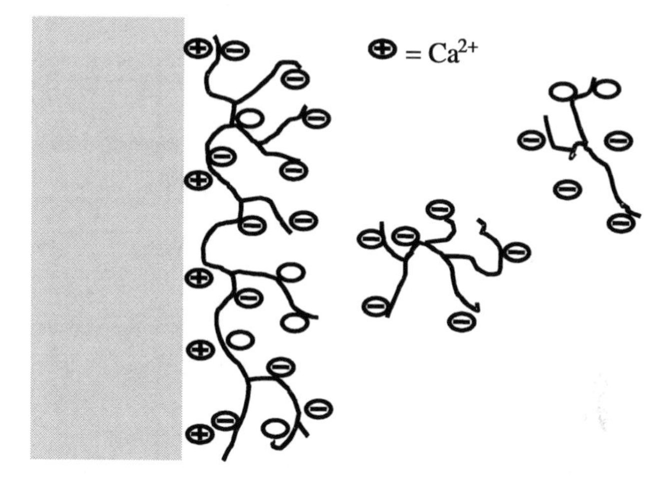

Figure 7.3 Superficial adsoption of superplasticizer molecules on a cement particle (after Jolicoeur et al. 1994).

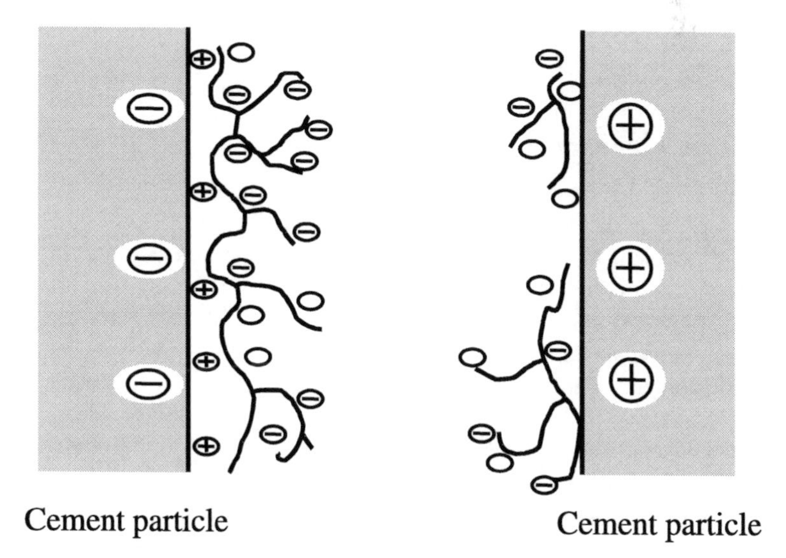

Figure 7.4 Electrostatic repulsion (after Jolicoeur et al. 1994).

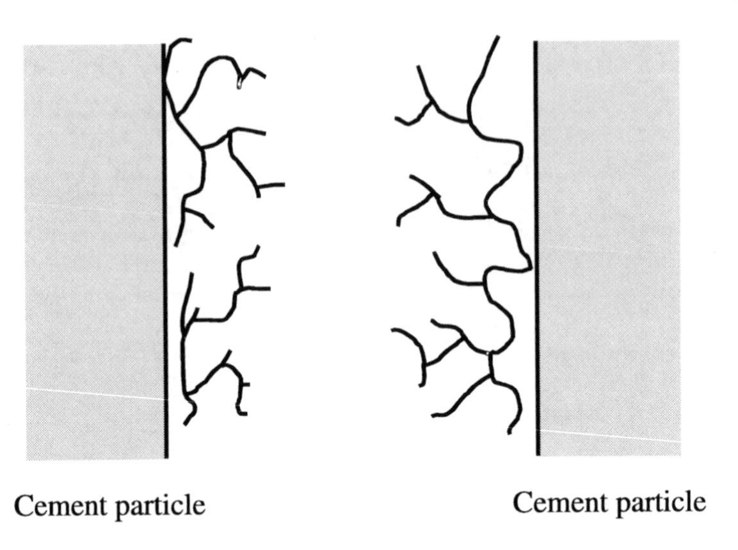

Figure 7.5 Steric repulsion (after Jolicoeur et al. 1994).

adsorption of the superplasticizers molecules on the surface of the cement particles prevents them from reacting with water. Not only is the peak delayed, but it is also not as high. This simple experiment shows that the action of superplasticizer molecules is not simply physical but that there is a chemical action taking place simultaneously. In the following pages, the interaction mechanism will be reviewed in more detail.

7.3.2 Lignosulphonates

For many years, lignosulphonates were practically the only dispersing agents used by the concrete industry: they were not expensive, and they were sufficiently efficient to satisfy the needs of the concrete industry because they were able to reduce by 5 to 8 per cent the amount of mixing water needed to obtain a given slump (Paillere et al. 1984; Dodson 1990). Most of the time, lignosulphonates were compatible with Portland cements; only very few compatibility problems were found to happen (Dodson and Hayden 1989; Ranc 1990).

Water reducers based on lignosulphonates can be mixed with polynaphthalene or polymelamine sulphonates when making concretes having water/binder ratios between 0.40 and 0.50, a range where superplasticizers are too efficient and increase too much the risks of segregation; this domain is presently reserved for the use of so-called 'mid-range' water reducers. Sometimes ligno-

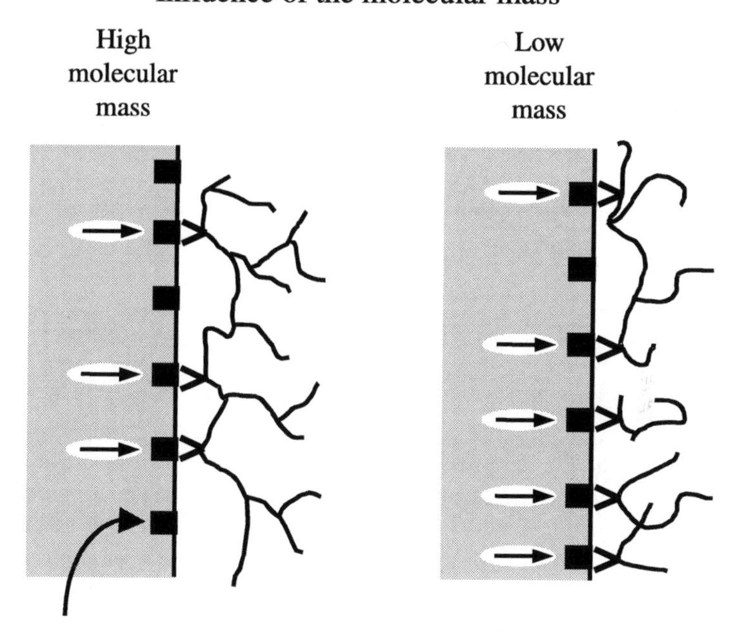

Figure 7.6 Inhibition of reactive sites (after Jolicoeur et al. 1994).

sulphonates are used in small amounts to correct the action of polynaphthalene sulphonate superplasticizers with Portland cements that are too reactive.

From a dispersion point of view, lignosulphonates are simply dispersing agents that are less efficient than synthetic polynaphthalene and polymelamine sulphonates or polycarboxylates, but they will be used as long as they are available, because they are not expensive and are adequately efficient in high water/cement ratio concrete ($W/C > 0.50$). Rixom and Mailvaganam (1999) explain this limited efficiency action by the fact that lignosulphonate molecules have a spheroidal shape (Figure 7.7) due to their strong reticulation, favoured by the inclusion of foreign ions trapped within the polymer chains.

Lignosulphonates are the by-product of the bisulfite process of pulp and paper production. When attacking the wood pulp with the hot acidic sulfite solution, the objective is to dissolve everything that is not cellulose in the wood in order to liberate the cellulose fibres. Of course, in a pulp and paper plant, this process is optimized not to produce the most efficient water dispersing agent for the concrete industry but, rather, to optimize the quality of the cellulose fibres. Therefore, industrial lignosulphonates have characteristics that are influenced by the nature of the wood that is treated (evergreen vs. deciduous

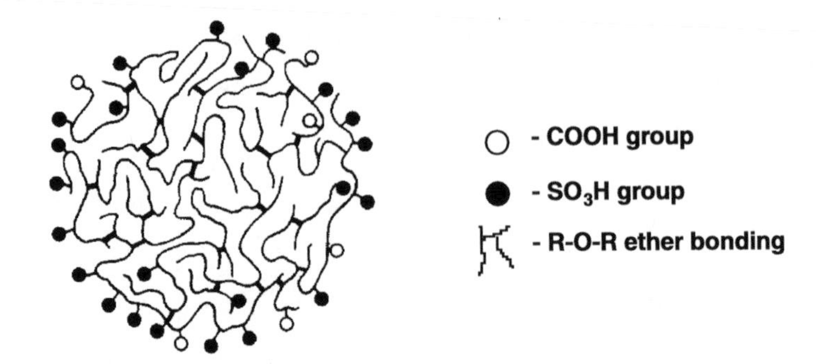

Figure 7.7 Micelle of lignosulphonate (according to Rixom and Mailvaganam 1999).

trees, species of wood) and by the season during which the trees were felled: the trees felled in winter contain no sap, trees felled in summer contain sap. Moreover, in summer the logs exposed to the sun's 'heat' due to the fermentation of sugars contained in the wood caused by bacterial action, wheras in winter this does not occur.

Of course, in the case of a pulp and paper plant that uses only one type of wood to produce only one type of paper, it is possible to produce a lignosulphonate having very constant characteristics and properties. This is not the case for all pulp and paper plants; the author has visited a pulp and paper plant which produces four different types of pulp by mixing different amounts of evergreen and deciduous trees. Some lignosulphonates produced by this pulp and paper plant cannot be used as dispersing agents and can only be burnt to recover some heat from them.

Lignosulphonates are usually marketed as their calcium salt, but some sodium salts are also available. Commercial lignosulphonates have a 40 ± 2 per cent solids content, and their colour is dark brown. Commercial lignosulphonates also contain surface-active compounds and sugars. The presence of these surface-active products results in the entrainment of a certain amount of air or, rather, as will be seen later, in the stabilization of their bubbles which are entrained during the mixing. The presence of sugars in lignosulphonates results in the retardation of setting and hardening, which is advantageous in hot climates but disadvantageous in cold climates.

It is possible to eliminate partially or totally these two types of secondary effects, but the process can be costly, depending on the amount of impurities and also on the way they are linked to the lignosulphonate chains (Mollah and al. 1995). Sometimes it is less expensive to add a defoamer and an accelerator to counteract the effects of the surface-active agent and sugar present in industrial lignosulphonates. In some cases, the elimination of surface-active agents is done very simply by agitating the lignosulphonate and by injecting compressed air. The foam that is formed is eliminated. However, this very simple technique does not work all the time, as sometimes the surface-active agent also entrains most of the lignosulphonate.

The presence of surface-active agents and sugars explains why, from a practical point of view, it is difficult to use a lignosulphonate dosage greater than 1 to 1.5 litres per m^3 of concrete, while it is possible to use up to 10 to 15 litres of polynaphthalene sulphonate before seeing any secondary effects. As will be seen dosages varying between 0.8 to 1.5 litres are sufficient to disperse the cement particles present in concretes having water/binder ratios greater than 0.50.

Some polynaphthalene or polymelamine sulphonates can be added to industrial lignosulphonates in order to improve their efficiency, but it increases their price.

In the future, the availability of lignosulphonate will decrease due to technological changes in the pulp and paper industry, which is moving away from chemical processing to thermo-mechanical processing.

7.3.3 Superplasticizers

Currently, six different types of superplasticizers are used (Bradley and Howarth 1986; Rixom and Mailvaganam 1999; Ramachandran et al. 1998):

1. sulphonated salts of polycondensates of naphthalene and formaldehyde known as 'naphthalene polysulphonates' or, more simply, 'polynaphthalene';
2. sulphonated salts of polycondensates of melamine and formaldehyde known as 'melamine polysulphonates' or, more simply, 'polymelamine';
3. lignosulphonates having a very low content of surface active agents and sugar;
4. polycarboxylates;
5. polyacrylates;
6. some companies are also marketing superplasticizers based on polyphosphonates and different copolymers.

Until recently polysulphonates of naphthalene and melamine types were the principal source of commercial superplasticizers, but recently polycarboxylates have begun to be used more and more often in spite of their higher price.

In their commercial form, superplasticizers are not inevitably composed of pure polymers; they are often the result of a secret cocktail containing some retarders or accelerators and even some lignosulphonate in order to decrease their production cost (Dodson 1990).

From an historical point of view, the dispersing properties of naphthalene polysulphonates were discovered and patented by Tucker in 1938, but they were not commercially exploited before the 1960s when the Japanese found that they could be used to make high slump concretes which could not be produced when using lignosulphonates. Almost simultaneously, in Germany, it was discovered that polymelamine sulphonates also had very effective dispersing properties (Meyer 1979).

From a practical point of view, for the first time, the admixture industry was using synthetic molecules specially designed for the concrete industry instead of using industrial by-products. Thirty years later, it cannot be said that the concrete industry is using superplasticizers to their full potential (Hewlett and Rixom 1977; Rosenberg and Gaidis 1989; Aïtcin et al. 1994). Progress must be made by the industry to apply what is already known in the field of superplasticizers, and while there are still problems that have not been totally solved, in many cases practical solutions to overcome these problems are known.

What has been encouraging in the recent history of superplasticizers is to see that purely technological preoccupations have resulted in the creation of a science of admixtures. In order to increase the efficiency of certain polymers in maintaining their effect on the properties of fresh concrete for 1.5 hours, without increasing their retarding effect, and the excessive entrainment of air, it has been necessary to invest in fundamental research to understand the physico-chemical action of superplasticizers on anhydrous and hydrated cement particles.

This research effort has resulted not only in interesting technological developments in the domain of concrete but also in the domain of the manufacturing processes of superplasticizers. Presently, superplasticizer manufacturers have better knowledge of the kind of polymers they have to produce and the kind of configuration that is favourable for an improved dispersing effect. For example, the molecular configuration of a polynaphthalene sulphonate (PNS) used to disperse cement particles is different from the architecture of a PNS used to disperse calcium hemihydrate during the fabrication of gypsum boards or from the PNS used to disperse the pitch found in pulp during the manufacturing of paper; or from the PNS used in the leather industry.

7.3.3.1 Polynaphthalenes (PNS)

The first step of the manufacturing process of a PNS consists in the sulphonation of one of the benzenic rings of naphthalene. This sulphonation is followed by a reaction of condensation using a formaldehyde molecule. Lastly, in the final step, the polysulphonic acid produced is neutralized, with or without a filtration step, depending on the base that is used during this neutralization (Ramachandran et al. 1998).

PNS is a brown liquid having a total solid content of about 40 to 41 per cent. PNS can be dried in spray driers. In dry form, PNS appears as a pale brown powder.

The manufacturing of a good PNS necessitates not only the use of good raw materials, as pure as possible, but also the full control of the process, particularly its thermodynamic parameters (Ramachandran et al. 1998).

The principal parameters that influence the efficency of a PNS are:

- the ratio between the number of sulphonate groups that are in a position β (2 o'clock) or α (12 o'clock). The greater the number of β sites the better (Figure 7.8);
- the ratio of the number of sulphonate groups per naphthalene rings; the closer this number to one the better (Figure 7.9);
- the degree of polymerization, which should be close to 9 or 10 in order to avoid too much reticulation and branching (Figure 7.10);
- the amount of active solids in the commercial solution which can be different from the total amount of solids.

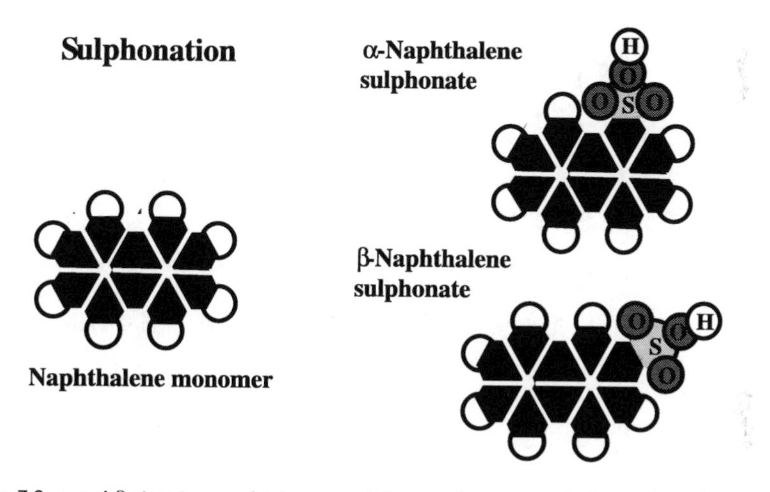

Figure 7.8 α and β sites in a naphtalenene sulphonate (courtesy of Martin Piotte).

Condensation

Formaldehyde

Naphthalene sulphonate

Naphthalene sulphonate

Water

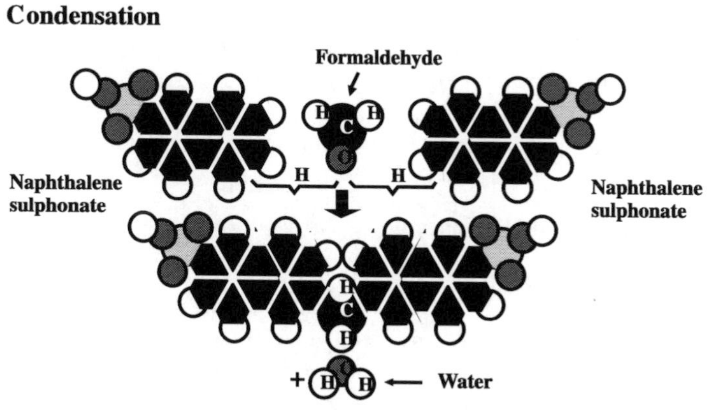

Figure 7.9 Polymerization of a polynaphtalene sulphonate (courtesy of Martin Piotte).

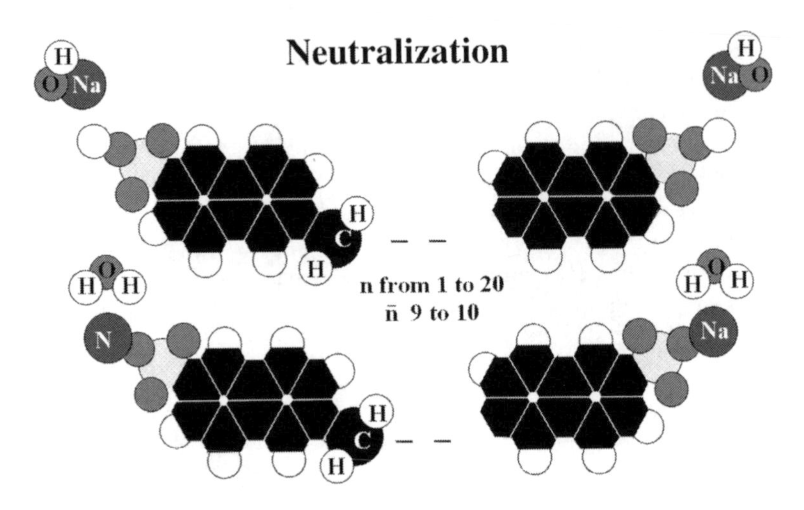

Figure 7.10 Neutralization of a polynaphtalene sulphonate with Na(OH) (courtesy of Martin Piotte).

Unfortunately, very few commercial technical data sheets include these parameters because commercial polynaphthalene superplasticizers are not pure products but 'cocktails' of different chemicals. Most of the time, technical data sheets indicate that the superplasticizer is a brown liquid containing 40 to 42 per cent solids, that it has a pH between 7.5 and 8.5 and a viscosity of 60 to 80 centipoises. Such a vague description can hide more fundamental differences between two commercial superplasticizers (Garvey and Tadros 1972; Childowski 1990; Ramachandran et al. 1998).

As the physico-chemical tests necessary to evaluate the fundamental parameters of a PNS are quite complex and long to perform, and, because some of them necessitate the use of expensive equipment for chemical analysis found only in university laboratories, the most economical way to evaluate the efficiency of a superplasticizer is to undertake rheological tests. Preferably these tests should be done with the cement that will be used with this superplasticizer (Piotte 1983; Piotte et al. 1995; Nkinamubanzi et al. 2000; Kim 2000).

7.3.3.2 Hydration in the presence of polysulphonate

Currently there is no accepted theory that explains in detail the action of a polysulphonate on a cement particle during the mixing of concrete and its hydration (Petrie 1976; Paillère and Briquet 1980; Buil et al. 1986; Uchikawa 1994; Ramachandran et al. 1998; Rixom and Mailvaganam 1999). However, the interaction cement/polysulphonate is better understood (Kim et al. 2000b; Nkinamubanzi et al. 2000).

When superplasticizers started being used, several researchers believed that the interaction was only physical. The action of superplasticizers on inert powder was studied (Foissy and Pierre 1990). Still other researchers preferred a chemical approach and studied the dissolution rate of the different ionic species in the presence of a superplasticizer (Andersen 1986; Odler and Abdul-Maula 1987; McCarter and Garvin 1989; Paulini 1990). Others studied the action of superplasticizer molecules on pure phases, hoping that the hydration of Portland cement could be deduced by summing up its individual action on each phase (Massazza and Costa 1980). None of these approaches worked very well, because of the strong interaction of sulphates when polysulphonates are used. It is recognized that superplasticizers interfere not only with the hydration of the cement phases but also with the dissolution rate of sulphates (Vernet and Cadoret 1992), as seen in Figure 7.11.

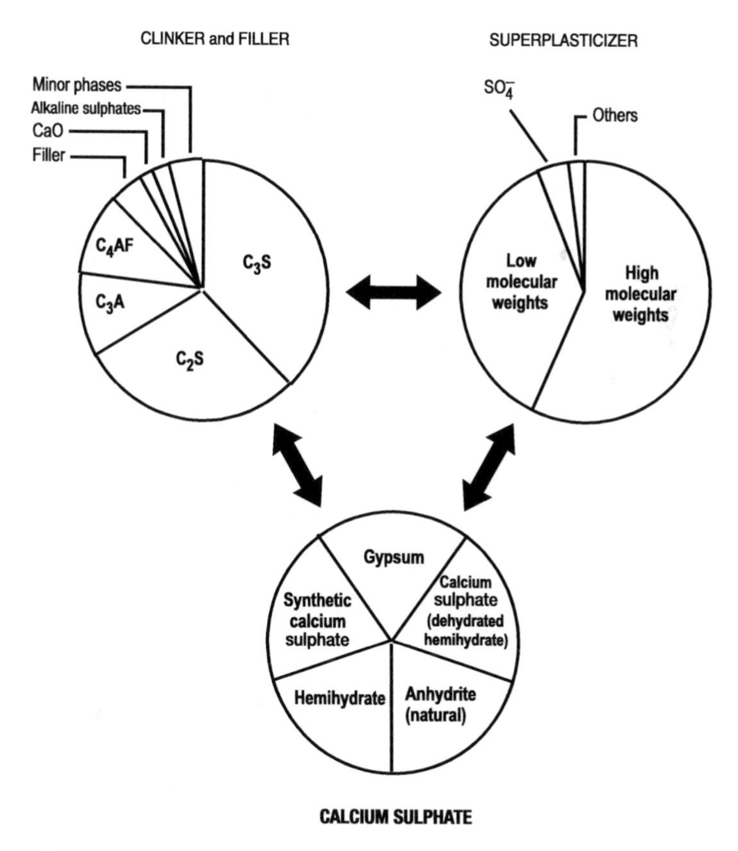

Figure 7.11 The complexity of the interaction of cement, superplasticizer and calcium sulphate (after C. Jolicoeur and P.-C. Aïtcin).

In Figure 7.12 it can be seen that according to the value of the ratio $\frac{SO_4^{2-}}{AlO_2^-}$ it is possible to face a 'normal' slump loss or a situation of false set or flash set. In a recent paper on the dissolution rate of calcium sulphate in the presence or in the absence of a superplasticizer, Vernet (1995) has shown that polysulphonate-based superplasticizers interact strongly with the hydration of calcium hemihydrate by favouring in some cases the precipitation of gypsum, which is the reason that polynaphthalene superplasticizers are used in the manufacturing of gypsum boards.

Due to these different approaches, when studying the superplasticizer–Portland cement interaction, we better understand why some superplasticizers are more efficient than others and why some cements are not compatible with certain superplasticizers (Buil et al. 1986; Hewlett and Young 1987; Andersen et al. 1988; Cunningham et al. 1989; Uchikawa et al. 1997; Fernon 1994a and 1994b; Jolicoeur et al. 1994; Vernet 1995; Huyhn 1996; Ramachandran et al. 1998; Jiang et al. 1999; Nkinamubanzi et al. 2000; Kim 2000).

For a given Portland cement, it is well established that the dosage of superplasticizer necessary to obtain a paste of a given fluidity increases with the specific surface area of the cement. It is known that superplasticizer molecules react preferentially with the interstitial phase, but that they can also be adsorbed by the C_3S. The adsorption of the superplasticizer by the C_3S has been directly observed on clinker-polished surfaces covered with drops of a polynaphthalene superplasticizer, where the sulphur of the sulphuric acid used during sulphonation was marked with radioactive sulphur (Onofrei and Gray 1989). Indirect observations have led to the same conclusion: as the superplasticizer dosage increases, it is observed in calorimetric studies that the development of the hydration reaction is delayed (Aïtcin et al. 1987). This retardation has also been observed by the MNR (Magnetic Nuclear Resonance) of the proton, and also in the field, when an abnormally excessive dosage of superplasticizers has been accidentally used. In one case the author was confronted with a retardation of greater than one day (Aïtcin et al. 1984).

In their experiment, Onofrei and Gray (1989) have clearly established that a part of the superplasticizer was combined in the hydrated interstitial phase. Luke and Aïtcin (1991) have also observed a modification of the rate at which ettringite was formed. This modification of the rate of the formation of ettringite was also accompanied by a change in its morphology. Hanna et al. (1989) and Khalifé (1991) have correlated the changes of the rheology of cement pastes with their C_3A and C_4AF contents. Carin and Halle (1991) and Vernet and Noworyta (1992) have found similar results. Aïtcin et al. (1991) have also reported that the use of a particular Portland cement containing less than 10 per cent of interstitial phase (3.6 per cent of C_3A and 6.9 per cent of C_4AF) was very economical in terms of superplasticizer when making high-performance concrete (HPC) having a low water/binder ratio. Using such a cement, it has been possible to make a HPC having a water/binder ratio as low

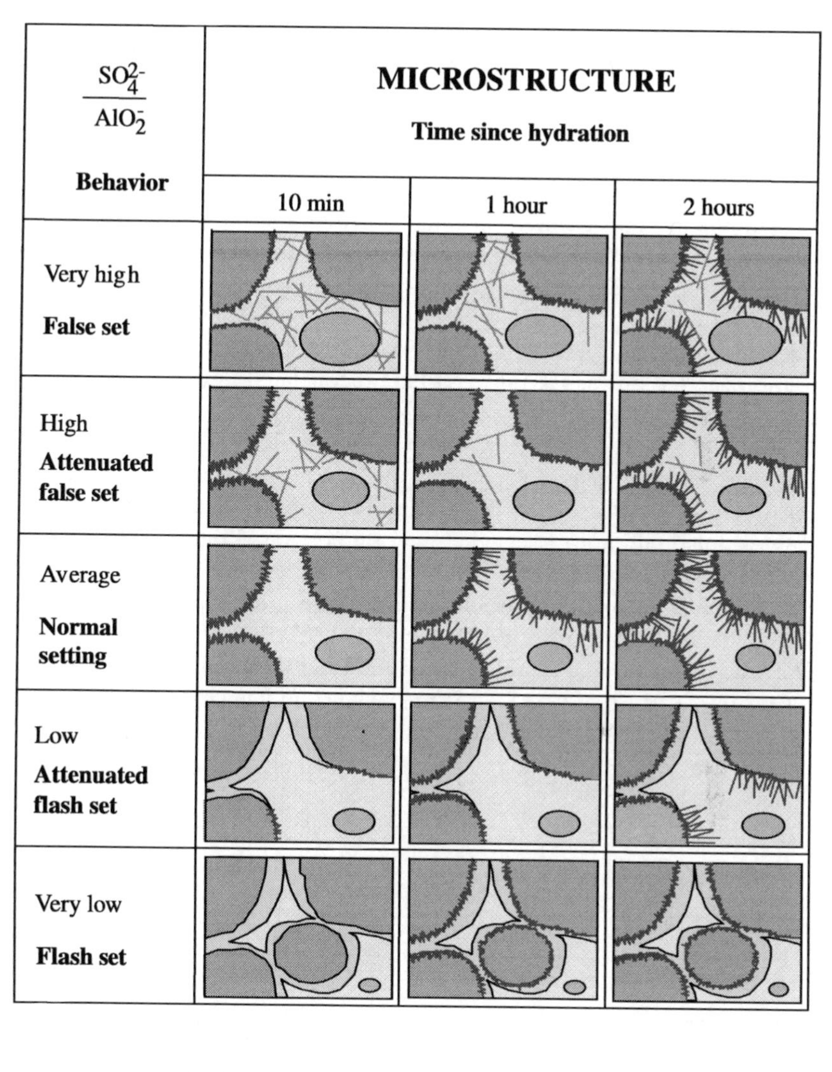

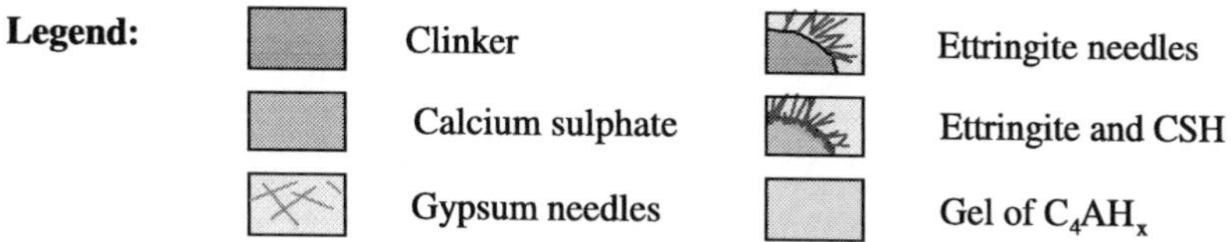

Figure 7.12 Influence of the $\frac{SO^{2-}{}_4}{AlO^-{}_2}$ concentration on the setting of cement (from C. Vernet).

as 0.17 and a slump of 230 mm one hour after mixing using only 100 litres of water and 28.6 litres of polynaphthalene sulphonate. The hardening of that concrete was not delayed: its twenty-four hour compressive strength was equal to 72 MPa.

7.3.3.3 The crucial role of calcium sulphate

As seen in the previous chapter, calcium sulphate is added to clinker to control its setting. Calcium sulphate can be considered as a retarder of clinker. In the presence of calcium sulphate, C_3A is transformed into ettringite ($3\ CaO \cdot Al_2O_3 \cdot 3CaSO_4 \cdot 32\ H_2O$). The first layer of ettringite covering the C_3A present on the surface of the cement particles is impervious and, therefore, blocks further hydration, so that during the dormant period concrete can be transported and placed without losing too much of its workability.

However, the action of calcium sulphate is not always as simple, because, as we have already seen in Chapter 5, there are different types of calcium sulphate currently used by the cement industry, and these different forms of calcium sulphate do not present the same rate of solubilization. Moreover, in Portland cement there are other forms of sulphates than calcium sulphate, for example, alkali sulphates which dissolve much more rapidly than calcium sulphate (Nawa and Eguchi 1987). Modern clinkers quite often are made with fuels rich in sulphur so they can contain significant amounts of arcanite (K_2SO_4), apithalite ($Na_2\ SO_4 \cdot 3\ K_2\ SO_4$) and calcium langbeinite ($2\ CaSO_4 \cdot K_2\ SO_4$). These alkali sulphates are deposited on the interstitial phase, but some of them can also be found at the surface of alite and belite crystals (Figure 5.30).

A few years ago, cement companies almost exclusively used gypsum as their source of calcium sulphate, but, as has been seen in Chapter 5, for economical reasons, modern Portland cements can contain either:

- Natural anhydrite ($Ca\ SO_4$);
- Gypsum ($Ca\ SO_4 \cdot 2\ H_2O$);
- Hemihydrate ($Ca\ SO_4 \cdot ½\ H_2O$);
- Totally dehydrated gypsum ($Ca\ SO_4$);
- Synthetic calcium sulphate ($Ca\ SO_4$).

The solubilization rates of each of these forms of calcium sulphate are not the same and can be modified in the presence of a polysulphonate, so that at a given time, the rate at which SO_4^{2-} ions are liberated and the C_3A rate of solubilization are no longer balanced (Luke and Aïtcin 1991). In certain cases, conditions favouring a flash set or a false set situation can occur, in spite of the fact that when that cement was tested at the cement plant in the absence of a superplasticizer it was correctly dosed in calcium sulphate. Therefore, in some circumstances, a cement and a polysulphonate that fulfil all the requirements of their acceptance standards are not compatible (Dodson and Hayden 1989). This kind of problem will occur more and more frequently in the future when making low water/cement mixtures if special attention is not paid to the solubilization rate of the calcium sulphate that is added in a cement (Tagnit-Hamou et al. 1992).

7.3.3.4 The compatibility between polysulphonate and cement

Following the recent research work done in Japan and at the Université de Sherbrooke (Nawa et al. 1991; Kim et al. 1999, 2000; Jiang et al. 1998, 1999, 2000; Nkinamubanzi et al. 1997, 2000) and by Pagé (Pagé et al. 1999), the compatibility between Portland cement and polysulphonate superplasticizers, or rather the robustness of their combination (Aïtcin et al. 2002), is better understood. Robustness is a general notion in engineering, concerning the influence of a variation of a parameter, ΔV, on a property P (Figure 7.13). All industrial processes aim for a robust process, in which a small variation ΔV of V does not result in the stopping of the production line because ΔV entrains an unacceptable variation ΔP of the property P, or too great a change in the quality of the final product such that the product is no within expected quality standards. A cement superplasticizer dosage is not robust when a variation of 0.1 per cent in its dosage has catastrophic effects on the rheology of a cement paste or of a concrete.

Recent research has shown that the quality of the superplasticizer is as crucial as the solubility rate of sulphates in low water/binder ratio made with a polynaphthalene superplasticizer. Presently, all polynaphthalenes and polymelamines superplasticizers on the market do not have the same efficiency and effective solids content, and, therefore, it is often necessary to test several commercial superplasticizers with a given cement in order to find the most compatible and robust combination.

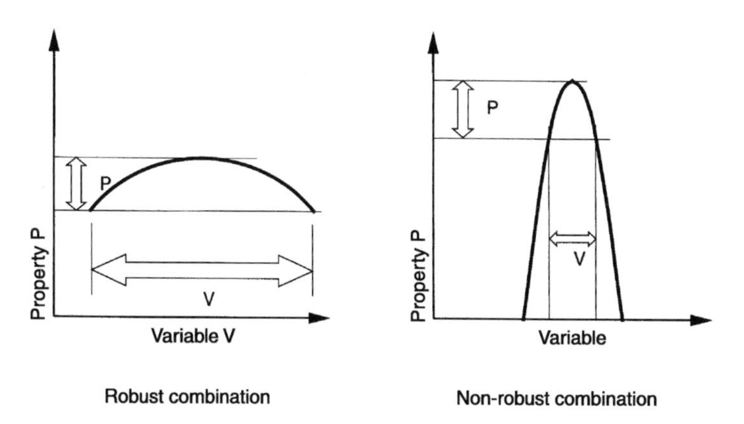

Figure 7.13 Notion of robustness.

Significant progress has been made in establishing the more important parameters of a Portland cement that influence its compatibility with pure polynaphthalene sulphonates. In the case of conventional Portland cement that contains between 6 and 10 per cent C_3A, it seems that an optimal amount of 0.4 to 0.6 per cent of soluble alkalis, expressed as Na_2O equivalent, is necessary to have a compatible and robust cement/superplasticizer combination as suggested by Fernon (1994a and 1994b), Nkinamubanzi et al. (2000) and Kim et al. (2000).

As soon as Portland cement enters into contact with water, if there are not enough SO_4^{2-} ions going into solution as compared to the amount of reacting C_3A, the sulphonate termination of the superplasticizer reacts with the active sites of the interstitial phase (C_3A and C_4AF) (Fernon 1994a and 1994b; Fernon et al. 1997; Pollet et al. 1997; Flatt 1999), instead of the SO_4^{2-} ions that should have been released by the calcium sulphate. In a high water/cement ratio concrete containing little cement and a lot of water, Portland cement calcium sulphate (and alkali sulphates) provides enough SO_4^{2-} ions. On the contrary, in low water/cement concretes when there are not enough SO_4^{2-} ions in solution, some superplasticizer molecules are trapped within the $C_4A\ H_x$ lamellae of the hexagonal hydrates and are no longer available to deflocculate cement particles. The lower the water/binder ratio, the richer the mix in C_3A and the higher the deficit in SO_4^{2-} ions, the larger the number of superplasticizer molecules trapped in the $C_4A\ H_x$ lamellae. This lack of SO_4^{2-} ions provided by calcium sulphates can be compensated by SO_4^{2-} ions coming from the dissolution of alkali sulphates, because alkali sulphates are not only more soluble than calcium sulphate, but they are also more rapidly soluble.

Studies done at the Université de Sherbrooke have shown that the Portland cements that present the poorest compatibility with polysulphonates are white cements and low alkali cements. However these studies have shown that

- there are some low-alkali cements that present an adequate compatibility or robustness with polysulphonates because their alkali sulphates go very rapidly into solution;
- the addition of some sodium sulphate can correct the situation as well as a delayed addition of the superplasticizer, or the addition of a small amount of retarder.

It can be useful that cement producers market cements with a low-alkali content in order to avoid alkali/aggregate problems, but it is not desirable to have cements with a too-low-alkali content, because these cement will be incompatible with polysulphonates so that they could only be used in high water/binder concretes which are not particularly durable. The research done at the Université de Sherbrooke has shown that soluble alkali sulphates play a significant role in the robustness of cement/superplasticizer combinations, which means that any deviation of superplasticizer dosage within a 0.2 to 0.3

per cent dosage range does not produce either a very rapid slump loss or an unacceptable bleeding and segregation.

The SO_3 content found in mill certificates of Portland cement is not sufficient for prediction of the compatibility of a Portland cement with a polysulphonate, because some of the SO_3 can also be trapped in the silicate phase, and not all the alkalis are combined with SO_3, as some are trapped in the C_3A. A simple method has been developed to find the total amount of rapidly soluble alkalis, which are the more important ones from a compatibility or robustness point of view (ATILH 1997).

The first tests conducted on the newly available polycarboxylates and polyacrylates have demonstrated the efficiency of this new family of superplasticizers. Their dosage at the saturation point can be two to three times smaller than the corresponding saturation point of a polysulphonate, but in certain cases some of the present formulations still lack robustness with some cements.

As some polycarboxylates have a strong natural tendency to entrain air, a detraining agent has to be added, and this can cause problems when the concrete must be air-entrained to make it freeze/thaw resistant.

Moreover, it seems that incompatibility problems have already been found in Japan (Yamada et al. 2001). **In fact it seems that the more compatible a cement is with a polynaphthalene the more incompatible it is with a polycarboxylate and vice versa.** Alkali sulphates seem to play a key role in the efficiency of polycarboxylates, and recent work tends to show that the optimum sulphate content for polycarboxylates is much lower than for polynaphthalenes (Saric-Coric 2002).

It is hoped that in the very near future our knowledge of the mode of action of polycarboxylates will improve and that the number of polymers presently tested will decrease. Consequently, until then, the testing of the efficiency of Portland cement/superplasticizer combinations cannot be avoided.

7.3.3.5 The limits of the testing of the rheology of cement paste

As it is impossible to determine theoretically the extent to which a particular cement is compatible, or robust, with a particular superplasticizer, and as the direct testing of this compatibility or robustness on concrete batches consumes a lot of time, materials and energy, the testing of the rheology of cement pastes will remain a fruitful first approach that will limit the hard work that will inevitably have to be done on concrete. These tests are based on the study of the rheology of cement grouts (Kantro 1980; de Larrard et al. 1996). When these methods are applied appropriately, they can be used to rapidly discriminate between efficient and non-efficient combinations. Experimental planning can also be done to improve the efficiency of this method (Rougeron and Aïtcin 1994; Fonollosa 2002).

However, it has been found that the testing of the rheology of a cement paste does not have a 100 per cent predictive value when dealing with concrete. In some cases, a much better rheology has been reported in a particular concrete than in a grout of the same W/C ratio, and in other cases it has been the reverse. But, it is certain that if a cement/superplasticizer combination does not work in a grout it will not work in a concrete.

In spite of the merit of these simplified methods, it is always necessary to end up making concrete batches in order to confirm the results obtained in grouts. It is always prudent to follow the slump retention of the concrete for 1 to 1.5 hours. In this case, the use of factorial experimental planning is also very helpful, because with a limited number of concrete batches, it is possible to obtain a maximum number of conclusions about the particular efficiency of the cement/superplasticizer combination.

The essential difference between a grout and a concrete is that a concrete also contains sand particles and coarse aggregates. It is possible to fluidify a cement paste much more in a grout than in a mortar because in the grout the liquid phase has to sustain in suspension only very small cement particles and not sand particles. Of course, gravity always leads to the sedimentation of cement particles, but the interactive forces created by the action of the superplasticizer molecules on the cement grains do not have to oppose very large gravity forces.

On the contrary, in the case of a concrete, the grout has to sustain cement particles in suspension within the mortar phase, and the mortar phase has to sustain the coarse aggregates in suspension. This is why, when making concrete, there is always an optimum superplasticizer dosage that has to be considered, because without any special precautions (the addition of a colloidal agent) a strong segregation can occur as soon as a 200 to 230 mm slump is obtained. This explain why it is very often necessary to add a colloidal agent when making self-compacting concrete. The use of a colloidal agent increases the robustness of the self-compacting concrete formulations.

Consequently, it is always risky to try to directly translate the results obtained on the rheology of a grout to the rheology of a concrete, because often the necessity of making a robust concrete from a rheological point of view limits somewhat the amount of superplasticizer that can be used.

Two methods are currently used to follow the rheological effect of a superplasticizer on a cement: the method known as the mini-slump method (Figure 7.14) and the Marsh cone method (Figure 7.15). The advantage of the mini-slump method is that it requires very little material (Figure 7.16), but the paste is tested in a 'quasi-static' mode (Figure 7.17). When using the Marsh cone method, it is necessary to use a larger amount of material (Figure 7.18), but the grout rheology is tested in 'dynamic' conditions (Figure 7.19). The use of one method rather than the other is a matter of personal choice. The simultaneous use of both methods is interesting because the grout is tested in different rheological conditions.

These two methods are currently used to determine the superplasticizer dosage at the *saturation point*, that is the superplasticizer dosage beyond which there is not any improvement of the rheology (Figure 7.20).

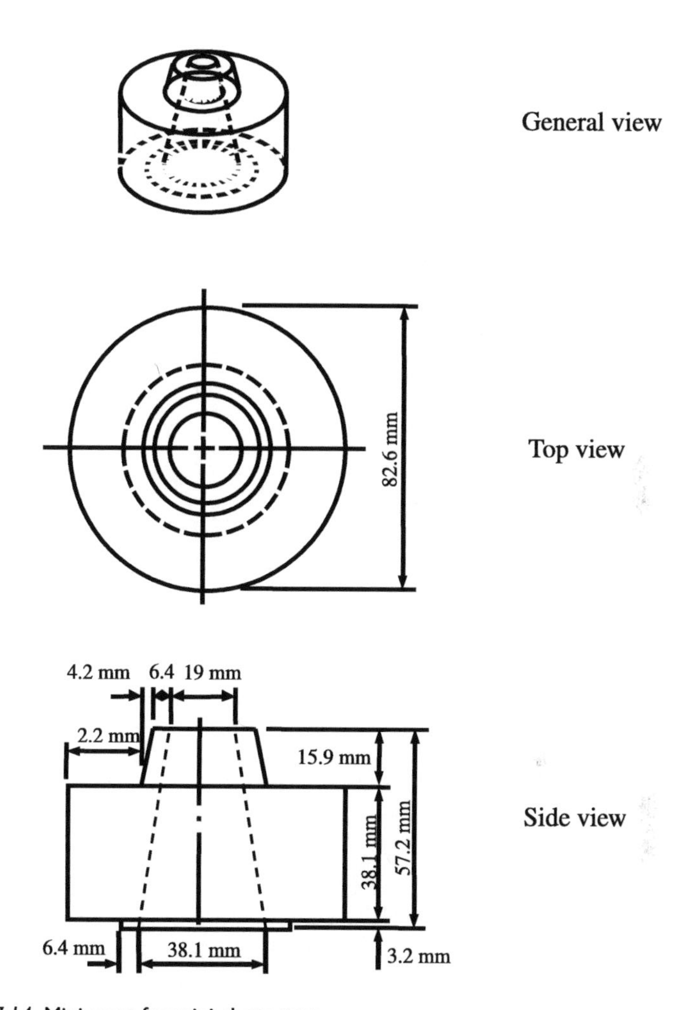

Figure 7.14 Mini-cone for mini-slump test.

These rheological tests show that presently it is impossible to manufacture polysulphonate or polycarboxylate superplasticizers that can be efficient with all Portland cements, of any water/binder ratio, because Portland cements present too broad a range of 'reactivity' and because the calcium sulphate used to control their setting presents too broad a range of solubility rate (Cunningham et al. 1989; Tagnit-Hamou et al. 1992; Tagnit-Hamou and Aïtcin 1993; Ramachandran et al. 1998; Jiang et al. 1999) (Figure 7.21).

But when a cement producer is careful of the nature of the calcium sulphate he is using, and especially of its solubility rate, he can produce Portland cements that are compatible with good polymelamine, polynaphthalene and polycarboxylate superplasticizers. When the sulphate solubility rate is not

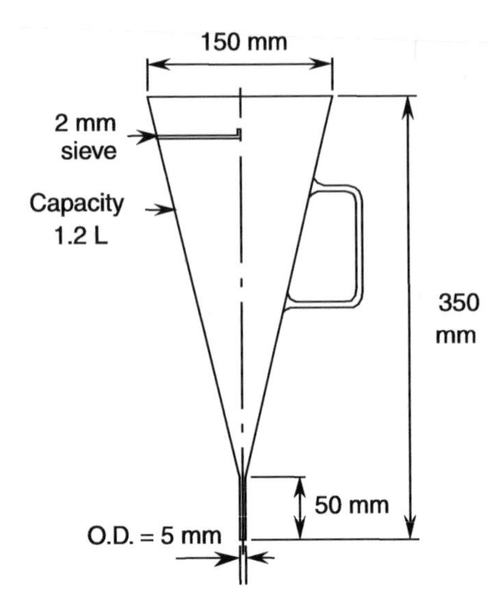

Figure 7.15 Example of Marsh cone.

well balanced with the C_3A reactivity, it is necessary to develop a special 'cocktail' in order to control concrete rheology, especially for low water/binder applications.

7.3.3.6 Some typical cases

Figure 7.22 presents the case of a particularly efficient cement/superplasticizer combination. The flow time at 5 minutes and 60 minutes for a 0.35 water/binder ratio grout is around 60 seconds (for 700 ml) for a large range of superplasticizer dosages starting at a dose of 0.6 per cent. At 60 minutes, the flow time is only 5 seconds longer for these dosages, which is remarkable. Figure 7.22 presents the influence of the cement fineness on the flow time. Of course, the greater the fineness, the greater the flow-time (Nawa et al. 1991). The chemical composition and the Bogue composition of this type 20 M Canadian cement used by Hydro-Quebec to build its dam, corresponds to an ASTM type II cement having a very low heat of hydration (Table 7.1).

INFLUENCE OF THE CEMENT TYPE

Figure 7.23 presents the flow-time of two cements produced in the same cement plant. One of these cements is a Type 10 Canadian cement (a US Type I cement that can contain up to 5 per cent limestone filler) and the other is a modified Type II cement (the previous Type 20 M) whose composition is given in Table 9.1. It is seen that the superplasticizer dosage at the saturation point of this

Figure 7.16 Mini-slump testing (courtesy of B. Samet).

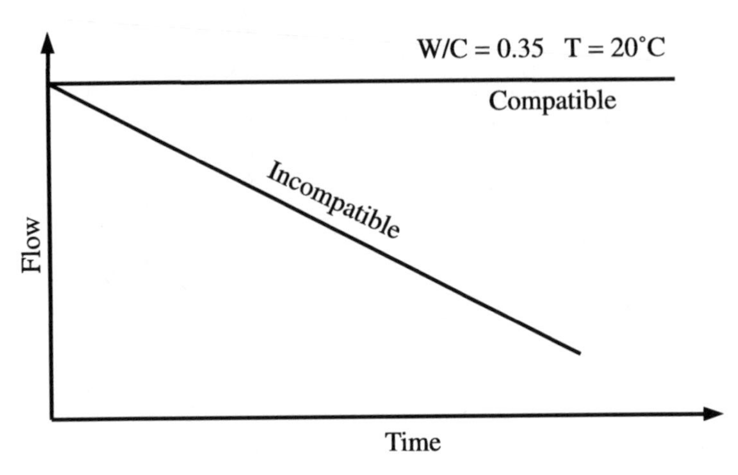

Figure 7.17 Mini-slump typical behaviour.

modified Type II cement, which has a very low C_3A and C_3S content, is much lower than that of the Type 10 cement; moreover, the flow time at 60 minutes is close to that at 5 minutes, which indicates a very good balance between the solubility rate of the sulphates and the reactivity of the C_3A.

A COMBINATION THAT PRESENTS A SIGNIFICANT SLUMP LOSS DURING THE FIRST HOUR

A Type I Portland cement has been tested with a good quality superplasticizer, and the slump loss is presented in Figure 7.24; the test was done on 0.35 water/binder concrete with a 0.8 per cent superplasticizer dosage. In this figure, it is seen that up until 45 minutes the slump loss is not very significant but that between 45 and 60 minutes the slump drops significantly, a situation that can create serious problems in the field.

This kind of behaviour can be improved by substituting a certain fraction of this cement by a mineral component or by adding a small amount of retarder or some sodium sulphate in the mix. The improvement of the rheology of concrete by the addition of a retarder is presented in Figure 7.25. In this particular case, a small addition of a retarder significantly improved the rheology of a concrete made with a blended cement containing silica fume when a 0.35 water/binder ratio was used. At 5 minutes, the presence of the retarder had no effect on the initial slump and kept the slump almost constant until 60 minutes, but after 60 minutes the slump started to decrease at the same rate as when the retarder was not used.

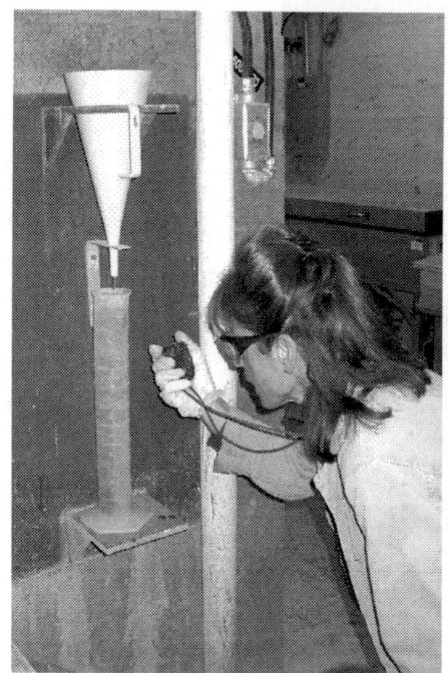

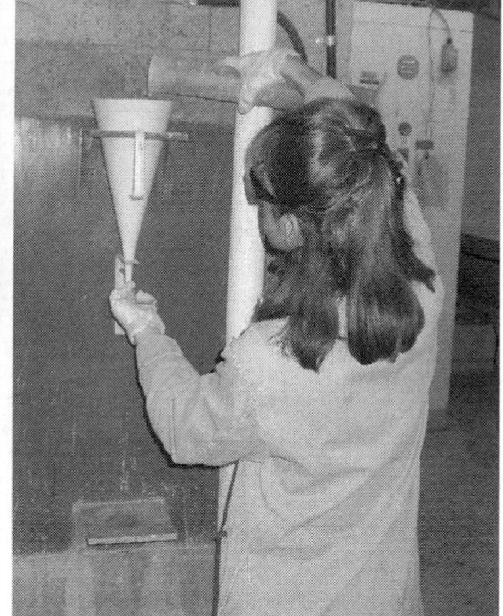

Figure 7.18 Marsh cone test.

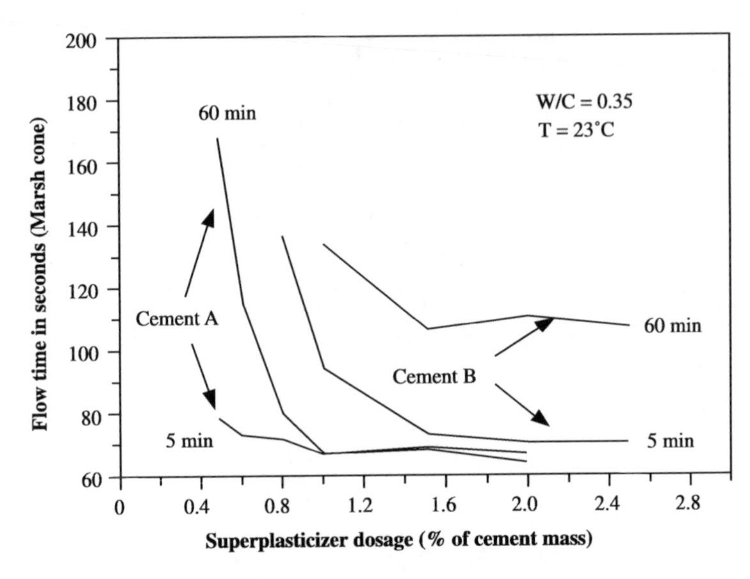

Figure 7.19 Examples of flow-time curves at 5 and 60 minutes for a compatible combination.

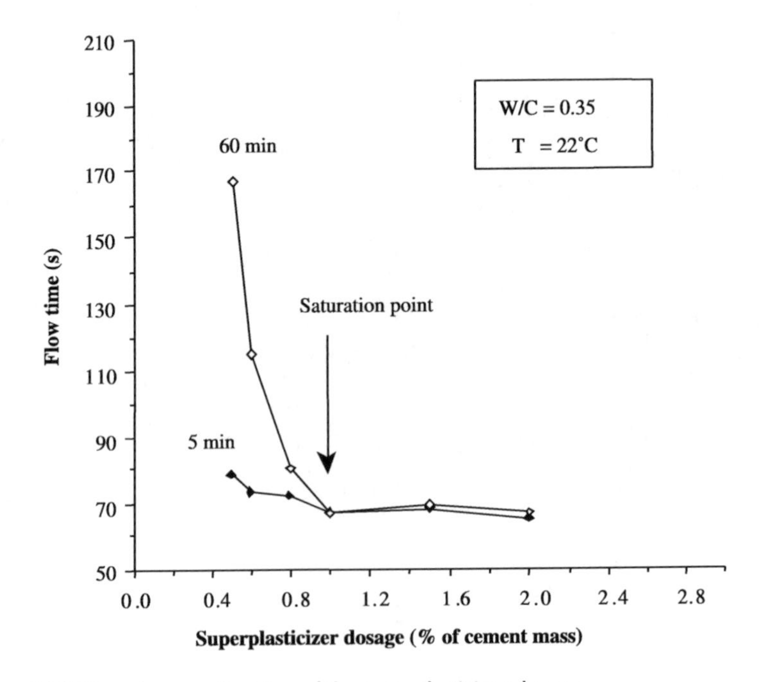

Figure 7.20 Flow time as a function of the superplasticizer dosage.

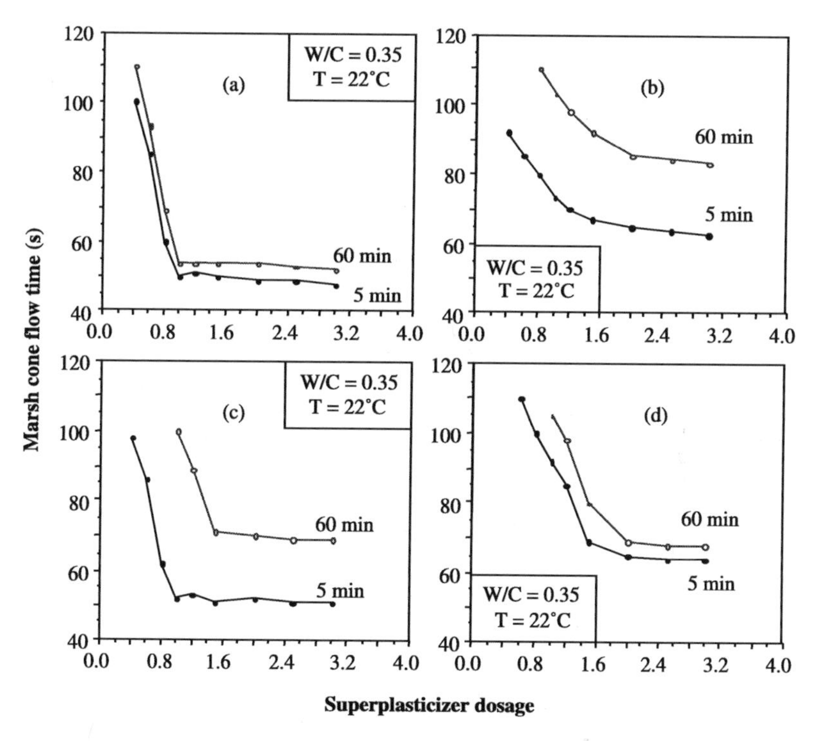

Figure 7.21 Different types of rheological behaviour.

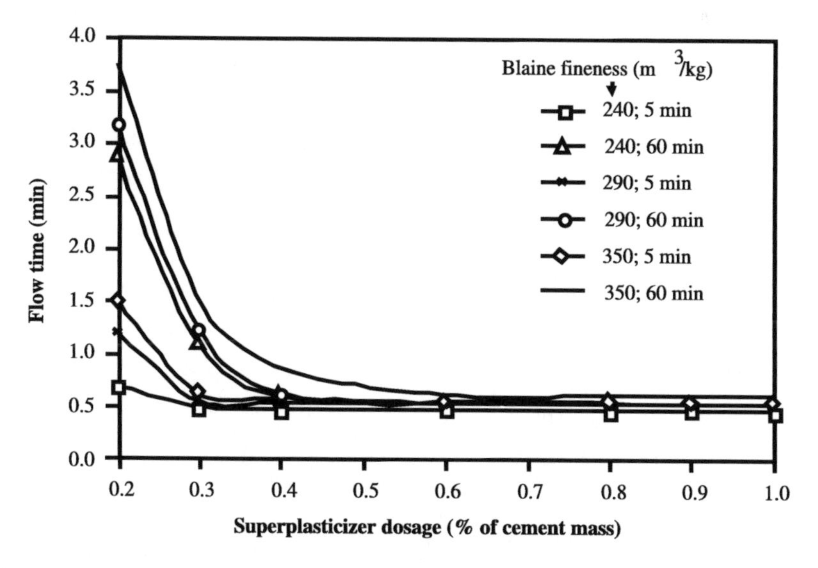

Figure 7.22 Effect of the fineness on the rheological behaviour.

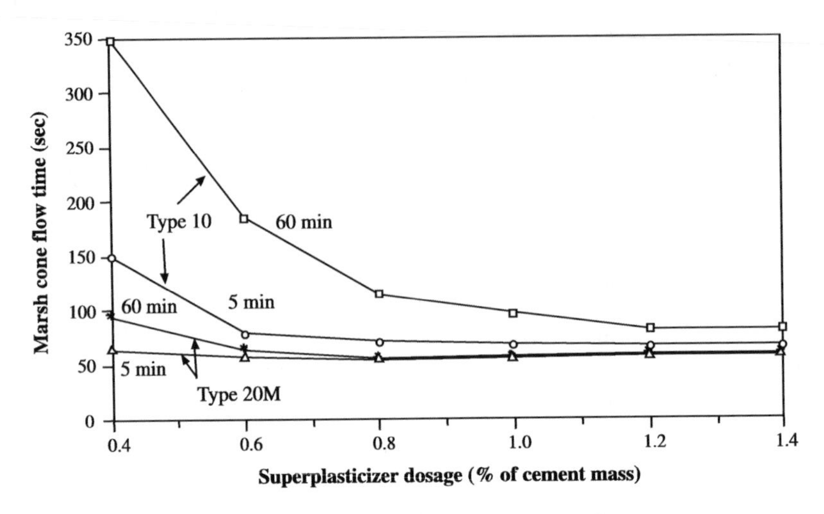

Figure 7.23 Comparison of flow time as a function of the type of cement.

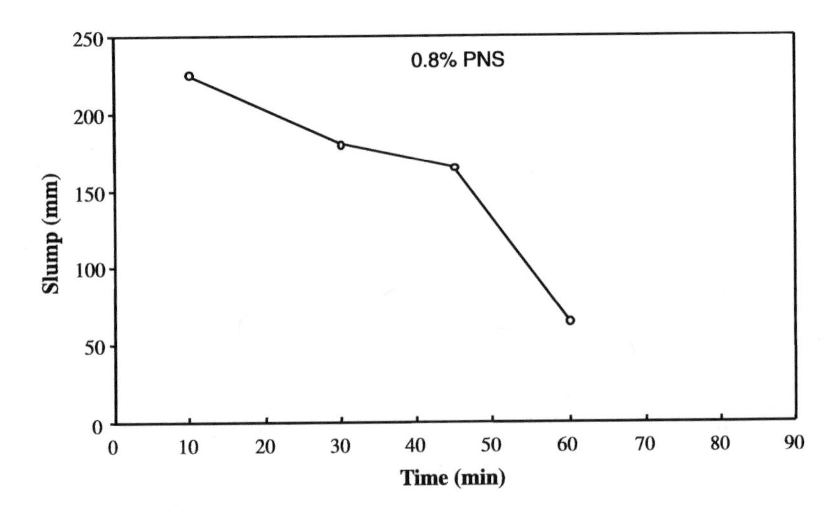

Figure 7.24 Example of slump loss.

7.3.3.7 Selection of a superplasticizer

When making high-performance concrete (HPC), the selection of the superplasticizer is a crucial step, just as crucial as the selection of the cement, because not all commercial superplasticizers have the same efficiency and react in the same way with all cements (Aïtcin et al. 1994; Uchikawa et al. 1997; Ramachandran et al. 1998). The initial slump increase that follows the introduction of the superplasticizer is not always maintained long enough (Daimon

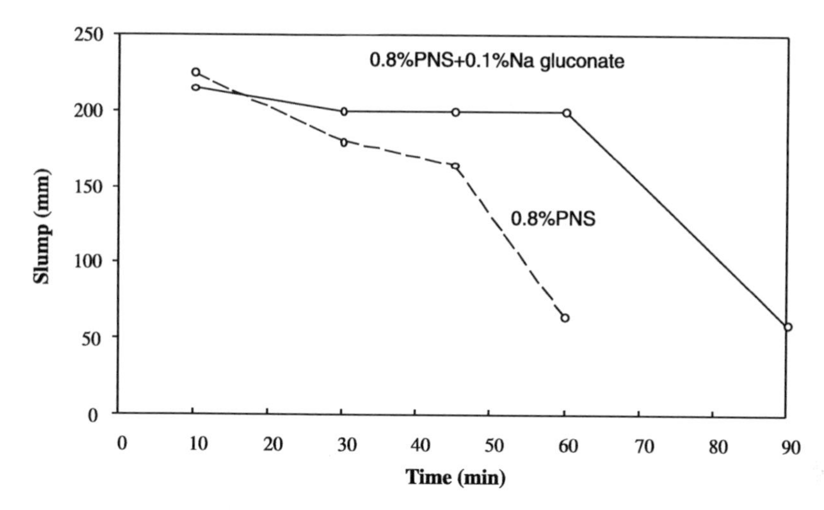

Figure 7.25 Improving slump retention with a retarder.

and Roy 1978; Piotte 1983; Singh and Singh 1989). This is due to the fact that the acceptance criteria of superplasticizers and cements have been developed separately, at a time when superplasticizers were essentially used to improve the rheology of concrete having water/binder ratios in the range of 0.50. Presently, superplasticizers are commonly used to reduce the amount of mixing water in flowable concretes having a low water/binder ratio.

(a) POLYMELAMINE SULPHONATES

Until recently, the production of polymelamine sulphonates was covered by a patent so that there was only one quality product on the market. But now polymelamine sulphonates are manufactured by several producers (Lahalih et al. 1988) and their quality can be more variable than previously, because not all manufacturers control the production parameters of polymelamine to the same degree.

Melamine superplasticizers are sold as transparent liquid solutions with a solids content of 22 per cent, but certain formulations are proposed with 30 or 40 per cent solids. Melamine superplasticizers are sold as sodium salt in a liquid or solid form.

When melamine superplasticizer users are asked why they are using a melamine superplasticizer their answer varies:

- Melamine superplasticizers do not retard cement hydration as much as naphthalene superplasticizers.
- As the solids content is lower (22 per cent), any accidental overdosage does not create a critical situation.

- Melamine superplasticizers do not entrain as much air as naphthalene superplasticizers.
- It is easier to obtain a stable air-bubble system.
- Melamine superplasticizers do not give a pale beige colour to white concretes.
- Melamine superplasticizers produce concrete surfaces free of air bubbles in pre-cast panels (although in Scandinavia some precasters maintain the contrary).
- The quality of the service and the consistency of polymelamines is better than that of polynaphthalenes.
- Some users candidly admit that they started to use polymelamine superplasticizers, and, as they had good results, they have not found any reason to change.

(b) POLYNAPHTHALENES

Polynaphthalene sulphonate water reducers were patented in 1938 (Tucker 1938), but they were not used extensively in the industry before the 1960s. When they start being used in the 1960s, the production of naphthalene superplasticizers was no longer covered by a patent, so several sources of polynaphthalene sulphonates were available and not all the manufacturers mastered their manufacturing process to the same degree of efficiency.

Polynaphthalene sulphonates are sold either in liquid form having a solids content between 40 and 42 per cent, or in a powder form. Sodium salt is more often used than calcium salt. There are some specific applications for which the use of the calcium salt is recommended, for example, in certain applications of reinforced concrete or pre-stressed concrete, particularly in the nuclear industry, where any trace of chloride in the concrete is absolutely forbidden. As typical industrial soda is made from sodium chloride, the sodium salt of polynaphthalene sulphonates always contain some traces of chloride, but this is not the case with calcium salt, because industrial limes are free from chloride.

There is another case when some specifiers require the use of a calcium salt, and that is when they want to absolutely avoid any risk of alkali/aggregate reaction. In such a case, it is evident that the use of a calcium salt is safer than the use of a sodium salt in spite of the fact that the amount of sodium ions brought by the superplasticizer is very small.

In any other cases, the use of a sodium salt is recommended because it is less costly: there is no need for a long filtration process after the neutralization of the sulphonic acid.

Polysulphonate superplasticizers have been extensively used in North America and in Japan to produce HPCs. When polynaphthalene users are asked why they are using a polynaphthalene they give the following answers:

- Polynaphthalenes have a higher solids content so that their use is more economical.
- It is easier to control the rheology of HPC because of the slight setting retardation they induce.
- The cost of a polynaphthalene can be negotiated with different producers.
- The quality of the service and the consistency of the superplasticizer are excellent.
- They started to use polynaphthalene, they had good results and did not see the necessity of changing.

(c) LIGNOSULPHONATE-BASED SUPERPLASTICIZERS

The dosage of most commercial lignosulphonates cannot be increased without increasing the development of excessive air-entrainment or retardation. However, if the lignosulphonate produced by the pulp and paper mill does not contain too many surface active compounds, or if these surface active components can be extracted easily by agitating the lignosulphonate and extracting the foam produced, and if the lignosulphonate does not contain either too much sugar, or if these sugars can be extracted easily, then the lignosulphonate can be used at a dosage similar to that of polynaphthalenes and polymelamines. Such lignosulphonates are more expensive than the lignosulphonates used as water reducers, but they are competitive on a price basis with polynaphthalenes and polymelamines.

Some admixture companies also sell commercial superplasticizers which are a mixture of polynaphthalene and lignosulphonate. It is difficult to simply differentiate such a mixture from a pure naphthalene base because both have a brown colour and both have a solids content between 40 and 42 per cent. It is therefore necessary to do an infrared analysis (Figure 7.26).

When they are making a HPC, some concrete producers like to first introduce a lignosulphonate water reducer at the beginning of the mixing, and at the end of the mixing a polynaphthalene, in order to take advantage of the slightly retarding action of the lignosulphonate and of its cheaper price. Moreover the lignosulphonate that combines with C_3A decreases the consumption of polynaphthalene molecules later on.

(d) POLYCARBOXYLATES

This is the last family of superplasticizers that has appeared on the market after many years of research to discover the most efficient polymers (Kinoshita et al. 2000; Ohta et al. 2000). Figures 7.27 to 7.30 present some molecular configurations that have been found to be efficient. This new type of superplasticizer that works essentially by steric repulsion (Uchikawa 1994) is very efficient on a solids basis because the dosage necessary to reach the saturation point is two to three times smaller than the dosage of a polynaphthalene.

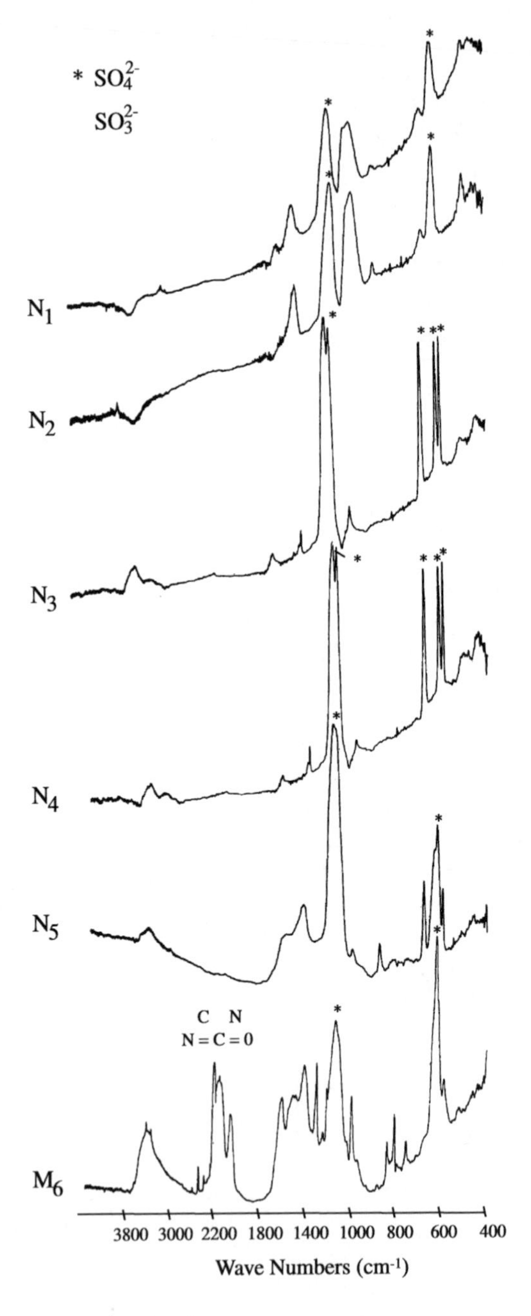

Figure 7.26 Infrared spectrographs.

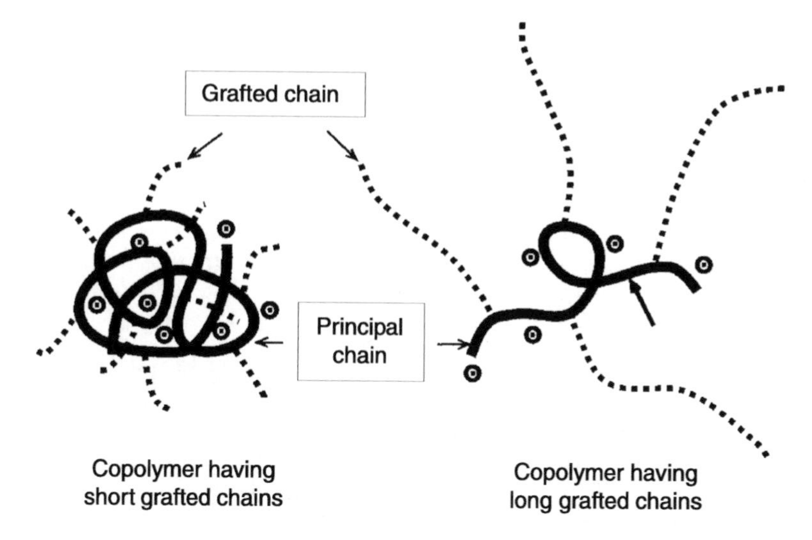

Figure 7.27 Influence of the length of the grafted chains on the morphology of a polymer (Ohta et al. 2000) (reprinted with permission of the American Concrete Institute).

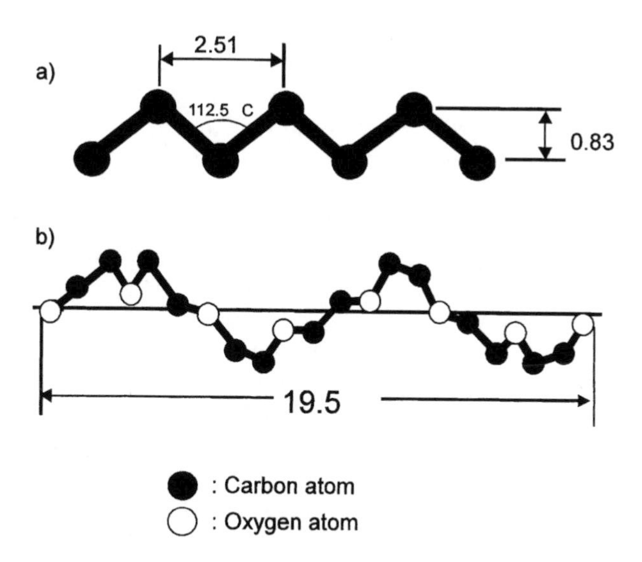

Figure 7.28 Examples of principal chains of polycarboxylates (Ohta et al. 2000) (reprinted with permission of the American Concrete Institute).

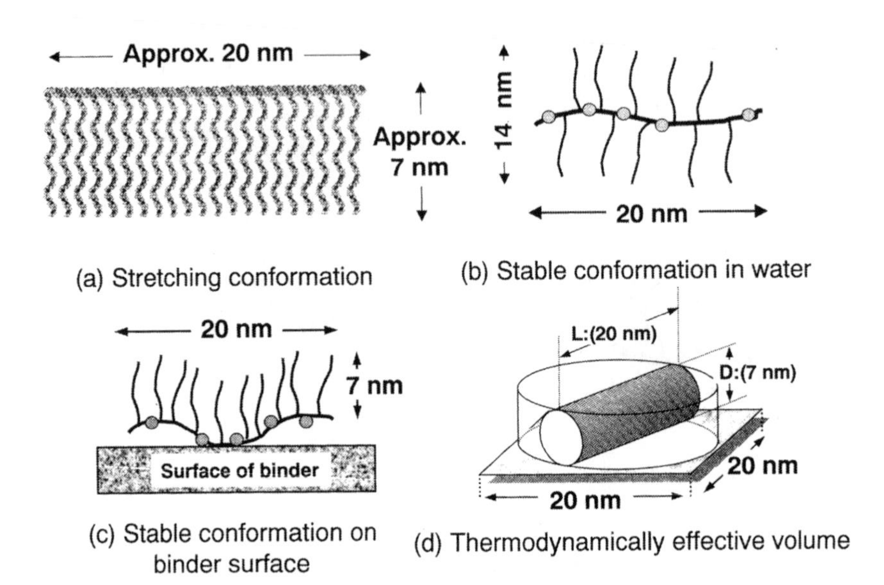

Figure 7.29 Effect of the configuration of polyacrylates (Ohta et al. 2000) (reprinted with permission of the American Concrete Institute).

However, the present formulation of polycarboxylates is not always very robust and on a performance/cost basis they are usually still more expensive than polynaphthalene. This lack of robustness can be explained as follows: the dispensing equipment found in modern concrete plants is quite precise, but what is not as precise is the consistency of the water content of the sand. For example, a variation of ± 0.5 per cent of the water content of sand results in a variation of 3 to 4 litres of mixing water, which is significant enough when using polycarboxylates.

Moreover, some polycarboxylates have a natural tendency to entrain air so that it is necessary to combine them with a costly air-detraining agent. The adjustment of the amount of detraining agent is critical when air entrained concretes have to be made.

Finally, it has been found that cases exist in which polycarboxylates are not compatible with some cements. In some cases it is the radical COO^- that reacts with the C_3A to form an organomineral compound that looks like a carboaluminate. The addition of calcium chloride has been proposed to eliminate this consumption of superplasticizer molecules. The rapidly soluble calcium chloride provides Cl^- ions that combine with C_3A to form chloraluminates. Usually an addition of sodium sulphate does not improve the rheology of grouts or concretes made with a polycarboxylate as it does with a polynaphthalene, because it leads to a desorption of the polycarboxylic molecules

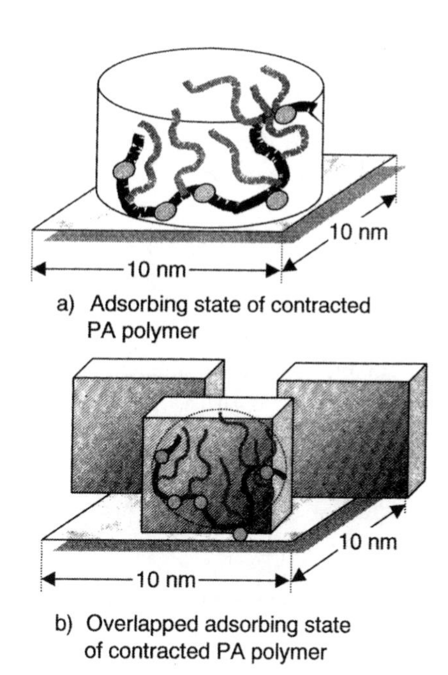

Figure 7.30 Adsorption of polyacrylates on the surface of cement grains (Ohta et al. 2000) (reprinted with permission of the American Concrete Institute).

from the surface of the cement particles. In the case of slag-blended cements, Saric-Coric and Aïtcin (2002) have found that the optimum dosage of sulphate was much lower in the case of polycarboxylates than in the case of polysulphonates.

In North America, polycarboxylates do not yet compete on performance/ price basis with polynaphthalenes, but, in the future when the knowledge of polycarboxylates equals that of polysulphonates, the situation could change. In Europe, superplasticizers are used essentially in the pre-cast industry, and precasters use very fine cements rich in C_3S and C_3A so that polycarboxylates are used more than polysulphonates (Nivesse 2002 pers. comm.).

(e) THE CONTROL OF THE QUALITY OF SUPERPLASTICIZERS

It is long, difficult and very costly to determine the particular architecture of a superplasticizer polymer (Piotte 1983). As stated previously, admixture companies provide a list of physico-chemical characteristics in their data sheets, but they are of little interest in determining the real efficiency of a particular superplasticizer. Moreover, commercial superplasticizers are quite often an elaborate 'cocktail' of several chemical compounds. Consequently, it is important to develop simple tests to check the consistency of a particular source of superplasticizer.

An infrared spectrum can be used to check the composition and the uniformity of a superplasticizer in the case of polysulphonates (Figure 7.26) (Khorami and Aïtcin 1989). An infrared spectrum can be considered as characteristic of any superplasticizer because all the details of the 'cocktail' or of the impurities present in the raw materials can be identified.

The determination of the average molecular weight of a polymer is not easy; neither is the determination of the number of α and β sulphonated sites in the case of polysulphonates. This is done by liquid chromatography for molecular mass determination and by magnetic resonance for the number or α and β sites. Therefore, a direct and qualitative measurement of these two very important characteristics of a superplasticizer is difficult, long and costly. Consequently, the simplest way to check the efficiency and consistency of a superplasticizer is to check its rheological efficiency on a reference cement.

7.3.3.8 Superplasticizer dosage

The following parameters have been found to control the rheology of a cement/superplasticizer combination:

- The water/binder ratio. It is better to check the efficiency of a superplasticizer at a water/binder ratio as close as possible to the one that will be used when making concrete (Hanna 1990; Khalifé 1991);
- The initial temperature of the water, which greatly influences the rate of hydration and the viscosity of the superplasticizer.
- The cement fineness (Uchikawa et al. 1992; Nawa et al. 1991; Stoïca 1996).
- The phase composition of the cement (Hanna 1990).
- The amount and quality of the sulphate phase and its dissolution rate (Kim 2000; Nkinamubanzi et al. 2000).
- The amount of alkali sulphates present in the cement (Nawa and Eguchi 1987; Kim et al. 2000a and 2000b; Jiang et al. 1998 and 1999).
- The intensity of the mixing.
- The mixing sequence (Baalbaki 1998).

As pointed out earlier, the intensity of mixing influences the superplasticizer dosage necessary to reach a certain fluidity in a grout. It is not certain that the dosage at the saturation point obtained on a grout will correspond exactly to the dosage necessary to obtain a given slump when mixing concrete. Usually, the dosage at the saturation point obtained on a grout that has been vigorously agitated is higher than the one necessary when making concrete in an industrial mixer. Therefore, it is *sometimes* better to start making the first concrete with a dosage equal to 80 per cent of the dosage at the saturation point and to adjust the final dosage according to the slump that has been obtained.

The optimum dosage of a superplasticizer is not always linked to the rheology. In a pre-cast plant, the most critical concrete parameter is its early strength. When a high early strength is desired, it is better to increase the amount of mixing water and, consequently, decrease the superplasticizer dosage (at a constant water/binder ratio). On the contrary, when long-term compressive strength is the critical factor, it is better to decrease as much as possible the amount of mixing water and, consequently, to increase the superplasticizer dosage (at a constant water/binder ratio), because the delaying action of the superplasticizer can be beneficial.

When rheological properties are more critical than strength, it is better to make concrete with the highest (safe) water/binder ratio in order to be sure to obtain the required strength and to adjust the rheology.

In many cases, the optimum dosage of a superplasticizer is a trade-off between the optimum dosages from a strength point of view and from a rheological point of view. When this exercise has to be done for the first time, it can be painful and long.

The optimum dosage can be determined using the three water/binder ratios method. Three concretes are made with low water/binder ratios, always using the same cement content and different amounts of mixing water. These three trial mixes give the strength and the slump that can be achieved in the three cases.

Another approach consists in making a first water/binder mix which will have the desired strength with a superplasticizer dosage equal to 1 per cent and to increase or decrease this dosage according to the slump that is obtained.

Usually, when operating systematically, the right dosage of superplasticizer should be obtained at the third or fourth trial batch.

7.3.3.9 Practical advice

(a) IS IT BETTER TO USE A SUPERPLASTICIZER IN POWDER FORM OR IN LIQUID FORM?

A superplasticizer must be introduced into concrete in liquid form in order to benefit from its action almost instantaneously. The dissolution in water of a superplasticizer in powder form takes longer. Of course, the use of a superplasticizer in a liquid form necessitates taking into account the amount of water brought by the superplasticizer (Aïtcin 1998).

(b) THE USE OF A RETARDER

If the reactivity of the interstitial phase is critical, particularly when the water/binder ratio is lower than 0.40, the simultaneous introduction, or better, the initial introduction, of a retarder can help to solve the slump loss problem as shown in Figure 7.25. It is, of course, very important to fine-tune the respective dosages of the superplasticizer and the retarder to counteract the too-rapid slump loss without lowering excessively the early strength of the concrete (Bhatty 1987). It is better to introduce the retarding agent as early as possible in the mixing sequence so that it can rapidly neutralize the most reactive sites; on the contrary, the superplasticizer should be introduced as late as possible.

During the 1970s some admixture companies tried systematically to incorporate a retarding agent into some of their superplasticizer formulations, but this initiative was not very successful, because not all Portland cements require the same amount of retardation. It was rapidly realized that it was far better to add a specific amount of retarding agent at the concrete plant.

(c) DELAYED ADDITION OF THE SUPERPLASTICIZER

Several researchers have demonstrated that the delayed addition of a superplasticizer is always beneficial from a rheological point of view (Malhotra 1978; Penttala 1993; Uchikawa et al. 1995; Baalbaki 1998; Flatt 1999). The slump loss can be reduced significantly when the addition of the superplasticizer is delayed as long as possible during the mixing sequence. Some concrete producers have even started using a sacrificial lignosulphonate-based water reducer at the beginning of the mixing. When the first SO_4^{2-} ions and lignosulphonate molecules start to react, there are less superplasticizer molecules to react with the C_3A. Consequently, it will take less superplasticizer to efficiently fluidify the cement paste (Ranc 1990). Flatt (1999), Fernon et al. (1997) and Uchikawa et al. (1997) suggest the formation of organomineral compounds.

(d) THE ADDITION OF A SMALL AMOUNT OF SODIUM SULPHATE

Nawa and Eguchi (1987) and Kim et al. (2000) have shown that the addition of a small amount of sodium sulphate that very rapidly provides soluble SO_4^{2-} ions can drastically improve the rheology of some cement pastes in which a polynaphthalene has been used. Of course, this technique works when the cement suffers from a lack of rapidly soluble sulphates.

(e) ADDITION OF A SMALL AMOUNT OF CALCIUM CHLORIDE

Calcium chloride is essentially known and used in the concrete industry as an accelerator, and consequently it is surprising to use it to improve the slump retention of some incompatible combinations, as has been proposed by

Hanehara et al. (2001), when polycarboxylate superplasticizers are used. This beneficial action of calcium chloride can be easily explained: the Cl^- ions that rapidly go into solution react with the C_3A to form chloroaluminates (Dodson 1990) so that less COO^- terminations get trapped in some kind of organomineral components when reacting with the C_3A.

7.4 Admixtures that modify hydration kinetics

It has been known for a long time that the kinetics of hydration can be accelerated or retarded without the use of heat. However, the combined effect of heat, or of a low temperature, are sometimes used to reinforce the effects of accelerators or retarders.

Some researchers classify calcium sulphate as the first retarder used to control the hydration of Portland cement, but gypsum and other forms of calcium sulphates are more traditionally considered as setting and hardening regulators. It would be better to say that calcium sulphate is a retarder of the setting and hardening of clinker as stated previously.

As Mindess and Young (1981) and Vernet (1984) have done, we will first treat the hardening accelerators and then the retarders because their modes of action are not the same.

7.4.1 Accelerators

There are several circumstances when a precaster or a constructor would want to accelerate the hardening of concrete, in order, for example, to shorten the time of form removal, to accelerate the construction process or the rotation of the moulds, or simply when the temperature is cold. Several measures can be undertaken: the concrete can be heated, or the water/binder ratio can be lowered, but it has also been known that at ambient temperature calcium chloride accelerates both the setting and hardening of Portland cement. According to Venuat (1984), Millant and Nicholas (1885) filed the first patent on the accelerating effect of calcium chloride. Also according to Venuat (1984), Candlot later was one of the first to carry out research on the effect of calcium chloride on Portland cerment hydration.

In spite of the merits of calcium chloride as an accelerator, its use can cause serious durability problems in reinforced and pre-stressed concretes. In fact, chloride ions can attack reinforcing rebars and cause swelling due to the formation of rust. This swelling is so destructive that reinforced concrete starts to fall apart. Therefore, in most countries, the use of calcium chloride is severely controlled. For example, in Canada, where calcium chloride is commonly used, the CSA A23.1 standard requires that the calcium chloride dosage be lower than 2 per cent by weight of cement. However, the use of calcium chloride is limited when the concrete is to be cured with low-pressure steam, or when different metals are cast within the concrete or when galvanized steel bars are used.

Moreover, the use of calcium chloride is totally prohibited in the cases of: pre-stressed concrete, external parking garages, when aluminum pieces are cast in the concrete, when sulphate reactive aggregates are used, when metallic hardeners are used to increase the abrasion resistance of floors and, of course, in mass concrete and in hot-weather concreting.

It is still amazing that, every year, thousand of tonnes of calcium chloride are added to concrete and we still don't totally understand the role it plays (Dodson 1990; Rixom and Mailvaganam 1999). After working for a long time on this question, Dodson reached the following conclusions:

1 Calcium chloride does not react with C_3S, it only accelerates its hydration. This effect results in an acceleration of the setting and hardening time.
2 Calcium chloride does not react with C_3A, except when there is not enough calcium sulphate.
3 The basic action of the action of calcium chloride has yet to be discovered.

Due to the potentially aggressive action of chloride ions, researchers have searched for accelerators which are less harmful. Unfortunately, up to now all the chemical compounds and molecules that have been tried present two major disadvantages when their accelerating action is compared to that of calcium chloride: they don't accelerate the hardening of Portland cement as efficiently as calcium chloride, and they are more costly.

After calcium chloride, the most efficient accelerator is calcium formate $Ca(HCOO)_2$, which has been studied extensively by Dodson (1990). However, Dodson recognizes that calcium formate is not as efficient as calcium chloride as an accelerator, and its mode of action is not explained any more precisely than that of calcium chloride. Rixom and Mailvaganam (1999) suggest that the acceleration of the hardening mechanism by calcium formate must be similar to that of calcium chloride.

Moreover, as calcium formate is not very soluble in water, it has to be introduced in powder form in concrete, which is not always practical and does not correspond to the usual way in which admixtures are introduced into concrete. It is certain that if one day a miraculous powder that accelerates concrete hardening without any harmful effect on concrete durability and which costs less than calcium chloride is discovered, concrete producers and admixture companies will find an easy and efficient way to introduce it into concrete. Presently, the higher cost of calcium formate prevents many calcium chloride users from switching to calcium formate.

Calcium nitrite $Ca(NO_2)_2$ is also an accelerator, but it is less efficient and more costly than calcium chloride. It is used instead as a corrosion inhibitor and as an anti-freezing admixture, as will be seen in Section 7.7.4.

There is also another type of accelerator, that is used: triethanolamine $N(CH_2\ CH_2\ OH)_3$ (TEA). According to Dodson (1990), TEA has been used in concrete as an accelerator since 1934 when it was found that a small amount of

TEA could compensate for the retarding action of lignosulphonates. Small amounts of TEA are often found in commercial superplasticizers containing some lignosulphonate. TEA is used at a dosage that does not go beyond 0.025 per cent of the cement mass. It must be remembered that for dosages smaller than 0.001 per cent, as well as at dosages higher than 0.5 per cent, TEA can act as a retarder. At a dosage of 1 per cent, TEA again acts as an accelerator. TEA is an admixture that must be dosed very precisely, because if it is not, the opposite effect can be obtained.

TEA is thought to solubilize iron ions of the interstitial phase and promote the hydration of C_4AF. The production of a small amount of rust, $Fe(OH)_3$, within the paste could result in a retarding effect.

As pointed out by Aïtcin and Baron (1996), in matters of acceleration of hydration, it is often forgotten that the most beneficial way is to simply lower the water/cement ratio. Presently, it is quite easy to obtain within 12 hours a compressive strength in excess of 20 MPa: it is only necessary to lower the water/binder ratio to the 0.30 to 0.35 range without any heating and/or without using calcium chloride. When lowering the water/binder ratio, cement particles get closer to each other in the cement paste so that the first C-S-H forms bridges very rapidly and very strongly bonds the cement particles together, which results in the development of early strength in concrete.

Of course, a fine cement can be used to obtain high initial strength, but this solution has many deleterious secondary effects: high thermal gradient, a higher shrinkage and a higher creep, three particularly harmful behaviours when dealing with concrete durability.

Presently, considering the different admixtures that can be used to easily decrease the water/binder ratio, the use of Portland cement with a fineness greater than 400 m^2/kg should be forbidden because these cements create more problems than they solve.

The preceding discussion does not mean that calcium chloride should be banned, except in very high water/binder ratio concretes. There are many industrial situations where the addition of a small amount of calcium chloride is not harmful, and it would be a pity to ban its use on such occasions.

7.4.2 Set accelerators

This type of accelerator is used in shotcrete or when a rapid hardening mixture is needed to block a water leak, for example. The chemical compounds most widely used in such cases are sodium silicate, potassium aluminate and aluminum silicate. In a shotcrete mixture, these types of admixtures reduce the rebound, that is, less material is lost due to the rebound of the fresh shotcrete, and also increase the thickness of material that can be sprayed in a vertical wall or on an horizontal one. Of course, the quality, impermeability and durability of such an accelerated shotcrete is less than that of a non-accelerated one. The setting of the shotcrete can be as short 3 to 12 minutes (Rixom and

Mailvaganam 1999), and a compressive strength of 3.5 to 7 Mpa can be reached very rapidly, but at the same time the twenty-eight-day compressive strength is reduced by 30 per cent.

The addition of sodium silicate should favour the precipitation of calcium silicate without increasing compressive strength, while the use of potassium aluminate results in a much stronger concrete, which is why this compound is usually used by the shotcrete industry. Potassium aluminate combines directly with calcium sulphate so that ettringite cannot be formed around C_3A grains, and, consequently C_3A hydrates rapidly as hydrated calcium aluminate.

The admixture industry uses different chemical compounds in conjunction with sodium silicate or potassium aluminate; these products are supposed to attenuate some of the deleterious secondary effects of sodium silicate and potassium aluminate (Rixom and Mailvaganam 1999).

7.4.3 Retarders

There are circumstances in which the setting and hardening of Portland cement has to be delayed in order to place concrete in more favourable conditions. This is, for example, the case when the weather is hot, or when the concrete has to be transported over a long distance, or when the delivery schedule is lengthened by traffic jams. It is also important to delay the setting of concrete when concrete is placed in difficult conditions, as in a mine when the placing rate is very slow. It is also essential to delay concrete setting and hardening when a large monolithic concrete base has to be built to support heavy vibrating equipment or submitted to severe shocks. This monolithic base is necessary to dissipate the vibratory mechanical energy, and the absence of cold joints within the base allow a good dissipation of the energy in the whole mass of the concrete.

According to Venuat (1984), the retarding action of sugar was discovered as early as 1909 and, later on, the retarding action of zinc oxide and phosphates was also discovered. Presently several chemical compounds are known to delay setting and hardening.

Before examining some of these compounds, an event that took place in Montreal in the 1970s will be related. One Tuesday morning in the early autumn, the engineer in charge of quality control of a ready-mix company received phone calls from his field inspectors. The concrete delivered late Monday morning and at the beginning of the afternoon had not yet hardened while the concrete delivered earlier in the morning and later in the afternoon had behaved normally. On Tuesday evening, after checking all the concrete-plant equipment, the engineer was still unable to find a good explanation for such strange behaviour. The most plausible explanation was that on Monday morning he had received a delivery of bad cement, but, when he complained to the cement plant, he was told that no other customer had faced such a problem on Monday. The rest of the week, everything went absolutely as normal.

The next Tuesday morning, the same phenomenon happened, and the rest of the week was normal. The inspectors told their customers that the concrete plant was facing a tricky problem, which the best specialists in concrete technology were unable to explain, but that this abnormal delay of setting and hardening was not a concern from a strength point of view, because the delayed concretes of the preceding week had a higher seven-day compressive strength than the usual concrete. For a third week in a row, the same phenomenon happened, but, on the Friday afternoon at a beer party celebrating some special event, the engineer in charge of quality control of the concrete learned from the yardman in charge of the reception and dispatch of the aggregates, that even if the company had problems with the quality of the delivered concrete, there would be no problems with the aggregate delivery, since the company had hired a new independent truck driver to transport the sand.

The yardman was very happy with this new truck driver, as he was a hard-working and trustworthy person, always on schedule. He was so hard-working that on weekends he was even trucking the wastes of a sugar plant. The quality-control engineer finally understood the origin of his problem. The first deliveries of sand on Monday morning were contaminated by the sugar left in the bin of the truck.

As there was still some sand from the previous Friday delivery left in the sand bin, the first concrete batches on Monday morning had a normal setting and hardening behaviour, but it was when the first sand delivered on Monday morning arrived at the bottom of the bin that the problem started. After the second sand delivery on Monday morning, the bin of the truck was clean of any trace of sugar and the concrete recovered its normal behaviour.

The truck driver was asked to carefully clean his bin with water before loading the first delivery of sand, which he did very conscientiously every Sunday night after transporting sugar wastes. The Tuesday problem no longer occurred. This anecdote illustrates how tricky it can be to make quality concrete.

Essentially, retarders temporarily block the hydration of C_3S and C_3A by delaying the precipitation of $Ca(OH)_2$, because it is the precipitation of $Ca(OH)_2$ that triggers the hydration reaction. Retarders can also cover cement particles with a thin layer that neutralizes the hydration of C_3S and C_3A (Venuat 1984). When all the retarder has been combined, the hydration reaction starts normally or at a slower rate. Usually it is observed that the twenty-eight day strength is greater than that of a non-delayed concrete.

When a retarder acts equally on C_3S and C_3A, it delays both setting and hardening, when it acts only on the C_3A it acts as a set retarder, and when it acts only on C_3S it delays hardening.

The principal commercial retarders are:

- lignosulphonates due to their sugar content;

- the salts of carboxylic acids that contain the active termination $- C = H_2 - OH$ such as sodium gluconate $CH_2OH\ (CHOH)_4\ COONa$;
- hydrocarbon $Cn(H_2O)_n$, such as corn syrup.

The salts of carboxylic acids have a much higher purity than lignosulphonates because there are used by the food and pharmaceutical industry. They are, of course, more expensive than lignosulphonates.

Other retarders are also used, but less frequently because they are more expensive, such as citric acid, phosphates, lead oxide, zinc oxide, lead nitrate, zinc acetate, etc. Although these compounds are used only rarely by the concrete industry, it is important to know about their retarding effect because it can explain certain strange behaviour in concretes that have been contaminated by them.

When concrete architectural panels are prefabricated using galvanized reinforcing bars (because the architect does not want to see any rust efflorescence at the surface of the panel), it is very important to check that the galvanized reinforcing bars are clear of a whitish film which can result from being stored outside for a long time. This whitish film is in fact zinc oxide (white rust). When galvanized rebars covered with zinc oxide are embedded in concrete, the hardening of the concrete around the reinforcing bars is delayed so that when the precast panels are demoulded and moved quite roughly to be stored in the yard, the concrete doesn't adhere to the reinforcing steel. Consequently, the panel acts as an unreinforced panel and it cracks during its handling. These cracks constitute preferential avenues for the penetration of aggressive agents so that when the galvanic protection disappears, the unprotected reinforcing bars rust. This result is exactly the opposite of what was desired.

7.4.4 *Cocktails of dispersing agent: accelerator or dispersing-agent retarder*

In their effort to decrease the price of their commercial admixtures, admixture companies have long been trying to use lignosulphonates as water reducers as much as possible. In order to compensate for the variability of commercial lignosulphonates and also to compensate for the secondary effects of lignosulphonates (presence of surface-active agents and sugars), admixture companies have been adding small amounts of pure products as corrective chemical, but as these pure products are more expensive, they are being used in smaller quantities. In some cases, lignosulphonates being sold as water/reducing admixtures are in fact a cocktail of several products. For example, when a lignosulphonate entrains too much air, some tributylphosphate is added at a dosage of less than 1 per cent of the solids contained in the lignosulphonate. Tributylphosphate is a strong defoaming agent. Dibuthylphtalate, insoluble alcohols and derivatives of silicone can also be used (Rixom and Mailvaganam 1999).

In order to compensate for the retarding action of the sugars contained in lignosulphonate, some triethanolamine (TEA) is used, most often at a dosage of up to 15 per cent of solids contained in the lignosulphonate. To accelerate setting and hardening, some calcium chloride or sodium nitrite can also be added.

Of course, the dosages of these additives are adapted so that the commercial water reducer or superplasticizer fits into one of eight or nine different categories of water reducers that are specified in the standards.

The author has worked with a particular lignosulphonate-based water reducer which was made essentially from very inexpensive and poorly performing lignosulphonate (which is why it was so inexpensive), containing five other components. In order to improve the dispersion of cement particles, some polynaphthalene superplasticizer was added. As this water reducer was to be used in a hot country, the admixture company wanted to delay somewhat the concrete setting, therefore, a very cheap corn syrup was added, but in order for a better control of the retarding action, some more expensive sodium gluconate was also added. Finally, in order to accelerate somewhat the hardening of the concrete, some triethanolamine was added, but, as it was expensive, some less expensive sodium nitrite was added.

Fortunately, the lignosulphonate did not entrain air, which would have necessitated the addition of a defoaming agent, perhaps because it had already been added at the pulp and paper plant. The admixture company was able to demonstrate to the author the number of pennies that were saved per litre of water reducer, and, to the great surprise of the author, this complex cocktail worked quite well with the Type I cement that was primarily used in the specific project in which he was involved.

7.5 Admixtures that react with one of the by-products of hydration reaction

The main by-products of the hydration reaction are portlandite and sulphoaluminates (ettringite and monosulphoaluminate) and also some alkalis.

If a very small amount of fine aluminum powder is added during mixing, it will react with the lime liberated by the hydration of C_3S and C_2S to produce small hydrogen bubbles in the fresh concrete during its hardening, which will result in a significant apparent volume increase. This reaction is used to produce cellular concrete used as insulating materials such as the products known as SIPOREX® or YTONG®. This volumetric expansion is also used in expansive mortars or grouts.

The production of hydrogen bubbles can also occur with some silica fumes which contain silicon metal. In two or three cases, accidents due to the explosion of hydrogen liberated by such a reaction have occurred. A very simple test has been described to find out if a particular silica fume contains too high an amount of silicon metal (Edwards-Lajnef et al. 1997).

The alkalis contained in cement can be the origin of an alkali aggregate reaction (AAR) with some potentially reactive aggregates. There are different ways to avoid AAR, the first one being to use a low alkali cement that contains less than 0.6 per cent of Na_2O equivalent, and the second one, of course, to use non-reactive aggregates, but these two solutions are not always easy to implement from an economic point of view. The problem can also be solved more economically by using a blended cement containing either a fly ash, a slag or some silica fume that fix the alkalis in the C-S-H. The use of lithium salt has also been recommended as well as the use of siloxanes and fluorisilicates. In this last case, it is the water-repellent effect of these chemical compounds that controls the alkali/aggregate reaction.

7.6 Air-entraining agents

Air-entraining agents are not foaming agents like the degraded proteins that are sometimes used when making lightweight concretes to improve acoustic insulating properties. The surface active agents that are introduced into concrete as air-entraining agents instead stabilize the air bubbles that are entrained during the mixing in the form of very tiny bubbles.

It is not the purpose of this paragraph to describe the beneficial effects of a network of microscopic air bubbles on the rheology of fresh concrete and on its durability to freezing and thawing cycles in the presence or absence of deicing salts.

To make a good air-entrained concrete, it is not only necessary to entrain a certain percentage of air, and measure it with an airmeter, but it is also absolutely necessary that the entrained bubbles be uniformly dispersed within the mass of concrete. A hole of 50 to 60 litres of air in the centre of a 1 m^3 cube does not protect concrete against freezing and thawing. The uniformity of the dispersion of the bubbles is controlled by measuring the spacing factor, $\bar{L}$, that represents roughly half the distance between two adjacent bubbles. According to Powers' theory, the spacing factor, $\bar{L}$, represents the distance that water must travel to freeze in the nearest bubble, so that the volume of water could increase during the freezing without rupturing the paste. Currently, however, we know that the freezing process is not as simple as that.

It is often forgotten that the presence of air bubbles in a concrete has a very beneficial effect on concrete rheology. The millions of tiny bubbles present in the mass of concrete improve drastically its workability, placeability and finishability. The absorptivity of concrete is also decreased, as the air bubbles act as 'air plugs', and, consequently, the impermeability of concrete is also greatly improved. Moreover, air bubbles dissipate the energy at the tips of a progressive crack so that air-entrained concretes are less prone to develop long and large cracks. Entrained air can also be used to correct the grain-size distribution of a coarse sand, or the harshness of a manufactured sand composed of particles having a very irregular shape. In high-performance concretes, a very small amount of entrained air drastically improves workability and decreases stickiness.

In a particular case of concreting in the Arctic where the only sand available had a fineness modulus of 4.0, the introduction of 8 to 10 per cent of entrained air greatly improved the workability and finishability of the concrete. This high amount of air was only added to improve concrete rheology. Freezing and thawing is not a problem in the Arctic, because concrete is almost always frozen, and, the annual rain fall corresponds to a desert climate, so that the concrete is almost always dry.

Sodium abietate extracted from the pitch of certain woods, lignosulphonates, sulphonated hydrocarbons, proteins and fatty acids were among the first air-entraining agents used. Vinsol resin, a very common air-entraining agent, is extracted from wood (the word Vinsol is in fact an abbreviation of **v**ery **insol**uble).

All of these anionic molecules present a hydrophilic end that combines with calcium ions in the interstitial solution to form insoluble calcium salts (Figure 7.31). This insolubility inhibits the coalescence of the bubbles so that the creation of larger bubbles is prevented. Air-entraining agents do not entrain air bubbles, they are only used to stabilize the very small bubbles trapped within the concrete during its mixing and those dissolved in the mixing water, those trapped in the dry cement or those present within the aggregates. The film that is formed around each bubble must be strong enough to resist internal and external pressures and last long enough to give the cement paste time to harden. Moreover, air-entraining agents must not alter the other properties of fresh and hardened concrete.

Of course, the presence of a certain volume of air within a concrete usually results in a decrease of compressive strength. There is a rule of thumb saying that each 1 per cent entrapped or entrained air modifies by 5 per cent the twenty-eight-day compressive strength, but this is not the case in low-strength concrete. In concrete having a compressive strength lower than 20 MPa, 5 to 6 per cent entrained air slightly increases compressive strength: concrete becomes more homogeneous, and the quality of the transition zone improves

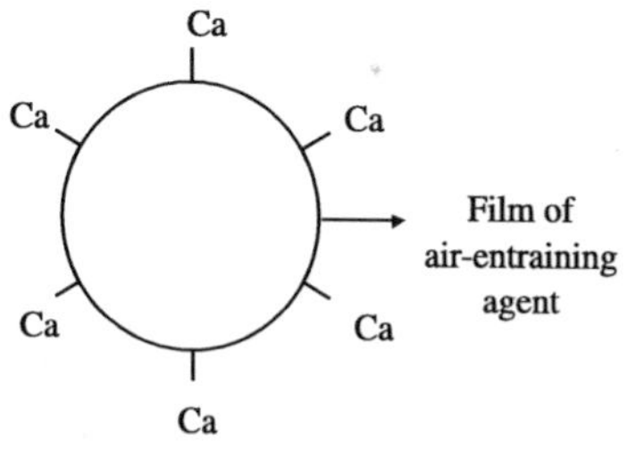

Figure 7.31 Air-entrained bubble (anionic type).

because the volume of the paste has been indirectly increased by 35 to 45 litres corresponding to the volume of the entrained air.

When a polished section of an air-entrained concrete is observed under a microscope, it is very easy to distinguish the air bubbles that have been stabilized by the air-entraining agent and to differentiate them from the coarse bubbles of entrapped air. Air-entrained bubbles are perfectly spherical, with uniform diameters and are dispersed homogeneously within the paste. The bubbles of entrapped air are much coarser and have irregular shapes.

According to the intensity of mixing, the grain-size distribution of the aggregates, particularly of the sand, the cement type and composition, the quality of the mixing water, etc., 1 to 2 per cent of air is normally entrapped in any concrete. This means that when a total air content of 6 per cent is measured with an airmeter in an air-entrained concrete, it can be considered that only about 4 to 5 per cent of this air is actually entrained air and thus 1 to 2 per cent is entrapped air.

Abundant literature exists on the principal factors that affect the characteristics of a network of entrained air in concrete, and these factors are listed in *Design and Control of Concrete Mixtures* by Kosmatka et al. (2002). Dodson (1990) cites several factors: the cement, the sand, the coarse aggregate, the water, the slump, the temperature, the mixer, the length of the mixing, vibration, pozzolans, oils, greases and other admixtures.

Before looking briefly at other types of admixtures, the author would like to re-emphasize that **the entrainment of air bubbles is not only beneficial for freezing and thawing, but, more importantly, entrained air greatly improves concrete rheology**. All the high-performance concrete made at Université de Sherbrooke contains between 4 to 5 per cent air, that is, 2 to 3 per cent entrained air (20 to 30 litres) in addition to the 1 to 2 per cent of entrapped air normally found in high-performance concretes.

7.7 Other types of admixtures

7.7.1 Colloidal agents

Colloidal agents were introduced long ago in masonry cements, as will be seen in Chapter 9, in fact ever since limestone has been substituted for lime in masonry cements. Masons need a ‘fat’ mortar which strongly retains its water and stays plastic long enough. Traditionally, masons (and good masons today) mixed half Portland cement and half lime when making their mortars. When cement producers started to propose to them a half and half mixture of limestone filler and Portland cement, cement companies were obliged to add colloidal agents to retain the mixing water to reproduce the fattiness of the mortar made from lime and Portland cement. Rixom and Mailvaganam (1999) classify colloidal agents into five broad categories.

1 Polymers soluble in natural water or synthetic polymers such as cellulose ethers, pregelanitized starch, polyethylene oxides, alginates, polyacrylamides, carboxivinylic polymers and polyvinyl alcohol.
2 Organic flocculates that can be adsorbed by cement particles and create interparticle attraction forces, such as the copolymers of styrene and carboxyclic groups and some natural gums.
3 Emulsions that increase the interparticle forces, such as acrylic emulsions.
4 Finely divided inorganic materials that have a great capacity for water retention, such as bentonite, silica fume, asbestos mine tailings.
5 Inorganic materials such as lime, kaolin, distomaceous earth or calcined pozzolanic material.

Kawai (1987) and Khayat (1993) have published a list of chemical products already used in making self-compacting concrete. These products are usually used at 1.5 per cent of the mass of cement.

Colloidal agents started being used in concrete at the end of the 1970s in order to increase the cohesion of concretes cast underwater, to avoid the washing out of the concrete when it was poured directly underwater. More recently, colloidal agents have been used extensively in self-compacting concrete where they contribute to the cohesiveness of the mix so that the concrete can level off without any segregation and bleeding, without the need for vibration.

Self-compacting concretes can also be qualified as 'silent' or 'noiseless' concretes, because they don't need to be vibrated. This characteristic is very important, because it allows the delivery of concrete at night in downtown areas without disturbing citizens during their sleep. Finally, in pre-cast plants, the use of self-compacting concrete results in an increase in productivity and also, in some countries, in a decrease in insurance rates because many fewer workers suffer from premature deafness.

7.7.2 *Water-repellent admixtures*

These admixtures are essentially stabilized suspensions of stearates that are incorporated in concrete to fill capillaries in the hardened concrete in order to reduce concrete absorptivity and capillarity. This type of admixture is commonly used during the fabrication of concrete blocks, bricks and paving blocks. During the vibration of these dry concretes, the presence of these water-repellent admixtures has a secondary beneficial effect: it favours a better compaction.

7.7.3 *Shrinkage-reducing admixtures*

This type of admixture is one of the latest to appear on the market. Shrinkage-reducing admixtures reduce the tensile strength developed within the menisci and the angle of contact of the menisci on the walls of the capillaries. This tensile strength and contact-angle reduction decreases the contraction forces which are the origin of shrinkage (plastic, autogenous or drying). This type of admixture can be very useful because it reduces the shrinkage forces (or rather delays them) at a moment when the concrete has not fully attained its tensile stress potential. When concrete is drying, of course, the menisci develop in capillaries that are smaller and smaller, and it takes more drying to develop an equivalent stress to the one developed in a capillary of given diameter when no shrinkage-reducing admixture has been used.

From a practical point of view, in any case, shrinkage-reducing admixtures stabilize shrinkage at a lower level than when no shrinkage-reducing admixture is used, because in the field, except for some desert places, concrete gets rewetted more or less regularly. In northern countries where concrete must be air-entrained, the use of the present shrinkage-reducing admixtures is not recommended because they tend to destabilize the air bubble system. Some researchers have suggested instead the use of shrinkage-reducing admixtures to impregnate already hardened concretes.

7.7.4 Anti-freeze admixtures

The technology of anti-freeze admixtures has been developed essentially in the former USSR, more particularly in Siberia. In Western countries, the protection of concrete against sub-zero temperatures has been undertaken by heating the concrete, using insulated forms and blankets, adding calcium chloride or even by building temporary shelters. For example, in Canada, when grouting a frozen rock, the rock is first heated with water vapour for hours or even days so that the 0°C front is moved deeply within the rock, such that when the grout is injected in the heated rock there is enough time for the grout to harden before the temperature of the rock recovers its sub-zero initial temperature.

Several anti-freeze admixtures have been used, which are essentially based on sodium nitrite or a mixture of sodium nitrite and nitrate. By using anti-freeze admixtures, it is possible to place concrete in temperatures down to –15°C. Using very high dosages concrete can even harden at –30°C (Hernandez 1989). Below this temperature, it is better to stop casting concrete, moreover, steel becomes very brittle at such a low temperature so that many mechanical problems that have nothing to do with concrete prevent the use of such a concrete.

7.7.5 Latexes

Latexes are introduced in concrete, mortars or grouts during mixing. Latexes are stabilized suspensions of small polymer chains. The latexes most widely used presently are copolymers of styrene butadiene (SBR), ethylene vinyl acetate (EVA), polyacrilic esters (PAE) and epoxy resins (PE). Latexes are commonly used in the USA in bridge decks.

The coagulation of the latex in parallel to hydration reaction results in the formation of two matrices, one constituting of the hydrated cement paste and the other one of the coagulated polymers closely linked to the cement matrix. The chloride ion permeability measured according to ASTM C1202, which is a measure of the conductivity of the concrete, demonstrates clearly the benefits provided by latexes in the reduction of the movement of ionic species. Latexes usually reduce chloride ion permeability in the same range as a 6 to 10 per cent silica fume addition. In order to reach a significant decrease in permeability of concrete, the solids content of the latex must be equal to 10 to 20 per cent of the mass of cement. Of course, the higher the amount of latex, the lower the resulting permeability. On the other hand, the addition of 15 to 20 per cent latex can double the cost of 1 m^3 of concrete. Moreover, it is difficult to exceed a dosage of 20 per cent due to a strong tendency to entrain air.

7.7.6 Foaming agents

Preformed foams can be introduced into concrete; they are essentially made from degraded proteins. These degraded proteins are also used as a foam in airports when planes have to make emergency landings. To be used in concretes and grouts, these proteins are modified to remain stable in an interstitial solution having a high pH, such as the interstitial solution found in grouts and concrete.

Foamed concretes are used to improve acoustic insulation, competing in this field with wood chips, polystyrene beads, vermiculite or perlite.

7.7.7 Corrosion inhibitors

A certain number of admixtures are presently sold as corrosion inhibitors, and they control, or at least delay, the corrosion of reinforcing bars. It would be dangerous to count only on these chemical compounds to solve corrosion problems. In fact, to be efficient, these corrosion inhibitors must be used in a high-quality concrete where all the usual means to increase concrete durability have already been used: low water/binder ratio, adequate cover over the reinforcing steel, adequate curing, control of plastic shrinkage, control of autogenous shrinkage and entrained air in the case of concretes that have to face freezing and thawing cycles.

In fact corrosion inhibitors provide a second level of protection. Rixom and Mailvaganam (1999) recognize two types of corrosion inhibitors according to their type of action: the anodic corrosion inhibitors and the corrosion inhibitors that have a compound action. The National Association of Corrosion Engineers of the United States of America use a classification comprising six types of corrosion inhibitors: those having a passivative oxidizing effect, a non-oxidizing passivative effect, a cathodic effect, a precipating effect, a mobilizing effect on oxygen, and the formation of a protective coating adsorbed at the surface of steel.

For example, anodic inhibitors of the nitrite type, chromates and mobyldates are passivators of oxidation which reinforce the protective coating of steel

$$2\ Fe^{++} + 20\ H^{-} + 2\ NO_2 \rightarrow 2\ NO + Fe_2O_3 + H_2O$$

Currently, from a practical point of view, nitrites are most often used in spite of the fact that numerous other types such as sodium benzoates, potassium or sodium chromates, sodium salts of silicates and phosphates, sodium fluophosphate, anilin and ferrocyanides are also used. This long list of chemical compounds demonstrates the importance of the problem of corrosion of reinforcing bars, and the efforts that are presently devoted to find a miraculous product that could solve this problem.

The miraculous means that can protect steel against corrosion is in fact a non-cracked low water/binder ratio covercrete which, in most environmental conditions, brings enough protection to steel. When using a high-performance concrete, it is only under very severe environmental conditions that the addition of a corrosion inhibitor should be mandatory.

7.8 Conclusion

It is always surprising to see the great variety of chemical products that are incorporated into concrete to improve one of its properties in the fresh or hardened state. The reason is simple: many people are interested in finding the magic compound that will drastically improve one or several properties of concrete, because several cubic centimetres of this compound multiplied by the number of cubic metres that are presently used in the world, represents thousands of cubic metres of this compound, and, of course, many dollars.

Admixtures are becoming increasingly essential components of modern concretes. Most of the recent advances in concrete technology are due almost exclusively to the progress made in the comprehension of the interaction mechanisms between cement and admixtures. In fact, the cost of admixtures is becoming more and more important when calculating the cost of 1 cubic metre of high-performance concrete, as can be seen in Figure 7.32.

The time is past when miraculous properties of miraculous molecules were found accidentally, and most of the recent polymers and polymer compounds have been developed purposely. Of course, lignosulphonates will continue being used (as long as they are available), but more and more synthetic molecules will be used. These molecules will have very specific effects on the properties of fresh and hardened concrete.

One day, cement companies will systematically add certain admixtures to their cements and hydraulic binders in spite of the failure of the introduction of air-entraining Portland cement in the USA some years ago. As will be seen in Chapter 11, the addition of a water reducer to all Portland cement and hydraulic binders produced in the world would reduce by 5 to 10 per cent the

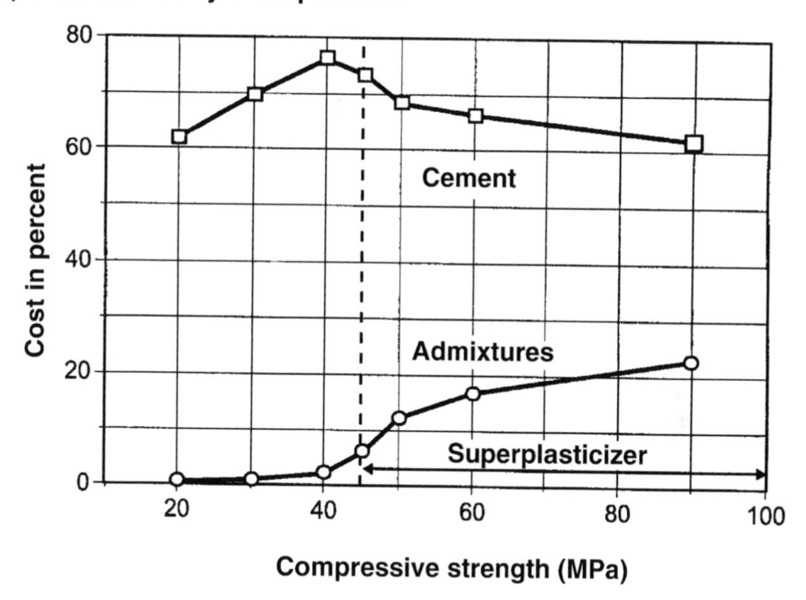

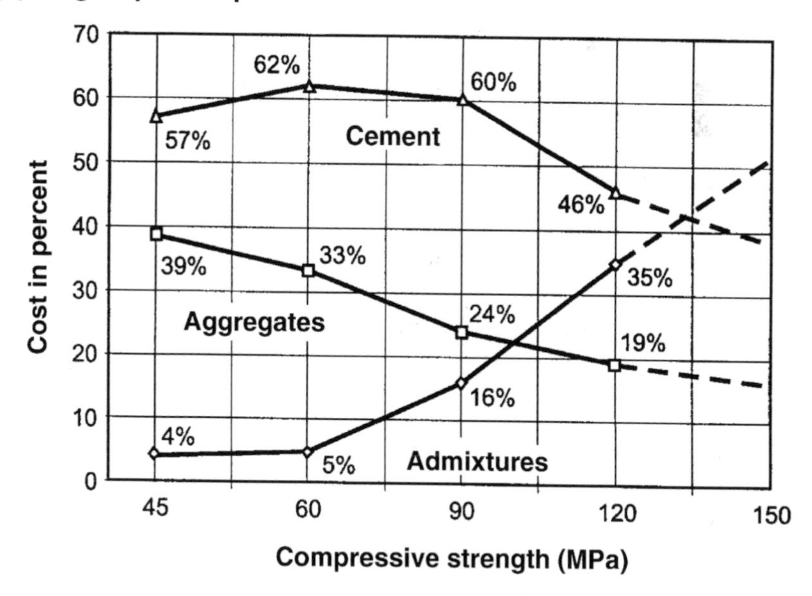

Figure 7.32 Importance of admixtures in modern concretes.

CO_2 emission of the cement industry, or else would result in the fabrication of more concrete for the same level of CO_2 emission. Therefore, in addition to improving the quality of concrete, the ecological impact of concrete would be decreased when using admixtures.

Finally, although the main mechanism of the interaction between hydraulic binders and admixtures are more or less well understood, there is still much progress to be made to develop admixtures that are better performing.

References

Aïtcin, P.C. (1998) *High performance concrete*, London: E & FN SPON.

Aïtcin, P.-C., Jiang, S.P., Kim, B.-G. and Petrov, N. (2002) 'L'Interaction ciment/superplastifiant: cas des polysulphonates', *Bulletin de Liaison des Laboratoires des Ponts et Chaussées*, 233: 8–97.

Aïtcin, P.-C. and Baron, J. (1996) 'Les adjuvants normalisés pour bétons', in J. Baron and J.-P. Ollivier (eds) *Les bétons, bases et données pour la formulation*, Paris: Eyrolles, pp. 87–131.

Aïtcin, P.-C. Jolicoeur, C. and MacGregor, J. (1994) 'Superplasticizers: How They Work and Why They Occasionaly Don't', *Concrete International*, 16 (5, May): 45–52.

Aïtcin, P.-C., Sarkar, S.L., Ranc, R., and Lévy, C. (1991) 'A High Silica Modulus Cement of High Performance Concrete', in S. Mindess (ed.), *Ceramic Transactions: Advances in Cementitious Materials*, Vol. XVI, Westerville, Ohio: American Ceramic Society, pp. 103–20.

Aïtcin, P.-C., Bédard, C., Plumat, M., and Haddad, G. (1984) *Very High Strength Cement for Very High Strength Concrete*, Fall Symposium, Boston, Mass.: Materials Research Society, pp. 202–10.

Aïtcin, P.-C., Sarkar, S.L., Regourd, M. and Volant, M. (1987) 'Retardation Effect of Superplasticizer on Different Cement Fraction', *Cement and Concrete Research*, 17 (6, December): 995–9.

Andersen, P.J. (1986) 'The Effect of Superplasticizers and Air-Entraining Agents on the Zeta Potential of Cement Particles', *Cement and Concrete Research*, 16 (6): 931–40.

Andersen, P.J., Roy, D.M. and Gaidis, J.M. (1988) 'The Effect of Superplasticizer Molecular Weight on its Adsorption and Dispersion of Cement', *Cement and Concrete Research*, 18 (6): 980–6.

Association Canadienne du Ciment Portland (1995) *Dosage et contrôle des mélanges de béton*, 6th edn, Montreal.

ASTM C494, Standard Specification for Chemical Admixtures for Concrete, Section 4, Construction, Vol. 04.02, Concrete and Aggregates.

Baalbaki, M. (1998) 'Influence des interactions du couple ciment/adjuvant dispersant sur les propriétés des bétons: influence du mode d'introduction des adjuvants', Ph.D. thesis, Université de Sherbrooke, Quebec.

Baussant, J.-B. (1990) 'Nouvelle méthode d'étude de la formation d'hydrates des ciments: applications à l'analyse de l'effet adjuvants organiques', Ph.D. thesis, Université de Franche-Comté.

Bhatty, J.I. (1987) 'The Effect of Retarding Admixtures on the Structural Development of Continuously Sheared Cement Paste', *International Journal of Cement Composites and Lightweight Concrete*, 9 (3, August): 137–44.

Black, B., Rossington, D.R. and Weinland, L.A. (1963) 'Adsorption of Admixtures on Portland Cement', *Journal of the American Ceramic Society*, 46 (8, August): 395–9.

Bradley, G. and Howarth, J.M. (1986) 'Water Soluble Polymers: The Relationships Between Structure, Dispersing Action and Rate of Cement Hydration', *Cement Concrete and Aggregate*, 8 (2, winter): 68–75.

Buil, M., Witier, P., de Larrard, F., Detrez, M. and Paillère, A.M. (1986) Physicochemical Mechanism of the Action of the Naphthalene Sulphonate Based Superplasticizers on Silica Fume Concretes, ACI SP–91, pp. 959–71.

Carin, V. and Halle, R. (1991) 'Effect of Matrix Form on Setting Time of Belite Cement which Contains Tricalcium Aluminate', *Ceramic Bulletin*, 70(2): 251–3.

Chatterji, V.S. (1988) 'On the Properties of Fresly Made Portland Cement Paste: Part 2, Sedimentation and Strength of Floculation', *Cement and Concrete Research*, 18: 615–20.

Childowski, S. (1990) 'Effect of Molecular Weight of a Polymer on the Structure of a Layer Adsorbed on the Surface of Titania', *Powder Technology*, 63: 75–9.

Cunningham, J.C., Dury, B.L. and Gregory, T. (1989) 'Adsorption Characteristics of Sulphonated Melamine Formaldehyde Condensates by High Performance Size Exclusion Chromatography', *Cement and Concrete Research*, 19 (6): 919–26.

Daimon, M. and Roy, D.M. (1978) 'Rheological Properties of Cement Mixes: 1: Methods, Preliminary Experiments, and Adsorption Studies', *Cement and Concrete Research*, 8 (6, November): 753–64.

de Larrard, F., Bosc, F., Catherine, C. and Deflorenne, F. (1996) 'La Nouvelle Méthode des coulis de l'AFREM pour la formulation des bétons à hautes performances', *Bulletin de Liaison des Laboratoires des Ponts et Chaussées*, 202 (March–April): 61–9.

Detwiler, R.J. and Mehta, P.K. (1989) 'Chemical and Physical Effect of Condensed Silica Fume in Concrete', Supplementary Papers, *Third CANMET/ACI International Conference on Fly Ash, Silica Fume, Slag and Natural Pozzolans in Concrete*, Trondheim, Norway, pp. 295–306.

Diamond, S. and Struble, L.J. (1987) 'Interaction Between Naphthalene Sulphonate and Silica Fume in Portland Cement Pastes', Materials Research Society, Fall meeting, Boston, Mass.

Dodson, V. (1990) *Concrete Admixtures*, New York: Van Nostrand Reinhold.

Dodson, V.H. and Hayden, T. (1989) 'Another Look at the Portland Cement/Chemical Admixture Incompatibility Problem', *Cement, Concrete and Aggregates*, 11 (1, summer): 52–6.

Edwards-Lajnef, M., Aïtcin, P.-C., Wenger, T., Viers, P., and Galland, J. (1997) 'Test Methods for the Potential Release of Hydrogen Gas from Silica Fume', *Cement Concrete and Aggregates*, 19 (2, December): 97–102.

Fernon, V. (1994a) 'Caractérisation des produits d'interaction adjuvants/hydrates du ciment', Journée technique adjuvants, les Technodes, Guerville, September.

Fernon, V. (1994b) 'Étude de nouveaux solides lamellaires obtenus par coprécipitation d'hydrate aluminocalcique et de sulphonate aromatique', Ph.D. thesis, Université d'Orléans.

Fernon, V., Vichot, A., Le Goanvic, N., Colombet, P., Corazza, F., and Costa, U. (1997) Interaction Between Portland Cement Hydrates and Polynaphthalene Sulphonates, ACI SP–173, pp. 225–48.

Flatt, R.J. (1999) 'Interparticle Forces and Superplasticizers in Cement Suspension', Ph.D. thesis, École Polytechnique de Lausanne.

Flatt, R.J. and Houst, Y.F. (2001) 'A Simplified View on Chemical Effects Perturbing the Action of Superplasticizers', *Cement and Concrete Research*, 31(8): 1169–76.

Foissy, A. and Pierre, A. (1990) 'Les Mécanismes d'action des fluidifiants', *Ciments, Bétons, Plâtres, Chaux*, 782: 18–19.

Fonolossa, P. (2002) Mechanical properties and microstructure of new matrices with ultra high performance, Ph.D. Thesis No. 1388 (in French), Université de Sherbrooke.

Garrison, E. (1991) *A History of Engineering and Technology*, Boca Raton, Fla.: CRC Press.

Garvey, M.J. and Tadros, T.F. (1972) 'Fractionation of the Condensates of Radium Naphthalene 2-Sulphonate and Formaldehyde by Gel Permeation Chromatography', *Colloid-Z.u., Z. Polymere*, 250(10): 967–72.

Hanna, É. (1990) 'La Rhéologie des bétons à haute performance', Master's thesis, Université de Sherbrooke.

Hanna, É., Luke, K., Perraton, D. and Aïtcin, P.-C. (1989) 'Rheological Behavior of Portland Cement in the Presence of a Superplasticizer', *Third CANMET/ACI International Conference on Superplasticizers and other Chemical Admixtures in Concrete*, Ottawa, pp. 171–88.

Hanna, E., Ostiguy, M., Khalifé, M., Stoïca, O., Kim, B.-G., Bédard, C., Saric-Coric, M., Baalbaki, M., Jiang, S., Nkinamubanzi, P.-C., Aïtcin, P.-C. and Petrov, N. (2000) 'The Importance of Superplasticizers in Modern Concrete Technology', *Sixth CANMET/ACI International Conference on Superplasticizers and other Chemical Admixtures in Concrete*, Nice.

Hattori, K. (1979) Experience with Mighty Superplasticizer in Japan, ACI SP–62, pp. 37–66.

Hernandez, P. (1989) 'Étude des scellements d'ancrages dans les massifs rocheux soumis à l'action du gel', Master's Thesis, Université de Sherbrooke.

Hewlett, P.C. (1998) *Lea's Chemistry of Cement and Concrete*, 4th edn, London: Arnold.

Hewlett, P.C. and Rixom, R. (1977) 'Superplasticized Concrete', *ACI Journal*, 74 (5, May): 6–12.

Hewlett, P.C. and Young, J.F. (1987) 'Physico-Chemical Interactions Between Chemical Admixtures and Portland Cement', *Journal of Materials Education*, 9(4): 389–436.

Huyhn, H.T. (1996) 'La Compatibilité ciment/superplastifiant dans les bétons à hautes performances: synthèse bibliographique', *Bulletin de Liaison des Laboratoires des Ponts et Chaussées*, 206 (November-December): 63–73.

Jiang, S.P., Kim, B.-G. and Aïtcin, P.-C. (1998) 'Some Practical Solutions Dealing with Cement and Superplasticizer Compatibility', in *Proceedings of the 4th Beijing International Symposium on Cement and Concrete*, October, Vol. I, pp. 724–9.

Jiang, S.P., Kim, B.-G., and Aïtcin, P.-C. (1999) 'Importance of Adequate Soluble Alkali Content to Ensure Cement/Superplasticizer Compatibility', *Cement and Concrete Research*, 29(1): 71–8.

Jiang, S.P., Kim, B.-G. and Aïtcin, P.-C. (2000) 'A Practical Method to Solve Slump Loss Problem in Superplasticized High-Performance Concrete', *Cement, Concrete, and Aggregates*, 22(1): 10–15.

Joisel, A. (1973) *Admixture for Cement: Physico Chemistry of Concrete and its Reinforcement*, n.p.

Jolicoeur, C., Nkinamubanzi, P.-C., Simard, M.-A. and Piotte, M. (1994) Progress in

Understanding the Functional Properties of Superplasticizer in Fresh Concrete, ACI SP–148, pp. 63–88.

Kantro, D.L. (1980) 'Influence of Water-Reducing Admixtures on Properties of Cement Paste: A Miniature Slump Test', *Cement, Concrete and Aggregates*, 2 (2, winter): 95–108.

Kawai, T. (1987) 'Non-Dispersible Underwater Concrete Using Polymers', *Marine Concrete*, International Congress on Polymers in Concrete, Brighton, UK, Chap. 11.5.

Khalifé, M. (1991) 'Contribution à l'étude de la compatibilité ciment/superplastifiant', Master's thesis, Université de Sherbrooke.

Khayat, K. (1998) 'Viscosity Enhancing Admixtures for Cement-based Materials: An Overview', *Cement and Concrete Composite*, 20: 171–88.

Khorami, J. and Aïtcin, P.-C. (1989) Physico-Chemical Characterization of Superplasticizers, ACI SP–119, pp. 117–31.

Kim, B.-G. (2000) 'Compatibility Between Cements and Superplasticizers in High-Performance Concrete: Influence of Alkali Content in Cement and of the Molecular Weight of PNS on the Properties of Cement Pastes and Concretes', Ph.D. thesis, Université de Sherbrooke.

Kim, B.-G., Jiang, S.P. and Aïtcin, P.-C. (1999) 'Influence of Molecular Weight of PNS Superplasticizers on the Properties of Cement Pastes Containing Different Alkali Contents', *Proceedings of the International RILEM Conference*, Monterrey, Mexique, pp. 97–111.

Kim, B.-G., Jiang, S.P. and Aïtcin, P.-C. (2000a) 'Effect of Sodium Sulphate Addition on the Properties of Cement Pastes Containing Different Molecular Weight PNS Superplasticizer', *Sixth CANMET/ACI International Conference on Superplasticizers and other Chemical Admixtures in Concrete*, Nice, October, pp. 485–504.

Kim, B.-G., Jiang, S.P. and Aïtcin, P.-C. (2000b) 'Slump Improvement Mechanism of Alkalis in PNS Superplasticized Cement Pastes', *Matériaux et Constructions*, 33 (230): 363–9.

Kim, B.-G., Jiang, S.P., Jolicoeur, C., and Aïtcin, P.-C. (2000c) 'The Adsorption Behavior of PNS Superplasticizer and its Relation to Fluidity of Cement Paste', *Cement and Concrete Research*, 30 (6): 887–93.

Kinoshita, M, Nawa, T., Iida, M. and Ichiboji, H. (2000) Effect of Chemical Structure on Fluidizing Mechanism of Concrete Superplasticizer Containing Polyethylene Oxide Graft Chains, ACI SP–195, pp. 163–79.

Kreijger, P.C. (1980) *Plasticizers and Dispersing Admixtures, Admixtures Concrete International*, The Construction Press, London, United Kingdom.

Lahalih, S.H., Absi-Halabi, H. and Ali, M.A. (1988) 'Effect of Polymerization Conditions of Sulphonated – Melamine Formaldehyde Superplasticizers on Concrete', *Cement and Concrete Research*, 18(4): 513–31.

Legrand, C. (1982) 'La Structure des suspensions de ciment', in J. Baron and R. Sauterey (eds) *Le béton hydraulique*, Paris: Presses de l'École Nationale des Ponts et Chaussées, pp. 99–113.

Luke, K. and Aïtcin, P.-C. (1991) 'Effect of Superplasticizer on Ettringite Formation', in S. Mindess (ed.) *Ceramic Transactions: Advances in Cementitious Materials*, Vol. XVI, pp. 147–66.

Malhotra, V.M. (1978) 'Effect of Repeated Dosages of Superplasticizers on Workability, Strength and Durability', First International Symposium on Superplasticizers in Concrete, Ottawa, Canada, May.

Mansoutre, S. (2000) 'Des suspensions concentrées aux milieux granulaires lubréfiés: étude des pâtes de silicate tricalciques', Ph.D. thesis, Université d'Orléans.

Massazza, F. and Costa, V.B. (1980) 'Effect of Superplasticizers on the C_3A Hydration', Seventh International Congress on Concrete Chemistry, Vol. IV, Paris, pp. 529–34.

McCarter, W.J, and Garvin, S. (1989) 'Admixtures in Cement: A Study of Dosage Rates on Early Hydration', *Matériaux et Constructions*, 22: 112–20.

Meyer, A. (1979) Experiences with the Use of Superplasticizers in Germany, ACI SP–62, pp. 21–36.

Mielenz, R.C. (1984) 'History of Chemical Admixtures for Concrete', *Concrete International*, 6 (4 April): 40–53.

Mindess, S. and Young, J.F. (1981) *Concrete*, Englewood Cliffs, NJ: Prentice-Hall.

Mollah, M.Y., Palta, P., Hers, T.R., Vempati, R.K. and Coche, D.L. (1995) 'Chemical and Physical Effects of Sodium Lignosulphonate Superplasticizer on the Hydration of Portland Cement and Solidification/Stabilization Consequences', *Cement and Concrete Research*, 25 (3): 671–82.

Nawa, T. and Eguchi, H. (1987) 'Effects of Types of Calcium Sulphate on Fluiding of Cement Paste', Review of the 41st General Meeting, The Cement Association of Japan, pp. 40–3.

Nawa, T., Eguchi, H. and Fukaya, Y. (1989) Effect of Alkali Sulphate on the Rheological Behavior of Cement Paste Containing a Superplasticizer, ACI SP–119, pp. 405–24.

Nawa, T., Eguchi, H., and Okkubo, M. (1991) 'Effect of Fineness of Cement on the Fluidity of Cement Paste and Mortar', *Transactions of JSCE*, 13 (August): 199–213.

Nkinamubanzi, P.-C. and Aïtcin, P.-C. (1998) Compatibilité ciment/adjuvant Phase II, Report 98-06, Paris: ATILH, 71 p.

Nkinamubanzi, P.-C., Baalbaki, M. and Aïtcin, P.-C. (1997) 'Comparison of the Performance of Four Superplasticizers on High-Performance Concrete', in P.K. Mehta (ed.) *Fifth CANMET/ACI International Conference on Superplasticizers and Other Chemical Admixtures in Concrete, Supplementary Papers*, Rome, October, pp. 199–206.

Nkinamubanzi, P.-C., Kim, B.-G., Saric Coric, M. and Aïtcin, P.-C. (2000) 'Key Cement Factors Controlling the Compatibility Between Naphthalene-Based Superplasticizers and Ordinary Portland Cements', *Sixth CANMET/ACI International Conference on Superplasticizers and other Chemical Admixtures in Concrete, Supplementary Papers*, Nice, October, pp. 33–54.

Odler, I. and Abdul-Maula, S. (1987) 'Effect of Chemical Admixtures on Portland Cement Hydration', *Cement and Concrete Aggregates*, 9 (1, summer): 48–48.

Ohta, A., Sugiyama, T. and Momoto, T. (2000) Study of Dispersing Effects of Polycarboxylate-Based Dispersant on Fine Particles, ACI SP-195, pp. 211–27.

Onofrei, M. and Gray, M. (1989), Adsorption Studies of 35s-Labelled Superplasticizer in Cement-Based Grout, ACI SP-119, pp. 645–60.

Ostiguy, M. (1994) 'Compatibilité ciment/superplastifiant: influence du type de ciment', Master's thesis, Université de Sherbrooke.

Pagé, M., Nkinamubanzi, P.-C., and Aïtcin, P.-C. (1999) 'The Cement/Superplasticizer Compatibility: A Headache of Superplasticizer Manufacturers', in J.G. Cabrera and R. Rivera-Villareal (eds) *The Role of Admixtures in High Performance Concrete: Proceedings of the RILEM International Conference*, Monterrey, Mexico, March 21–26, pp. 48–56.

Paillère, A.M. (1982) 'Les Adjuvants', in J. Baron and R. Sauterey (eds) *Le Béton hydraulique*, Paris: Presses de l'École nationale des Ponts et Chaussées, pp. 69–82.

Paillère, A.M., Alègre, R., Ranc, R. and Buil, M. (1984) Interaction entre les réducteurs d'eau-plastifiants et les ciments, Lafarge/Laboratoire Central des Ponts et Chaussées Report, Paris, France, pp. 105–8.

Paillère, A.M. and Briquet, P. (1980) 'Influence des résines de synthèse fluidifiantes sur la rhéologie et la déformation des pâtes de ciment avant et en cours de prise', Seventh International Congress on Concrete Chemistry, Paris, Vol. III, pp. 186–91.

Paulini, P. (1990) 'Reaction Mechanisms of Concrete Admixtures', *Cement and Concrete Research*, 20: 910–18.

Penttala, U.E. (1993) 'Effects of Delayed Dosage of Superplasticizer on High-Performance Concrete', in I. Holland and E. Sellevold (eds) *Proceedings of the International Conference on High-Strength Concrete,* Lillehammer, Oslo: Norwegian Concrete Association, pp. 874–81.

Petrie, E.M. (1976) 'Effect of Surfactant on the Viscosity of Portland Cement: Water Dispersions', *Industrial Engineering Chemistry, Prod. Res. Dev.*, 15(4): 242–9.

Piotte, M. (1983) 'Caractérisation d'un naphthalène sulphonate: influence de son contre-ion et de sa masse moléculaire sur son interaction avec le ciment', Ph.D. thesis, Université de Sherbrooke.

Piotte, M., Bossanyi, F. Perreault, F. and Jolicoeur, C. (1995) 'Characterization of Poly (Naphthalene Sulphonate) Salts by Ion-Pair Chromatography and Ultrafiltration', *Journal of Chromatography A.*, 704(2): 377–85.

Pollet, B., Germaneau, B., and Defossé, C. (1997) 'Fixation des adjuvants de type poly-naphtalène ou polymélamine sulphonate dans les mortiers et bétons', *Matériaux et Constructions*, 30 (204): 627–30.

Powers, T.C. (1956) 'Structure and Physical Properties of Hardened Portland Cement Paste', *Journal of the American Ceramic Society*, 4 (January): 1–6.

Ramachandran, V.S. (1995) *Concrete Admixtures Handbook: Properties, Science and Noyes, New Technology*, 2nd edn, Park Ridge, NJ: Noyes Publications.

Ramachandran, V.S., Beaudoin, J.J. and Shilua, Z. (1989) 'Control of Slump Loss in Superplasticized Concrete', *Matériaux et Constructions*, 22: 107–11.

Ramachandran, V.S., Malhotra, V.H., Jolicoeur, C. and Spiratos, N. (1998) *Superplasticizers: Properties and Applications in Concrete*, Ottawa: Materials Technology Laboratory, CANMET.

Ranc, R. (1990) 'Interactions entre les réducteurs d'eau-plastifiants et les ciments', *Ciments, Bétons, Plâtres et Chaux*, 782: 19–20.

Rixom, R. and Mailvaganam, N. (1999) *Chemical Admixture for Concrete*, 3rd edn, London: E & FN Spon.

Rosenberg, A.M. and Gaidis, J.M. (1989) 'A New Mineral Admixture for High-Strength Concrete', *Concrete International*, 11 (4, April): 31–6.

Rougeron, P. and Aïtcin, P.-C. (1994) 'Optimization of the Composition of a High-Performance Concrete', *Cement Concrete and Aggregates*, 16 (2, December): 115–24.

Singh, N.B. and Singh, A.C. (1989) 'Effect of Melment on the Hydration of White Portland Cement', *Cement and Concrete Research*, 19: 547–53.

Stoïca, O.M. (1996) 'La Production d'éléments fabriqués en béton de poudres réactives', Master's thesis, Université de Sherbrooke.

Tagnit-Hamou, A. and Aïtcin, P.-C. (1993) 'Cement and Superplasticizer Compatibility', *World Cement*, 24 (8, August): 38–42.

Tagnit-Hamou, A., Baalbaki, M. and Aïtcin, P.-C. (1992) 'Calcium-Sulphate Optimization in Low Water/Cement Ratio Concretes for Rheological Purposes', *Ninth International Congress on Concrete Chemistry*, New Delhi, India, Vol. V, pp. 21–5.

Tucker, G.R. (1938), US Patent 2 141 569, Concrete and Hydraulic Cement, 5 p.

Uchikawa, H. (1994) 'Hydration of Cement and Structure Formation and Properties of Cement Paste in the Presence of Organic Admixture', *Conférence en hommage à Micheline Moranville-Regourd*, edited by Béton Canada, Faculté des sciences appliquées, Université de Sherbrooke, Québec, Canada, October, 55 p.

Uchikawa, H., Hanehara, S. and Sawaki, D. (1997) 'The Role of Steric Repulsive Force in the Dispersion of Cement Particles in Fresh Paste Prepared with Organic Admixtures', *Cement and Concrete Research*, 27 (1): 37–50.

Uchikawa, H., Hanehara, S., Shirosaka, T. and Sawaki, D. (1992) 'Effect of Admixture on Hydration of Cement, Adsorptive Behavior of Admixture and Fluidity and Setting of Fresh Cement Paste', *Cement and Concrete Research*, 22 (6): 1115–29.

Uchikawa, H., Sawaki, D. and Hanehara, S. (1995) 'Influence of Kind and Added Timing of Organic Admixture on the Composition, Structure and Properties of Fresh Cement Paste', *Cement and Concrete Research*, 25 (2): 353–64.

Venuat, M. (1984) *Adjuvants et traitements: techniques modernes d'amélioration des ouvrages en béton*, n.p.

Vernet, C. (1995) Mécanismes chimiques d'interactioins ciment-adjuvants, CTG Spa. Guerville Service Physico-Chimie du ciment, January, 10 p.

Vernet, C. and Cadoret, G. (1992) 'Suivi en continu de l'évolution chimique et mécanique des BHP pendant les premiers jours', in Y. Malier (ed.) *Les Bétons à hautes performances: Caractérisation, durabilité, applications*, Paris: Presses de l'École Nationale des Ponts et Chaussées, pp. 115–56.

Vernet, C. and Noworyta, G. (1992) Interaction des adjuvants avec l'hydratation du C_3A: points de vue chimique et rhéologique, personal communication, 56 p.

Vichot, A. (1990) 'Les Polyméthylenenaphtylsulphonates: modificateurs de la rhéologie', Ph.D. thesis, Université Pierre et Marie Curie, Paris VI.

Yamada, K., Takahashi, T., Hanehara, S. and Matsuhisa, M. (2000) 'Effect of the Chemical Structure on the Properties of Polycarboxylate-Type Superplasticizer', *Cement and Concrete Research*, 30 (2): 197–207.

Yamada, K. and Hanehara, S. (2001) 'Interaction Mechanism of Cement and Superplasticizers: The Roles of Polymer Adsorption and Ionic Conditions of Aqueous Phase', *Concrete Science and Engineering*, 3 (11): 135–45.

Young, J.F. (1983) 'Slump Loss and Retempering of Superplasticized Concrete: Final Report', Civil Engineering Studies, Illinois Cooperative Highway and Transportation, University of Illinois at Urbana Champaign, Urbana, Illinois, March.

Chapter 8

Cementitious materials other than Portland cement

Supplementary cementitious materials, mineral components, Portland cement additions

8.1 Terminology

Although the terms 'clinker' and 'Portland cement' are quite well accepted, there are several other expressions used to designate the different materials that are blended with Portland cement clinker in cement plants or directly with Portland cement in concrete batching plants to produce hydraulic binders, that is, binders that harden with and in water. This is also true in languages other than English; for example, in French, the terminology is purposely vague: Baron and Ollivier (1995) make the following distinctions:

- Secondary constituents are products incorporated at a level not greater than 5 per cent. It is not necessary to be precise about their nature, and the cement can still be sold as Portland cement.
- Additives are products added in very small amounts during grinding. They do not modify the essential characteristics of the cement. For example, grinding aids are additives.
- Other constituents, which are products combined with Portland cement clinker to produce blended cements. Slag, silica fume, natural pozzolans, fly ashes, shale, calcined clays, limestone and siliceous fillers are some of the other constituents that can be found in blended cements.

This vague terminology results from the long tradition of usage of these products by the French cement industry.

The European standard ENV 197-1 is very specific, with each of the different products that can be blended with Portland cement clinker, they are designated by a symbol. For example, the letter K is used to designate clinker, S slag, P natural pozzolan, Q industrial pozzolan, V silicoaluminous fly ashes, W silicocalcic fly ashes, T calcined shale, L limestone filler, D silica fume and F fillers (in general) (Dreux and Festa 1995; Baron and Ollivier 1995; Jackson 1998).

In North America, different general expressions are used to designate all of these products. This lack of uniformity in the terminology comes from the fact that blended cements were not produced in North America in significant amounts before 1998. Before 1998, these products were introduced directly into the concrete in concrete plants. Presently in the literature these products can be designated by some of the following general expressions: mineralized admixtures, cementitious additions, cementitious materials, mineral components, supplementary cementitious materials, etc. ASTM standards C311 and C618 use the expression 'mineral admixture' but in the book *Design and Control of Concrete Mixtures*, the Portland Cement Association (Kosmatka et al. 2002) uses the expression 'supplementary cementitious materials'.

In this book we will favour the expression 'cementitious materials' in spite of the fact that some fillers which are presently blended with Portland cement have little or no cementitious properties by themselves. The expression 'mineral component', which is quite often found in some commercial literature has not been used because slags, fly ashes, silica fume, natural pozzolans are glassy or amorphous products. The expression 'supplementary cementitious materials' has not been kept because it is unnecessarily long. Therefore, in the following we will use the expression 'cementitious material', but in most cases we will be precise about the exact nature of the cementitious material.

The addition of cementitious materials when making concrete is not a new idea as has already been seen in Chapter 2. In fact, the first concretes, made long before the discovery of Portland cement, were made by mixing lime with natural cementitious materials or calcined clays. More recently, the first use of a slag dates back to the beginning of the twentieth century: in Germany, slag was blended with Portland cement to produce the first actual blended cement (Josephson et al. 1948; Moranville-Regourd 1998).

The use of cementitious materials in a specific location has been either favoured or fought by cement companies. For a cement producer, it is easy to perceive cementitious materials as direct competitors of Portland cement, especially when they are used directly in concrete plants. In such a case, they can be perceived as products that dilute Portland cement. However, they can be also perceived as materials that improve certain characteristics of fresh or hardened concrete (Regourd 1983b; Malhotra 1987; Malhotra and Mehta 1996) and as a source of profit.

Generally speaking, in Chapter 11, we will see that in the years to come, the use of cementitious material will increase, because it is the easiest way for the cement industry to decrease drastically its CO_2 levels of emission while adding value to coproducts of other industries. Adding value is another positive step in the implementation of a sustainable development policy (Uchikawa 1996; Nkinamubanzi and Aïtcin 1999; Malhotra 2000).

In Figure 8.1, most of the cementitious materials that are presently used by the cement and concrete industries have been placed in a CaO–SiO_2–Al_2O_3 ternary diagram in order to show the great variety of products with diverse chemical compositions that can be used as cementitious materials. This graph also shows that not all materials containing CaO, SiO_2 and Al_2O_3 can be used as cementitious materials.

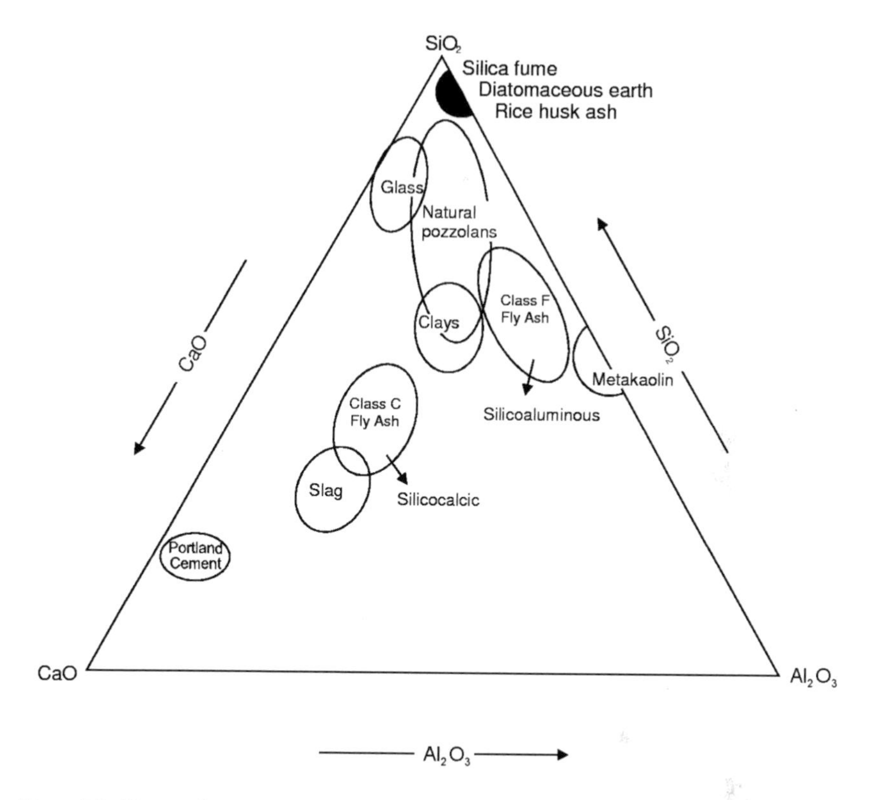

Figure 8.1 Chemical composition of the principal cementitious materials.

In order to simplify the presentation of the most important materials presently used as cementitious materials, they have been regrouped into three categories: the slags, the pozzolanic materials and the fillers.

These three categories will be examined in more detail in the following paragraphs. The number of references will be limited, and it is suggested that those who would like to further their knowledge on the subject consult the series of the Proceedings of the six International Conferences organized by ACI/CANMET on this subject. These conferences were held in Ottawa, Madrid, Istanbul, Trondheim, Milwaukee and Bangkok. These proceedings have been published as ACI SP 79, 91, 132, 114, 153 and 178, and each time a supplementary volume was published by CANMET.

Please note that it is always very important to remember, when reading a paper on cementitious materials, that it is dangerous to generalize the results and the conclusions of a paper. In spite of the fact that cementitious materials can be classified in broad categories, based on similar behaviour and effects when introduced into cement or concrete, no two cementitious materials are identical in their properties and effects. Moreover, the quality of cementitious materials can vary over time, because they are essentially coproducts or natural products having a chemical composition that is not as constant as that of a Portland cement. Portland cement is an industrial product produced with care, under the control of prescriptive standards.

8.2 Blast furnace slag

There are many type of industrial slags, but in this paragraph we will consider only the one that is used in large scale by the cement and concrete industries: the blast furnace slag produced during the fabrication of pig iron.

8.2.1 Fabrication process

In the literature, the following expressions are found to describe the coproduct of the fabrication of pig iron that is used as a cementitious material by the cement and concrete industries: granulated slag, quenched slag, vitreous slag and granulated ground blast furnace slag (GGBFS). In the following pages, this cementitious material will simply be called slag.

The first use of vitreous slag as a cementitious material dates back to 1868 as a result of the work done by E. Langen in Germany (Papadakis and Venuat 1966).

When producing pig iron in a blast furnace, all the impurities of the iron ore and the clay used to pelletize it, as well as all the impurities contained in the metallurgical coke, are melted and float on top of the heavier melted pig-iron bed. These impurities are collected at the bottom of the blast furnace (Figure 8.2). The liquefied slag is quenched as soon as it is tapped from the furnace. This quenching is usually done in modern plants by projecting pressurized water in gullies in which the molten slag is running. The slag can also be poured into large water basins or projected into the air after being quenched with a small amount of water. When the slag is quenched in water, it is transformed into a sand whose colour can vary from pale beige to dark brownish beige. When it is quenched in air, it is transformed into 10 to 20 mm pellets having a pale or dark beige colour.

From a mineralogical point of view, quenched slag is amorphous, and it can also be described as vitreous or a glassy material.

Blast furnace slag can also be left to cool down slowly in air in special pits where it crystallizes essentially as melilite (Figures 8.2 and 8.3). Melilite is a solid solution of ackermanite $2CaO \cdot 2SiO_2 \cdot MgO$ (C_2S_2M) and gehlenite $2CaO \cdot SiO_2 \cdot Al_2O_3$ (C_2SA). Crystallized slag does not have any cementitious properties and can only be used as an aggregate when making concrete, or as part of the raw meal, as will be seen in Chapter 11.

Melted slag can also be transformed into a lightweight aggregate or mineral wool (Josephson et al. 1949; Moranville-Regourd 1998).

Figure 8.2 Crystallized slag.

Figure 8.3 Crystallized slag or air-cooled slag under polarized light in an optical microscope. The prismatic white crystals are melilite crystals.

The iron-ore pellets which are currently introduced at the top of modern blast furnaces are much purer than the crude iron ores which were used fifty years ago, but they always still contain some impurities. Iron-ore pellets are usually made by binding concentrated iron-ore particles with clay. As a result, less slag is currently produced when making pig iron as compared to fifty years ago.

From a process point of view, it is necessary to have some slag on top of the layer of melted pig iron to purify the pig-iron droplets that are passing through the layer of molten slag (Figure 8.4).

As the melting temperature of the impurities contained in the iron-ore pellets and the metallurgical coke can be quite high, as compared to the melting temperature of the pig iron, steel producers add fluxing agents (Figure 8.5) to correct the chemical composition of these impurities and to produce a slag whose chemical composition corresponds in the ternary phase diagram $CaO–SiO_2–Al_2O_3$ to a composition whose melting temperature is one of the two lowest possible in the ternary diagram $CaO–SiO_2–Al_2O_3$ (Osborn and Muan 1960; Moranville-Regourd 1998).

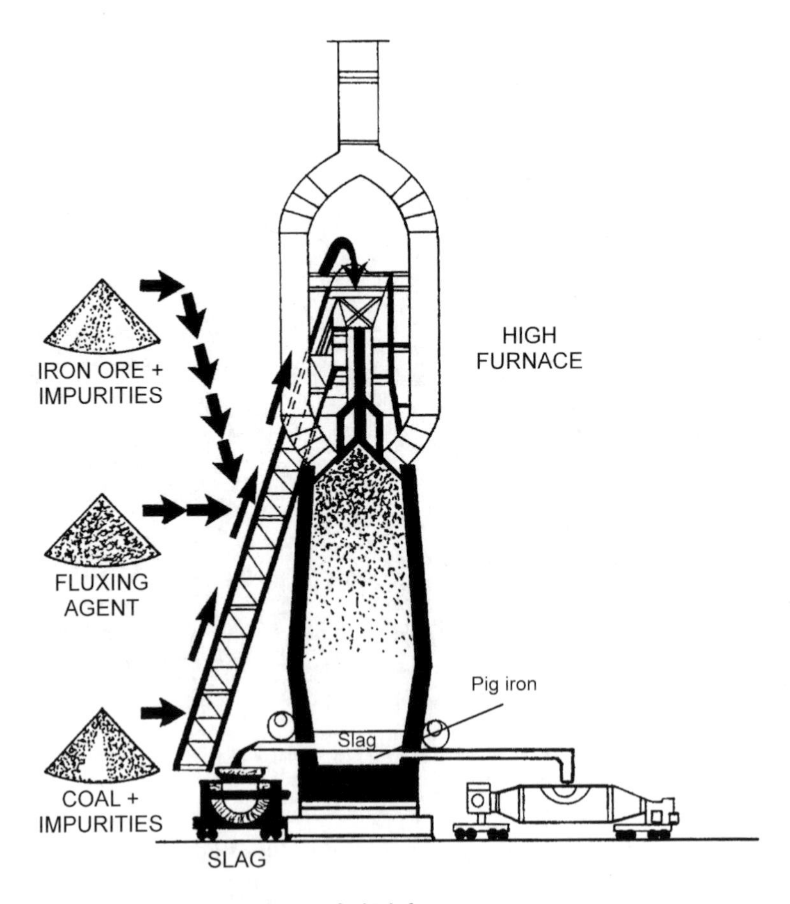

Figure 8.4 Schematic representation of a high furnace.

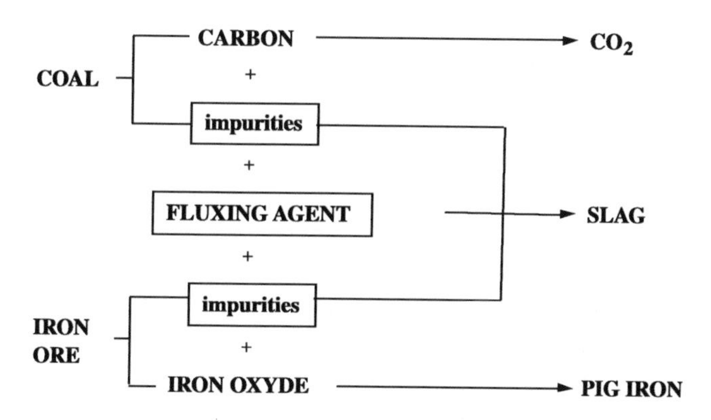

Figure 8.5 Schematic representation of the production of slag.

In the $CaO–SiO_2–Al_2O_3$ phase diagram, there are two eutectic compositions that have low melting temperatures E_1 and E_2 (Figure 8.6). E_1 has a chemical composition rich in SiO_2 that would produce a pig iron too rich in SiO_2. Steel producers prefer, for the quality of their pig iron, to adjust the composition of their slag to E_2 eutectic which has a slightly higher melting temperature. Therefore, blast-furnace slags have a chemical composition that varies within narrow limits. Their chemical composition shows that they can contain different amounts of Al_2O_3 and MgO depending on the type of fluxing agent that was used as shown in Table 8.1. Usually, this variation does not have a dramatic influence on the reactivity of the slag as a cementitious material. In the $CaO–SiO_2–Al_2O_3$ diagram (Figure 8.1), it is seen that slags have a composition quite close to the composition zone of Portland cements.

Quenched slag (Figure 8.7) displays hydraulic properties when it is activated by Portland cement, calcium sulphate and alkalis. At twenty-eight days a slag-blended cement is as strong as a pure Portland cement when used at the same water/binder ratio. The early strength increase of slag cement is, of course, slower than that of Portland cement (Moranville-Regourd 1998).

It is easy to check the amorphous state of a quenched slag through an X-ray

Table 8.1 Chemical composition of some slags

	French slags	*North American slags*
SiO_2	29 to 36	33 to 42
Al_2O_3	13 to 19	10 to 16
CaO	40 to 43	36 to 45
Fe_2O_3	< 4%	0.3 to 20
MgO	< 6%	3 to 12
S^-	< 1.5%	–

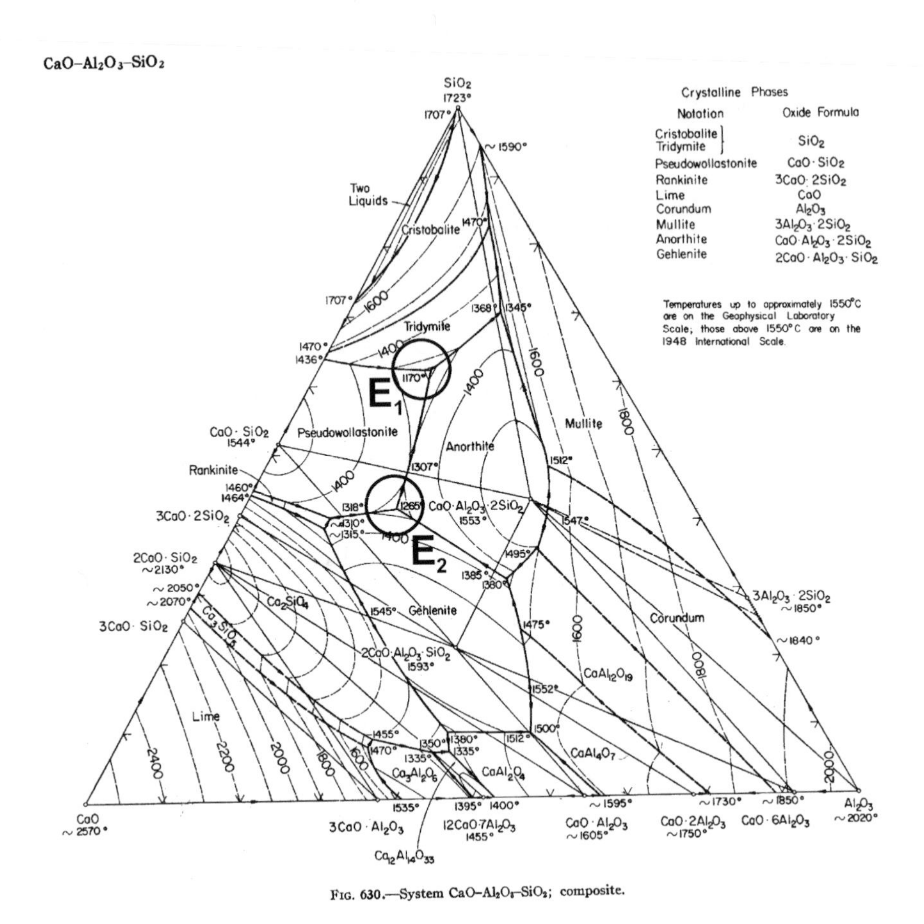

Figure 8.6 Position of the two eutectics having the lowest temperature of melting in a $CaO–SiO_2$ phase diagram.

diffractogram. The diffractogram does not present any peaks but rather a hump located at a 2θ angle around 30°. This angle corresponds to the principal peak of melilite (Figure 8.8).

The best slags are 'hot' slags which were quenched when they were at a high temperature. The ions in the solid slag are quite disorganized, making the slag quite reactive. Hot slags have a pale colour (Aïtcin 1968). 'Cold' slags have a darker colour and are usually not as reactive. Slags whose quenching temperature was between the liquidus and solidus temperatures contain some minute crystals of melilite (Figure 8.9). The closer a slag's quenching temperature to its solidus temperature, the more melilite crystals it has in it and the darker it is.

Slag can be blended directly with clinker and calcium sulphate to produce blended cements, but slag can also be introduced directly into concrete mixers

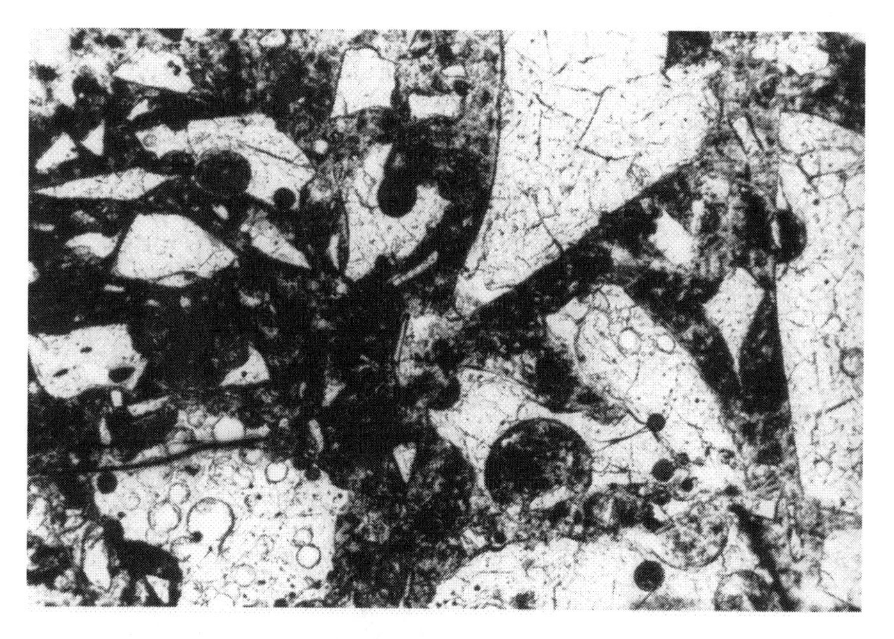

Figure 8.7 In white, vitreous slag particles. Note the angular shape of the slag particles and their porosity. This slag was quenched at a high temperature because no crystals are visible. Such slag is called a hot slag.

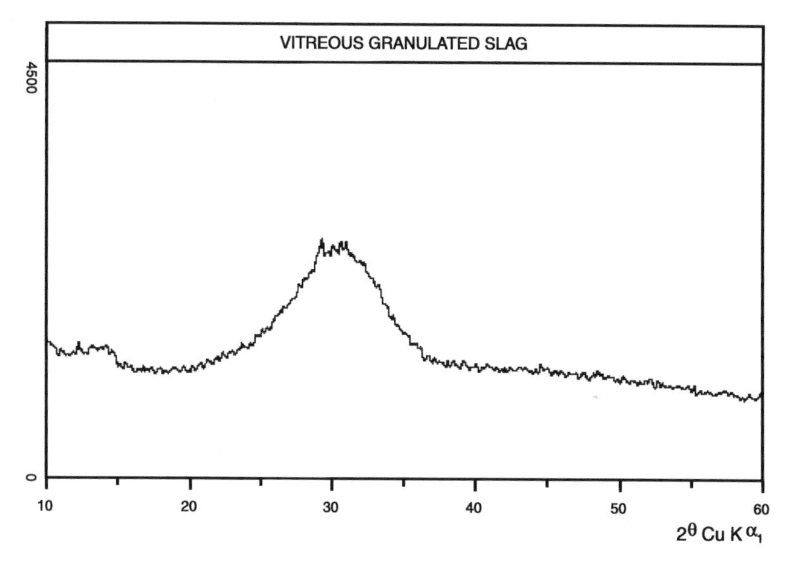

Figure 8.8 X-Ray diffractogram of a hot slag after its quenching.

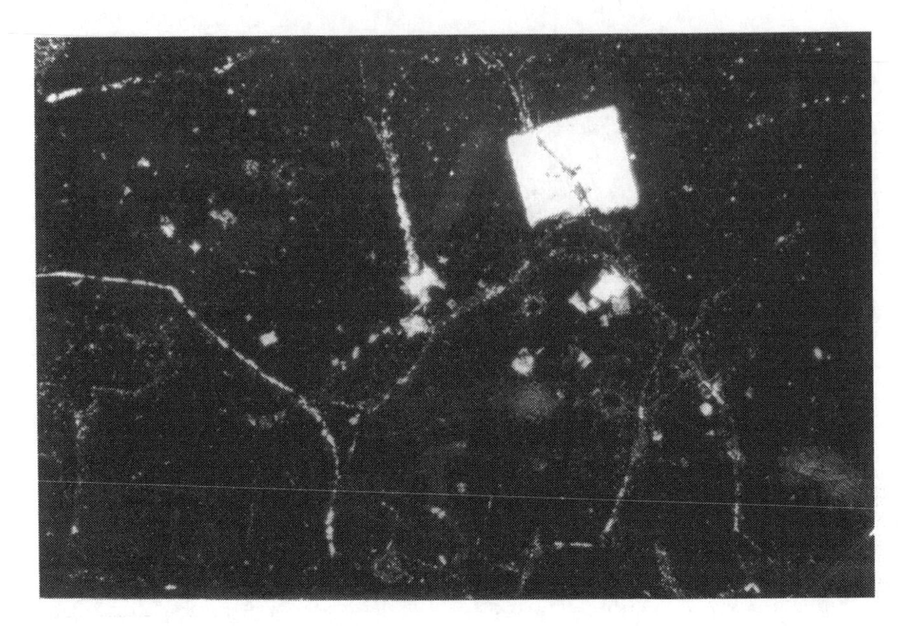

Figure 8.9 Melitite crystal in a slag particle after quenching. This slag was a cold slag quenched under a temperature of a liquidus.

to produce slag concrete. In order to avoid rheological problems, it is better to introduce an appropriate amount of calcium sulphate with the slag (Saric-Coric and Aïtcin 2003).

Slag (quenched or not) can also be used as a raw material during the preparation of the raw meal when making Portland cement and results in a decrease of CO_2 emission. The lime it brings to the raw meal during the fabrication of Portland cement decreases the amount of limestone necessary to adjust the composition of the raw meal, or more precisely, this lime has already released its CO_2 into the atmosphere at the steel plant (see Chapter 11).

8.2.2 Slag hydration

Slag is neither a hydraulic binder, nor a pozzolanic material (Tanaka et al. 1983; Moranville-Regourd 1998). When slag is mixed with water, it does not harden nor does it combine directly with the lime liberated by the C_3S and C_2S to form secondary C-S-H. Slag can be activated by lime and also by calcium sulphate, potash or soda, which act as catalysts in the attack of the glass (Figure 8.10). This kind of chemical attack is called alkaline activation.

When a finely ground slag is mixed with water, few Ca^{2+} ions get into solution and, instead an impervious 'acid' gel is formed around slag particles. This 'acid' gel temporarily blocks the dissolution of the slag. However, the lime

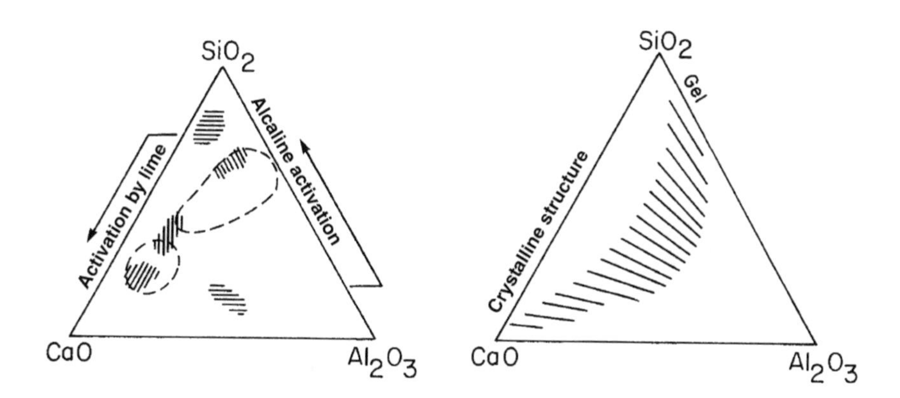

Figure 8.10 Principle of the different activation types for the supplementary cementing materials (according to Idorn).

liberated by the hydration of C_3S and C_2S reacts with this 'acid' layer of gel to form C-S-H and C_4AH_{13}. When slag is activated by calcium sulphate, C-S-H and ettringite are formed.

Portland cement is a good catalyst for slag activation because it contains the three main chemical components that activate slag: lime, calcium sulphate and alkalis. Sulfate ions dissolve the first layer of hydrate formed around slag particles when they entered into contact with water. The result of this reaction is the formation of a coarse layer of hydrate that is less impervious than the 'acidic' gel and does not oppose further penetration of water. When slag particles are in the presence of a saturated solution of Ca^{2+} ions, slag hydration accelerates, and well-crystallized hydration products are formed within the slag particle. A hydrate 'skeleton' is thus formed as shown schematically in Figure 8.11. Slag somewhat delays the hydration of the C_3S when 'pumping' Ca^{2+}, and it also delays the hydration of C_3A and C_4AF.

8.2.3 Effect of slag on the main characteristics of concrete

Of course, the effect of slag on the properties of a blended cement are proportional to the level of substitution. European standards consider several levels of substitution: from 6 to 20 per cent and from 21 to 35 per cent in the so-called CPJ blended cements, but also 36–65 per cent, 66–80 per cent and 81–95 per cent in blast-furnace cements. Some ternary blends have also been standardized. Apart from Portland cement, they can contain 20–39 per cent and 40–64 per cent of slag.

In North America, only three types of slag-blended cement have been standardized in the ASTM C980 standard: class 80, 100 and 120 slag-blended cements; the higher the grade, the greater the cube strength. For example, a

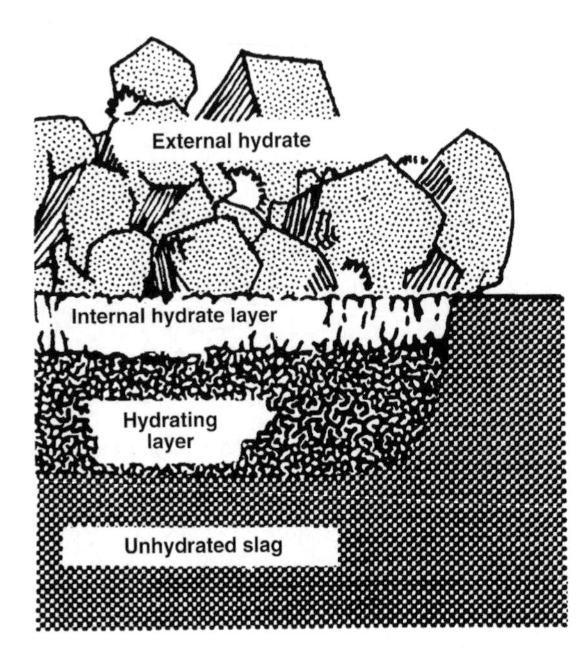

Figure 8.11 Schematic representation of the hydration of slag, according to Tanaka et al. (1983). Reprinted with permission from the American Concrete Institute.

120-grade slag has an activity index of 115 when it is mixed in equal parts with Portland cement. The activity index, at a given age, represents the ratio between the cube strength of a mortar that contains slag and the cube strength of a pure Portland cement mortar of the same water/binder ratio.

The effect of slag is also proportional to its fineness. Generally speaking, slags are ground finer than Portland cement to increase their reactivity. When slag and Portland cement clinker are ground together, it is the clinker that is ground finer because it is not as hard as slag.

Generally speaking the addition of slag:

- lowers early age strength (24 hours), particularly in cold weather;
- lowers the amount of heat that is initially liberated, but, in order to reach a significant decrease of the heat of hydration it is necessary to reach a degree of substitution that is higher than 50 per cent (such a decrease of heat is particularly advantageous in hot countries and for mass concrete [Bramforth 1980]);
- increases the ratio of tensile strength/compressive strength;
- increases the imperviousness of concrete;
- very efficiently protects concrete against swelling when potentially reactive aggregates (to alkalis) are used;

- increases the resistance of concrete to sulphates;
- increases the resistance of concrete to sea water.

The addition of a certain percentage of fly ash in ternary blended cements (cements containing Portland cement, slag and fly ash) is very advantageous, as there is a kind of synergitic effect between the slag and the fly ash that has not yet been clearly explained. A synergitic effect that is even more advantageous is obtained when, instead of adding a fly ash, 5 to 10 per cent silica fume is blended in the slag cement (Baalbaki et al. 1992; Nkinamubanzi et al. 1998).

Very recently some cement companies have started to market quaternary blended cement containing slag, fly ash and silica fume (Isaia 1997).

8.3 Pozzolans

The second great family of cementitious materials used in concrete is pozzolans. This general term is used to designate natural as well as industrial coproducts that contain a certain percentage of vitreous silica (Figure 8.12). This vitreous (amorphous) silica reacts at ambient temperature with the lime produced by the hydration of C_3S and C_2S to form C-S-H similar to that produced by the direct hydration of C_3S and C_2S (Venuat 1984; Malhotra 1987; Malhotra and Mehta 1996).

ASTM C618 proposes the following definition for the word pozzolan: 'siliceous, or siliceous and aluminous materials, which in themselves possess little or no cementitious value but will, in finely divided form and in the presence of moisture, chemically react with calcium hydroxide at ordinary temperatures to form compounds possessing cementitious properties'.

When a pozzolanic reaction takes place, the lime produced during C_3S and C_2S hydration is transformed into calcium silicate hydrate. This can be considered as the weakest link of the hydration products of Portland cement in terms of mechanical properties and durability (lime can be easily leached out). Therefore, a pozzolanic reaction improves the intrinsic quality of the cement paste and, consequently, the quality of concrete (Massazza 1998). It is only in the field of reinforcing bar passivation that this consumption of lime is not as beneficial, because it results in a decrease in the pH of the interstitial solution. But, in decreasing the permeability of the concrete, a pozzolanic reaction also decreases the vulnerability of reinforced concrete to the penetration of aggressive agents.

Schematically a pozzolanic reaction can be written in the following manner:

pozzolan + portlandite → calcium silicate hydrate

Therefore, hydration of a blended cement containing a pozzolan can be written as follows:

Portland cement + pozzolan + water → C-S-H + sulphoaluminates

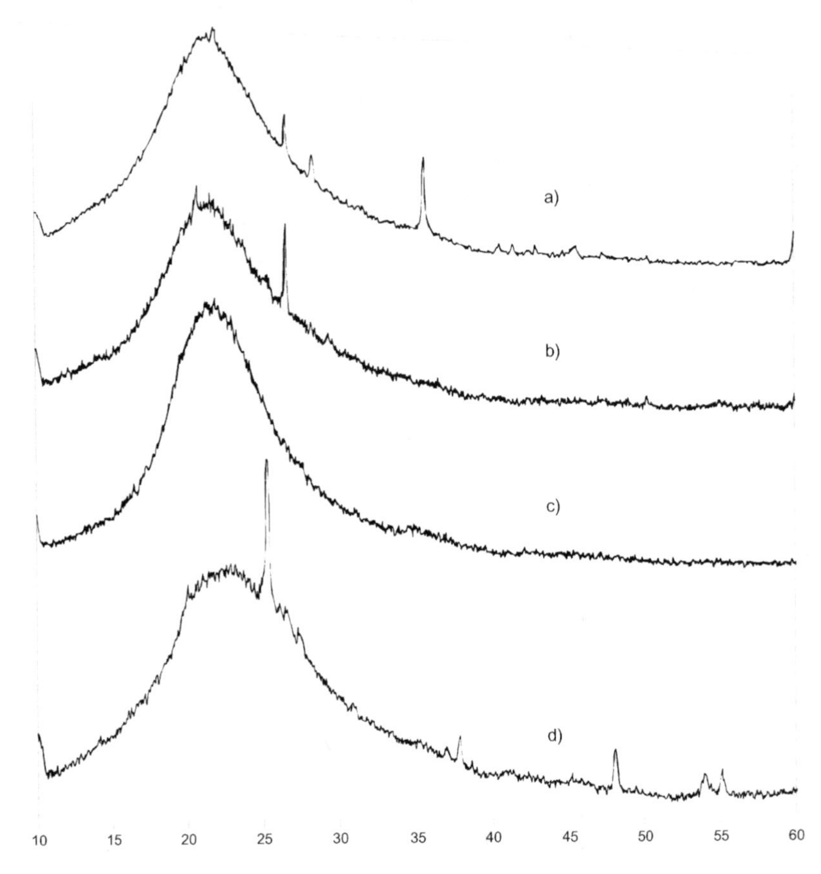

Figure 8.12 Typical diffractograms of various forms of amorphous silica used as supplementary cementitious materials: (a) silica fume; (b) rice husk ash; (c) diatomaceous earth; (d) metakaolin.

Of course, in the real world, a pozzolanic reaction never develops to the extent that all the lime liberated by C_3S and C_2S hydration is consumed. In fact, the biggest particles of vitreous silica, or the ones that are partially crystallized, stay unreacted or partially reacted. Even in the case of silica fume, which is a very fine reactive pozzolan, 100 times finer than Portland cement, it is possible to see unreacted silica-fume particles in old concretes.

At ambient temperature, the development of a pozzolanic reaction is much slower than the rate of Portland cement hydration, but a water-cured concrete that contains a pozzolan has a strength that increases and a permeability that decreases with time. Of course, the fineness of the pozzolan, its degree of amorphousness and the temperature of the concrete influence significantly the strength increase and permeability decrease of concrete.

In the following sections pozzolans will be divided into two main categories: artificial and natural pozzolans. The first category includes fly ashes, silica fumes, calcined clays and shales, metakaolin and rice-husk ash. Volcanic ashes, volcanic tuffs, trass and zeolites are the natural pozzolans that will be briefly presented. It is well known that heating to a temperature between 500 and 800°C improves the pozzolanic reactivity of natural pozzolans, but this is rarely done due to the costs associated with this heating (Malhotra and Mehta 1996).

8.3.1 Fly ashes

The term 'fly ash' is used to describe the fine particles that are collected in the dedusting system of power plants that burn coal or lignite (Figure 8.13). Fly ashes are in fact composed of all the fused and unfused residues that are present in the pulverized coal or lignite that is injected in the burners (Figure 8.14). These fused particles are quenched when they leave the flame, and they solidify in the shape of vitreous glass spheres (Figures 8.15 and 8.16) but a certain number of coarse impurities can pass so rapidly through the flame that they are not fused (Figure 8.17).

As all the coals and lignites that are burned in power plants do not contain the same impurities, the fly ashes that are recovered in the dedusting system have a chemical and phase compositions that vary over a wide range, as shown in Table 8.2 and Figure 8.17. Moreover, the combustion of coal or lignite is not always completed to 100 per cent; therefore, fly ash usually contains a certain percentage of unburned carbon, usually at a content lower than 6 per cent of its mass. Finally, the amount of unburned carbon can vary within a day, as it is usually lower when the power plant is producing the maximum energy during peak hours, but greater during the night when the power plant produces less energy.

Therefore, the term 'fly ash' covers a large family of powders having approximately the same grain-size distribution as Portland cement and containing more or less vitreous particles. There are some fly ashes which are entirely crystallized and therefore do not have any pozzolanic action (Figures 8.18 and 8.19). There are fly ashes with a very low carbon content, others with a high one, some with a constant carbon content, others with a variable carbon content. The carbon content of a fly ash is very important, particularly when using admixtures, because carbon particles can adsorb preferentially some admixtures, particularly air-entraining agents.

In an effort to classify fly ashes into broad categories, they are now regrouped into two or three main classes, but it should be remembered that there are fly ashes that close the gap between the ideal characteristic of each fly ash class.

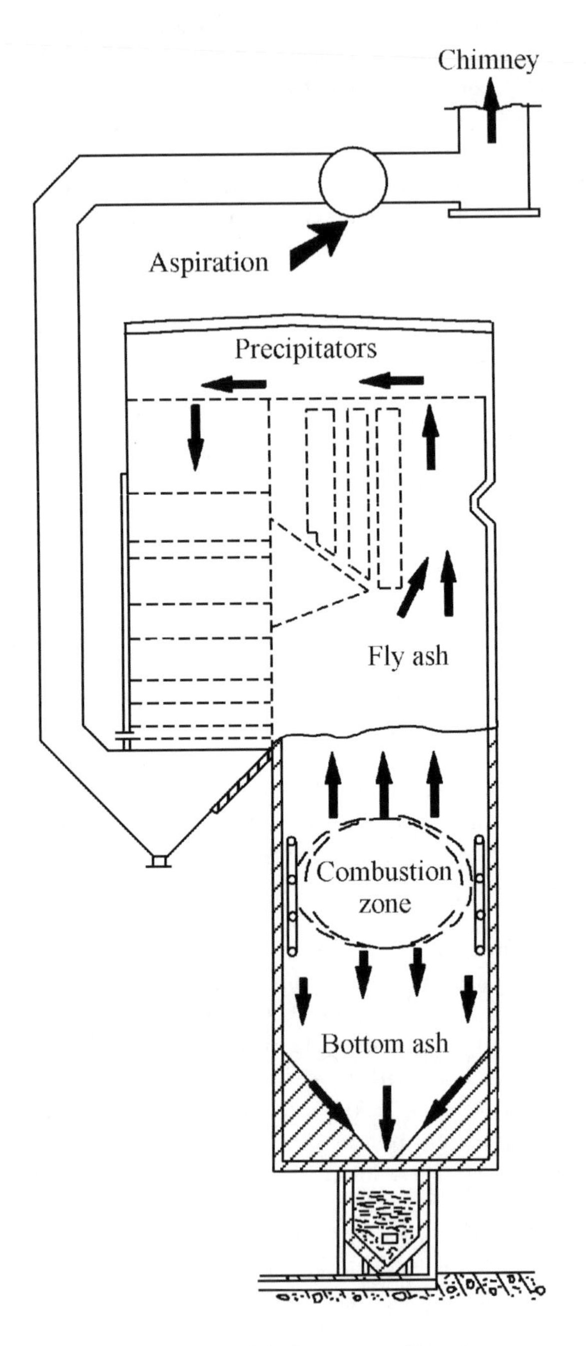

Figure 8.13 Schematic representation of the formation of fly ashes.

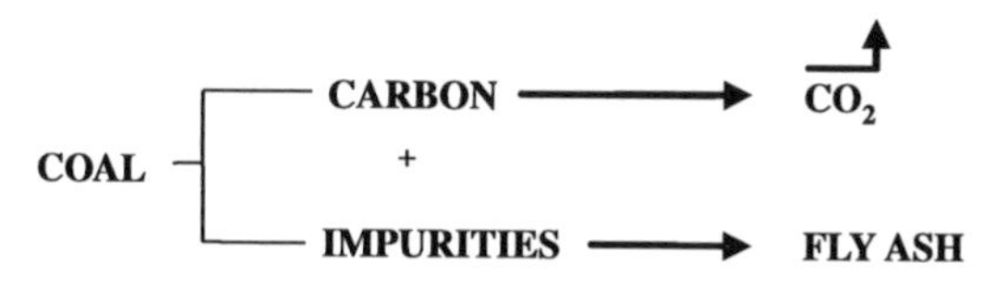

Figure 8.14 Schematic representation of the production of fly ash.

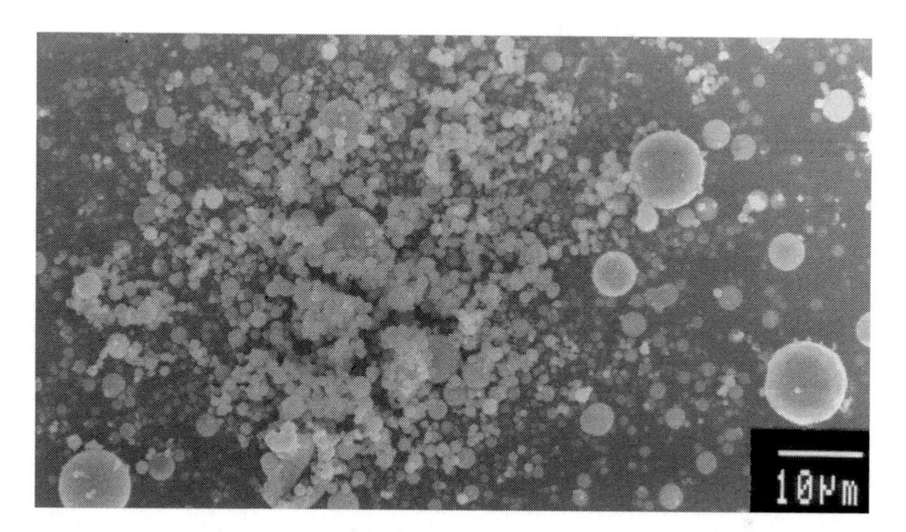

Figure 8.15 Spherical particles of fly ashes.

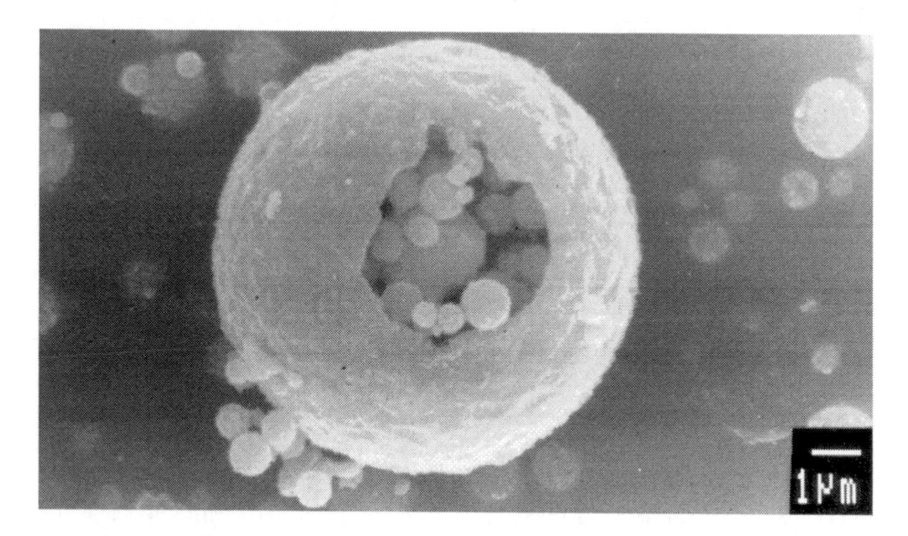

Figure 8.16 Plerosphere containing cenospheres in a fly ash.

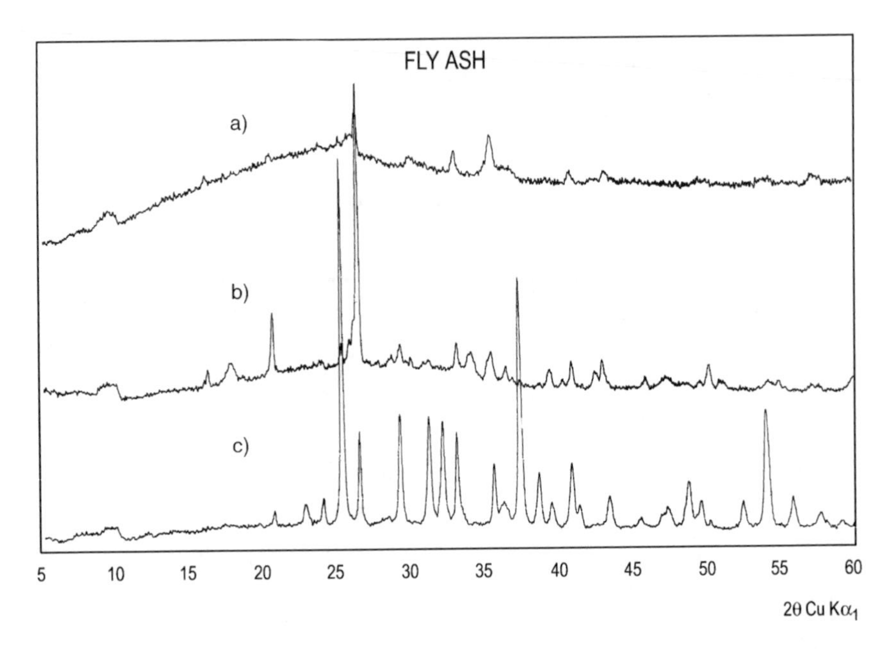

Figure 8.17 Typical X-Ray diffractogram of different types of fly ash: (a) Class F or silicoaluminous; (b) Class C or silicocalcic; (c) sulphocalcic fly ash totally crystallized.

Table 8.2 Chemical composition of some fly ashes

	Class F Silicoaluminous		*Class C Silicocalcareous*	*Sulpho-calcareous*	*Sulpho-calciareous*
SiO_2	59.4	47.4	36.2	24.0	13.5
Al_2O_3	22.4	21.3	17.4	18.5	5.5
Fe_2O_3	8.9	6.2	6.4	17.0	3.5
CaO	2.6	16.6	26.5	24.0	59
MgO	1.3	4.7	6.6	1.0	1.8
Na_2O equiv.	2.2	0.4	2.2	0.8	–
SO_3	2.4	1.5	2.8	8.0	15.1
Loss on ignition	2.0	1.5	0.6	–	–
$SiO_2 + Al_2O_3 + Fe_2O_3$	90.7	74.9	60.0	59.5	22.5
Free lime	–	–	–	–	28.0

ASTM C618 standard recognizes two main families of fly ashes based on the value of the sum of SiO_2, Al_2O_3 and Fe_2O_3:

- If the value of $SiO_2 + Al_2O_3 + Fe_2O_3$ is greater than 70 per cent, the fly ash is said to be a Class F fly ash.
- If the value of $SiO_2 + Al_2O_3 + Fe_2O_3$ is lower than 70 per cent and the CaO content higher than 10 per cent, the fly ash is said to be a Class C fly ash.

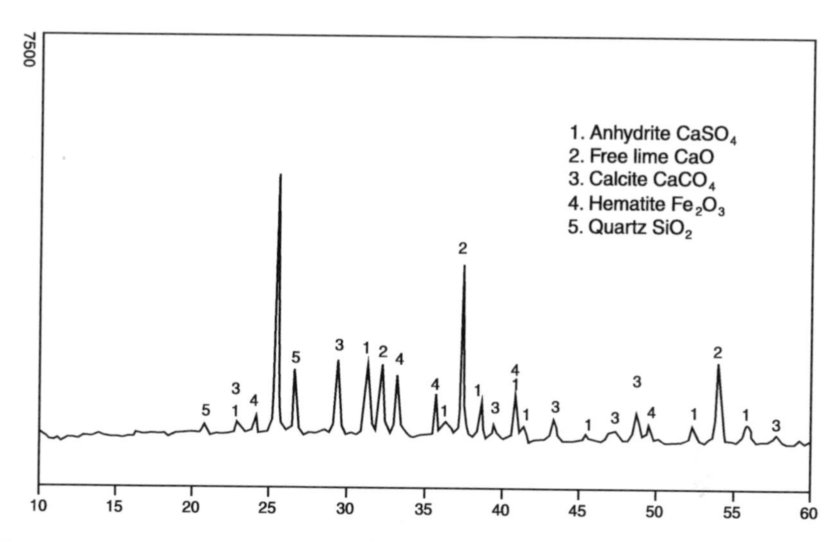

Figure 8.18 X-ray diffractogram of a totally crystallized fly ash.

Figure 8.19 Coarse crystallized fly ash particle (courtesy of I. Kelsey-Lévesque).

In France, fly ashes are classified as silicoaluminous fly ashes (Class F) and silicocalcareous fly ashes (Class C). There is another class of fly ash called sulphocalcareous fly ashes which correspond to fly ashes that are rich in sulphur and calcium.

Silicoaluminous fly ashes (Class F fly ashes) are truly pozzolanic, silico-calcareous ones are more or less so. Some Class C fly ashes have been found to have cementitious properties by themselves, but others are definitely not pozzolanic et all (Aïtcin et al. 1986).

This long introduction on the terminology and classification of fly ashes has been made only to show that it is always dangerous to generalize the results obtained with a particular fly ash, because two fly ashes of the same class can be quite different. Of course, in the same family there are many fly ashes that present behaviours which do not deviate very much from an average behaviour (Aïtcin et al. 1986; Swamy 1993).

Before presenting the main effects resulting from the substitution of a certain mass of Portland cement by an equivalent mass of fly ash in a concrete, it is interesting to speak about a new type of fly ash that is starting to be available on the market. These fly ashes are sold as beneficiated fly ashes or as micro fly ashes.

Up until very recently, it was out of the question to spend money to try and improve the quality of a fly ash, as the only objective was to get rid of it. But at the moment some fly ashes are sold with a very low carbon content varying between 1.5 to 2 per cent, and the coarse particles of some fly ashes are eliminated with cyclones so that most of the non-reactive crystallized particles are no longer contained in the fly ash. The coarse particles that are extracted by cyclones are essentially composed of quartz, ion ore and alumina and can be used to correct the chemical composition of the raw meal. Of course, such fly ashes that are clean (low carbon content) and that do not contain any non-reactive coarse crystalline particles are more reactive than the initial fly ash and can be sold for a much better price to the cement and concrete industries.

Moreover, in certain power plants, dedusting precipitators are composed of several stages so that particles of different cleanliness and particle size are collected separately and can be commercialized separately. For example, two Australian power plants are producing two beautiful and efficient fly ashes known as pozzofume and kaolite. Pozzofume is a micro fly ash having an average particle size not greater than 2 μm, which is in fact a very active pozzolan. Unfortunately, the power plant produces only 2000 to 3000 tonnes of pozzofume every year. This very reactive pozzolan is mixed with another fly ash which is somewhat coarser (average particle size between 10 and 20 μm) and which has a chemical composition similar to metakaolin. The mixture of these two very particular fly ashes results in a super fly ash that is only surpassed by silica fume as a pozzolanic material.

The reactivity of a fly ash can be increased by grinding. The creation of a fresh fractured surface increases the reactivity of the fly ash because it exposes more unbalanced ions, but the advantage of the spherical form of the fly ash from a rheological point of view is thus lost (Naproux 1994).

Now that fly ashes are considered as a coproduct of coal firing with an added value, it is certain that the quality of the fly ashes available will increase, or some cement companies will invest money before introducing them into their blended cements.

8.3.2 Silica fume

Silica fume is the byproduct of the fabrication of silicon metal, ferrosilicon alloys or occasionally of the fabrication of zirconium (Figure 8.20). These metals and alloys are produced in arc electric arc furnaces where, for example, in the case of silicon, quartz (SiO_2) is reduced to the metallic state Si. During the process SiO vapors are formed within the arc, and when these vapors escape from the top of the furnace and enter into contact with air, they are oxidized and form tiny vitreous SiO_2 vitreous particles, a hundred time finer than Portland cement particles (Figure 8.21). These very fine particles are collected in dedusting systems (Figure 8.22) (Aïtcin 1983; Fidjestøl and Lewis 1998). A few other particles are also collected in the dedusting system: very fine untransformed quartz particles, carbon particles, graphite particles from the electrodes and, occasionally, some wood chips (when silicon is produced). Usually these 'impurities' count for a very small percentage of the silica fume collected in the dedusting system. Silica fumes can also contain a small amount of silicon metal that can react with the lime liberated during the hydration of C_3S and C_2S to form hydrogen (Edwards-Lajnef et al. 1997).

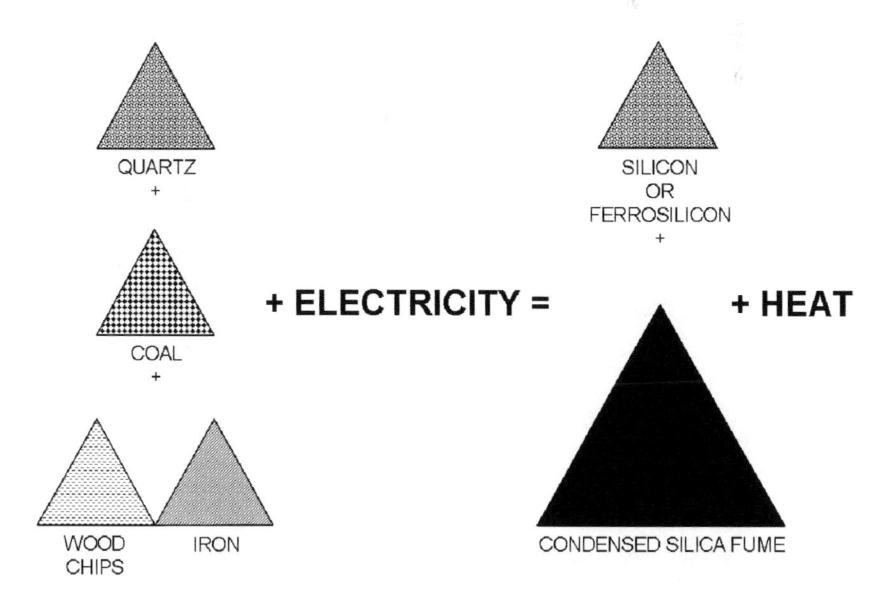

Figure 8.20 Principle of the production of silica fume.

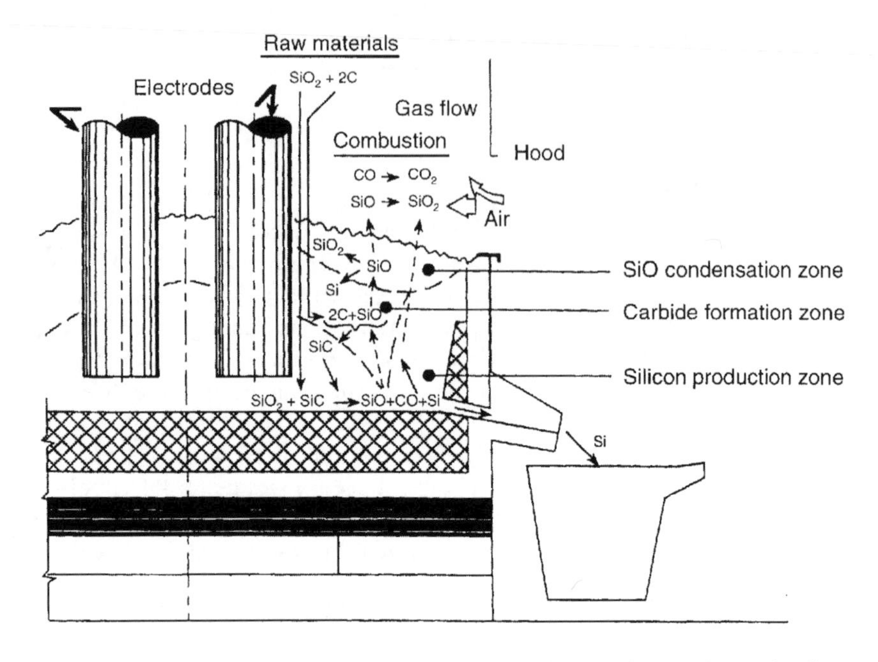

Figure 8.21 Schematic representation of an electric arc furnace that produces silica fume.

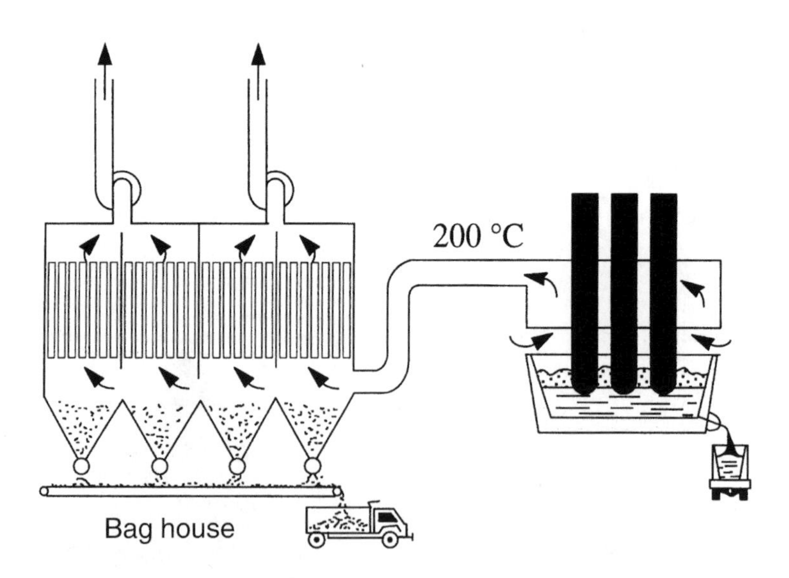

Figure 8.22 Collecting silica fume in the 'bag house'.

The average diameter of amorphous silica particles is about to 0.1 μm (Figures 8.23 and 8.24). Silica-fume particles are usually grey in colour, darker or lighter according to their carbon and iron content.

An X-ray diffractogram of a silica fume presents a hump centred at the location of the main peak of cristobalite (one form of crystallized SiO_2 stable at high temperature) (Figure 8.26). The specific gravity of silica fume is of the order of 2.2, which is the usual specific gravity of vitreous silica. Its specific surface area, which has to be measured by nitrogen adsorption, varies over a wide range from 10000 to 25000 m^2/kg, which is 6 to 10 times the specific surface area of a Portland cement when it is measured by nitrogen adsorption.

As the materials introduced in the arc furnace to produce silicon or ferrosilicon have quite a constant chemical composition, the silica fumes that are collected in the dedusting system also have a constant chemical composition (Table 8.3) (Pistilli et al. 1984; Pistilli et al. 1984).

Some arc furnaces are equipped with a heat-recovery system so that the exhaust gas leaves the furnace at a temperature of 800°C. The silica fume collected from a furnace equipped with a heat-recovery system is whitish because it does not contain any carbon or graphite particles. The silica fume collected during the fabrication of zirconium is also white. In conventional furnaces, the hot gas has to be diluted with fresh air in order to lower the average temperature of the gas to a temperature lower than 200°C. It is important to lower the temperature of the exhaust gas in order to avoid the burning of the filter bags.

Silica fumes are very reactive fine pozzolans (Figure 8.27) that can have, in addition to their pozzolanic effect, a filler effect (Regourd 1983a; Detwiler and

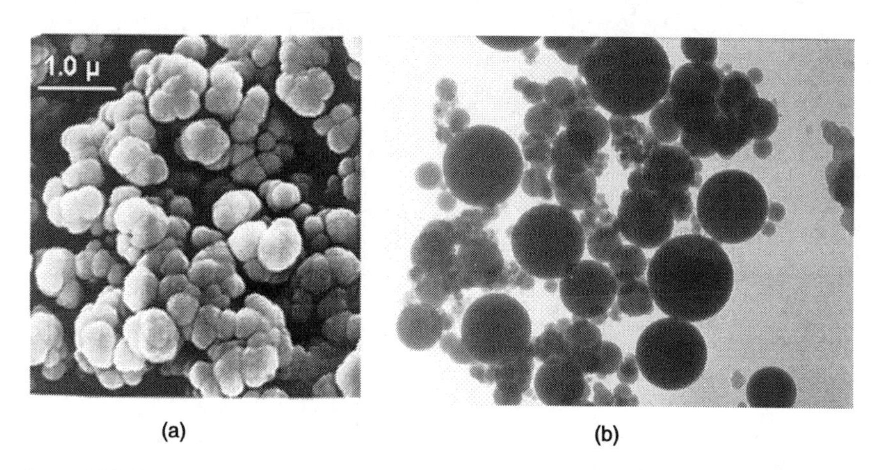

(a) (b)

Figure 8.23 Silica fume as seen in an electronic microscope: (a) scanning electronic microscope, silica fume particles are naturally agglomerated in an as-produced silica fume; (b) transmission electronic microscope, dispersed individual particles.

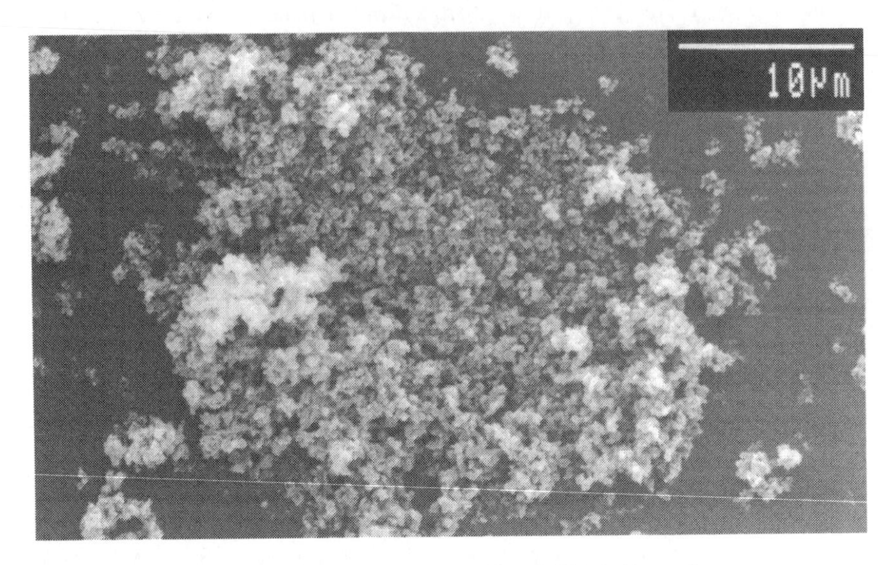

Figure 8.24 As produced silica fume (courtesy of Arezki Tagnit-Hamou).

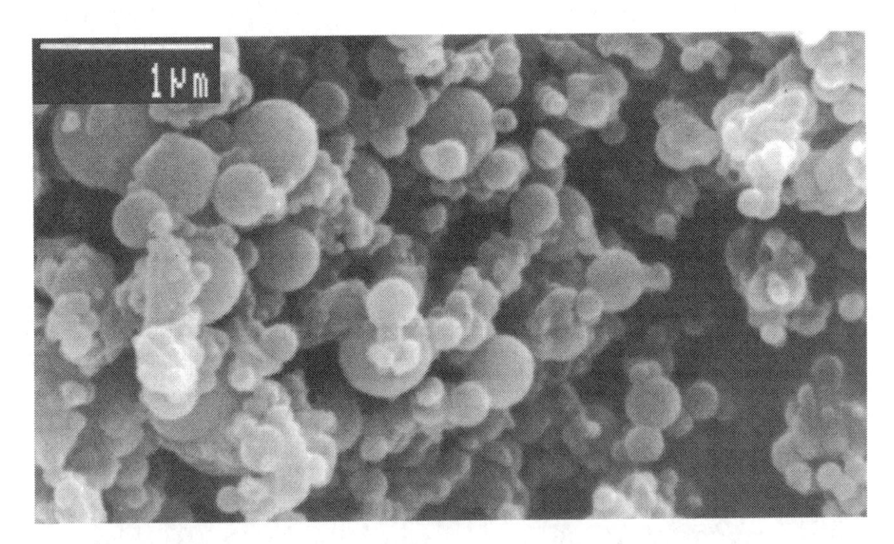

Figure 8.25 Enlarged view of the preceding figure. It is seen the silica-fume particles have a diameter lower than 1 µm (courtesy of Arezki Tagnit-Hamou).

Mehta 1986; Regourd et al. 1986; Mehta 1987; Goldman and Bentur 1993a; Sellevold and Justnes 1993; Fidjestøl and Lewis 1998). Silica-fume particles can also play the role of nucleation sites for very tiny portlandite crystals ($Ca(OH)_2$). These tiny portlandite crystals are not seen under an electron

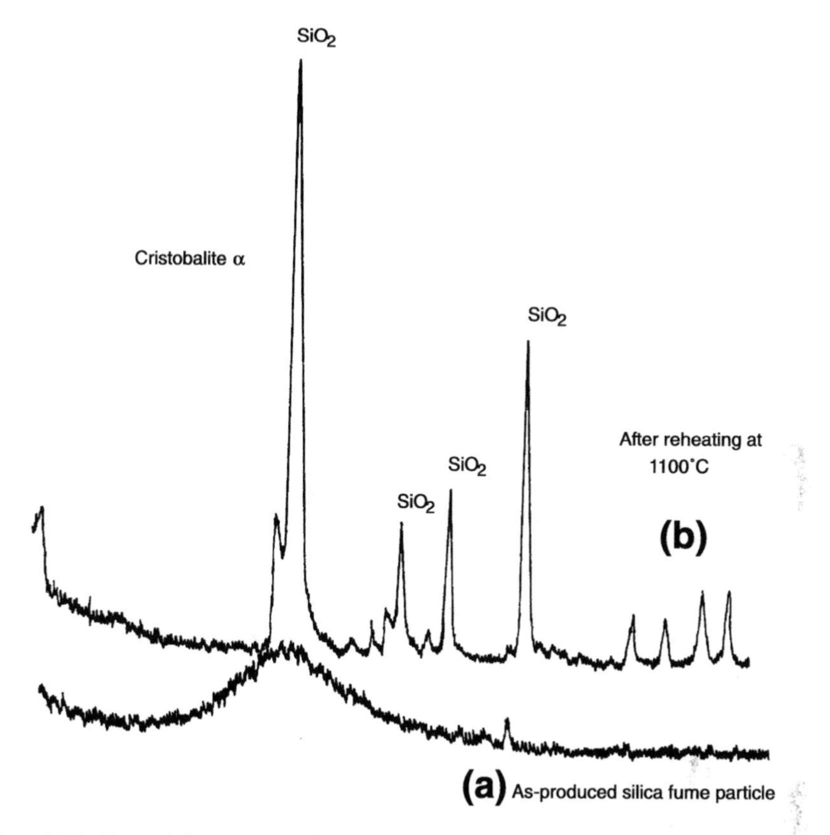

Figure 8.26 X-ray diffractogram of a silica fume particle: (a) as-produced; (b) after its reheating at 1100°C. After the reheating the silica crystallized as cristobalite. The hump found in the as-produced silica fume corresponds to the main pick of cristobalite which indicates than in a silica fume particle silica tetrahedrons are organized as in cristobalite α on a short-range distance.

microscope or in an X-ray diffractogram, but they can be observed when carrying out a differential thermal analysis. When following the hydration of a silica-fume blended cement it is seen that, at the beginning, some portlandite is present (dehydration peak at 450°C) and that this peak disappears with time when the portlandite reacts with silica-fume particles to form some C-S-H.

The introduction of silica fume into concrete decreases bleeding and modifies the microstructure of the hydrated cement paste which looks quite massive and 'amorphous' (Figure 8.28). Moreover, it is observed that the transition zone around the coarse aggregates is much more compact than that observed during the hydration of pure Portland cement (Regourd 1983a).

Table 8.3 Chemical composition of some silica fumes

	Grey production of silicium	*Grey production of ferrosilicium*	*White*
SiO_2	93.7	87.3	90.0
Al_2O_3	0.6	1.0	1.0
CaO	0.2	0.4	0.1
Fe_2O_3	0.3	4.4	2.9
MgO	0.2	0.3	0.2
Na_2O	0.2	0.2	0.9
K_2O	0.5	0.6	1.3
LOI	2.9	0.6	1.2

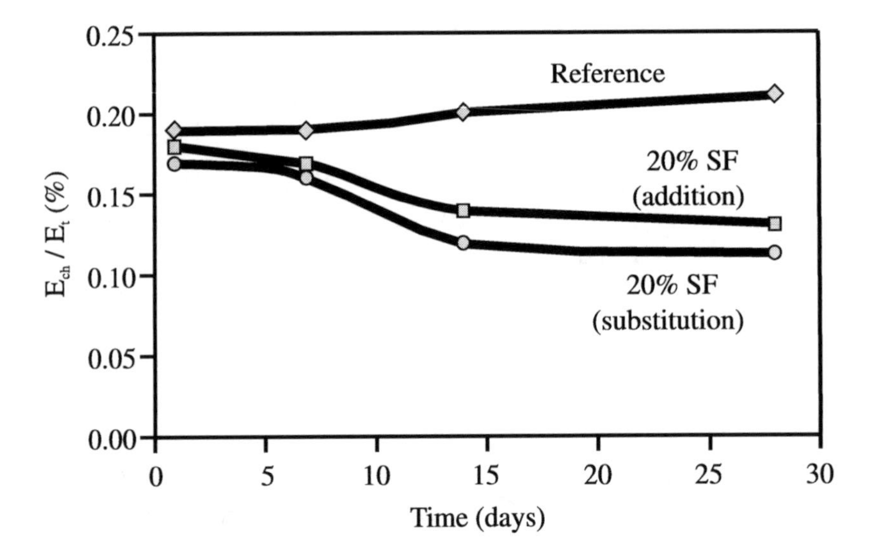

Figure 8.27 Mass loss with and without silica fume.

Silica fumes are marketed in four different forms (Holland 1995):

- those produced with a density of 300 to 500 kg/m^3;
- in an agglomerated form;
- as a slurry;
- blended with cement in binary, ternary or quaternary blended cements.

The cost of silica fume is usually quite expensive when compared to the price of Portland cement so its use is usually limited to the production of high-performance concrete, concrete facing severe environmental conditions or grouts for injection works (Carles-Gibergues et al. 1989; de Larrard et al. 1990; Mehta and Aïtcin 1990; Gjørv 1991; Cail and Thomas 1996). In Quebec, the

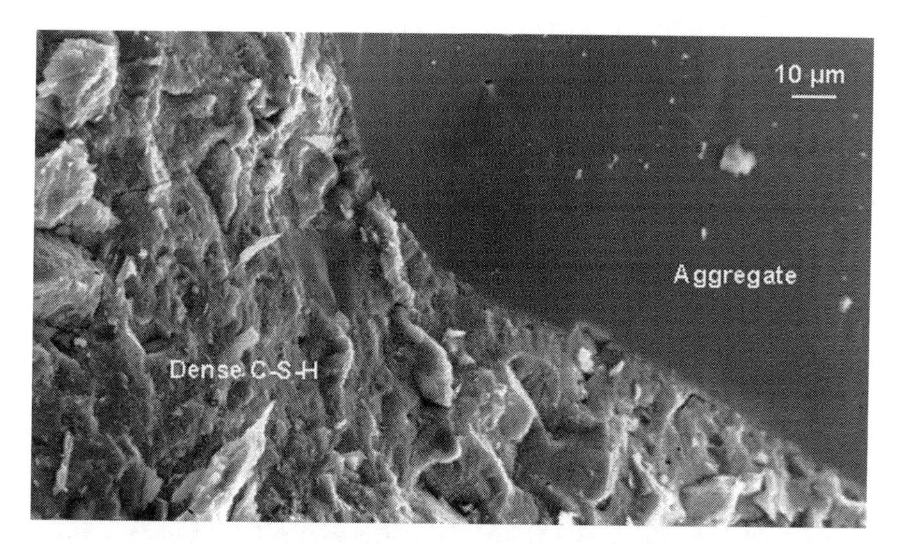

Figure 8.28 Compact C-S-H around a concrete aggregate in a silica fume concrete. The absence of transition zone can be noted.

three cement manufacturers have been producing, since 1983, about 100 000 tonnes per year of a blended cement containing 7 to 8 per cent silica fume. Since the year 2000, this binary blended cement has been transformed into a ternary blended cement containing 5 per cent silica fume and fly ash, or slag, or fly ash and slag. In Iceland, silica fume is also blended with Portland cement.

8.3.3 Calcined clays and shales

The use of calcined clay in concrete dates back to the Phoenician times as seen in Chapter 2. When a clay or a shale is calcined at a temperature of 700 to 750°C, the clay is dehydrated and its crystalline structure is totally disorganized. Silicon tetrahedra become active so that they can react at ambient temperature with the lime liberated by the hydration of C_3S and C_2S (Massazza 1998). However, the addition of calcined clay or shale increases the water demand in a concrete.

Metakaolin ($2\ SiO_2 \cdot Al_2O_3$) is the result of the calcination of kaolin, a particular clay mineral $2\ SiO_2 \cdot Al_2O_3 \cdot 2H_2O$ (Figure 8.29).

8.3.4 Rice husk ash

The husk that protects rice grains against rain has a siliceous skeleton representing about 20 per cent of its mass. When rice husks are burnt at 700 to

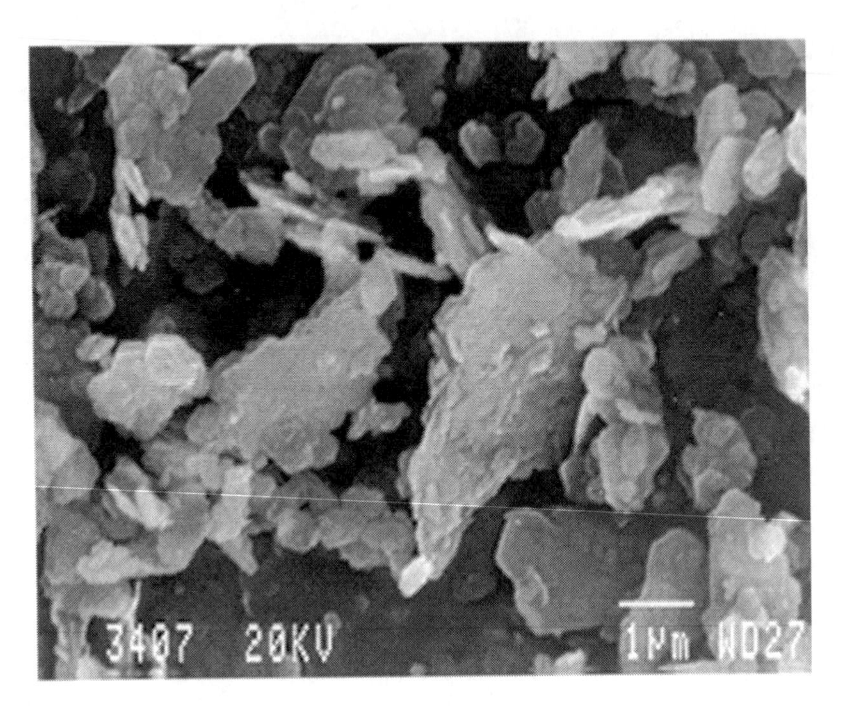

Figure 8.29 Metakaolin particles (courtesy of Martin Cyr).

750°C, the ashes are therefore essentially composed of vitreous silica. Calcined rice-husk ashes are more or less dark according to the unburned carbon they contain. This vitreous silica presents excellent pozzolanic properties. However, due to their very particular texture and morphology, rice-husk ashes increase the water demand of the fresh concrete (Figure 8.30) (Malhotra and Mehta 1996; Massazza 1998).

8.3.5 Natural pozzolans (Figure 8.31)

Dreux and Festa (1995) define natural pozzolans as materials usually of volcanic origin, or sedimentary rocks that are essentially composed of reactive silica in a proportion greater than 25 per cent and of alumina and iron oxide. Massazza defines natural pozzolans as pyroclastic rocks essentially vitrified or transformed into a zeolite, or more generally as inorganic materials, that harden under water when they are in the presence of lime or of the lime liberated by Portland cement hydration.

As seen in Chapter 2, natural pozzolans were in fact used as the first binders developed in antiquity.

When the pozzolanicity of a rock or of a natural earth has to be tested, it is necessary to determine whether this material contains some vitreous silica. It

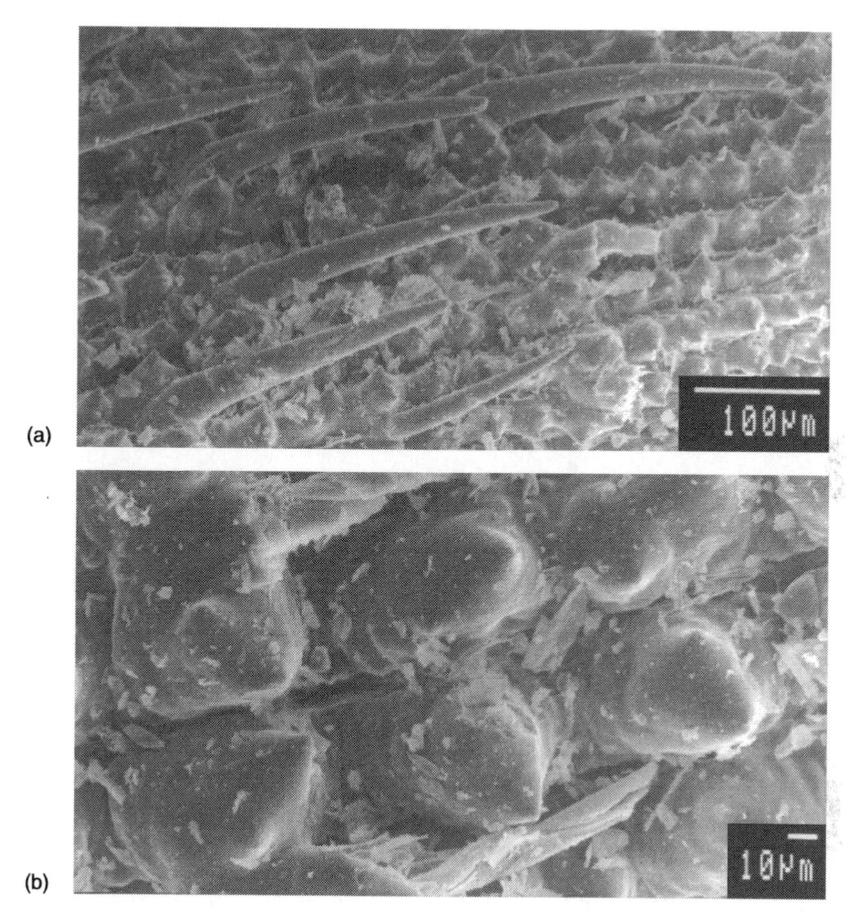

Figure 8.30 (a) Rice-husk ash; (b) enlarged view (courtesy of Arezki Tagnit-Hamou and Irène Kelsey-Lévesque).

is, therefore, necessary to make a chemical analysis *and* an X-ray diffractogram. Moreover, several standard tests exist to evaluate the pozzolanicity of a material. Generally speaking, these tests consist of substituting a certain amount of Portland cement by the same mass of pozzolan and then accelerating the pozzolanic reaction by heating the tested samples in the presence of water or in a sealed container. When the accelerated curing is finished, the cube strength of the mortar containing the pozzolan is compared to the cube strength of pure Portland cement cured in the standard way.

Massazza (1998) provides the chemical composition of several pozzolans and explains in detail the specific chemical reactions that are developed within blended cements containing these pozzolans. In Tables 8.4 and 8.5 the chemical composition of some calcined clay and natural pozzolans can be found.

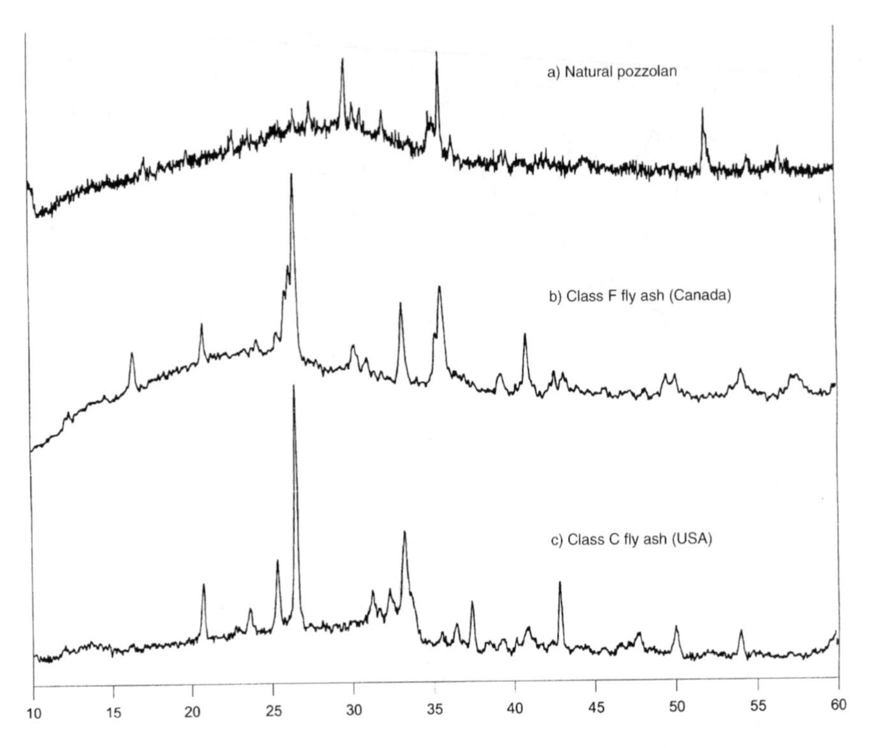

Figure 8.31 RX diffractogram of a natural pozzolan and two fly ashes.

Table 8.4 Chemical composition of some pozzolanic materials (Bogue 1952)

	SiO_2	Al_2O_3	Fe_2O_3	CaO	MgO	K_2O	Na_2O	SO_3	*Loss on ignition*
Santorin earth	63.2	13.2	4.9	4.0	2.1	2.6	3.9	0.7	4.9
Rhenan trass	55.2	16.4	4.6	2.6	1.3	5.0	4,3	0.1	10.1
Punice	72.3	13.3	1.4	0.7	0.4	5.4	1,6	tr.	4.2
Calcined gaize	83.9	8.3	3.2	2.4	1.0	n.a.	n.a.	0.7	0.4
Calcinated clay	58.2	18.4	9.3	3.3	3.9	3,1	0.8	1.1	1.6
Calcined shale	51.7	22.4	11.2	4.3	1.1	2,5	1.2	2.1	3.2
Blastfurnace slag[1]	33.9	13.1	1.7	45.3	2.0	n.a.	n.a.	tr.	n.a.

Notes

[1] Blastfurnace slag is not considered as a pozzolan anymore.

n.a. = non-available; tr. = trace

Pozzolans are commonly used in Italy, Greece, Turkey, Morocco, Mexico and Chile. During the twenty-first century, the use of natural pozzolans will increase, at least in countries where natural pozzolans can be found of the

Table 8.5 Chemical composition of some natural pozzolans and calcined clays (Papadakis and Venuat 1966)

Oxides	*Pozzolans*			*Trass*				
	Italy Latium segni	*Greece Santorinl Island*	*Canary island*	*Rhenan*	*Romanian*	*Calcined Gaize*	*Calcined clay*	*Pumice*
SiO_2	48	65	47	55	62	84	58	55
$Al_2O_3 + TiO_2$	22	13	20	16	12	8	18	22
Fe_2O_3	9	6	3	4	2	3	9	3
CaO	7	3	4	3	6	2	3	2
MgO	3	2	0.5	1	1	1	4	0.5
$Na_2O + K_2O$	5	6.5	9	9	3	–	4	11
SO_3	0.5	0.5	–	–	–	0.7	1	–
LOI	5	4	16	10	14	0.8	2	6

quantity and quality required. In several developing countries, natural pozzolans could prevent the production or importation of a significant amount of Portland cement clinker, which would otherwise be necessary to fulfill the needs of the construction industry. The use of natural pozzolan will decrease the amount of CO_2 generated during the production of clinker. Each time a developing country uses its own natural pozzolan to produce the hydraulic binders needed to develop its infrastructure, it will be able to make all the concrete it needs without increasing the emission of greenhouse gases. This economical and ecological contribution of natural pozzolans can be easily enhanced by using appropriate admixtures.

Massazza (1998) and Malhotra and Mehta (1996) provide data on the effects of some natural pozzolans on concrete properties, but it should be pointed out that, as in the case of fly ashes, it is always dangerous to make generalizations. However, the basic chemical principle explaining the binding properties of natural pozzolans always remains the same:

silica + lime → C-S-H.

8.3.6 Diatomaceous earth (Figure 8.32)

Diatomaceous earths are also called Tripoli or kiselghur, and are composed of the siliceous skeleton of microscopical algae that lived in fresh or in sea water. The examination of diatomaceous earth under an electron microscope shows some different types of patterns (Figure 8.32). It is easy to check that diatomaceous earths are composed of vitreous silica when doing an X-ray diffractogram as seen in Figure 8.12. In this case, silicon tetrahedra are more or less organized over a short as in cristobalite.

Figure 8.32 (a) Diatomaceous earth; (b) enlarged view (photos taken by I. Kelsey-Lévesque, courtesy of Arezki Tagnit-Hamou).

Rice husk ash and diatomaceous earth are made of a very reactive silica, as reactive as the one found in silica fume, but their large porosity increases their water demand. Such pozzolans should be used saturated, which is convenient in concrete plants, but not in making blended cement. Presently, this is their principal limitation when making concrete.

8.3.7 Perlite (Figure 8.33)

Rhyolites are volcanic rocks rich in silica (Bensted 1998; Sims and Brown 1998). Perlite is obtained when heating a rhyolitic rock which is transformed into a spongy mass that absorbs a lot of water. Presently perlite is used as an ultra-lightweight aggregate to make insulating concretes (thermal and acoustic

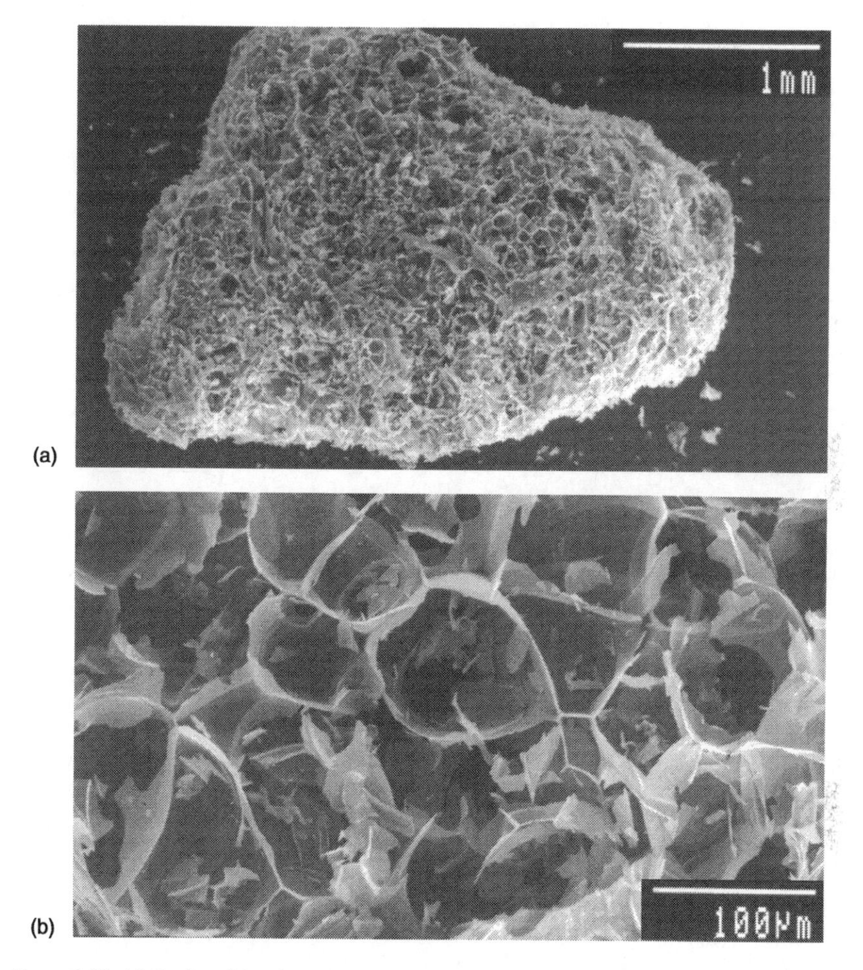

Figure 8.33 (a) Perlite; (b) enlarged view (courtesy of Arezki Tagnit-Hamou and I. Kelsey-Lévesque).

insulation). The very high absorptivity of perlite limits its use as a pozzolan in concrete. In a few cases ground rhyolitic rocks are used as a natural pozzolan.

8.4 Fillers

In Europe and in Canada it is possible to introduce 5 per cent limestone filler in a Portland cement and sell this cement as 'pure' Portland cement. In Europe limestone filler can be incorporated at a 35 per cent level in blended cement of the CPJ category. In the USA the addition of limestone filler is forbidden. The only difference between a US Type I cement and a Type 10 Canadian cement is that the Type 10 Canadian cement can contain up to 5 per cent limestone filler.

Some researchers maintain that limestone fillers are not totally inert fillers because calcium carbonate can react with C_3A to form a carboaluminate. Moreover, Goldman and Bentur (1993b) and Venuat (1984) have shown that calcium carbonate accelerates somewhat the hydration of C_3S. In addition, limestone filler particles are active as nucleation sites that favour the growth of C-S-H.

Limestone fillers can be used to make economical self-compacting concrete (Ghezal and Khayat 2001) or even high-performance concrete (Nehdi et al. 1996).

8.5 Producing blended cement or adding cementitious materials directly in the concrete mixer?

Cementitious materials can be blended with Portland cement clinker in cement plants or added directly into concrete mixers at the batching plant. In fact, when a concrete containing cementitious material has been thoroughly mixed it does not make any difference if the cementitious material was blended with Portland cement clinker at the cement plant, or if the cementitious material was added directly in the mixer. Each method of using cementitious materials has its own advantages and disadvantages (Uchikawa et al. 1987).

When a cementitious material is added in a cement plant to make a blended cement, its quality has been checked, its dosage is quite precise, the cementitious material is thoroughly mixed with the Portland cement, but the addition rate is fixed. Concrete producers do not need an additional silo, nor do they need to be concerned about the control of the quality and consistency of the characteristics of the cementitious material.

In order to be more flexible in the dosage rate of the cementitious material, some cement producers have developed blending units where they can blend 'à la carte' blended cement or 'ciment du jour' according to the specifications of each of their customers.

Presently we have the technological facilities to make these 'à la carte' blendings and to control their quality. This kind of marketing should be available for customers in the near future to fulfil some of their very specific needs. For example, the percentage of cementitious materials could be increased in summer, for mass concrete applications, or for very high performance applications, while it could be decreased for winter applications, or pre-cast applications where a very high early strength is required, etc.

When the cementitious material is added at the concrete plant, it is necessary to have a special silo devoted to the storage of the cementitious material, to modify the scales, to take care of the cementitious material delivery schedule, to check the quality and consistency of the cementitious material, etc., but the substitution rate can be modified from one batch to another to take into account the needs of the customer.

There is no universal rule at present that favours one mode of utilization of cementitious material over another. In France and Belgium cementitious materials are blended at the cement plant; in Germany slag is blended at the cement plant but fly ash at the concrete plant; and in the USA and Canada both methods are presently used. In Canada, ternary or quaternary blended cements (Portland cement + slag + silica fume; Portland cement + fly ash + silica fume; Portland cement + slag + fly ash + silica fume) are blended at the cement plant.

8.6 Effects of cementitious materials on the principal characteristics of concrete

The introduction of a cementitious material into concrete modifies the properties of the fresh and hardened concrete. It is not the intention of the author to review in minute detail the effect of each cementitious material on the properties of the concrete, as this kind of development fits better in a book devoted to concrete rather than in a book dealing with hydraulic binders. Readers are referred to specialized books or papers (Regourd 1983a; Malhotra 1987; Malhotra and Ramezanianpour 1994; Fidjestøl 1998; Massazza 1998).

According to the type and morphology of their individual particles, some cementitious materials increase the water demand in order to get a given slump. When the water demand is increased, the situation can be partially or totally overcome by using some superplasticizer, but the extra cost of the superplasticizer will have to be taken into account when pricing the economical performance of the cementitious material. Usually, the water demand is decreased when using fly ash, but this is not always true, and the addition of slag is rather neutral on the water demand except when slags are ground very fine. The water demand is increased when using silica fume due to the extreme fineness of their particles, and is also increased in the case of rice husk ash and diatomaceous earth due to the morphology of the particles. It is difficult to be very specific about the water demand of natural pozzolans because there are so many different types.

In the case of silica fume it is interesting to describe a very specific behaviour when a superplasticizer is used. At the beginning of the mixing the mix looks quite dry and suddenly it becomes flowing. At that time the superplasticizer particles have not only deflocculated the cement particles, but the silica fume particles have started to displace the water trapped between the cements flocs. The water that is liberated makes the mixture flow. Very low water/binder ratio concrete can be placed easily (high-performance concretes, reactive powder concretes), but it is, necessary to wait until deflocculation occurs, and this deflocculation time depends on the shear capacity of the mixer that is used.

When using admixtures in concrete containing some cementitious material it is often necessary to modify their dosage according to their carbon content. Usually carbon particles trap organic molecules, diverting them from their intended action in the mixture. This is particularly true for fly ashes that have a high loss on ignition, essentially due to the presence of unburned carbon particles. Another drawback of the presence of unburned carbon in fly ashes is the variability of this carbon content, which makes it quite tricky, for example, to obtain a stable and reproducible bubble system in entrained air concrete.

Usually, it is easier to maintain the slump of a concrete containing cementitious materials for 1 hour to 1.5 hours because cementitious materials are less reactive than the Portland cement they replace in the mix. In the case of slag it can be necessary to add an amount of calcium sulphate that is roughly equal to that which was contained in the Portland cement that has been replaced (Saric-Coric and Aïtcin 2003), because slag consumes some SO_4^{2-} ions when it starts reacting during the first hour and a half following the introduction of water in the mix.

The amount of heat that is developed by blended cements is significantly lower when the dosage of cementitious material is greater than 50 per cent. When the substitution rate is in the order of 15 to 25 per cent, there is, of course, a slight decrease in the intensity of the peak temperature registered in the concrete. However, this is not large enough to have positive consequence in sufficiently decreasing the thermal gradients within the concrete and in reducing potential thermal cracking. Usually this slight decrease of the maximum temperature reached by the concrete is due to the fact that blended cements are ground somewhat finer in order to make them more reactive.

Some cementitious materials modify the colour of a concrete. Silica fume and rice husk ashes tend to darken concrete colour due to the presence of very finely divided carbon particles, but, of course, this is not true when using a 'white' silica fume obtained in a furnace equipped with a heat recovery system (Aïtcin 1983). Fly ashes and slags have a tendency to give paler concretes. In the case of slags it should be mentioned that under its pale grey skin the concrete has rather a greenish tone (Aïtcin 1973). The origin of this greenish tone has not yet been explained in a satisfactory manner (Saric-Coric 2002).

Of course, generally speaking, the early strength resistance of a concrete that contains some cementitious materials is lower than that of a plain Portland cement concrete, except in the case of silica fume. According to the reactivity of the cementitious material the decrease of the early and short term strength can be noticed until seven, fourteen or twenty-eight days. This is a major drawback for constructors, especially in cold weather, but in hot climates, where the concrete is placed at a high temperature, it is not a significant disadvantage, because heat activates the reactivity of the cementitious material.

As far as long term strengths are concerned, they can be adjusted to the desired level, if the water/binder ratio is correctly adjusted and if is the *concrete is adequately water cured*. It is very important to point out that the actual performance of a concrete containing a cementitious material depends to a significant extent on the availability of water, because in order to see the development of a pozzolanic reaction it is absolutely necessary to have *not only vitreous silica but also water*. This point is too often forgotten.

Cementitious materials do not seem to significantly affect shrinkage. Of course the particles of the cementitious material are less reactive but, as blended cements are usually ground finer the shrinkage behaviour is not very greatly affected.

Concretes containing cementitious materials are usually more durable than plain Portland cement concretes having the same water/binder ratio. The transformation of portlandite ($Ca(OH)_2$) into secondary C-S-H fixes lime into the more stable calcium silicate hydrate, which is more difficult to attack or leach out.

When tested under the same testing procedure as plain Portland cement concretes, concretes containing some slag tend to display a lower durability to slow freeze/thaw cycles in the presence of deicing salts. This is essentially due to the fact that when these slag concretes are exposed to the first freeze/thaw cycles and deicing salts they have not yet reached the same degree of maturity as plain Portland cement concrete, and, that their surface has not been sufficiently carbonated (Saric-Coric and Aïtcin 2000).

Duchesne and Bérubé (2001) have clearly demonstrated that one of the best ways to fight potential alkali aggregate reaction is to use fly ash and slag. Glassy fly ash and slag particles are activated by the rapidly soluble alkalis present in Portland cement (alkaline activation of glass) so that these alkalis are trapped in the early C-S-H developed during the development of the pozzolanic reaction. Carles-Gibergues (2000) maintains that slag and pozzolanic materials delay only by several years the development of the alkali aggregate reaction when accelerated tests are done at 38°C in perfectly saturated conditions. The question that can now be raised is this: How many years of curing at ambient temperature in a non-saturated atmosphere corresponds to one year of curing at 38°C in perfectly saturated conditions with an aggregate that is so reactive that any engineer, with a little common sense, would never use?

8.7 Blended cements

Over the years' several typical blended cements have been developed to meet the needs of the construction industry. These blended cements have been standardized in certain countries. European standards are particularly rich in the matter of blended cement categories (Pliskin 1993; Dreux and Festa 1995; Baron and Ollivier 1995; Jackson 1998), ASTM Standard C595 is not as rich in these categories.

8.8 Conclusion

In order to reach its objectives in terms of the reduction of its greenhouse gas emissions and to contribute significantly to the implementation of a sustainable development policy, the cement industry has no other choice but to implement the use of cementitious material by providing the construction industry with blended cements. The cement industry should now instead be called the hydraulic binder industry.

In addition to these first two ecological objectives the use of cementitious materials will help the concrete industry, because when cementitious materials are used in combination with appropriate admixtures concrete usually becomes more durable when it has an appropriate water/cement ratio and when it is cured appropriately. *Pozzolanic reaction needs vitreous silica and lime to develop, but also water, which is often forgotten.* The curing of concretes containing blended cements is crucial in the development of the cementitious potential of cementitious materials.

The easiest way to be sure that concrete is well cured in the field is to specifically pay contractors to do it; when water curing becomes an easy source of profit, contractors become rather zealous. Since the city of Montreal has adopted such a policy it enjoys almost crack-free structures. The cost of the implementation of such a water-curing policy has increased the cost of construction of urban infrastructures by 0.1 to 1 per cent, depending on the type of construction, its location and its importance. Contractors are not asked to flood their concrete with water. They are only asked to provide enough water to appropriately cure their concrete. Water curing is a key factor for the successful use of cementitious material.

References

Aïtcin, P.-C. (1968) 'Sur les propriétés minéralogiques des sables de laitier de haut fourneau de fonte Thomas et leur utilisation dans les mortiers en bétons', *Revue des Matériaux de Construction* (May): 185–94.

—— (1983) *Condensed Silica Fume*, Quebec: Éditions de l'Université de Sherbrooke.

Aïtcin, P.-C., Autefage, F., Carles-Gibergues, A. and Vaquier, A. (1986) Comparative Study of the Cementitious Properties of Different Fly Ashes, ACI SP–91, pp. 91–114.

ASTM C311, Standard Test Method for Sampling and Testing Fly Ash or Natural Pozzolans for Use as a Mineral Admixture in Portland Cement Concrete.

ASTM C595, Standard Specification for Blended Hydraulic Cements.

ASTM C618, Standard Specification for Fly Ash and Raw or Calcined Natural Pozzolan for Use as a Mineral Admixture in Portland Cement Concrete.

ASTM C989, Standard Specification for Ground Granulated Blast-Furnace Slag for Use in Concrete and Mortars.

ATILH (1997) *Guide pratique pour le choix et l'emploi des ciments*, Paris: Eyrolles.

Baalbaki, M., Sarkar, S.L., Aïtcin, P.-C. and Isabelle, H. (1992) 'Properties and Microstructure of High-Performance Concretes Containing Silica Fume, Slag and Fly Ash',

Fourth CANMET/ACI International Conference on Fly Ash, Silica Fume, Slag and Natural Pozzolans in Concrete, Istanbul, May, pp. 921–42.

Baron, J. and Ollivier, J.-P. (1995) *Les bétons: bases et données pour leur formulation*, Paris: Eyrolles.

Bensted, J. (1998) 'Special Cements', in P.C. Hewlett (ed.) *Lea's Chemistry of Cement and Concrete*, London: Arnold, pp. 779–835.

Bramforth, P.B. (1980) 'In Situ Measurement of the Effect of Partial Portland Cement Replacement Using Either Fly, Ash or Ground Granulated Blast-Furnace Slag on the Performance of Mass Concrete', *Proceedings of the Institution of Civil Engineers*, 69 (2, September): 777–800.

Cail, K. and Thomas, H. (1996) 'Development and Field Applications of Silica Fume Concrete in Canada', CANMET-ACI Intensive Course on Fly Ash, Slag, Silica Fume, Other Pozzolanic Materials and Superplasticizers in Concrete, Ottawa, Canada, April.

Carles-Gibergues, A. (2000) Personal communication.

Carles-Gibergues, A., Ollivier, J.-P. and Hanna, B. (1989) 'Ultrafine Admixtures in High Strength Pastes and Mortars', in V.M. Malhotra (ed.) *Third CANMET/ACI International Conference on Fly Ash, Silica Fume, Slag and Natural Pozzolans in Concrete*, Trondheim, Proceedings. Detroit, Mich.: American Concrete Institute, 2 vols, Vol. II, pp. 1101–16.

de Larrard, F., Gorse, J.F. and Puch, C. (1990) 'Efficacités comparées de diverses fumées de silice comme additif dans les bétons à hautes performances', *Bulletin de Liaison des Laboratoires des Ponts et Chaussées*, 168 (July–August): 97–105.

Detwiler, R.J. and Mehta, P.K. (1986) 'Chemical and Physical Effects of Silica Fume on the Mechanical Behavior of Concrete', *ACI Materials Journal*, 86 (6): 609–14.

Detwiler, R.J. and Mehta, P.K. (1989) 'Chemical and Physical Effects of Condensed Silica Fume in Concrete', *Third CANMET/ACI International Conference*, Supplementary Papers, Trondheim: 295–306.

Dreux, G. and Festa, J. (1995) *Nouveau guide du béton et de ses constituants*, Paris: Eyrolles.

Dron, R. and Voinovitch, I.A. (1982) 'L'Activation hydraulique des laitiers, pouzzolanes et cendres volantes', in J. Baron and R. Sauterey (eds) *Le Béton hydraulique*, Paris: Les Presses de L'École Nationale des Ponts et Chaussées, Chapter 8.

Duchesne, J. and Bérubé, M.-A. (2001) 'Long-Term Effectiveness of Supplementary Cementing Materials Against Alkali-Silica Reaction', *Cement and Concrete Research*, 31 (7): 1057–63.

Edwards-Lajnef, M., Aïtcin, P.-C., Wenger, F., Viers, P. and Galland, J. (1997) 'Test Method for the Potential Release of Hydrogen Gas from Silica Fume', *Cement, Concrete and Aggregates*, 19 (2, December): 64–9.

Fidjestøl, R. and Lewis, R. (1998) 'Microsilica as an Addition', in P.C. Hewlett (ed.) *Lea's Chemistry of Cement and Concrete*, London: Arnold, pp. 675–708.

Ghezal, A. and Khayat, K.H. (2001) 'Statistical Design of Self-Consolidating Concrete with Limestone Filler', Workshop on Self-Compacting Concrete, University of Tokyo, Japan, October.

Gjørv, O.E. (1991) 'Norwegian Experience with Condensed Silica Fume in Concrete', CANMET/ACI International Workshop on the Use of Silica Fume in Concrete, Washington, DC, April, pp. 49–63.

Goldman, A. and Bentur, A. (1993a) 'Effects of Pozzolanic and Non-reactive Fillers on the Transition Zone of High-Strength Concrete', in J.-C. Maso (ed.) *RILEM International Symposium on Interfaces in Cementitious Composites, Toulouse, Proceedings*, London: E & FN Spon, pp. 53–62.

Goldman, A. and Bentur, A. (1993b) 'The Influence of Micro Fillers on Enhancement of Concrete Strength', *Cement and Concrete Research*, 23 (8): 962–72.

Gutteridge, W.A. and Dalziel, J.A. (1990) 'Filler Cement: The Effect of the Secondary Component on the Hydration of Portland Cement, Part II, Fine Hydraulic Binders', *Cement and Concrete Research*, 20 (6): 853–61.

Holland, T.C. (1995) Specification for Silica fume for Use in Concrete, ACI SP–154.

Isaia, G.C. (1997) 'Synergic Action of Fly Ash Ternary Mixtures with Silica Fume and Rice Husk Ash: Pozzolanic Activity', in H. Justnes (ed.) Tenth *International Congress on the Chemistry of Cement*, Gothenburg, comptes rendus, Amarkai AB, 4v., Vol. 4, 4iv005, 8 p.

Jackson, P.J. (1998) 'Portland Cement: Classification and Manufacture', in P.C. Hewlett (ed.) *Lea's Chemistry of Cement and Concrete*, London: Arnold, pp. 25–94.

Josephson, G.W., Sillers, F. and Runner, D.G. (1949) *Iron Blast Furnace Slag: Production, Processing, Properties and Uses*, Washington, DC: Bureau of Mines.

Jouenne, C.A. (1984) *Traité de céramique et matériaux minéraux*, Paris: Septima.

Khayat, K., Hu, C. and Laye, J.M. (2000) 'Study on Control Technique of Workability of Self-Compacting Concrete', *Proceedings of the International Conference: High Performance Concrete – Workability, Strength and Durability*, Vol. II, Hong Kong University of Science and Feelinology, Kowloon, China, pp. 659–67.

Kosmatka, S.H., Kerkoff, B., Panarese, W.C., MacLeod, N.F. and McGrath, R.J. (2002) *Design and Control of Concrete Mixtures*, Michigan: Portland Cement Association.

Malhotra, V.M. (ed.) (1987) *Supplementary Cementing Materials for Concrete*, Ottawa: Canadian Government Publishing Centre.

—— (2000) 'Role of Supplementary Cementing Materials in Reducing Greenhouse Gas Emissions', in O. Gjørv and K. Sakai (eds) *Concrete Technology for a Sustainable Development in the 21st Century*, London: E & FN Spon, pp. 226–35.

Malhotra, V.M. and Mehta, P.K. (1996) *Pozzolanic and Cementitious Materials*, Amsterdam: Gordon & Breach.

Malhotra, V.M. and Ramezanianpour, A.A. (1994) *Fly Ash in Concrete*, Ottawa CANMET Natural Resources Canada.

Massazza, F. (1998) 'Pozzolana and Pozzolanic Cements', in P.C. Hewlett (ed.) *Lea's Chemistry of Cement and Concrete*, London: Arnold, pp. 471–632.

Mehta, P.K. (1987) 'Studies on the Mechanisms by which Condensed Silica Fume Improves the Properties of Concrete', *International Workshop on Condensed Silica Fume on Concrete*, Ottawa, pp. 1–17.

Mehta, P.K. and Aïtcin, P.-C. (1990) 'Principles Underlying Production of High-Performance Concrete', *Cement, Concrete, and Aggregate*, 12 (2): 70–8.

Moranville-Regourd, M. (1998) 'Cement Made from Blastfurnace Slag', in P.C. Hewlett (ed.) *Lea's Chemistry of Cement and Concrete*, London: Arnold, pp. 633–74.

Naproux, P. (1994) 'Les Microcendres (cendres volantes traitées) et leurs emplois dans les bétons hydrauliques', Ph.D. thesis, INSA de Toulouse.

Nehdi, M., Mindess, S., and Aïtcin, P.-C. (1996) 'Optimization of High-Strength Limestone Filler Cement Mortars', *Cement and Concrete Research*, 26 (6, June): 883–93.

Nkinamubanzi, P.-C. and Aïtcin, P.-C. (1999) 'The Use of Slag in Cement and Concrete in a Sustainable Development Perspective', WABE International Symposium on Cement and Concrete, Montréal, pp. 85–110.

Nkinamubanzi, P.-C., Baalbaki, M., Bickley, J. and Aïtcin, P.-C. (1998) 'Slag on HPC', *World Cement*, October: 97–103.

Nonat, A. (2005) The structure of C-S-H, Cement Wapno Beton, vol.10, No 21 pp 65–73

Osborn, E.F. and Muan, A. (1960) Phase Equilibrium Diagrams of Oxyde Systems, Plate 1. Published by the American Ceramic Society and the Edward Orton, Jr., Ceramic Foundation, *Phase Diagrams for Ceramists*, Vol. 1, The American Ceramic Society (1964), Figure 630, p. 219.

Papadakis, M. and Venuat, M. (1966) *Fabrication et utilisation des liants hydrauliques*, 2nd edn (self-published).

Pistilli, M.F., Rau, G., and Cechner, R. (1984) 'The Variability of Condensed Silica Fume from a Canadian Source and its Influence on the Properties of Portland Cement Concrete', *Cement, Concrete and Aggregates*, 6 (1): 33–7.

Pistilli, M.F., Winterstein R., and Cechner, R. (1984) 'The Uniformity and Influence of Silica Fume from a US Source on the Properties of Portland Cement Concrete', *Cement, Concrete, and Aggregates*, 6, (2): 120–4.

Pliskin, L. (1993) *La Fabrication du ciment*, Paris: Eyrolles.

Regourd, M. (1983a) 'Pozzolanic Activity of Condensed Silica Fume', in Aitcin, P.-C. (ed.) *Condensed Silica Fume*, Quebec: Les Éditions de l'Université de Sherbrooke, pp. 20–4.

Regourd, M. (1983b) 'Caractérisation et activation des produits d'addition', Main report Thema III, *Ninth International Congress on Chemistry of Cement*, Rio de Janeiro, Vol. I, pp. 199–229.

Regourd, M., Mortureux, B. and Hornain, H. (1986) Use of Silica Fume as Filler in Blended Cements, ACI SP–79, Vol. 2, pp. 747–764.

Saric-Coric, M. (2002) 'Superplasticizer-Slag Interaction in Slag Blended Cement: Properties of the Concrete', Ph.D. thesis, Université de Sherbrooke.

Saric-Coric, M. and Aïtcin, P.-C. (2003) 'Influence of Curing Conditions on the Shrinkage of Blended Cements Containing Various Amounts of Slag', *ACI Materials Journal*, 100 (6): 477–84.

Sellevold, E.J. and Justness, H. (1993) 'High Strength Concrete Binders. Part B: Non-Evaporable Water, Self-Desiccation and Porosity of Cement Pastes with and without Condensed Silica Fume', in V.M. Malhotra (ed.) *Fourth CANMET/ACI International Conference on Fly Ash, Silica Fume, Slag, and Natural Pozzolans in Concrete*, Istanbul, 1992. Proceedings, American Concrete Institute, 2 v., Vol. 2, pp. 891–902 (SP–132).

Sims, I. and Brown, B. (1998) 'Concrete Aggregates', in P.C. Hewlett (ed.) *Lea's Chemistry of Cement and Concrete*, London: Arnold, pp. 902–1011.

Swamy, R.N. (1993) 'Fly Ash and Slag: Standards and Specifications. Help or Hindrance?' *Matériaux et Constructions*, 26 (164, December): 600–13.

Tanaka, H., Totani, Y. and Saito, Y. (1983) Structures of the hydrated glossy blast-furnace in concrete, ACI SP–79, pp. 963–977.

Uchikawa, H., Uchida, S. and Okamura, T. (1987) 'The Influence of Blending Components on the Hydration of Cement Minerals and Cement', *Review of the 41st General Meeting*, The Cement Association of Japan, Tokyo, Japan, pp. 36–9.

Venuat, M. (1984) *Adjuvants et traitements: Techniques d'amélioration des ouvrages en béton*', n.p.

Chapter 9

Special Portland cements and other types of hydraulic binder

9.0 Introduction

In this chapter, several special Portland cements and other types of hydraulic binders are presented. These hydraulic binders have a chemistry that is totally different from that of Portland cement and they are presently used for very specific applications. This is why they are presented in a separate chapter in order to emphasize their specificity and what makes them different from usual Portland cement or blended cements. As they are produced in much smaller quantities than Portland cement, their production and marketing costs are much higher than those of Portland cement. Moreover, they usually have to be transported over long distances so that their market price can be three to five times higher than that of regular Portland cement, and, even eight to ten times more expensive in the case of some aluminous cement.

First, a few very specific Portland cements will be presented, followed by aluminous cements and calcium sulphoaluminate cements and finally other industrial hydraulic binders will be briefly described. Appendix X discusses pigments used to colour concrete and a phenomenon known as efflorescence. Efflorescence is very unappealing when it appears on coloured concrete (Neville 2002a and 2002b).

9.1 Special Portland cements

The expression 'special' Portland cement is used to identify Portland cements made from clinkers that are different from usual Portland cement in one or several ways. These special Portland cements are manufactured to fulfil certain technological or esthetic needs. However, in essence they are still Portland cements and their binding properties involve the reaction of their silicate and interstitial phases. Table 9.1 presents the chemical composition of some of these special Portland cements.

Table 9.1 Comparison of the chemical composition of some special Portland cements

	Gray	*White*	*Low alkali content*	*Oil well*	*Type 20M**
SiO_2	20.9	21.9	21.2	21.3	23.1
CaO	63.2	66.5	66.4	63.3	62.9
Al_2O_3	4.1	5.6	5.0	2.9	3.2
Fe_2O_3	2.8	0.3	3.9	4.1	4.7
MgO	2.7	1.0	1.3	4.1	1.8
Na_2O	0.32	0.08	0.09	0.17	0.12
K_2O	0.95	0.13	0.11	0.68	0.40
Na_2O equiv.	0.94	0.16	0.16	0.62	0.38
SO_3	3.3	3.1	2.4	4.1	2.4
P.A.F.	2.4	2.5	0.5	1.8	0.63

* Cement with very low heat of hydration used by Hydro Québec for the construction of its dams.

9.1.1 White Portland cement

White Portland cement is simply a Portland cement that is made from raw materials that contain very little iron (Freedman 1971). The Fe_2O_3 content of a white cement is usually lower than 0.8 per cent, which means that its interstitial phase is almost exclusively composed of C_3A. Consequently, white Portland cement must be fired at a higher temperature than grey Portland cement. White Portland cements must be ground in grinding mills that do not generate too many iron particles; balls made from very hard steel alloys are presently used. Apart from its white colour and high 'reactivity', white Portland cement has no other special characteristic.

From a rheological point of view, white Portland cements are very reactive because their interstitial phase is almost exclusively composed of C_3A. White cements usually react poorly with polysulphonate superplasticizers because they also have a low alkali content. Therefore, with white cement, it is better to use polycarboxylate superplasticizers or to add an adequate amount of retarder or sodium sulphate (Kim 2000).

White Portland cements are used to manufacture white or coloured architectural panels, white or coloured bricks, blocks, paving stones or paving bricks. In order to colour them, pigments are added (see Appendix X).

9.1.2 Buff cement

Some years ago, Texas architects complained that grey Portland cement didn't fit well with the brown or red bricks they were using nor with the reddish or brownish colour of Texas soil. They also complained about the lack of consistency of the coloured mortars proposed to masons by concrete producers and consequently, they asked cement companies if it would be possible to manufacture a buff cement which would not need to be coloured.

It is possible to change the colour of clinker by quenching it rapidly before it leaves the kiln, much more rapidly than is usually done to fix the C_3S and C_2S into their active phases. It is only necessary to spray water onto the hot clinker when its temperature is still around 1400°C. All the iron contained in the clinker remains in the Fe^{3+} form. This is exactly the same phenomenon that can be observed when quenching blast furnace slag. Crystallized slag obtained after a slow cooling is dark grey, while when the same slag is quenched it becomes a pale brown or yellowish sandy material, depending on its iron content and the temperature at which it was quenched so-called 'cold slags' are dark brown after quenching, while 'hot slags' can be whitish or pale yellow.

A buff cement was manufactured in Texas for several years. Apart from its colour, this cement was similar to grey Portland cement as far as its rheological and mechanical characteristics were concerned.

However, quenching a hot clinker with water as it exits a cement kiln is very harsh for the refractory lining. This is why buff clinker was always manufactured at the end of a production period, just before a maintenance shutdown. The production of buff clinker was therefore costly, and as the market was not very large, cement companies stopped producing it. A few cement companies now offer coloured Portland cements in which pigments are blended to a white Portland cement, but unfortunately these coloured cements don't reach the warmth of the buff colour obtained by direct quenching of the clinker.

9.1.3 Oil well cements

Although petroleum standards recognize the existence of nine different types of oil well cements, from Class A to H plus one of Class J, it can be said that in general, these cements are not very different from ASTM Type V cement. They are coarse cements (280 to 350 m^2/kg Blaine fineness), their C_3S content is low and they contain very little C_3A. These cements are ground without any grinding aids because drilling companies prefer to buy non-altered ground clinkers to which they can add the admixtures they want without worrying about a secondary reaction between the grinding aids and the additives contained in the cement and admixtures they are introducing into their grouts. They prefer to buy a virgin cement to which they can incorporate the specific additives they need.

The grouts injected in oil wells have to withstand very high temperatures and very harsh and sometimes aggressive environments only found in oil fields; moreover, the grouts must have a rheology that is easy to control. John Bensted (1998), an oil well specialist. has written a whole paragraph on the subject (Paragraph 14.2) in *Lea's Chemistry of Cement and Concrete*.

9.1.4 Shrinkage compensating cements

Shrinkage compensating cements are hydraulic binders that swell when they begin to hydrate as well as during their initial hardening. Theoretically, it is possible to adjust the amplitude of this swelling so that it is equal to the future drying shrinkage of the concrete. Therefore, theoretically, a concrete made with such a cement does not display any shrinkage when it dries. Three types of expansive cements, another name used for these cements, are standardized. In each of these cements, expansion is obtained through different chemical reactions.

The first type of expansive cement is Type K cement. It is a blended cement made of Portland cement, tetracalcic trialuminosulphate ($C_4A_3\bar{S}$), calcium sulphate ($C\bar{S}$), and free lime (CaO). The second type is Type M cement. It contains Portland cement, calcium aluminate and calcium sulphate. Finally, Type S cement contains Portland cement, a high dosage of tricalcium aluminate (C_3A) and calcium sulphate. In each case, the initial expansion mechanism is the same: expansion is caused by the formation of large amounts of ettringite in the hardening concrete. When the matrix becomes rigid enough, expansion stops.

When concretes made with expansive cements dry, they begin to shrink, and as previously said, theoretically, the initial expansion can be adjusted so that the final deformation after drying is equal to zero. From a practical point of view, one will almost always ends up some residual shrinkage or expansion, but nothing comparable to the 500 to 600 μm/m usually observed on conventional concretes after drying.

In reinforced concrete, initial expansion is restrained by the reinforcing steel so that concrete becomes naturally pre-stressed.

Shrinkage compensating cements are used to manufacture pre-stressed prefabricated elements because they limit prestressing losses due to drying shrinkage. But from a practical point of view, shrinkage compensating cements present a rheology that is difficult to control because of their high aluminate content. This explains partially why expansive cements are not used very frequently. In the USA, only 50 000 to 60 000 tonnes are produced every year, which represents 0.5 to 0.6 per cent of the total cement production.

9.1.5 Regulated set cements

Regulated set cements are Portland cements whose setting time can be adjusted from 1 to 2 minutes to 60 minutes. These cements also present very rapid strength increase. This rapid hardening is obtained by doping the clinker with some fluorine. Under the action of this fluorine, C_3A is transformed into $C_{11}A_7 \cdot CaF_2$. In this hybrid notation, $C_{11}A_7$ represents 11 CaO • 7 Al_2O_3 and CaF_2 represents calcium fluorine (Lawrence 1998).

Regulated set cement can be manufactured directly in regular cement kilns or by intergrinding an ordinary Portland clinker with an appropriate amount of $C_{11}A_7 \cdot CaF_2$. This last component is more reactive than C_3A and can give rise to a flash set phenomenon if there is not enough calcium sulphate in the blended cement. The setting time of regulated set cements can be lengthened by an addition of citric acid.

In spite of the fact that a decrease in the setting time of concrete can be advantageous in certain circumstances, it must be said that regulated set cements have never been able to attract users, because from a practical point of view, they do not offer enough flexibility. Even in a pre-cast plant, it is difficult to maintain a very tight schedule all day long and sometimes it is good for concrete not to harden too fast. Regulated set cements were used for a certain time by the US Army. During the Vietnam war, it was used to repair and build airstrips. In civilian applications, it was never successful. There are, however, still one or two cement plants in the world that produce regulated set cements.

Presently, high performance roller-compacted concrete advantageously replaces regulated set cement for building industrial slabs or repairing concrete highways and air strips.

9.1.6 Masonry cement

In the previous chapter, we saw that colloidal agents have been used for a long time to make masonry cements. Cement companies developed masonry cement to replace the usual mixture of hydrated lime and Portland cement masons used to prepare in the field. Roughly speaking, masonry cements are made by blending 50 per cent Portland cement clinker with 50 per cent limestone filler. Masonry cement is ground much finer than Portland cement and various admixtures are introduced in it to obtain a plasticity and a water retention similar to the 50–50 per cent mixes of hydrated lime and Portland cement. Of course, masonry cement cannot be used to make concrete.

Masonry cement represents between 3 and 7 per cent of the total cement consumption in North America and between 15 and 20 per cent in countries where mortars are very much used.

9.1.7 Air-entrained Portland cements

ASTM C150 defines three types of air-entraining cements. At the moment, these cements are practically no longer manufactured in North America because it is difficult always to entrain the right amount of air when making concrete. In fact, there are many factors other than the air-entraining admixture dosage that influence the final air content of a concrete: sand, mixing, water quality, mixing sequence, air-entraining admixture dosage in determining the final characteristics of the air bubble system entrained in a given mix. From a practical point of view, it is better to adjust the final air entraining agent dosage when making a particular concrete in a particular mixing plant with the particular materials used to make the concrete.

9.1.8 Low alkali Portland cements

Low alkali cements can be classified as special Portland cements. They are simply Portland cements with a low Na_2O equivalent content, usually lower than 0.60 per cent, which represents the usual upper limit for a cement to be classified as a low alkali Portland cement. Low alkali cements are simply Portland cements made with raw materials that have a low alkali content. The alkali content of the raw materials can be somewhat lowered by incorporating calcium chloride into the raw mix.

Some low alkali cements have a very low alkali content, as low as 0.20 per cent or even lower, and some cement companies are promoting them as the best cements to avoid the alkali/aggregate reaction. Such a very low alkali content is not so advantageous from a practical point of view, more specifically from a rehological point of view, because it results in serious incompatibility problems with most sulphonate based water reducers or superplasticizers. Very low alkali cements contain very little alkali sulphates so that they cannot very rapidly provide SO_4^{2-} ions in the interstitial solution. SO_4^{2-} ions control the reactivity of the interstitial phase (C_3A and $C_4A\bar{F}$ to a lesser extent) when they are used to make low W/B concretes. With low alkali cements, it is only possible to make porous concretes having a W/B ratio greater than 0.50. In such concretes containing a high amount of mixing water, the calcium sulphates that were added during the final grinding process rapidly provide enough SO_4^{2-} ions to control C_3A hydration. But, these porous and pervious concretes are not the best to control the development of an alkali/aggregate reaction when potentially reactive aggregates are used to make concrete, because they favour the penetration of water within the concrete, the third essential condition to favour alkali/aggregate reaction.

To control alkali/aggregate reaction, it is better to use low W/B concretes made with a cement slightly richer in alkalis (0.45 to 0.60 per cent Na_2O equivalent) or blended cements. In Section 7.3.3.4 of Chapter 7, it has been pointed out that from a rheological point of view there exists an optimum soluble alkali content of around 0.4 to 0.5 per cent when low W/B concrete have to be made using polysulphonates.

It is not so certain that in the future it will be necessary to make low alkali cements anymore, because blended cements will be available almost everywhere, and it is recognized that blended cements constitute a very efficient way of counteracting alkali/aggregate reaction. Blended cements trap most of the alkalis present in the cement in a very safe way.

If a cement producer has the raw materials allowing the fabrication of a cement having a 0.4 to 0.6 per cent Na_2O equivalent content, he can produce such a cement and promote it is a good low alkali cement, but if his raw materials bring more than 0.6 per cent Na_2O equivalent content, it would be better for him to blend his clinker with a fly ash or a slag to manufacture a safe cement as far as alkali/aggregate reaction is concerned. In such a case, alkalis are very useful when dealing with blended cements because they favour the activation of fly ashes, slag, and silica fume through a chemical process known as alkaline activation.

9.1.9 Microcements

Microcements are usually blended cements that have been ground very fine. Their maximum particle size can be as low 10 to 15 μm or even less than 10 μm. Microcements are used to make grouts that are injected into rocks, soils or even fissured concrete. Microcements have a Blaine fineness as high as 800 to 1 000 m^2/kg. Usually, the Portland cement content of a microcement is low, around 25 per cent, and the remaining 75 per cent is a finely ground slag, including an appropriate amount of calcium sulphate, in order to control adequately the rheology of the grouts to be made. As injections are often done in difficult working conditions in mine galleries, it is out of the question to use a 'nervous' microcement.

9.2 Aluminous cements

Aluminous cements have a totally different chemistry from that of Portland cement. The first aluminous cement was developed by Bied in 1908 in the Lafarge laboratories, but its industrial production began in 1918 (Papadakis and Venuat 1966).

It is out of question in this section to present in full detail the pecularities of these cements, nor to present their refractory applications, which presently represent their main industrial field of applications; only certain civil engineering applications will be briefly presented. Many references can be found in Adam Neville's books (1963 and 2000) and in a chapter by Scrivener and Capmas (1998) in the latest edition of *Lea's Chemistry of Cement and Concrete*.

As can be seen in Tables 9.2 and 9.3 and Figure 9.1, aluminous cements definitely have chemical and phase compositions that are completely different from those of Portland cement. Aluminous cements are essentially composed of hydraulic calcium aluminates. But, aluminous cements still pertain to the

Table 9.2 Comparison of the chemical composition of a Portland cement and an aluminous cement

	SiO_2	CaO	Al_2O_3	Fe_2O_3	Na_2O	K_2O	SO_3
Portland cement	21	63	4	3	0.2	0.5–1	2.5–3.5
Aluminous cement	5	40	40	15	–	–	–

Table 9.3 Comparison of the phase composition of a Portland cement and an aluminous cement

	Portland		*Aluminous*	
Essential constituents	C_3S	50–60 %	CA	60 %
	C_2S	10–30 %		
Other constituents	C_3A	0–10 %	C_2S	10 %
	C_4AF	5–15 %	C_2AS (gehlenite)	5–20 %
Minor constituents	CaO	1–2 %	$C_{12}A_7$	5–20 %
	Periclase	1–2 %	FeO	
	Alkali sulphates	1–2 %	Ferrite	
			Pleocerite	

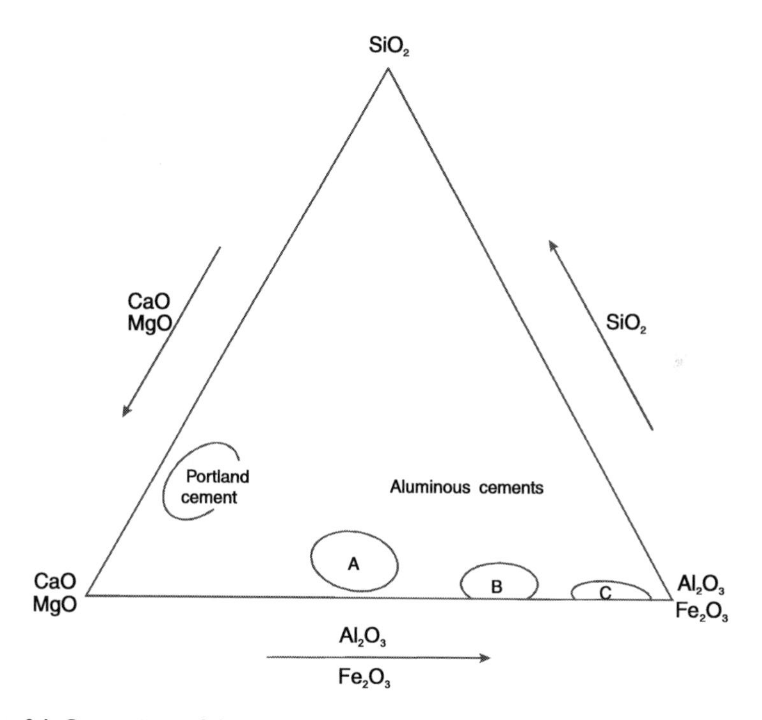

Figure 9.1 Comparison of the composition of Portland cements and aluminous cements: A: low alumina content (Ciment Fondu); B: medium alumina content; C: high alumina content.

family of hydraulic binders because they react with water and harden under water in the same way as Portland cement.

Aluminous cements that have an alumina content greater than 50 per cent are quite expensive and are used almost exclusively in refractory applications. They will not be considered in this paragraph. The only aluminous cements that will be considered are aluminous cements that have an alumina content between 30 and 40 per cent, such as Ciment Fondu marketed by Lafarge. Such aluminous cements have been and are still used in a very few civil engineering applications. These cements can be also used as low-duty refractory cements for applications where the service temperature never rises above 1000 to 1100°C.

Although the first patents related to aluminous cements date back to 1888, it is only after the developmental work carried out by Bied that these cements started to be manufactured on an experimental basis just before the First World War. Their first field application was developed by the French and English armies towards the end of the First World War. It was even considered a military secret, because it allowed English and French artilleries to be operational 24 hours after casting of the concrete bases on which the recoil cannon were anchored, while it took three or four days for the Germans to be ready to fire their guns anchored on Portland cement bases.

Ciment Fondu is made through a process that is completely different from that of Portland cement: it consists of melting an adequate mixture of limestone and ferruginous bauxite ($Al(OH)_3$ mixed with iron oxide) in order to form calcium aluminate (CA), as can be seen in Figure 9.2. When cooled, Ciment Fondu ingots are ground to a fineness similar to that of Portland cement.

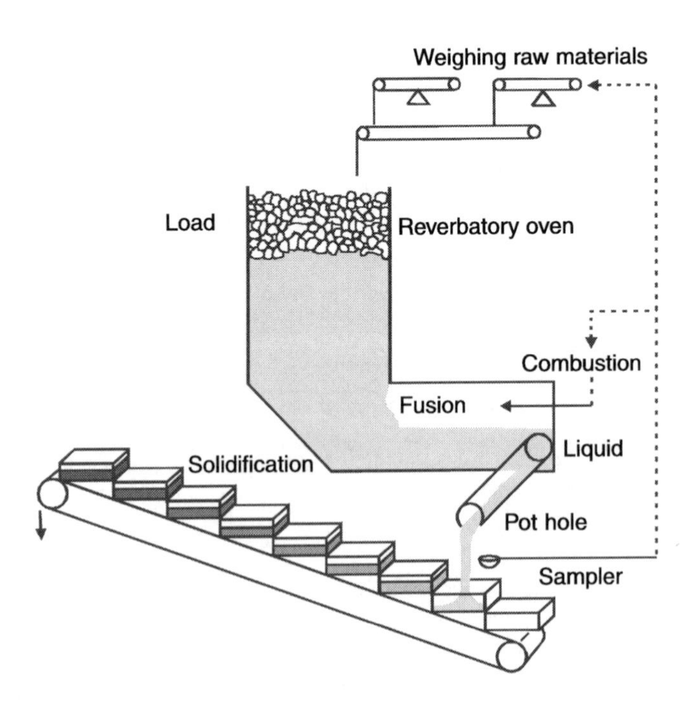

Figure 9.2 Principle of fabrication of an aluminous cement.

Ciment Fondu is used to make concrete in the same way as Portland cement, using the same mixing equipment. However, its hydration reaction is totally different. According to the temperature at which hydration proceeds, different calcium aluminate hydrates can be formed (Capmas and Scrivener 1998; Neville 2000).

if	$T < 10°C$	CA is transformed into CAH_{10}
if	$10 < T < 25°C$	CA is transformed into $C_2A\ H_8$ and AH_3
if	$T > 25°C$	CA is transformed into $C_3A\ H_6$ and AH_3

Hydration in ambient conditions results in the formation of a mixture of CAH_{10} and C_2AH_8.

When an aluminous cement is mixed with water, a dormant period lasting 2 or 3 hours is usually observed, as in the case of Portland cement, but as soon as hydration reaction starts, it proceeds very intensely and rapidly, with two concomitant consequences: a rapid hardening and the release of a large amount of heat. These two consequences have made the use of aluminous cements attractive in certain types of civil engineering applications. Moreover, other pecularities of aluminous cement resulted in their use in very special applications, for example, in moderately acidic industrial environments such as in milk factories and pulp and paper mills, because aluminous cements do not produce lime when they hydrate, but rather alumina which is more stable in acidic environments. Finally, another characteristic that resulted in certain civil engineering applications is their good resistance to sulphate attack, again because they do not produce lime during hydration.

The high amount of heat liberated during the first 24 hours is particularly interesting in arctic conditions because such concretes require only a very light heat protection.

For a long time, aluminous cements were the only cements with which a compressive strength of 20 to 40 MPa could be reached within 24 hours, which is why they were used in pre-casting. However, this is no longer the case, because such strengths, and even higher ones, can be reached by lowering the W/B ratio when using usual Portland cement and superplasticizers.

During the 1960s and 1970s in England, aluminous cements were commonly used in precasting operations (Neville 1963) because it was not necessary to heat concrete or add calcium chloride (with their deleterious consequences on the durability of reinforced concrete). The use of aluminous cement allowed the daily reutilization of the forms.

However, the use of aluminous cements in civil engineering presents a major problem known as the *conversion reaction*. C_2AH_8 and CaH_{10} formed at a temperature lower than 25°C are metastable compounds that are transformed into the more stable C_3AH_6, however with a significant decrease in strength due to a reduction of the solid volume of the hydrates. When this reaction is developing, it results in an increase of concrete porosity and microcracking of the hydrated cement paste. If this strength decrease is not taken into account during the design of the structural element, it can create serious problems.

When ambient temperature increases above 25°C, the *conversion reaction* accelerates. It is also accelerated inside precast elements cast at ambient temperature due to the high amount of heat released by aluminous cements. During their hydration, the temperature inside the concrete increases above 25°C. Moreover, it has been observed that when some $C_3A\ H_6$ has been formed, it continues to be formed even at a temperature below 25°C, a temperature at which its formation should stop, theoretically. Compressive strength losses are very rapid, more specifically when the W/C ratio is greater than 0.40.

On the contrary, when heat curing of aluminous cement is carried out with a low pressure water vapour, $C_3A\ H_6$ is formed directly and the conversion reaction does not occur. The only strength loss observed is due to the acceleration of the hydration rate, a phenomenonm which is also observed with Portland cement.

It is possible to take into account this strength decrease when designing structural elements built with aluminous cements, but it must be certain that the ambient temperature will not rise above 25°C (Scrivener and Capmas 1998; Neville 2000; Neville 2003a and 2003b; Neville 2004a and 2004b).

As a result of a series of structural failures, mostly in England, aluminous cement has been banned for permanent work in many countries (Neville 1963). One of the most dramatic failures occurred in a school where the roof collapsed during class only 15 minutes before recess, killing about thirty children. The pre-cast beams supporting the roof of this school were made of aluminous cement and the conversion reaction developed due to leakeage of water vapour from the pipes used to heat the school. These pipes were hidden in the roof-space along the concrete beams.

Aluminous cements are still used in civil engineering when very fast hardening mortars or paste have to be used to make emergency repairs, to stop water leakage and any time it is useful to use a very fast hardening hydraulic binder.

When Portland cement (40 to 80 per cent) is mixed with an aluminous cement (20 to 60 per cent), the setting time of the mixture is drastically shortened and can be as low as 5 minutes. Such conditions correspond to a flash set situation. However, such a mixture does not have great strength nor durability.

Presently, the use of aluminous cements is decreasing in civil engineering because of legal restrictions, their high price, and because high early strength concrete can be made using low W/B ratio concretes made with Portland cement.

9.3 Calcium sulphoaluminate cements[1]

9.3.1 Introduction

Calcium sulphoaluminate cements were developed in China around 1975 by the China Building Materials Academy (Lawrence 1998; Ambroise and Péra 2005). It is a new family of hydraulic cements where binding properties are no longer related to the formation of calcium silicate hydrate (C-S-H) but rather to the formation of ettringite.

Calcium sulphoaluminate cements are obtained through the calcination of a mixture of limestone, gypsum and bauxite at a temperature of around 1350°C, 100–200°C lower that that necessary to form Portland cement clinker. Consequently, their fabrication is less energy intensive than that of Portland cement.

Initially these cements were used primarily in China to manufacture self-stressed concrete pipes due to their slightly expansive properties. According to Péra and Ambroise (2005), 1 million tonnes of calcium sulphoaluminate cements were produced in China. In the USA, small quantities of calcium sulphoaluminate clinker are produced to make Type K expansive hydraulic cement (ASTM C845).

9.3.2 Composition and hydration of calcium sulphoaluminate cements

Table 9.4 presents a comparison of typical compositions of Portland cement and calcium sulphoaluminate cements to emphasize the essential differences between these two families of cements. Calcium sulphoaluminate cements contain essentially C_2S, C_4AF, $C_4A_3\bar{S}$ ($4CaO \cdot 3Al_2O_3 \cdot SO_3$ also called yeelimite) and gypsum. Lawrence (1998) describes in details the pyroprocess that results

Table 9.4 Comparative compositions of Portland and calcium sulphoaluminate cements

Component	*Cement*	
	Portland	*Sulphoaluminate*
C_3S	40–70	0
C_2S	0–30	10–60
C_3A	2–15	0
C_4AF	0–15	0–40
$C_4A_3\bar{S}$	0	10–90
Gypsum	3–6	15–30

1 Jean Péra, from the INSA of Lyon, must be thanked for his valuable help in writing this section, because the author has no experience with these particular types of cement.

in the formation of these different minerals. To decrease the energy consumption during the production of calcium sulphoaluminate cements the influence of different mineralisers is presently being studied (Lawrence 1998).

As has been seen in Chapter 6, in Portland cements the hydration of C_3S and C_2S results in the formation of calcium silicate hydrate (C-S-H) and portlandite (CH) and the hydration of C_3A and C_4AF in the formation of ettringite and/or calcium monosulphoaluminate. The hydration of calcium sulphoaluminate cements results in the formation of ettringite ($C_6A\bar{S}_3H_{32}$) according to the following reactions:
without any lime:

$$C_4A_3\bar{S}+2C\bar{S}H_2+34H \rightarrow C_6A\bar{S}_3H_{32}+AH_3 \quad (1)$$

or in the presence of lime:

$$C_4A_3\bar{S}+8C\bar{S}H_2+6CH+74H \rightarrow 3C_6A\bar{S}_3H_{32} \quad (2)$$

The microstructure of the ettringite produced according to these two reactions depends on the presence of lime. According to Mehta (1973) the ettringite produced by reaction (1), in the absence of lime, is not expansive and generally results in high early strength (Beretka et al. 1997). The ettringite produced according to reaction (2), in the presence of lime, is expansive (Su et al. 1992). It is this expansion that is used in expansive or shrinkage compensating cements.

9.3.3 Applications of calcium sulphoaluminate cements

The preceding hydration reactions point out the following specific characteristics of these cements that will define their principal field of applications:

- formation of ettringite as the binding material
- absence of lime

with the following practical consequences:

- limited total shrinkage or even slight expansion
- absence of efflorescences
- good durability in the presence of sulphates
- non-reactivity with glass fibres.

The present principal applications of calcium sulphoaluminate cements are the production of self-stressed concrete elements (ACI 223); self-levelling floors, rapid hardening mortars and cold weather concrete products. New markets could also develop rapidly in glass-fibre cement composites (Ambroise and Péra 2005), fireproof protective coatings (Péra and Ambroise 2005) and in the domain of the encapsulation of hazardous wastes (Péra et al. 2005). However, the long-term durability of calcium sulphoaluminate cements is not yet well documented in the literature.

9.4 Other types of cementitious systems

In spite of its controversial nature, Davidovits' work (1987) had the merit to bring back (reactualize) the interest on other hydraulic cementitious systems like Portland and aluminous cements. Two international conferences have been held in Brno on this subject in 2002 and 2005 (Bilek and Kersner 2002, 2005).

As a result of this new trend, it is interesting to see that already in 2005 a 0 per cent Portland cement binder, equivalent to a CEM 1 52.5 Portland cement, has been already introduced in the market in Belgium, France, and Netherlands. According to Gebauer et al. (2005) this new hydraulic binder is an improved version of the old supersulphated cement used in these countries since 1932. The main differences between the old supersulphated cement and the new 0 per cent Portland binder are the elimination of Portland cement, the use of anhydrite and the use of some alkali rich cement kiln dust (CKD). The binding properties of this new hydraulic binder rely exclusively in the formation of massive and very stable ettringite crystals (Mehta 1973).

When making concrete with such a 0 per cent Portland cement binder, high early compressive strength is obtained through a combination of physical and chemical activations: increase of the fineness of the slag (500 to 650 m^2/kg), drastic decrease of the interparticle distance in the cement paste by decreasing the W/B ratio with a superplasticizer, use of an alkali activation to accelerate the dissolution of the slag and use of anhydrite which provides just on time the necessary sulphate ions to form the ettringite crystals. As these ettringite crystals are formed in an interstitial liquid particularly low in Ca^{++} ions they are very stable.

Another type of very interesting hydraulic binder is also presently developed in Canada (Tagnit-Hamou and Laldji 2004). It has been already tested successfully in different experimental projects. This new binder is a mixture of Portland cement and Calcifrit (Tagnit-Hamou and Laldji 2004; Laldji et al. 2004), a new hydraulic binder obtained when processing spent pot liner (SPL) from the aluminum industry. Every six to eight years the electrolytic cells, in which aluminum is extracted from alumina, have to be rebuilt so that the production of 100 tonnes of aluminum generates every year the 'production' of 2 tonnes of SPL. These SPLs are essentially composed of refractory bricks, graphite, cryolite, etc. From a chemical point of view the typical chemical composition is:

SiO_2	23.0 per cent
Al_2O_3	31 per cent
CaO	8.26 per cent
Na_2O	11.0 per cent
F	8.7 per cent

They also contain cyanids that classify them as a hazardous waste.

The company NovaPb has developed a patented process that transform SPL into a non-hazardous slag or binder having typically the following chemical composition: 31.7 per cent SiO_2, 23.4 per cent Al_2O_3, 14.6 per cent CaO, 9.4 per cent Na_2O, 1 per cent K_2O, 9.4 per cent CaF. This slag is quenched, finely ground so that it can be blended with Portland cement.

Recently a ternary cement composed of 70 per cent Portland cement clinker, 20 per cent CalCiFrit, 5 per cent of silica fume and 5 per cent of gypsum was developed and used successfully in different experimental projects. It is interesting to note that this new performing binder contains two industrial products that for many years have been considered as waste and hazardous products. This is a positive consequence of the development of a sustainable conciousness in the cement and concrete industry.

9.5 Other types of cements

There are many other types of cement apart from those based on the hydration of calcium silicate and/or calcium aluminates. These cements are used, in much smaller quantities than Portland cement or aluminous cement, in very specific applications. Readers interested in these special cements can find a whole chapter on that topic, written by John Bensted, in *Lea's Chemistry of Cement Portland and Concrete* (Bensted 1998).

9.5.1 Sorel cement

Sorel cement is a magnesian cement based on magnesium oxychloride. Magnesium oxide (magnesia) is mixed with a concentrated solution of magnesium oxychloride. The chemical reactions that occur are quite complex and the first compounds formed react with the carbon dioxide found in the air. The mechanical performance of this type of cement is remarkable, but it tends to depend on the firing temperature of magnesia. Sorel cements were widely used in the Soviet Union, while in North America and Europe, applications have been quite limited due to the volumetric instability of the concretes made with such a cement and to their poor performance under freezing and thawing cycles.

9.5.2 Oxysulphate magnesium-based cement

This is a cement in which magnesium sulphate or phosphate are added to magnesium chloride.

9.5.3 Other cements

Cements based on zinc oxychloride, aluminum oxychloride, silicophosphate, sodium metaphosphate, calcium phosphate, zinc phosphate, magnesium phosphate, ammonium phosphate and magnesia, ammonium tripolyphosphate and magnesia, aluminosilicates and magnesia, etc., have been developed and are used in very small quantities.

Finally, two other cements that are manufactured only in the Soviet Union can be added to this long list, one of which is alinite. Alinite is obtained by introducing calcium chloride as a source of lime in the raw meal. The advantage of such a cement is its low processing temperature, which is decreased by 300 to 500°C compared with the production of Portland cement. This cement can be made at a firing temperature as low as 1150°C. Alinite can be mixed at a substitution rate of up to 20 per cent to Portland cement without causing any problems (Lawrence 1998).

There exists a second type of cement similar to alinite, known as belinite, in which magnesia is introduced in the raw meal. This cement does not have very interesting hydraulic properties (Lawrence 1998).

9.6 Conclusion

None of these special cements can compete with Portland cement in most of its applications. These very special cements are instead used in small niche markets where one of their particular properties can overcome their high price or when they can be used to make more durable concretes in very particular environments.

The very large consumption of Portland cement and its poor ecological performance, constitute very strong motivations to finding new types of binders, of which the most recent was the geopolymer developed, or rediscovered, by Davidovits, named Pyrament by Lonestar Cement. This will probably not be the last cement to be discovered. However, it must be admitted that up to now, very few of these binders have been able to compete against Portland cement.

References

ACI 223 'Standard Practice for Shrinkage-Compensating Concrete', Manual of Concrete Practice, American Concrete Institute.

Ambroise, J. and Péra, J. (2005) 'Durability of Glass-Fibre Cement Composites: Comparison Between Normal Portland and Calcium Sulphoaluminate Cements', *Third International Conference on Composites in Construction, Lyon, France*, Hamelin Editions.

Anonymous (1970) 'General Portland's Trinity Division Unveils Tan-Colored "Warmtone" Cement', *Modern Concrete*, 14 (7): 3.

—— (1977) 'Les efflorescences', *Le Bâtiment-Bâtir*, 5: 43–6.

ASTM C845, Standard Specification for Expansive Cement.

Bensted, J. (1998) 'Special Cements', in P.C. Hewlett (ed.) *Lea's Chemistry of Cement and Concrete*, 4th edn, London: Arnold, pp. 779–835.

Beretka, J., Sherman, N., Marrocoli, M., Pompo, A, and Valentu, G.L. (1997) 'Effect of Composition on the Hydration Properties of Rapid-hardening Sulphoaluminate

Cements', in H. Justness, A. B. Armakai and Congrex (eds) *Proceedings, Tenth International Congress on the Chemistry of Cement, Gothenburg, Sweden*, Vol. II, pp. 2029–37.

Bilek, V. and Kersner, Z. (eds) (2002) *Non-Traditional Cement and Concrete I*, Brno: Edice Betonove Stavitelski.

—— (2005) *Non-Traditional Cement and Concrete II*, Brno: Edice Betonove Stavitelski.

Capmas, A. and Verschaeve, M. (1996) 'Le Ciment d'aluminates de calcium et les règles de formulation des bétons faits avec ce ciment', in J. Baron and J.-P. Ollivier (eds) *Les Betons, bases et données pour leur formulation*, Paris: Eyrolles, pp. 23–45.

Davidovits, J. (1987) 'Ancient and Modern Concretes: What Is the Real Difference?', *Concrete International*, 9 (12, December): 23–35.

Deichsel, T. (1982) 'Efflorescence-origins, Causes, Counter-measures', *Betonwerk + Fertigteil-Technik*, 10: 590–7.

Detwiler, R.J. and Mehta, P.K. (1989) 'Chemical and Physical Effects of Condensed Silica Fume', in *Concrete, Supplementary Paper at the Third CANMET/ACI International Conference on Fly Ash, Silica Fume, Slag and Natural Pozzolans in Concrete*, Trondheim, pp. 295–306.

Dutruel, F. and Guyader, R. (1974) *Évolution de la teinte d'un béton et étude des mécanismes qui s'y rattachent*, Technical publication No. 11 by Centre d'Études et de Recherche de l'Industrie du Béton Manufacturé.

—— (1975) *Influence de l'étuvage sur l'évolution de la teinte d'un béton*, CERIB Technical Publication.

—— (1977) 'Influence de la carbonatation associée à l'étuvage et des conditions de stockage sur la teinte d'un béton', *Revue des Matériaux de Construction*, 704 (January): 27–30.

Freedman, S. (1971) *White Concrete*, Spokie, Ill.: Portland Cement Association.

Gebauer, J., Ko, S.-C., Lerat, A. and Roumain, J.-C. (2005) 'Experience with a New Cement for Special Applications', *Second International Symposium on Non-Traditional Cement and Concrete*, Brno, pp. 277–83.

Kim, B.-G. (2000) 'Compatibilité entre les ciments et les superplastifiants dans les bétons à haute performance et la teneur en alcalins du ciment et de la masse moléculaire des PNS sur les propriétés rhéologiques des pâtes de ciment et des bétons', Ph.D. thesis, Université de Sherbrooke.

Kroone, B. and Blakey, F.A. (1968) 'Some Aspects of Pigmentation of Concrete', *Constructional Review*, 41 (7): 24–8.

Laldji, S, Fares, G. and Tagnit-Hamou, A. (2004) 'Glass Frit in Concrete as an Alternative Cementitious Material', *International RILEM Conference on the Use of Recycled Materials in Buildings and Structures, RILEM Proceedings, PRO40*, Vol. 2, pp. 953–962.

Lawrence, C.D. (1998) 'The Production of Low-Energy Cements', in P.C. Hewlett (ed.) *Lea's Chemistry of Cement and Concrete*, 4th edn, London: Arnold, pp. 421–70.

Mehta, P.K. (1973) 'Mechanism of Expansion Associated with Ettringite Formation', *Cement and Concrete Research*, 3 (1): 1–6.

Neville, A.M. (1963) 'A Study of Deterioration of Structural Concrete Made with High-alumina Cement', *Proceedings of the Institute of Civil Engineers, London*, 25 (July): 287–324.

—— (2000) *Les Propriétés des bétons*, Paris: Eyrolles, Paris.

—— (2002a) 'Efflorescence: Surface Blemish or Internal Problem, Part I, The Knowledge', *Concrete International*, 24 (8): 86–90.
—— (2002b) 'Efflorescence: Surface Blemish or Internal Problem, Part II, Situation in Practice', *Concrete International*, 24 (9): 85–8.
—— (2003a) 'Draft Standard for High-alumina Cement', *Concrete*, 37 (7, July–August): 44–5.
—— (2003b) 'Should High-Alumina Cement Be Re-Introduced Iinto Design Codes?' *The Structural Engineer*, 81 (23/24): 35–40.
—— (2004a) 'Draft Standard for High-alumina Cement: Should it Tell Us How to Make Concrete?' *Materiales de Construcciones*, 54 (273): 89–90.
—— (2004b) 'Revised Guidance on Structural Use of High-Alumina Cement', *Concrete*, 38: 60–2.
Papadakis, M. and Venuat, M. (1966) *Fabrication et utilisation des liants hydrauliques*, 2nd edn (self-published).
Péra, J. and Ambroise, J. (2005) 'Thermal Behavior of Materials Based on Calcium Sulphoaluminate Cement', *Raimundo Rivera International Symposium on Durability of Concrete, Monterrey, Mexico*, pp. 451–68.
Péra, J., Ambroise, J., and Reysson, S. (2005) 'Immobilization of Waste by Calcium Sulphoaluminate Cement', personal communication.
Rabot, R., Coulon, C. and Hamel, J. (1970) 'Étude des efflorescences', *Annales de l'Institut Technique du Bâtiment et des Travaux Publics*, 23rd year, July-August, No. 271–2, pp. 70–84.
Scrivener, K.L. and Capmas, A. (1998) 'Calcium Aluminate Cements', in P.C. Hewlett (ed.) *Lea's Chemistry of Cement and Concrete*, 4th edn, London: Arnold, pp. 709–78.
Su, M., Kurdowski, W. and Sorrentino, F. (1992) 'Development in Non-Portland Cements, Proceedings', *9th International Congress on the Chemistry of Cement, New Delhi, India*, Vol. 1, pp. 317–54.
Tagnit-Hamou, A. and Laldji, S (2004) Development of New Binder Using Thermally-Treated Spent Pot Liners from Aluminum Smelters, ACI SP-219, pp. 145–159.
Venuat, M. (1984) *Adjuvants et traitements*, n. p.

Chapter 10

The art and science of high-performance concrete

10.1 Introduction

This chapter will examine how the application of the recent scientific knowledge developed in the field of hydraulic binders and admixtures resulted in the development of what is presently called high-performance concrete (HPC). It should be pointed out that the development of HPC was not simply the fruit of a scientific approach; as has very been often the case, the science and the art of HPC were developed simultaneously. At present, due to the knowledge accumulated, the trend of discoveries and development is changing, for example, self-compacting concretes and reactive powder concretes (RPCs) were the result of a scientific approach. However, as has been pointed out in the previous chapters, there is still scientific progress to be made in the field of hydraulic binders and concrete.

This chapter reviews certain themes previously developed, in order to integrate them in a field application that points out the present state of the art in the area of HPC and the general directions in which scientific and technological aspects have to be developed.

It is not pretentious to assert that recent developments in HPC constitute a giant step for the better use of concrete. Concrete is becoming a 'high-tech' material having improved characteristics and a much longer durability than a 15 to 30 MPa concrete. It is also very easy to demonstrate that HPCs are more ecological than 15 to 30 MPa concretes if both are considered from a sustainable development perspective when using socio-economic and environmental criteria different from the exclusively economic ones currently considered.

In HPC, Portland cement, cementitious materials, aggregates, and admixtures are used to their best advantage to make very compact, long-lasting concretes that can be used to build concrete structures having a long life cycle.

Of course, concrete is not an eternal material, no more than granite if durability is considered on a geological time scale. In nature, granite is transformed into quartz grains and clay particles; the bonding elements of a concrete will always end their structural life as limestone, clay and calcium sulphate – the three basic materials used to make hydraulic binders. As hydraulic binder specialists, the only thing we can do, and must do, is to try to delay as much as possible this unavoidable transformation. If we succeed, we will have made a valuable contribution in the quest to improve our material conditions with respect to our environment.

10.2 What is high-performance concrete?

The concrete that was known as high-strength concrete in the late 1970s, is now referred to as high-performance concrete, because it has been found to be much more than simply stronger: it displays enhanced performance in such areas as durability and abrasion resistance. Although widely used, the expression 'high-performance concrete' is often criticised as being too vague, even as having no meaning at all. What's more, there is no simple test for measuring the performance of concrete.

HPC can be defined as an engineered concrete in which one or more specific characteristics have been enhanced through the selection and proportioning of its constituents. This definition is admittedly vague, but it has the advantage of indicating that there is no one single type of HPC, but rather a family of new types of high-tech concretes whose properties can be tailored to specific industrial needs.

While we could devise a slightly more technically rigorous definition of HPC that would still be quite simple, such as it being a concrete with a low water/cement (W/C) (or rather a low water/binder [W/B] ratio), in the 0.30 to 0.40 range, this would still be inexact because HPCs with a W/B ratio in the 0.20 to 0.30 range have been used.

This definition can be technically refined by stating that a HPC is a concrete in which autogenous shrinkage can develop due to a phenomenon called self-desiccation, when the concrete is not water-cured. This technical jargon, however, does little to clarify things, because very few people are familiar with the terms 'self-desiccation' and 'autogenous shrinkage'.

Since there is no single best definition for the material that is called HPC, some researchers prefer to define it as a low W/B concrete with an optimized aggregate/binder ratio to control its dimensional stability and which receives an adequate water curing.

I would prefer to define HPC as concrete having a low W/C or W/B ratio, usually between 0.30 and 0.40.

10.3 Water/cement or water/binder ratio?

Both expressions were deliberately used above, either alone or together, to reflect the fact that the cementitious component of HPC can be Portland cement alone or any combination of Portland cement and cementitious materials, such as: slag, fly ash, silica fume, metakaolin, rice husk ash, and fillers such as limestone. Ternary systems are increasingly being used to take advantage of the synergy of some cementitious materials to improve concrete properties in the fresh and hardened states and to make HPC more economical. Some quaternary systems have even started to be marketed.

The use of cementitious materials when making HPC presents several advantages:

- technical, by modifying certain properties of the fresh and hardened concrete;
- economical, by decreasing the unit price of HPC;
- ecological, by providing added-value to certain industrial by-products and decreasing the emission of greenhouse gases;
- socio-economical, by putting in place a politic of sustainable development.

In spite of the fact that most HPC mixtures contain at least one mineral component, which should favour the use of the more general expression W/B ratio, the W/B and W/C ratios should be used alongside one another. This is because most of the cementitious materials that go into HPC are not as reactive as Portland cement, which means that most of the early properties of HPC can be linked to its W/C ratio while its long-term properties are rather linked to its W/B ratio.

It must be emphasized that the development of HPC technology has taught us what Feret and Abrams expressed in their original formulae for giving the compressive strength of a concrete mixture: concrete compressive strength is closely related to the compactness of the hardened matrix. HPC has also taught us that the coarse aggregate can be the weakest link in concrete when the strength of hydrated cement paste is drastically increased by lowering the W/B ratio. In such cases, concrete failure can start to develop within the coarse aggregate itself. As a consequence, there can be exceptions to the W/B ratio law when dealing with HPC. In some areas, decreasing the W/B ratio below a certain level is not practical because the strength of the HPC will not significantly exceed the aggregate's compressive strength. When the compressive strength is limited by the coarse aggregate, the only way to get higher strength is to use a stronger aggregate, or to make a concrete that does not contain any coarse aggregate: an RPC.

10.4 Concrete as a composite material

Standard concrete can be characterized solely by its compressive strength because that can be directly linked to the W/C ratio, which still is the best indicator of paste porosity. Most of the useful mechanical characteristics of concrete can then be linked to compressive strength with simple empirical formulas. This is the case with elastic modulus and the modulus of rupture (flexural strength), because the hydrated cement paste and the transition zone around coarse-aggregate particles constitute the weakest links in concrete. The aggregate component (especially the coarse aggregate) contributes little to the mechanical properties of ordinary concrete. As the strength of the hydrated cement paste increases in HPC, the transition zone between the coarse aggregate and the hydrated cement paste practically disappears (Figures 10.1 and 10.2). Since there is proper stress transfer under these conditions, HPC behaves like a true composite material, as shown by Baalbaki et al. (1991). Stress-strain curves of HPC are influenced by the stress-strain curve of the coarse aggregate, as seen in Figure 10.3.

To express this in another way, the elastic modulus (i.e., the rigidity) of a HPC can be tailored to the specific need of the designer simply by selecting the appropriate coarse aggregate. Nilsen and Aïtcin (1992) have been able to make 100-MPa concrete with elastic moduli varying from 27 to 60 MPa. Therefore, the relationships between modulus and strength found in most codes have no predictive value with respect to HPC. Recently, however, Baalbaki (1997) proposed two simple models that take into account the coarse aggregate's elastic characteristics so that the elastic modulus of any type of concrete can be calculated. One of the Baalbaki models is presented in Figure 10.4 and the predictive value of his model is presented in Figure 10.5.

10.5 Making high-performance concrete

HPC cannot be made using a casual approach. All ingredients must be carefully selected and checked because their individual characteristics significantly affect the properties of the final product. What has been said about the coarse aggregate also holds true for the cement, the cementitious materials, the sand, the superplasticizer and the other admixtures.

Once the materials have been carefully selected, their proportions must also be determined meticulously. Particular attention must be paid to water content. Even seemingly insignificant volumes of water present in the aggregates or admixtures must be accounted for. Increasing the performance of concrete is so difficult and expensive; ruining it is as easy as including a little too much water.

Compressive strengths of from 50 to 75 MPa can usually be achieved fairly easily with most cements. On the other hand, experience has shown that it is more difficult to control the rheology long enough to place a 200-mm slump HPC an hour or more after mixing, due to the potential for incompatibility and lack of robustness between the cement and superplasticizer.

10.6 Temperature rise

There is a belief firmly rooted in the concrete community that the heat developed in any concrete is a direct function of its cement content. This does not always hold true because Portland cement by itself does not develop heat. Since Portland cement develops heat only as a result of hydration, it should be rather said that the heat developed in a concrete is a direct function of the

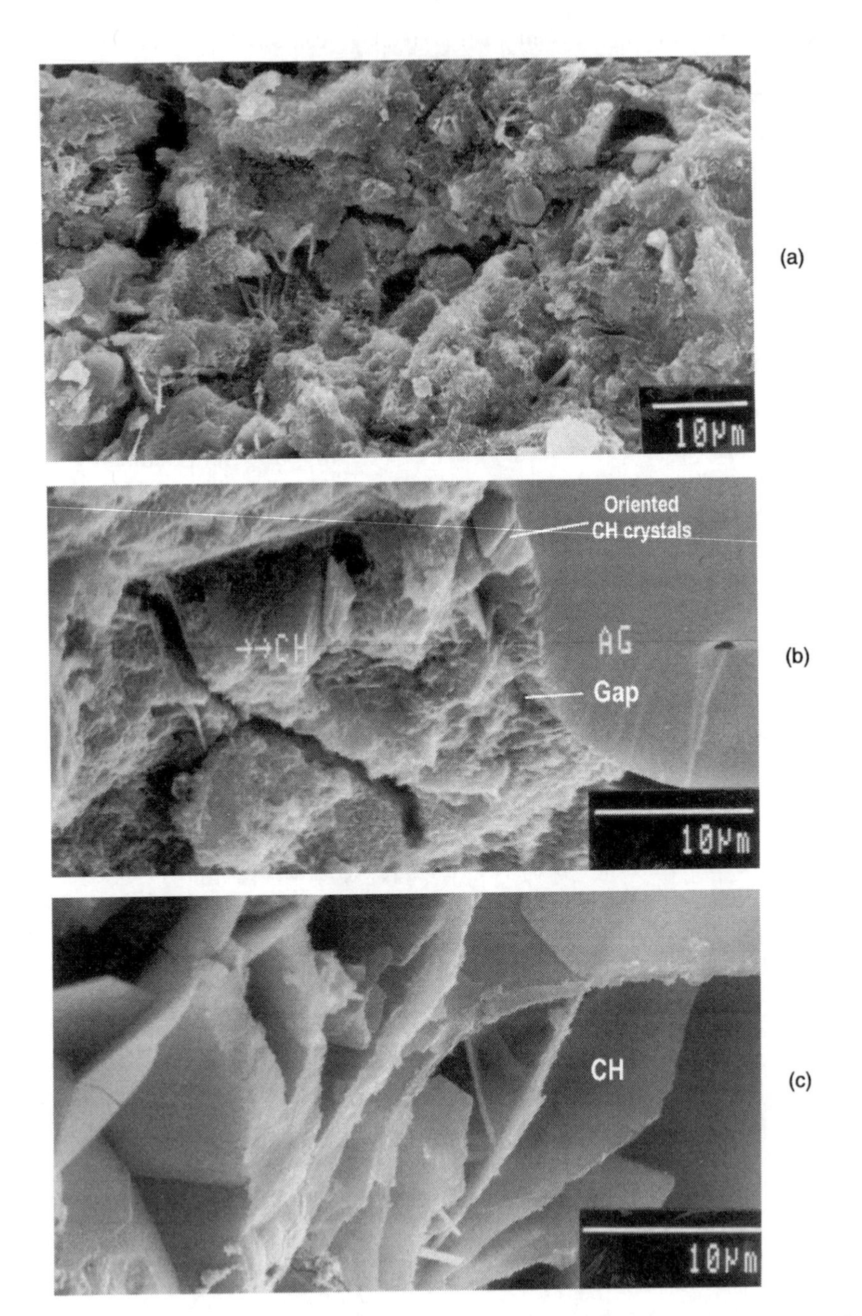

Figure 10.1 Microstructure of high water/cement ratio concrete: (a) high porosity and heterogeneity of the matrix; (b) orientated crystal of $Ca(OH)^2$ on aggregate; (c) CH crystals (courtesy of Arezki Tagnit-Hamou).

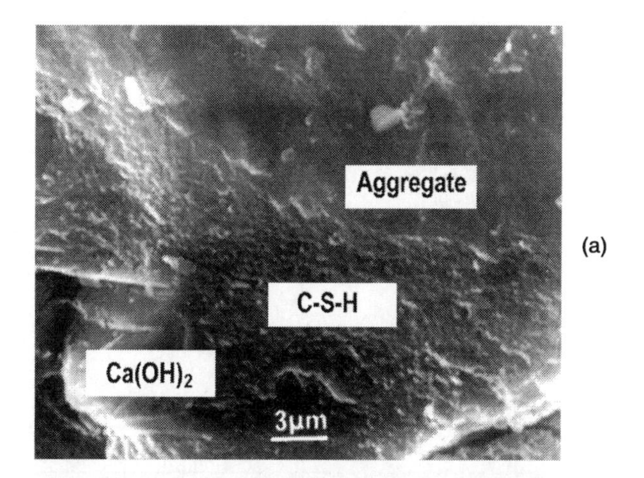

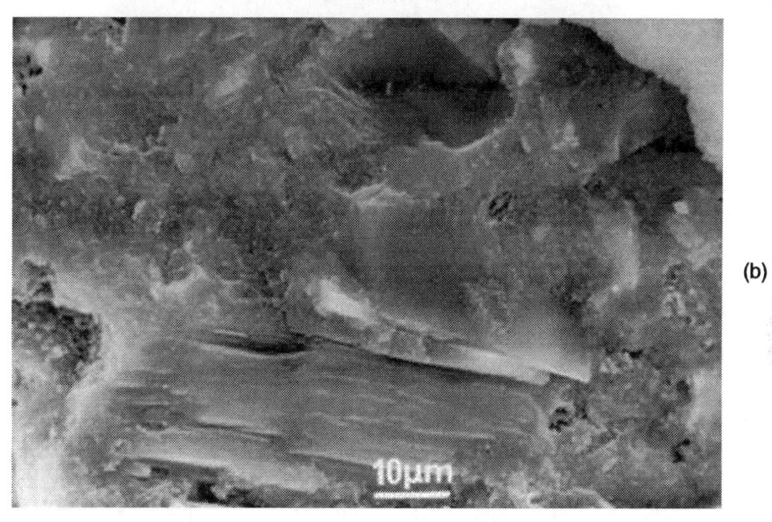

Figure 10.2 Microstructure of a high-performance concrete: low porosity and homogeneity of the matrix (courtesy of Arezki Tagnit-Hamou).

amount of cement that is hydrating and not a direct function of the total amount of cement contained in the concrete.

The amount of cement hydrating during the first hours can be limited by the amount of cement in the mix, such as in mass concrete. But, cement hydration can also be affected by the use of a retarder, a high dosage of superplasticizer or insufficient water to hydrate all the cement in the mix. The latter describes the usual situation in HPC.

The temperature variation in a concrete due to this development of heat is usually positive (temperature rise), but it can be negative (temperature decrease

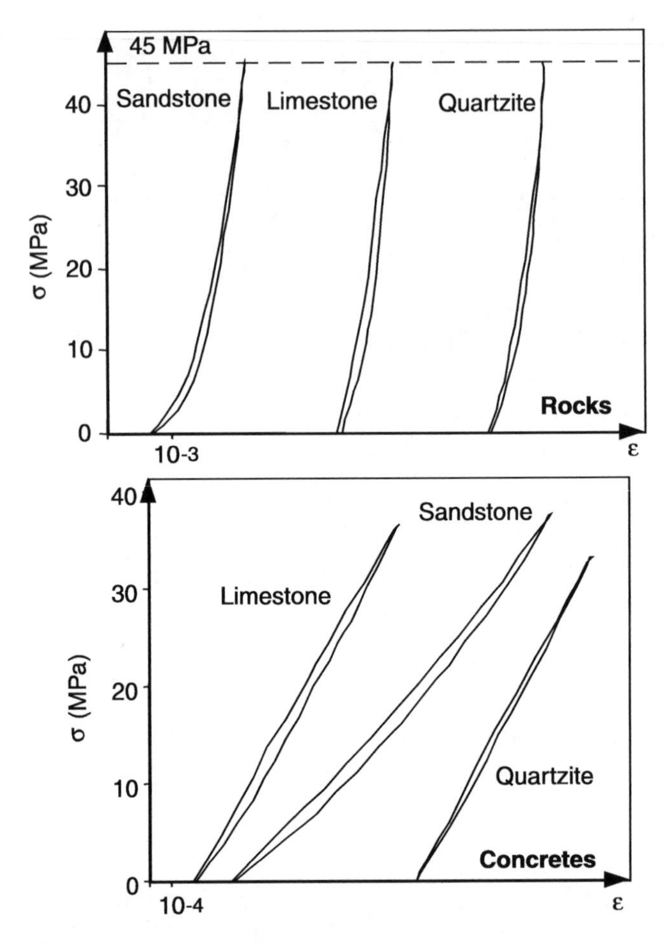

Figure 10.3 Stress-strain curves of different rocks and of high-performance concretes made with crushed coarse aggregates made out of these rocks (Baalbaki 1997, with the permission of Walid Baalbaki).

in winter conditions) or almost zero (when the amount of heat generated within the concrete equals the heat losses through the surface of the concrete and through the forms).

Consequently, it is not absolutely true that HPC develops greater heat of hydration than plain concrete. In fact, it can be totally false in the case of a particular structural element or under specific conditions. This has been confirmed by the work of Cook et al. (1992), in which three similar columns measuring 1 × 1 × 2.4 m were cast with three different concretes with compressive strengths of 30, 80 and 120 MPa. The temperature of the concrete was monitored at several places within the columns. The temperature rise at the

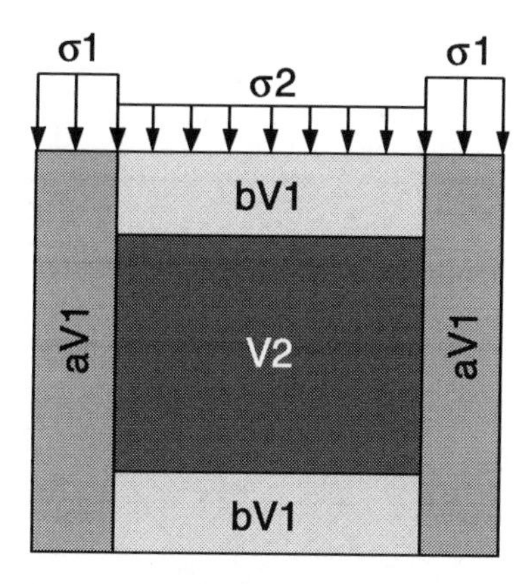

Figure 10.4 W. Baalbaki model for concrete.

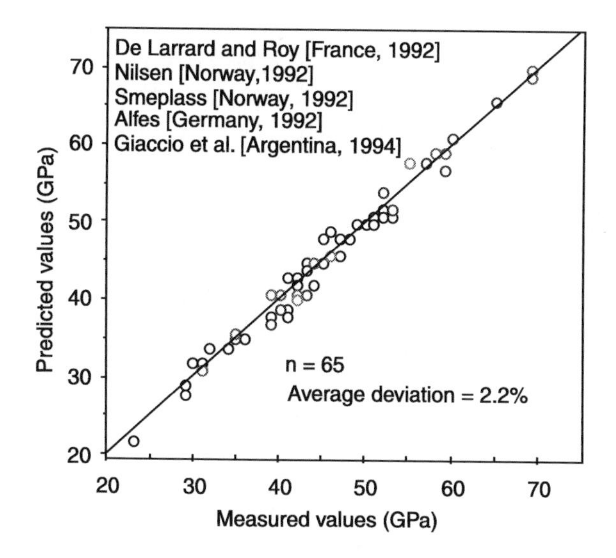

Figure 10.5 Predictive value of W. Baalbaki model for concrete.

centre of each column, where the temperature was always the highest, was nearly identical in the three columns, despite the fact that the cement content varied in the three mixes from 355 kg/m^3 for the 30-MPa concrete up to 540 kg/m^3 for the 120-MPa concrete as shown in Figure 10.6.

Faced with such results, many engineers are surprised to learn that a plain

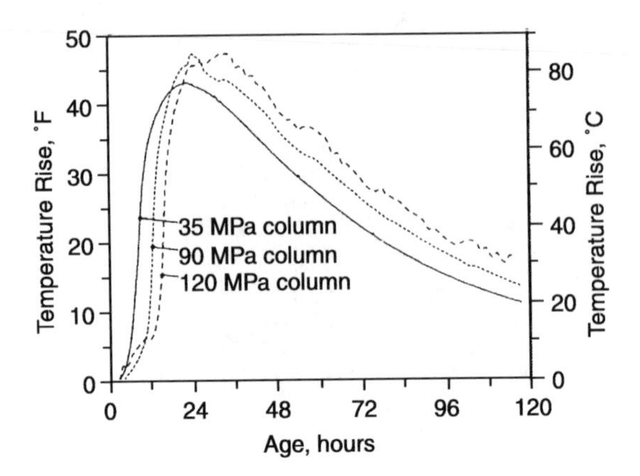

Figure 10.6 Temperature increases at the centre of each column.

concrete can develop as much heat and temperature rise as HPC, because this statement goes against the long-standing myth that the temperature rise in a concrete is directly proportional to the cement content.

Indeed, in this example the 120-MPa concrete evidenced the lowest temperature rise because less cement hydrated during the first 30 hours. Less cement had the chance to hydrate in the 120-MPa concrete for several reasons: (1) less water was used to make the concrete; (2) more superplasticizer was used (it is well known that naphthalene superplasticizers act as retarders when used at high dosages); (3) a retarder was used; and (4) the hydrates formed in the early stage of hydration in HPC are so compact that hydration kinetics are controlled by diffusion of water through the hydrates rather than by chemical processes of dissolution and precipitation, which controls kinetics when there is plenty of water in the mixture.

Moreover, as shown by Lachemi et al. (1996), temperature rise is not uniform through the structure and the maximum temperature is not reached at the same time. The temperature rise depends not only on the amount of heat developed within the concrete but also on the thermodynamic conditions at the boundaries. Of course, the more massive the structural element, the higher the temperature rise; the higher the ratio of exposed surface to the volume of concrete, the lower the temperature rise. A temperature decrease can be observed under severe winter conditions. as shown by Lessard et al. (1994) in the reconstruction of the sidewalk entrance and by Lessard (1995) during the construction of a bridge.

Using finite-element modelling, Lachemi et al. (1996) studied the influence of ambient temperature and concrete temperature on the temperature rise in some structural elements they monitored during a field experiment. Their main findings were that the highest temperature recorded was obtained at the highest ambient temperature when a hot concrete mixture was cast, but the more critical conditions for the development of high thermal gradients were achieved when a hot concrete was cast on a cold day and then allowed to cool.

10.7 Shrinkage

If water curing is essential to develop the potential strength of cement in plain concrete, early water curing is crucial for HPC in order to avoid the rapid development of autogenous shrinkage and to control concrete dimensional stability, as explained below.

Cement paste hydration is accompanied by an absolute volume contraction that creates a very fine pore network within the hydrated cement paste. This network drains water from coarse capillaries, which start to dry out. If no external water is added during curing, the coarse capillaries empty of water as hydration progresses, just as though the concrete were drying. This phenomenon is called self-desiccation. The difference between drying and self-desiccation is that when concrete dries water evaporates to the atmosphere, while during self-desiccation water stays within concrete (it only migrates towards the very fine pores created by the volumetric contraction of the cement paste) (Aïtcin et al. 1997). It is these tensile stresses developed within the concrete that are transformed into a reduction of the apparent volume that is described in terms of shrinkage.

In ordinary concrete with W/C greater than 0.50, for example, there is more water than required to fully hydrate the cement particles. A large amount of this water is contained in well-connected large capillaries so that the menisci created by self-desiccation appear in large capillaries where they generate only very low tensile stresses. Therefore, the hydrated cement paste barely shrinks when self-desiccation develops. Moreover if a 0.50 cement paste is hydrating in the presence of an external source of water, the water drained by the very fine porosity created during the volumetric contraction of Portland cement is immediately replaced in the coarse capillaries by the same volume of external water, consequently no menisci are formed and there is no autogenous shrinkage.

In the case of HPC with a W/B of 0.30 to 0.40 or less, significantly more cement and less mixing water have been used, so that the pore network is essentially composed of fine capillaries. When self-desiccation starts to develop as soon as hydration begins, the menisci rapidly develop in small capillaries if no external water is added. Since many cement grains start to hydrate simultaneously in HPC, the drying of very fine capillaries can generate high-tensile stresses that shrink the hydrated cement paste. This early shrinkage is referred to as autogenous shrinkage (of course, autogenous shrinkage is as large as the drying shrinkage observed in ordinary concrete when these two types of drying develop in capillaries of the same diameter).

But, when there is an external supply of water, the capillaries do not dry out as long as they are connected to this external source of water. The result is that no menisci, no tensile stress and no autogenous shrinkage develop within a HPC that is constantly water cured since its setting. However, as soon as the network of pores and capillaries gets disconnected from the external water source autogenous shrinkage starts to develop. This disconnection of the porous system is favoured by the fact that when a cement particle is hydrated by water from the external source this time there is an increase of the absolute volume of the unhydrated cement particle and therefore a disconnection in the capilary pore system (Figure 10.7).

Thus an essential difference between ordinary concrete and HPC is that ordinary concrete exhibits practically no autogenous shrinkage, whether it is water-cured or not, whereas HPC can experience significant autogenous shrinkage if it is not water-cured during the hydration process. Autogenous shrinkage does not develop in HPC if the capillaries are interconnected and have access to external water, but, when the continuity of the capillary system is broken, then and only then does autogenous shrinkage start to develop within the hydrated cement paste of a HPC, as shown in Figure 10.7.

Drying shrinkage of the hydrated cement paste begins at the surface of the concrete and progresses more or less rapidly through concrete, depending on the relative humidity of the ambient air and the size of capillaries. Drying in ordinary concrete is therefore rapid because the capillary network is well connected and contains large capillaries. Drying shrinkage in HPC is slow because the capillaries are very fine and soon get disconnected.

A major difference between drying shrinkage and autogenous shrinkage is that drying shrinkage develops from the surface inwards, while autogenous

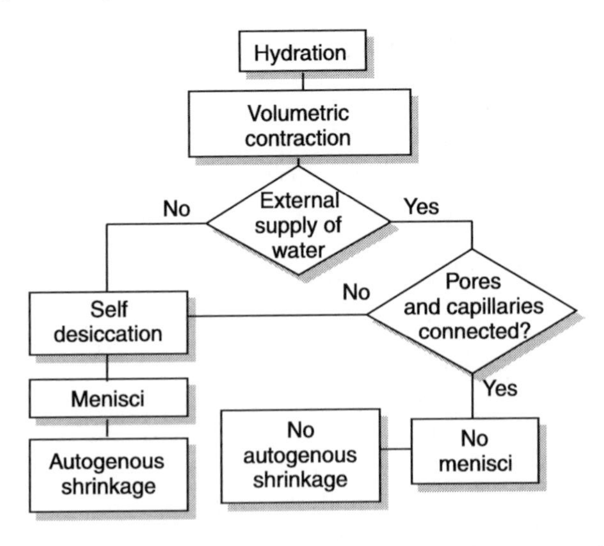

Figure 10.7 Influence of curing conditions on the occurence of autogenous shrinkage.

shrinkage is homogeneous and isotropic insofar as the cement particles and water are well dispersed within the concrete.

Thus there are considerable differences between ordinary concrete and HPC with respect to their shrinkage behaviour. The cement paste of ordinary concrete exhibits rapid drying shrinkage progressing from the surface inwards, whereas HPC cement paste can develop a high isotropic autogenous shrinkage when not water cured. This difference in the shrinkage behaviour of the cement paste has very important consequences for concrete curing and concrete durability.

Although the shrinkage of the hydrated cement paste is a very important parameter with respect to concrete volumetric stability, it is not the only one. A key parameter is the amount of aggregate and, more specifically, the amount of coarse aggregate. Too often it is forgotten that the aggregates do more than simply act as fillers in concrete. In fact, they actively participate in the volumetric stability of concrete when they restrain the shrinkage of the hydrated cement paste: concrete shrinkage is always much lower than that of a cement paste having the same W/C. It is common knowledge that concrete shrinkage can be easily reduced by increasing the coarse-aggregate content; the shrinkage of the hydrated cement paste stays the same, but it is more restrained, so that the volumetric stability of the concrete is increased. Restraining the shrinkage of hydrated cement paste by modifying the coarse-aggregate skeleton may or may not produce a network of microcracks, depending on the intensity of the tensile stresses developed by this process with respect to the tensile strength of the hydrated cement paste.

10.8 Curing

HPC must be cured quite differently from ordinary concrete because of the difference in shrinkage behaviour described above. If HPC is not water-cured immediately following placement or finishing, it is prone to develop severe plastic shrinkage because it is not protected by bleed water, and later on develops severe autogenous shrinkage due to its rapid hydration. While curing membranes provide adequate protection for ordinary concrete (which is insensitive to autogenous shrinkage), they can only help prevent the development of plastic shrinkage in HPC but they have no value in inhibiting autogenous shrinkage.

The critical curing period for any HPC runs from placement or finishing up to two or three days later, and the most critical period is usually from 12 to 36 hours, as shown in Figure 10.8. In fact, the short time during which efficient water curing must be applied to HPC can be considered a significant advantage over ordinary concrete. Those who specify and use HPC must be aware of the dramatic consequences of missing early water curing. Initiating water curing after 24 hours is too late because, most of the time, a great deal of autogenous shrinkage has already occurred and, by this time, the microstructure is already

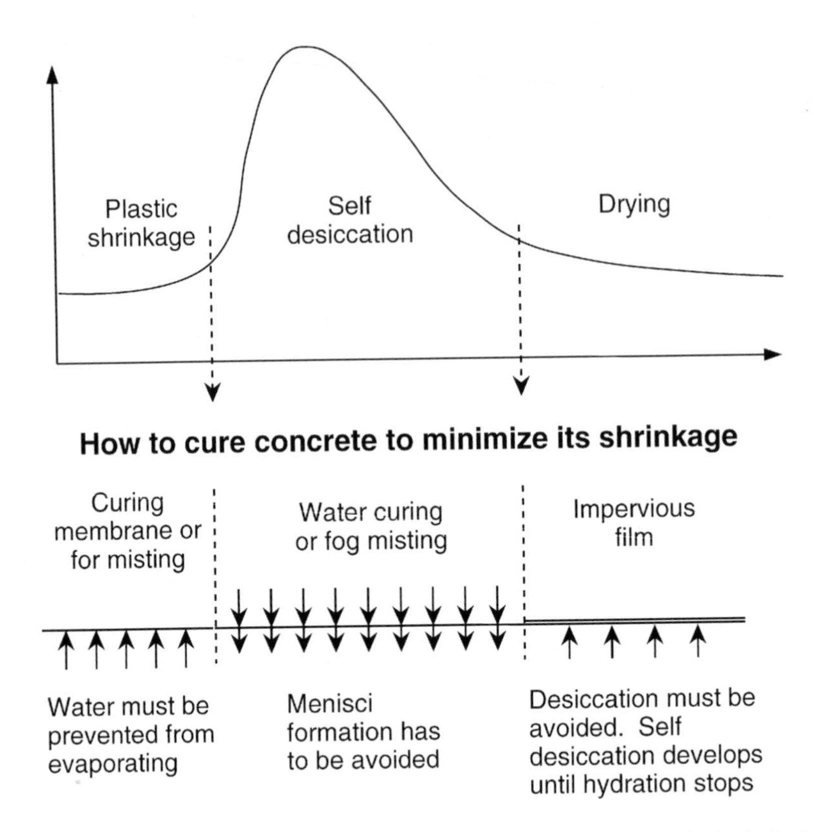

Figure 10.8 The most appropriate curing regimes during the course of the hydration reaction.

so compact that external water has little chance of penetrating very deep into the concrete.

Water ponding or fogging are the best ways to cure HPC; one of these two methods must be applied as soon as possible, immediately following placement or finishing. An evaporation retarder can be applied temporarily to prevent the development of plastic shrinkage. If, for any reason, water ponding or fogging cannot be implemented for seven days, then the concrete surface should be covered with wet burlap (hessian) or preferably a pre-wetted geotextile. The burlap or the geotextile must be kept constantly wet with a soaker hose and protected from drying by a polyethylene sheet in order to ensure that at no time during the curing period is the concrete allowed to dry and experience any autogenous shrinkage.

Moreover, it is observed that when any concrete is water cured during setting it does not shrink, but rather swells. Figure 10.9 illustrates the effect of early water curing on the volumetric change of concrete. The water curing can be

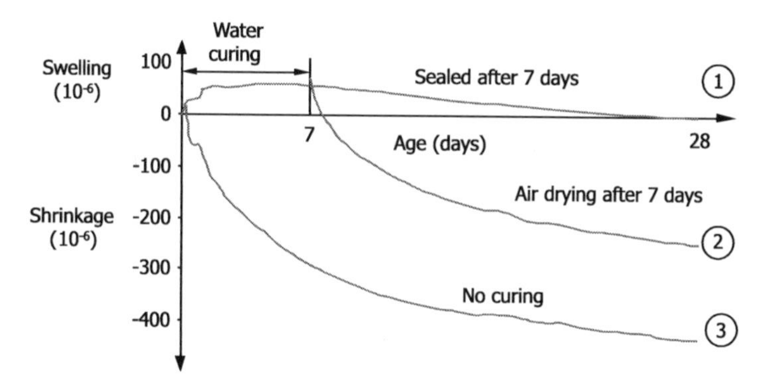

Figure 10.9 Length changes according to different curing regimes for the 0.35 W/C concrete.

stopped after seven days because most of the cement at the surface of concrete has hydrated and any further water curing has little effect on the development of shrinkage. After seven days of water curing, HPC experiences slow drying shrinkage due to the compactness of its microstructure, and autogenous shrinkage has already dried out the coarse capillaries pores. Even then, theoretically the best thing to do is to paint HPC or to use a sealing agent so that the last remaining water can be retained to contribute to hydration. There is no real advantage to painting or sealing a very porous concrete because it is impossible to obtain an absolutely impermeable coating; painting or sealing HPC, however, can be easy and effective. Figure 10.10 illustrates the curing

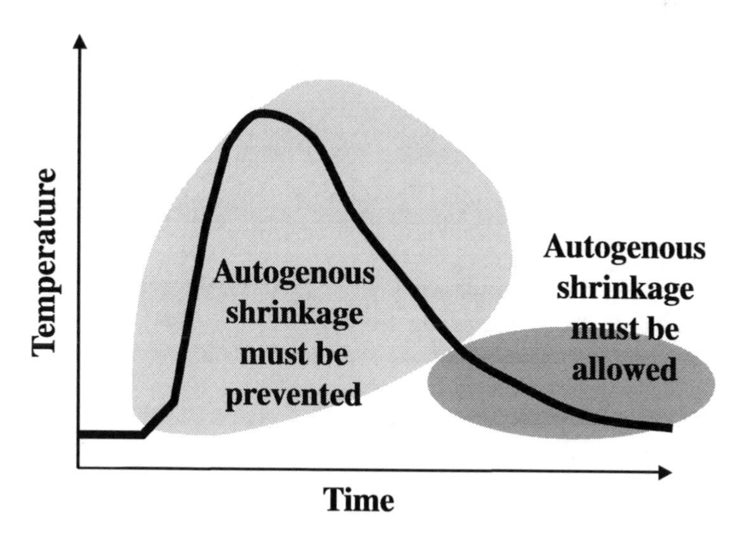

Figure 10.10 Control of autogenous shrinkage during concrete curing.

conditions that will or will not result in the development of autogenous shrinkage.

Partial replacement of coarse aggregate by an equivalent volume of saturated lightweight aggregate has been used to counteract autogenous shrinkage internally (Weber and Reinhardt 1997). The saturated lightweight aggregate particles act as small water reservoirs throughout the mass of concrete; they can be emptied into the very fine pores created by hydration reaction. Therefore, the water in the lightweight aggregate particles is drained along with that contained in the fine capillaries of the HPC. The menisci developed within the cement paste are not as small, which means lower tensile stress and less autogenous shrinkage.

Recently Jensen and Hansen (2001) proposed the use of super-absorbent polymers (SAP), of the type found in diapers, to store some curing water during mixing. This water is 'entrained' to control the development of autogenous shrinkage. One kilogramme of SAP can store more than 50 kg of water (even 200 kg of deionized pure water). When these 50 kg of water are entirely used to fill the porosity created by the absolute volumetric contraction they leave a network of 50 litres of gas bubbles corresponding to a 5 per cent air content. If the diameter of the SAP particles is well adjusted, an air-entrained network having the right spacing factor is achieved at the same time and the concrete is protected against the action of freezing and thawing cycles.

It is somewhat too early to see if the concept of entrained water by SAP works efficiently in the field, but the technology will be developed, and it will be a revolution in the art of making concrete, because during its mixing the necessary water to fight self-desiccation and autogenous shrinkage would already have been introduced in the concrete.

SAP could also be used in usual concrete to fight drying shrinkage as a source of curing water or to delay the development of drying shrinkage. One major cause of concrete cracking would be eliminated, that is the lack of water curing of concrete at an early age.

As autogenous shrinkage is a function of the development of tensile stresses within the menisci, another means to decrease autogenous shrinkage is to decrease the surface tensile stress developed within the menisci or the angle of contact of the interstitial solution. This is achieved through the use of shrinkage reducing admixtures.

But it is well known that concrete is never cured properly in the field, if spite of the fact that it is always written into the specifications that the contractor has to cure concrete. Contractors are not curing concrete for a very simple reason: they are not specifically paid for this, and, therefore, concrete curing is always perceived by them as an unprofitable activity or even a source of expense and therefore a waste of time. But, when contractors are specifically paid to water cure concrete, they do it as they do it for any other item that is paid. For six years now, the City of Montreal (1999) and the Department of Transportation of Quebec are requesting unit prices for each item directly related to an early water curing. Since the initiation of this new policy on the early water curing of concrete, it is amazing to see how zealous contractors are in the matter of water curing. For them water curing is now seen as a source of profit. From the first experiences in that matter it has been found that the cost of an early water curing varies from 0.1 to 1 per cent. A very modest price when considering the improved durability of the concrete structures that are built like that.

Therefore, the best way to be sure that HPCs are properly and efficiently cured in the field is to specifically pay contractors to cure concrete.

10.9 Durability

10.9.1 General matters

The durability of a material in a particular environment can only be established over time, so it is difficult to predict precisely the longevity of HPC because we do not have a track record for HPC exposed to very harsh environments for more than five to ten years, except perhaps for some North Sea offshore platforms. It must be remembered that the first uses of high strength concrete in the late 1960s and early 1970s were indoor applications, mainly in columns in high-rise buildings, which is not a particularly severe environment. Outdoor applications of HPC only date from the late 1980s and early 1990s, which means that not enough time has gone by to properly assess the real service life of any HPC structures under outdoor conditions.

Based on years of experience with ordinary concrete, we can safely assume that HPC is more durable than ordinary concrete. Indeed, the experience gained with ordinary concrete has taught us that concrete durability is mainly governed by concrete porosity and the harshness of the environment (Aïtcin 1994).

It is easy to assess the harshness of any environment with respect to HPC because hydrated cement paste is essentially a porous material that contains some freezable water. Assessment involves simply examining how the environment affects each of these characteristics.

On the other hand, it is not always simple to assess how easily aggressive agents will penetrate concrete. For example, water flow through a 0.70 W/C concrete is easy to measure, but water flow almost stops in a 0.40 W/C concrete, regardless of the thickness of the sample and the amount of pressure applied. The gas permeability is also difficult to measure. Sample preparation, particularly drying, significantly influences gas permeability. Therefore, the critical question remains how to appropriately assess the permeability of a concrete with a low W/B and a very compact microstructure.

Despite all the criticism levelled at it, the so-called 'Rapid chloride-ion permeability test' (AASHTO T-277) gives a fair idea of the interconnectivity of the fine pores in concrete that are too fine to allow water flow. Experience has revealed good correlation between the water permeability and rapid chloride ion permeability for concrete specimens with a W/C greater than 0.40. Chloride-ion permeability is expressed in coulombs, which corresponds to the total amount of electrical charge that passes through the concrete sample during the 6-hour test when subjected to a potential difference of 50 volts.

When the rapid chloride-ion permeability test is performed on concrete samples with lower W/C, the number of coulombs passing through the sample decreases. It is easy to achieve a chloride-ion permeability of less than 1 000 coulombs for a HPC containing about 10 per cent silica fume and having a W/B around 0.30. The only other way to achieve this would be with latex-modified concrete, which would be much more costly. Much lower chloride-ion permeability values can be achieved if the W/B is reduced below 0.25. Values as low as 150 coulombs have been reported, far lower than the 5 000 to 6 000 coulombs reported for ordinary concrete (Gagné et al. 1993).

The rapid chloride-ion test also reveals that the connectivity of the pore system decreases drastically as the W/B decreases, making the migration of aggressive ions or gas more difficult in HPC than in its plain counterpart. The author believes that this is the best indication that the service life of HPC should exceed that of ordinary concrete in the same environment. It is difficult to determine the number of years by which the service life would be extended because the predictive models developed for ordinary concrete cannot readily be extrapolated to include HPC. However, it can be said that some HPC structures will outlast the average life span of a human being.

10.9.2 Durability in a marine environment

10.9.2.1 Nature of the aggressive action

Sea water by itself is not a particularly harsh environment for plain concrete, but a marine environment can be very harmful to reinforced concrete due to the multiplicity of aggressive agents that it can face. In a marine environment, a concrete structure is essentially submitted to four types of aggressive factors:

1. chemical factors related to the presence of various ions dissolved in the sea water or transported in the wet air;
2. geometrical factors related to the fluctuation of the sea level (tides, storms, etc.);
3. physical factors such as freezing and thawing, wetting and drying, etc.;
4. mechanical factors such as the kinetic action of the waves, the erosion caused by sand in suspension in the sea water, floating debris and even floating ice in northern seas.

It is the combination of these different factors that can be harmful to reinforced concrete structures. In the following, we will review very briefly the nature of each attack in order to show how HPC is the best concrete fitted to resist not only each of these particular factors, but also their combined action.

10.9.2.2 Chemical attack on concrete

Several submerged plain concrete blocks and structures exposed for nearly 100 years in different marine environments are still in relatively good condition. The only chemical limitation usually recommended for a cement to be used in a marine environment is related to its C3A content, which should not be greater than 8 per cent.

Figure 10.11 represents the different successive altered zones found in a concrete exposed to sea water for several years: carbonation, formation of brucite, of monochloroaluminate, and sulphate attack with the formation of gypsum, ettringite or even thaumasite. Each of these chemical mechanisms is well known and explained in specialized books (Duval and Hornain 1992; Mehta and Monteiro 1993; Neville 1995).

Sea water is very harmful to reinforced concrete. Once chlorine ions have reached the reinforcing steel level, resulting in a rapid spalling of the covercrete, it is easier for chlorine ions to reach the second level of rebars, and so on. The only way to avoid, or to retard as long as possible, the corrosion of the rebars by chlorine ions is:

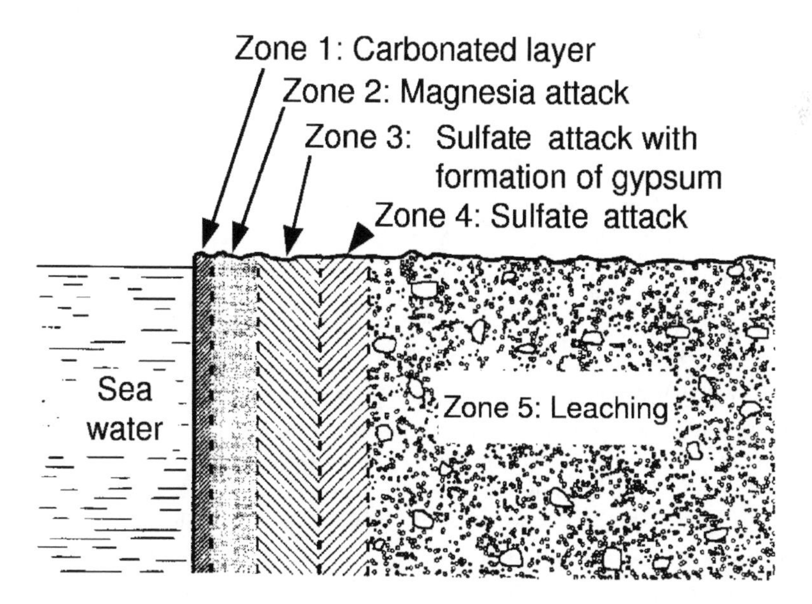

Figure 10.11 Schematic representation of the different altered layers found in a concrete marine structure (with the permission of H. Hornain).

- to specify a very compact and impervious concrete, and place and CURE it correctly, and;
- to increase the concrete cover.

The development of all of these mechanisms of aggression is closely related to the facility with which aggressive ions can penetrate concrete. Therefore it is obvious that a very dense and impervious matrix, such as the one found in HPC, constitutes the best protection that can presently be offered against a marine environment. HPC has been used very successfully for more than twenty years to build offshore platforms and, more recently, to build two major bridges for which the owner had requested a 100-year life cycle: the Confederation Bridge in Canada and the Tago Bridge in Lisboa, Portugal. It is interesting to point out that these two bridges have been built in a BOOT mode (Build, Own, Operate and Transfer) by consortia of contractors that will have to maintain these two bridges during the entire concession time.

In that respect, it is interesting to note that in the case of these two bridges, the concrete cover has been extended to 75 mm to meet the 100-year life cycle requirement.

It is also very important to point out that it is not sufficient to specify a Type V cement or a slag cement to obtain a concrete that will resist a harsh marine environment. The curing of this concrete is as important as the selection of an appropriate cementitious system. The rapid deterioration of the precast elements of the Dubai causeway in the Arabian peninsula is a good example of what must *not* be done.

10.9.2.3 Physical attack

There is still controversy about the necessity of entraining air in HPC to make it freeze-thaw resistant. In Canada, any exposed HPC must be air entrained, which is the case of the concrete of the Confederation Bridge. In Norway, HPC can contain only a small amount of air, but more to facilitate its placing and finishing rather than to improve its freeze-thaw resistance.

10.9.2.4 Mechanical attack

In this case also the compactness and the high strength of the matrix of a HPC offers a good protection to the abrasive action of the sand of the debris, or even from floating ice. In the case of the Confederation Bridge, the concrete used to build the conical part of the piles that deflects the ice loads in the tidal zone was a 90 MPa air-entrained HPC. This concrete is thought to be able to resist the tidal freezing and thawing cycles in winter and the abrasive action of the floating ice which is particularly severe in the Northumberland Strait, due to the presence of changing currents associated with the tides and winds.

10.9.2.5 Conclusion

It is obvious that HPC is the material of choice fitted to resist the harsh environment that is found in coastal areas particularly well. It would be a great mistake to take advantage of the improved durability of the concrete cover to reduce its thickness; on the other hand, it would be wise to increase it during the next millennium if we want to start to implement an efficient sustainable development policy in civil engineering.

10.10 Freeze-thaw resistance

This subject has always been and remains controversial (Aïtcin et al. 1998). Some researchers and engineers are saying that HPC does not need entrained air to be freeze-thaw resistant. Canadian standard CSA A23.1 is imperative on this subject: the HPCs made in Canada must contain entrained air. First of all, no single test can be used to ascertain if a particular concrete is resistant to freezing and thawing. Standards such as ASTM C666 propose more than one procedure for determining freeze-thaw resistance, and selecting the proper procedure is not always straightforward. Second, the freezing and thawing rate can vary over a large range when these tests are performed, and the variation in rate can influence the test results. Third, an arbitrary value for the durability factor is usually specified to distinguish a freeze-thaw resistant concrete from the one which is not. Finally, there is the issue of how many freeze-thaw cycles a concrete must resist in order to be declared freeze-thaw resistant.

In North America, freeze-thaw resistance of concrete is assessed using Procedure A (freezing and thawing in water) of ASTM C666. If the durability factor of concrete is still above 60 per cent after 300 cycles, the concrete is said to be freeze-thaw resistant. Because this test takes too long to perform (usually more than ten weeks), several other criteria giving a more rapid assessment of freeze-thaw resistance have been developed and correlated to the ASTM C666 test. This is the case, for example, with the use of entrained air spacing factor as a freeze-thaw acceptance criterion. Measuring the spacing factor of a particular concrete is not so easy, but it can be done within a week or less. To illustrate this point, Canadian Standard CSA A23.1 specifies that ordinary concrete can be classified as freeze-thaw resistant if its average spacing factor is less than 230 mm with no individual values higher than 260 μm. When this criterion was adopted by the CSA Committee, it was noted that the value also protected ordinary concrete from the scaling action of deicing salts. This fact has generally been forgotten.

Experience has proven that this criterion is not valid for HPC. HPCs with spacing factors as high as 350 μm, and even 425 μm in one case, were found to resist 500 freeze-thaw cycles. Therefore, the 2000 formulation of CSA A23.1 states now that when a HPC has a W/B ratio lower than 0.36 its average spacing factor has to be lower than 260 μm with no individual values greater than 300 μm in order to be said to be freeze-thaw resistant. It is still not clear how many cycles HPC should withstand before it is considered freeze-thaw resistant. A recent study carried out by Aïtcin et al. (1998) involving different HPCs with the same W/C but with spacing factors varying from 190 μm to 425 μm revealed an inverse relation between the spacing factor and the number of freeze-thaw cycles to failure. For instance, it took almost 2000 freeze-thaw cycles to fail a HPC with a spacing factor of 190 μm when tested in accordance with ASTM C666 Procedure A. The freeze-thaw resistance of HPC is destined to remain controversial for some time because there is no consensus on how freeze-thaw durability should be tested.

Two HPC mixtures were used to reconstruct two entrances to a restaurant in Sherbrooke (Lessard et al. 1994). The non-air-entrained concrete used to build one entrance failed to meet both CSA A23.1 and ASTM C666 criteria; the air-entrained concrete passed ASTM C666, but failed CSA A23.1 criteria. After twelve winters, during which it can be assumed that the concretes averaged fifty freeze-thaw cycles annually in 'saturated conditions' with exposure to deicing salts, the difference between the two concretes is almost imperceptible.

10.11 The fire resistance of HPC

For many years, the fire resistance of HPC has been a controversial subject, some reports saying that HPC performed as well as usual concrete, others the reverse (Diederich et al. 1993; Sanjayan and Stocks 1993; Felicetti et al. 1996; Jensen et al. 1996; Noumowe et al. 1996; Chan et al. 1996; Koylou and England 1996; Breïtenbücker 1996; Jensen and Aarup 1996; Phan 1997). Following the first fire that occured in a HPC structure, the Chunnel fire (Acker et al. 1997; Demorieux 1998), and from different studies in progress in several countries it is clear that the fire resistance of HPC does not seem to be as good as that of usual concrete, but that it is not as bad as some alarming reports have suggested. As is the case for any concrete, HPC is one of the safest construction material as far as fire resistance is concerned.

As construction details, as well as material resistance by itself, could greatly influence the fire resistance of a structural element, it is presently impossible to give very simple rules that should govern the design of HPC structures that could be exposed to a more or less severe fire. Presently, some models have been developed that can be used to forecast the structural consequence of a fire on the safety of a HPC structure. Moreover, several promising avenues are presently being investigated to improve the fire resistance of HPC itself.

Instead of reviewing in detail the controversial literature on the fire resistance of HPC, three brief presentations on the actual fire resistance of HPC and usual concrete will be done. The first one will deal with the violent fire that occured in the Chunnel, based on reports by Acker et al. (1997) and J.-M. Demourieux (1998), the second one will deal with the big conflagration that occured in the Düsseldorf Airport and the third one will present the latest fndings of the Brite-Euram HITECO BE-1158 research project (Neck 1999).

10.11.1 The fire in the Channel Tunnel

The fire of a truck in the Channel Tunnel did not surprise the safety department of the tunnel administration. The occurence of such a fire had been forecast because each day a truck burns in France and another in England (Acker et al. 1997). If a fire was to happen in the tunnel, the engineer of the train was asked to speed up to get the train out of the tunnel as soon as possible.

It had also been forecast that some hydraulic jacks used to support the access ramp used by the trucks to get on the railway platform could be loosened and create a risk of derailment for the train. In such a case, the engineer could be asked to stop the train and tighten the hydraulic jack. But, what had not been forecast was that the two incidents could happen simultaneously so that the engineer could receive two conflicting orders at the same time for which no priorization had been included in the safety procedures.

Facing two contradictory orders, the engineer decided on 11 November 1996 to stop the train at 18 km from the French entrance, fortunately in the driest zone of the tunnel from a water seepage point of view. In this area, the blue chalk through which the tunnel was excavated was the most impervious of all the rock formations. When the moles were excavating this area they registered their all time speed record and the workers asked for some water to be fogged in the excavated rock to get rid of the fine dry chalk dust generated by the moles.

It is difficult to imagine the damage that could have occurred to the tunnel if the fire had taken place a few kilometres further in a fault area where it would not have been possible to take advantage of the imperviousness of the chalk layer.

The fire was particularly violent as far as its maximum temperature and length are concerned. The maximum temperature that was reached was in the order of 700 to 1 000°C and the total duration of the fire was about 10 hours. The concrete lining in the area of the fire was composed of pre-cast concrete elements having a design strength of 70 to 80 MPa, using HPC having a water/binder ratio of 0.32 (Demorieux 1998).

In the most intense zone, the concrete lining was severely damaged and would not have been able to counteract the hydrostatic pressure for which it had been designed. An extensive survey done after the fire on the unaffected zone has shown that the in-place concrete had an average compressive strength between 70 and 80 MPa, a modulus of rupture between 7 and 8 MPa, and an elastic modulus between 37 and 44 GPa.

As far as the damaged zone is concerned, it is 480 m long, involving about 300 lining rings 400 mm thick. It could be divided into six parts that displayed different degrees of damage. In the two farthest zones from the centre of the fire, which were 80 m and 100 m long on both sides, the concrete lining had been only slightly damaged and no steel reinforcement was apparent.

In the next two adjacent zones, which were 70 m long, the concrete lining was more severely damaged over a thickness of between 50 to 100 mm. In this zone, the first reinforcing level could be seen here and there. The more severely damaged section was observed on the upper part of the lining.

The central zone where the fire was the most intense could be divided into two parts, a 50 m long area very severely damaged and another one 90 m long slightly less damaged. In the most severely damaged zone the concrete lining was completely destroyed all over its thickness (400 mm), more specifically in the 3 to 8 o'clock zones. In the less damaged part, the residual concrete had a thickness of between 50 and 200 mm except in the lowest part where its thickness was still 350 mm. In the less damaged zone, the concrete lining had a residual thickness of about 200 mm in the upper part and of about 350 mm in the lowest part.

It was observed that in this zone the reinforcing steel arrangement played a very important role because the concrete, which was prevented from falling down by the reinforcing bars, protected the interior concrete. Therefore in such zones, the steel reinforcing bars played a key role in protecting the interior concrete. In many places in the central zone, the reinforcing steel was highly deformed due to the severity of the fire. Under the effect of the heat, the restrained dilatation of the lining generated large transverse and horizontal stresses in the damaged concrete so that numerous 45° fissures could be observed at the edges of the precast elements. Concrete was litterally hollowed out at the centre of the reinforcing mesh over a more or less thick depth depending on its location from the centre of the fire.

Concrete spalled in small pieces having an average thickness of about 10 mm. Some of theses pieces were no bigger than a coin. The concrete lining was always more damaged in its upper zone than at its bottom end on the track.

In the two less damaged zones, concrete had spalled over greater surfaces in some places. It was possible to see that it happened frequently where nylon spacers had been used during precasting to correctly place the reinforcing steel. Most probably the pressure of the gas generated during the burning of the nylon spacers was responsible for this spalling.

All the tests done in the non-damaged part of the lining in the fire zone have shown that the residual concrete remaining in the lining was almost intact in the upper part as well as in the lower part of the section. SEM observations have shown that the residual concrete had not been altered significantly, no permanent strains could be seen even at the surface of the residual concrete. In all cases the residual concrete was altered in a very thin layer, and according to petrographical examinations, the maximum temperature reached by the concrete in that area was lower than 700°C.

The observations and test results conducted on the concrete of the track and at the lower part of the lining resulted in the decision to keep them in place during the reconstruction. Based on all the observations done and the results obtained, it was decided to rebuild the lining using a wet shotcrete after a careful cleaning of the damaged concrete.

Other fires have occurred in several tunnels in Europe after the fire of the

Channel Tunnel. These fires have shown both the advantages and the limits of concrete in such situations but also, and more importantly, the great danger of fume intoxication by the asphalt that was used as a road surfacing agent and its reinforcing effect on the fire. *It is absolutely urgent to remove asphalt from the tunnels and to replace it by a concrete pavement.*

10.11.2 The Düsseldorf Airport fire

On 16 April 1996 a devastating conflagration broke out at the Düsseldorf Airport, killing seventeen people by smoke inhalation and injuring several hundred. The fire was attributable to improper use of combustible insulating material and the plastic cover of the cables laid in the hollow ceiling space (Neck 1999).

The fire was triggered by some welding work which had been performed and developed unnoticed for a long time so that the smoke had time to spread into all parts of the building in the hollow ceiling and through the ventilation system.

The highest temperature to which the concrete ceiling was exposed has been estimated at 1000°C. At such a temperature the 25 MPa concrete spalled, but from a structural point of view the building elements were not damaged. Moreover, it was found that harmful substances such as dioxins did not penetrate into the concrete so that the concrete structure was still in a serviceable condition.

Since the owners and operators intended to erect an extended and modernized airport facility, they decided to demolish the burnt concrete structure.

10.11.3 Spalling of concrete under fire conditions

It is difficult to make a direct comparison between these two major conflagrations but it can be pointed out that in both cases, very fortuituously, the initial cause of the fire was the burning of some polystyrene, that the maximum temperature reached has been estimated at 1000°C, and that the HPC of the Chunnel and the usual concrete of Düsseldorf Airport both spalled.

Of course, the thickness of the spalling is a function of the maximum temperature that is reached during the fire and of the duration of exposure to this temperature, as has been seen in the various zones of the fire area in the Chunnel.

According to J.-M. Demourieux, a standard ISO 834 fire would have to cause a 30-mm thick spalling in the lining of the Chunnel.

10.11.4 The Britl–Euram HITECO BE-1158 Research Project

The preliminary conclusions of this research project, financed by the European Community, were presented on 9 March 1999 at a meeting of the French Civil

Engineering Association meeting. All the tests were done in Finland at the VTT Laboratories, one of the best equipped European laboratories for fire studies (Cheyrezy and Beloul 1999).

The conclusion of this presentation is: the experimental study undertaken on a 60 MPa HPC without silica fume and a 90 MPa high-performance concrete with silica fume have shown an excellent fire resistance, except for a small column that was heavily loaded.

In actual structures more favourable conditions are found:

- The columns have a bigger size than the one tested.
- The loading is the service loading and not the maximum loading.

The FIREXPO software can be used to predict the thermomechanical behaviour of any structural element.

10.12 The future of HPCs

HPC is not a passing fad. It is here to stay, not only because of its high strength but also because of its durability. For example, an outdoor concrete parking garage can be built with 20 MPa concrete under current codes. Its columns and slabs would be somewhat bigger than if an 80 MPa concrete had been used. But the life cycle of the garage will be very short in an environment as severe as that in Eastern Canada or in a sea coastal area, because 20 MPa concrete cannot adequately protect the reinforcing steel against corrosion from chlorine ions. The service life would be somewhat longer in a less severe environment, but would still be short due to carbonation.

Therefore, at the dawn of the twenty-first century, it is not particularly difficult to predict that the use of HPC will increase in order to extend the service life of concrete structures exposed to severe environments (Duval and Hornain 1992). The durability of a concrete structure depends on several factors, one of which is the durability of the concrete itself. As the durability of concrete is essentially linked to its permeability, HPC, with its compact microstructure and very low permeability, should obviously be more durable than ordinary concrete. It must be emphasized, however, that good concreting practice, including good curing, are essential to creating a durable structure. It would be a pity if improper practice and poor curing resulted in a structure with impervious concrete in between the cracks.

We still do not know how to make HPC with low permeability but without the high strength. Therefore, designers have to learn to take advantage of the extra strength provided by low W/B concrete. One day, we may be able to make durable concrete of lower strength.

Another reason that will lead to a greater use of HPC in the twenty-first century is society's greater interest in ecological concerns (Kreijger 1987). Many others share my opinion that we cannot continue the wasteful use of our natural resources and energy that characterized the nineteenth and twentieth centuries. HPC is more ecological than ordinary concrete. HPC, with its more compact microstructure, provides the sought property with less material.

Moreover, the twenty-first century will be the century of recycling. Recycling paper, cardboard, aluminium, steel and even plastic has already become quite common in many countries. The trend towards recycling has already reached the concrete field. We must remember, however, that recycling a material leads to a product of lower quality. This is because the original purity of the initial materials are lost when different additives are combined in order to enhance the final performance of the product. Concrete production has not reached the same degree of sophistication as rubber manufacturing, although concrete is becoming more complex every day.

Another new development is reactive-powder concrete (RPC), which can have a compressive strength of about 200 MPa in an unconfined state and as high as 350 MPa when confined in thin steel tubes (see Figure 10.38). Moreover, the addition of steel fibres to RPC produces a material with modulus of rupture as high as 25 to 35 MPa in addition to high compressive strength (Richard and Cheyrezy 1993; Bonneau et al. 1996). RPC will pose no threat to the HPC market for years to come. Once we are able to produce and use RPC and to design with it at an industrial scale, HPC will have already won over a significant part of the market of ordinary concrete.

Now that 1000-MPa Portland cement-based materials is possible to make, only a pessimist could refuse to see the future of HPC.

10.13 Some HPC structures

In this section some HPC structures will be presented in order to illustrate some recent uses of HPC. The selection of the structures was made on an arbitrary manner.

10.13.1 Monuments in HPC

Among the monuments and prestigious structures built with a HPC, the Grande Arche de la Défense in Paris was one of the first applications of the use of HPC in France (Figure 10.12a). The top beams that are linking the two sides of the arch were built with a HPC and not steel.

The Sagrada Familia Cathedral of Barcelona was designed by Gaudí as a structure working essentially in compression with catenaries and shapes inspired from nature. Presently some of the structural elements are built with HPC which accelerate the construction path. According to the present chief architect, the Sagrada Familia should be completed between 2020 and 2025.

Figure 10.12a The Arche de la Défense in Paris (courtesy of Pierre Richard).

10.13.2 Skyscrapers

Figure 10.13 presents the evolution of the height of concrete skyscrapers during the second half of the twentieth century. This passed from 100 metres to more than 800 metres in less than fifty years.

The Water Tower Place building of Chicago is the first skyscraper that was designed with a high strength concrete. The columns of the first floor were designed with a 60 MPa concrete in 1970, which was really an outstanding strength at that time. The columns of the upper storeys were built with concretes having decreasing compressive strength and an adjusted amount of reinforcing steel. Such a solution resulted in significant savings during the construction: the forms could be reused at every floor, as the finishing of each floor was absolutely identical.

The same concept was used to build the columns of the Petronas Towers in Kuala Lumpur and is presently used in Goïania in Brazil to build some story condominiums. In this last case the use of HPC to build the columns and a 30 MPa concrete for the floors results in 20 per cent saving on the structural part of the condominiums.

Figure 10.12b The use of high-performance concrete in the Sagrada Familia in Barcelona.

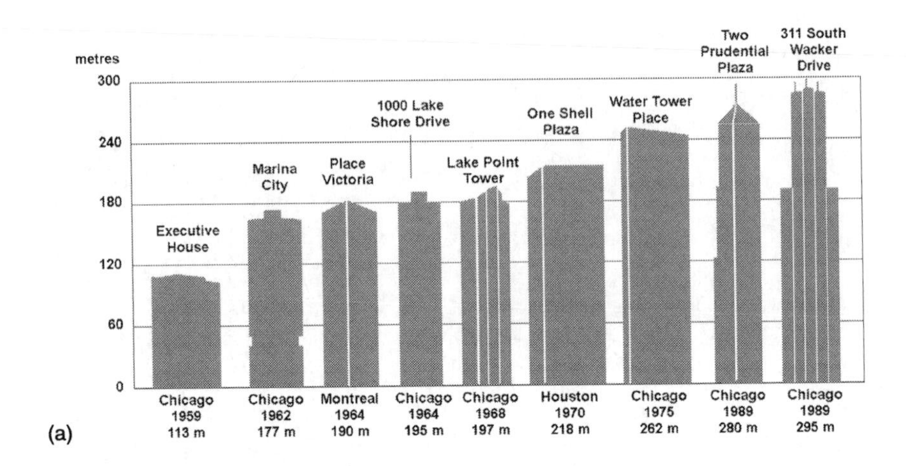

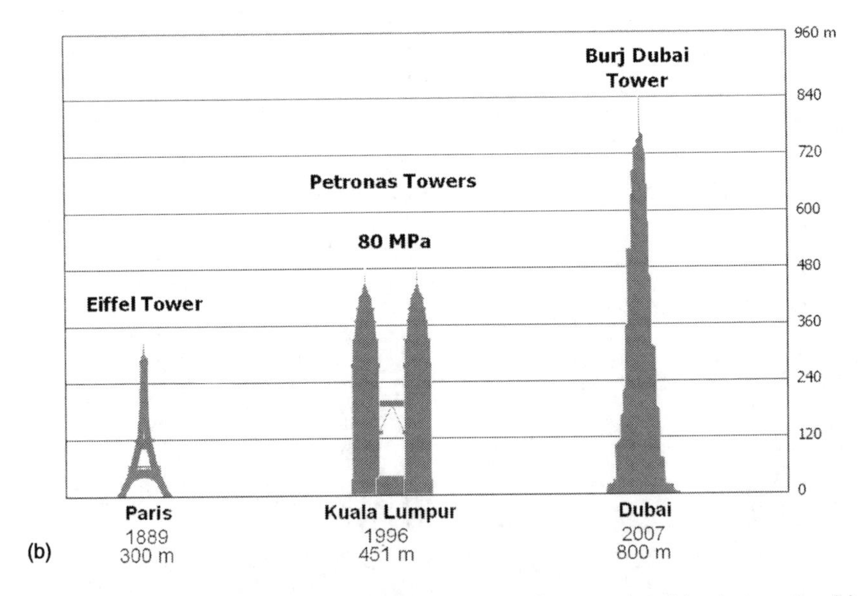

Figure 10.13 (a) Evolution of the height of concrete skyscrapers in North America; (b) future project that is manned to be built by an Indian religious sect at the centre of gravity of India.

10.13.2.1 The Petronas Towers in Kuala Lumpur

Who could have predicted that at the end of the twentieth century the higher high-rise building in the world would be built in Kuala Lumpur in Malaysia? With its 300 metres, the Eiffel Tower looks like a dwarf besides the twin towers that are 451 metres high.

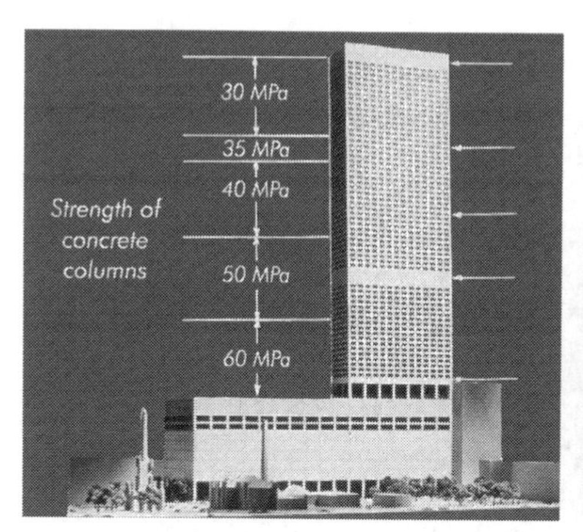

Figure 10.14 The Water Tower Place, high-rise building of Chicago built in 1970 (courtesy of J. Albinger).

10.13.2.2 High-rise buildings in Montreal

The first picture represents La Laurentienne (L) building in which the first HPC column in Canada was cast in 1984. The monitoring of this column is always under progress.

The second picture represents one of the latest high-rise building built in Montreal with a 60-MPa concrete. It contains IBM Canada headquarters. It is built just by the side of the La Laurentienne (L) building.

10.13.2.3 High-rise buildings in Canada

Scotia Plaza (SC) was the first high-rise building built with a HPC in Canada. It was built using jumping forms with a slag HPC having a compressive strength of 93 MPa. This HPC was entirely pumped. During its construction, the ambient temperature varied from –20°C to +35°C. In summertime the HPC was cooled using liquid nitrogen.

10.13.2.4 A very special American high-rise building

Among the numerous skyscrapers built with HPCs, the Two Union Square building of Seattle is special. Not only was it built with a 120-MPa concrete, but it was also designed with a confined concrete. The rigidity of this high-rise building was obtained by filling four large steel tubes with a 120-MPa concrete.

Figure 10.15 Petronas Towers in Kuala Lumpur (photo by Richard Gagné).

Smaller confined concrete sections was used at the periphery of the building to increase its rigidity. This building is so rigid that the people working on the top floor have the same comfort as the ones working on the first floor. (Seattle is competing with Chicago for the title of the City of the Winds in the USA.) Recently, the new extension of Toronto airport was built using the same design. A 60-MPa self-compacting concrete *was pumped up* from the bottom to the top of each column.

10.13.3 HPC bridges

Around 1990, HPC started to be used in outdoor applications in environments harsher than the inside of a high-rise building, though HPC had been used to build offshore platforms somewhat earlier.

Figure 10.16 La Laurentienne and IBM Canada headquarters in Montreal.

It took about ten years to see designers and owners specifying the use of HPC to build bridges. The following pictures represent some bridges built with HPC that are familiar to the author. There are many HPC bridges now built in many countries, some modest, other prestigious.

Figure 10.17 Scotia Plaza in downtown Toronto (courtesy of John Bickley).

10.13.3.1 Joigny Bridge in France

Joigny Bridge was the first deck built with a HPC in France. External post-tensioning was used, and as can be seen in the bottom picture the two holes can be used to add two new post-tensioning cables if the French army or Électricité de France had to use this bridge to pass an exceptional load. These two holes could also be used to replace any defective cable.

10.13.3.2 Small bridges

Air-entrained HPC (60-MPa) was used to build small prefabricated bridges and a pedestrian bridge in the Montreal area in 1982 and 1983.

In the case of the small bridge it was necessary to rapidly replace a previous small bridge that was destroyed by a spring flood when a large amount of ice was piled against the bridge.

In the case of the pedestrian bridge over Highway A15 it was necessary to launch it without interrupting the traffic, that is, the time slot allocated for the launching was between midnight and 6 o'clock in the morning. The picture was taken 15 minutes before the reopening of the traffic.

Figure 10.18 Two Union Square of Seattle, the first use of confined concrete in steel tube (courtesy of Weston Hester).

10.13.3.3 Portneuf Bridge

This was the first bridge entirely built with a HPC in Quebec. The beams were prefabricated and the deck cast in place (150 m^3 of 60-MPa, air-entrained concrete).

Figure 10.19 Joigny Bridge (courtesy of Y. Malier).

The bridge section is quite thin in order to leave as much clearance as possible beneath the bridge to allow the maximum space for ice during spring flood. A HPC design using external post-tensioning was preferred to a steel design for aesthetic reasons.

Figure 10.20 The first two prefabricated bridges built with high-performance concrete in Quebec in the early 1990s.

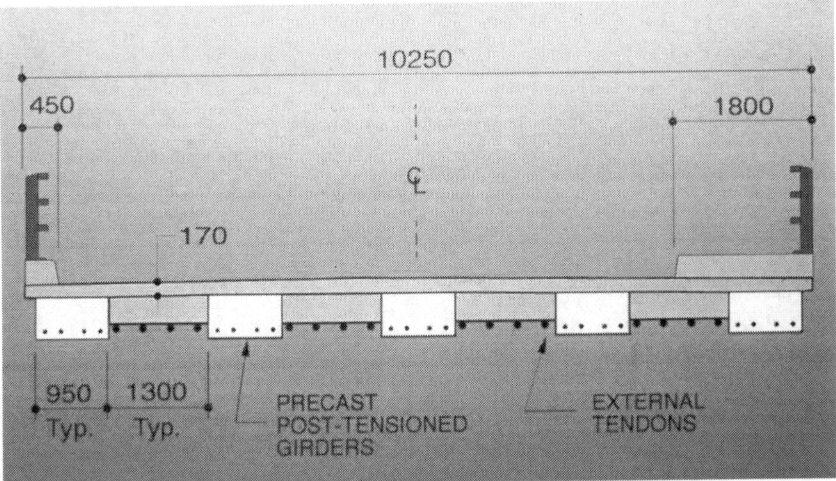

Figure 10.21 First cast in place bridge deck built with a HPC in North America (1992).

The concrete of the bridge deck was delivered by bin in order to be sure that it kept its original air-entrained system and that the spacing factor was lower than 230 μm, the required spacing factor for concrete at that time in Canada to insure freeze-thaw resistance.

Presently, as noted earlier, the spacing factor requirement has been relaxed to an average of 260 μm without any individual value greater than 300 μm.

Figure 10.22 Placing the high-performance concrete in Portneuf Bridge.

Therefore all HPC used to cast in place bridge decks is pumped. If after pumping the spacing factor does not meet the new requirement the contractor has to show that his concrete can sustain successfully 300 freezing and thawing cycles using Procedure A (freezing and thawing in water) of ASTM C666 standard.

10.13.3.4 Montée St-Rémi viaduct

This entirely cast-in-place HPC bridge was part of highway project where two other similar viaducts had to be built by the same contractor. In order to make a fair evaluation of the competitiveness of a HPC design, the Department of Structures of the Department of Transportation of the Province of Quebec decided to build the three viaducts using three different designs. The longer viaduct (3 in Figure 10.23) and the shorter one (1) were built using a 35 MPa concrete, one with precast beams, the other entirely cast in place. The intermediary bridge (2) was designed and built using a 60-MPa, air-entrained concrete. After cost normalization to eliminate the slight design difference it was found that the HPC bridge (2) had cost 5 per cent less than the two other viaducts (initial cost).

In this case too the bridge design involved external post-tensioning. The HPC was cast on the night of the 23 June. In order to lower the initial temperature of the concrete, up to 60 kg of ice per cubic metre was used. This viaduct was fully instrumented and monitored during and after its casting. Numerical simulations have been done to see the influence of the initial temperature of the concrete and of the ambient temperature on the development of thermal and stress gradients.

10.13.3.5 The Normandie Bridge in France

The concrete part of this bridge was built with a 60 MPa concrete. The cross-section of the bridge deck was designed to minimize the effect of the wind rather than static loading conditions. The concrete was controlled using the sand box method developed by Boulay and de Larrard (usual sulphur capping is replaced by a layer of sand contained in a box attached to the ends of the specimens).

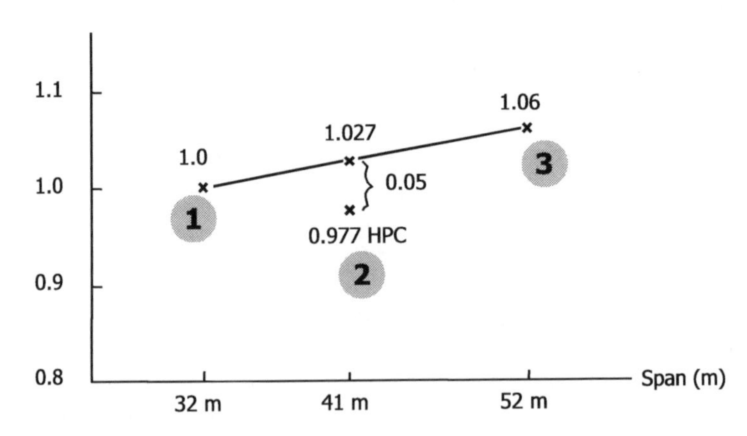

Figure 10.23 Normalized cost of 1 m² of bridge deck of the three viaducts of Montée St-Rémi near Mirabel Airport near Montreal.

Figure 10.24 The high-performance concrete viaduct of Montée St-Rémi near Mirabel Airport near Montreal.

10.13.3.6 The Confederation Bridge in Canada

This bridge was entirely prefabricated using essentially a 60 MPa, air-entrained concrete. It is the longest bridge in Canada (13 km), linking Prince Edward Island and the Province of New Brunswick. In the tidal area the pillars have a conical section to deviate the icebergs and the floes that are waved back and forth by strong currents. The concrete used in this conical part is a 90 MPa concrete particularly resistant to ice abrasion. The prefabricated beams were 190-m long, and 60-m drop-in slabs increased the span between two pillars

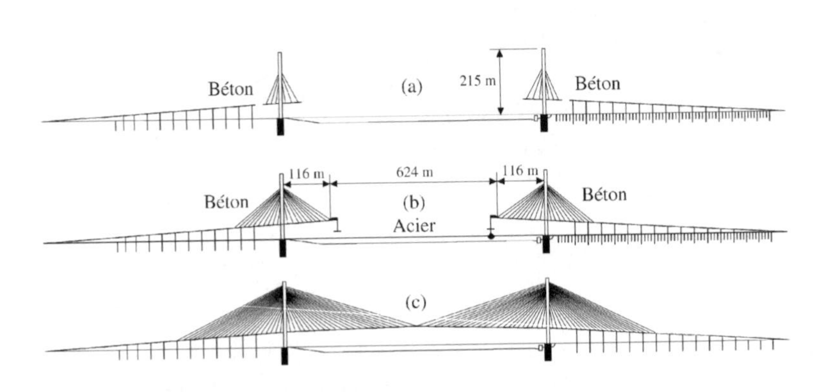

Figure 10.25 The Normandie Bridge.

(a)

(b)

Figure 10.26 (a) The Confederation Bridge; (b) the pre-cast yard.

to 250 m. These huge precast beams were launched by a special catamaran (the Svanen) except near the shore where the water was not deep enough.

10.13.4 Hibernia offshore platform

The expensive mechanical unit of this offshore platform (US$3 billion) was built over a huge gravity base of US$700 million. It is presently in operation in the Grand Banks area where it can resist the impact of the largest iceberg to be found in that area.

(a)

(b)

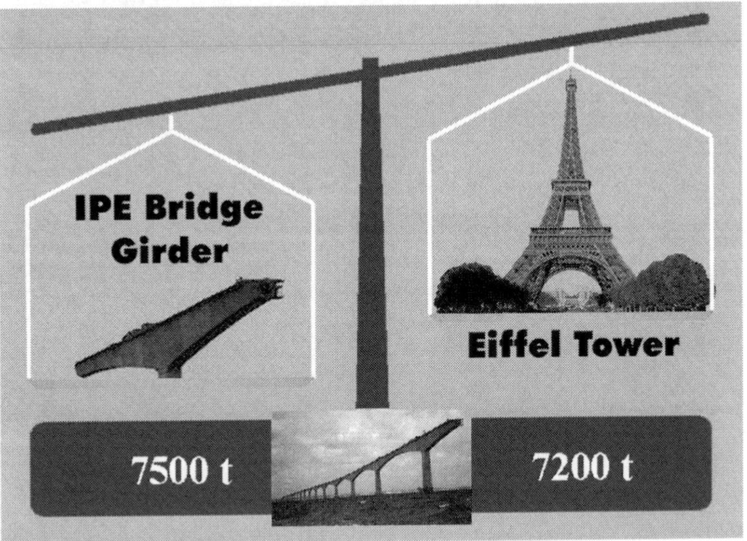

Figure 10.27 The launching of a prefabricated girder.
(a) The Svanen in action. This huge catamaran (100 × 100 × 100 m) is the strongest crane in the world. The maximum load that it lifts is 800 tonnes (7500 tonnes for the concrete elements + 500 for the handling head). The tension of all the cables is balanced by a computer. (Photo courtesy of G. Tadros.)
(b) The modular pre-cast beam made to build the bridge was by itself heavier than the entire Eiffel Tower (7200 tonnes) or than the 120000 inhabitants of Prince Edward Island.

Figure 10.28 Construction of Hibernia platform: (a) construction of the base; (b) construction of the four shafts.

10.13.5 Miscellaneous uses of HPC

10.13.5.1 Reconstruction of an entrance of a McDonald's restaurant

It would be an error to consider that HPC use is limited to prestigious or huge civil engineering projects. HPCs can be used in the day-to-day life to solve

(a) (b)

Figure 10.29 (a) Towing Hibernia platform; (b) Hibernia platform in operation (with permission of Exxon Mobile).

local problems, like the reconstruction of the two entrances of one of the two McDonald's restaurants in Sherbrooke.

As a McDonald's restaurant makes 80 per cent of its business between Thursday noon and Sunday midnight, the owner of the McDonald's restaurant told the City of Sherbrooke that he had no problem to operate his restaurant with his third entrance from Monday morning to Thursday noon. The two entrances that had to be rebuilt (Figure 10.30, 1) and (2) were demolished on Monday, the base was levelled, compacted and forms placed on Tuesday. A 60 MPa concrete was cast on Wednesday morning. Entrance (1) was built with a 60 MPa, non-air-entrained concrete and entrance (2) with a 60 MPa, air-entrained concrete. The traffic was reopened on Thursday at 11:00 a.m.

After ten winters it is difficult to see any difference in the behaviour of the two concretes.

For the City of Sherbrooke, the HPC solution represented an extra cost of US$12.50 per linear metre of pavement, when compared to a 30 MPa solution which would have closed the two entrances to direct use for a week.

Figure 10.30 McDonald's restaurant on King West street in Sherbrooke.

Figure 10.31a Placing the entrained-air concrete.

10.13.5.2 Piglet farm

Because it has a very low porosity, HPC offers a major advantage over usual concrete in agribusiness from a sanitation point of view. The use of HPC

Figure 10.31b Placing the non-air entrained concrete.

reduces considerably the proliferation of bacteries, germs and viruses. Moreover it is easier to clean and disinfect the surface of a HPC which results in a drastic decrease of the development of epidemics.

In a piglet farm where a HPC was used to build the floor and the walls, after three years of operation the owner had not faced any epidemic and had recovered his extra costs. A diarrhoea epidemic means that the owner is obliged to inject the piglets with antibiotics so that he cannot export his meat to Japan, where zero-antibiotic meat is sold for 50 per cent over the average market price. Moreover after each epidemic he was obliged to disinfect his whole piggery and feed the piglets until they recovered the 5–10 pounds lost in 24 hours.

10.13.6 Special HPCs

10.13.6.1 Self-compacting HPCs

HPC technology has already evolved into a series of very specialized concretes for niche markets. In self-compacting HPC the standard mix design is altered by the incorporation of additional fine material, a viscosity agent and some superplasticizer in order to place concrete without any vibration.

Such a concrete was used to build the reaction wall in the structural laboratory of the civil-engineering department of the Engineering Faculty of the Université de Sherbrooke. Such a concrete was selected because it was a 'silent'

Figure 10.32 Use of high-performance concrete floor in a piglet farm.

concrete that could be placed during working hours without disturbing the 2000 persons working in the faculty building. Otherwise the contractor would have been forced to cast the 240 m^3 of concrete on weekends between 10:00 p.m. on Saturday night and 6:00 a.m. on Monday morning.

10.13.6.2 Roller-compacted HPC

There are circumstances where instead of placing a fluid concrete it is more advantageous to cast a 'dry' concrete and compact it with rollers with the same type of equipment as that is used to place asphalt. This is the case for industrial slabs, streets and highways. This placing technique is very rapid and very economical. In the Province of Quebec, more than 100000 m^2 of such slabs are made with a 60-MPa concrete having a 6 to 7 MPa flexural strength due to its high coarse aggregate content.

10.14 Reactive powder concrete

Reactive Powder Concrete (RPC) was developed in France by Pierre Richard, the former Scientific Director of the company Bouygues. RPC does not contain any

Figure 10.33 The reaction wall in the structural laboratory of the Université de Sherbrooke.

Table 10.1 Scale factor between an armed concrete and an RPC

	Aggregate	*Reinforcement*	
	∗ *max (mm)*	∗ (μμ)	*Length (mm)*
RPC	0.2	0.2	12
Concrete	20	20	1200

coarse aggregate; the average size of its 'coarse' particles is 200 μm, i.e. the size of a grain of sand. So, why call such a mortar 'reactive powder concrete' when, from a grain size distribution point of view, RPC is a mortar and not a concrete. In fact when such a mortar contains specific steel fibres, these fibres work like reinforcing bars in a common concrete due to a 100-scale effect (Table 10.1).

Therefore, this fibre-reinforced mortar works from a mechanical point of view like a reinforced concrete, from which Pierre Richard devised the name 'reactive powder concrete'.

Reactive powder was developed to overcome the inherent weakness of most HPCs when they reach a compressive strength of 150 MPa. At such strength levels, it is observed that most of even the strongest 'normal' concrete aggregates fail within the matrix, leading to concrete failure. Attempts to reduce the maximum diameter and the intrinsic strength of the coarse aggregate have proven to be successful but also very prohibitive. Therefore, why not remove the

Figure 10.34 Mixing and placing a roller compacted high-performance concrete at the Domtar pulp and paper plant in Windsor Mills near Sherbrooke.

Figure 10.35 (a) The two layers of the Domtar slab; (b and c) aspect of the 96 000 m² slab.

coarse aggregate particles to make a very strong mortar having a very low W/B ratio, in which the coarse particles are not in contact with each other, in order to avoid stress concentration at the points of contact? This was Pierre Richard's idea when he developed RPC.

10.14.1 Reactive powder concrete concept

The concept of RPC is based on the three following principles:

1 An increase in the homogeneity of the material due to the elimination of the coarse particles, a limitation of the amount of sand to prevent sand particles from being in contact with each other in the hardened material, an improvement in the mechanical properties of the hydrated cement paste and the elimination of the transition zone at the interface of the hydrated paste/aggregates.
2 An increase of the compactness of the matrix by optimizing the grain-size distribution of all the powders that are used when making RPC and, when possible, by pressing during the hardening.
3 A refinement of the microstructure of the hardened paste by an appropriate heat treatment.

In the following, we will review very briefly these three concepts.

10.14.1.1 Increase of homogeneity

The elimination of the coarse aggregate and its substitution by a sand results in a decrease in the size of the microcracks of mechanical, thermal or chemical origin which are linked, in a traditional concrete, to the presence of the coarse aggregates that can be considered as rigid inclusions in an homogeneous matrix. This reduction of the size of the microcracks results in an improvement of the compressive and tensile strengths.

It is important to limit the amount of sand in the matrix to prevent sand particles, which act as rigid inclusions in the matrix, from entering into contact with each other, i.e. to limit interparticle contact as much as possible. With each sand grain behaving like an individual inclusion, and not as part of a densely packed aggregate skeleton, there are no longer any contact points creating stress concentrations. Moreover, when shrinkage develops, the matrix is free to contract as much as necessary because it is no longer restrained by the aggregate skeleton.

The increase of the granular compactness by using powders having complementary grain-size distributions, and the use of a large amount of ultrafine silica fume particles, eliminates the presence of a transition zone between the sand particles and the paste. As a consequence, the stress transfer between the paste and the aggregates is improved (Figure 10.36).

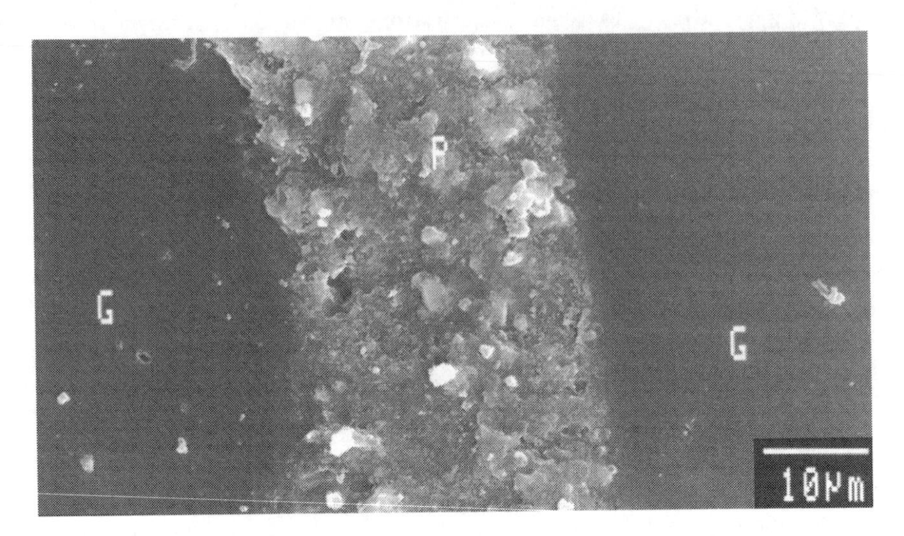

Figure 10.36 Microstructure of an RPC. One can notice:

- first the granular loosening, the two cement grains (G) being clearly separated by a quite large band of hydrated paste (P);
- the absence of a transition zone between aggregates and hydrated cement paste;
- the extreme compactness of the hydrated cement paste.

(Courtesy of Arezki Tagnit-Hamou)

Table 10.2 Typical composition of an RPC

Materials	*kg/m*^{3}
Cement	705
Silica fume	230
Crush quartz	210
Sand	1010
Superplasticizer	17
Steel fibers	140
Water	185
Water/binder	0.26

10.14.1.2 Increase of compactness

The optimization of the compactness of the granular matrix minimizes the interparticle voids and the amount of mixing water. Table 10.2 gives the composition of a typical RPC similar to the one used to build the Sherbrooke pedestrian bikeway bridge. As previously mentioned, when it is possible, RPC compactness can be increased by pressing fresh RPC, which results in the elimination of some of the mixing water, of some air bubbles and even some chemical shrinkage.

10.14.1.3 Improvement of the microstructure by thermal treatment

RPC must be heated to 70 to 90°C in a water bath, or by steam curing, for two days in order to obtain the maximum strength from all the powders it contains. However, this water curing must be done only *after two days of curing at ambient temperature* in order for hydration to develop sufficient lime. This heat curing results in an acceleration of the pozzolanic reaction between silica fume and the lime liberated by cement hydration and in a significant increase of the compressive strength.

Further heat treatment up to 400°C may be applied. It results in a complete transformation of the nature of the hydrates that were formed and in significant strength increases. Compressive strengths up to 500 MPa, and even 800 MPa when a metallic powder is used as sand replacement, have been achieved in the laboratory.

10.14.1.4 Improvement of the ductility of reactive powder concrete

RPC matrices have a purely elastic behaviour until they rupture in a brittle mode. RPC ductility can be improved by the addition of steel fibres (Figure 10.37), but not just any kind of steel fibre. The steel fibres must have a diameter/length ratio in accordance with the scale effect which was developed previously. As seen in Figure 10.38, a 140 kg/m^3 dosage (1.8 per cent in volume) of 12-mm long fibres provides a pseudo-ductile behaviour. Figure 10.39 represents an electron microscope view of the fracture surface of a piece of fibre reinforced RPC.

RPC can also be confined in a steel tube which results in a significant increase of its compressive strength and ductility. A maximum compressive strength up

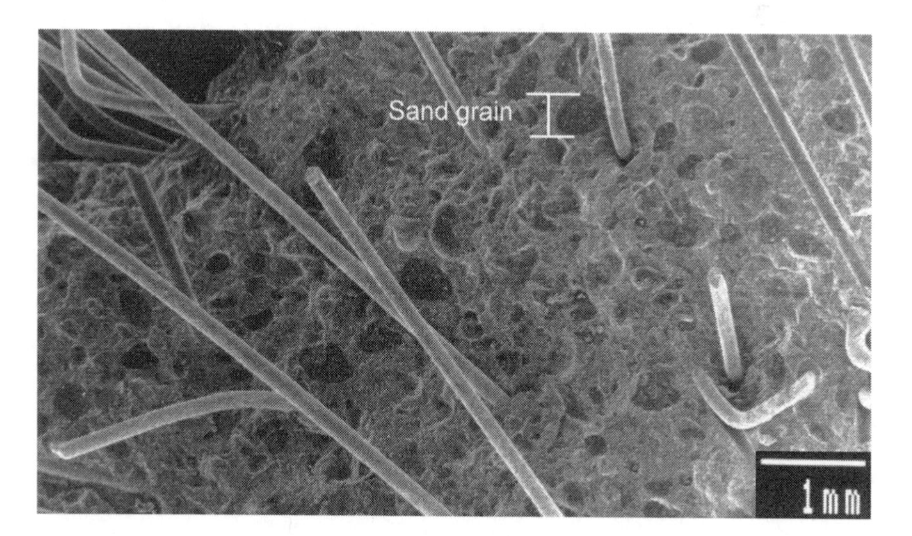

Figure 10.37 Failure surface for a fibered RPC (courtesy of Arezki Tagnit-Hamou).

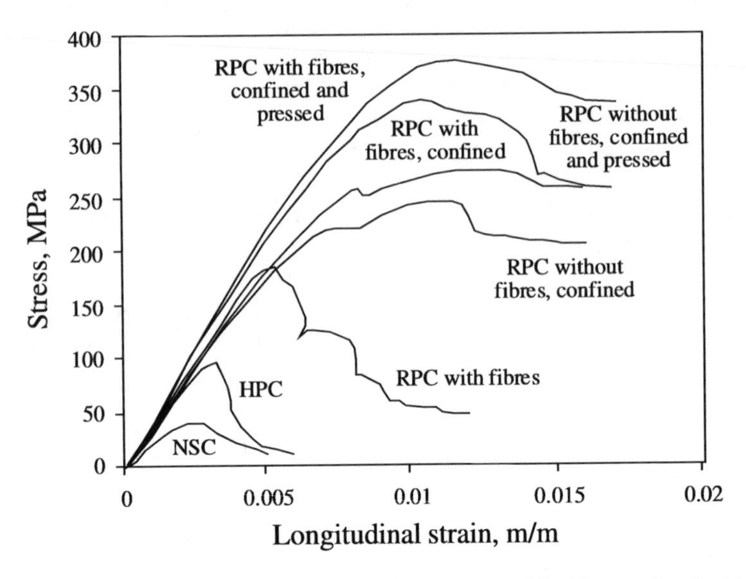

Figure 10.38 Compressive stress-strain curves for different RPCs fabricated at the Université de Sherbrooke (courtesy of Eyrolles).

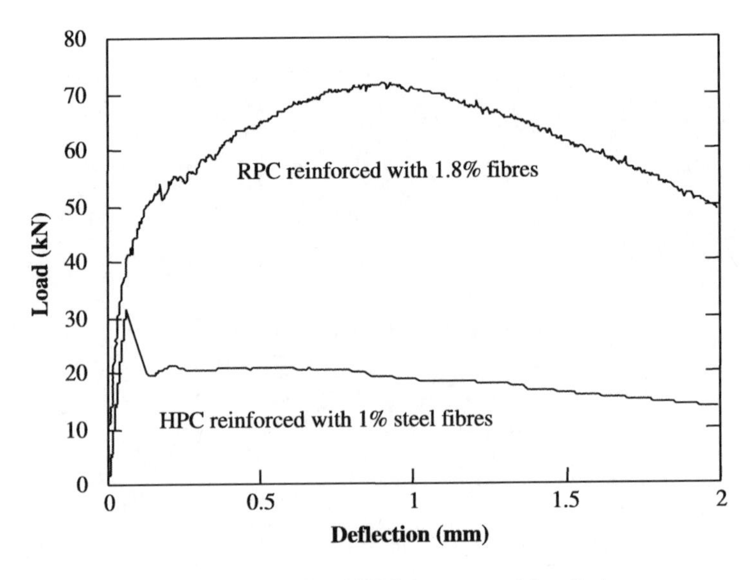

Figure 10.39 Load-deflection for RPC and HPC (courtesy of Eyrolles).

to 375 MPa, associated with a strain of slightly more than 1 per cent, can be achieved.

10.14.2 The Sherbrooke pedestrian bikeway

This 60m-long RPC structure has been built from Six pre-cast post-tensionned elements weighing on average 18 tonnes each, almost like a steel structure.

This structure is a pre-stressed concrete open-web-space truss that does not contain any passive reinforcement. The main tensile stresses are counter-balanced by post-tensioning, whereas some secondary tensile stresses are directly resisted by the steel fibres.

The cross-section is composed of two 380 × 320 mm bottom chords and a 30 mm upper slab comprising transverse stiffening ribs. The total height of the truss is 3.0 m and the overall width 3.3 m (Figure 10.40).

Truss diagonals are made of RPC confined in thin stainless steel tubes, joined to the top and bottom chords through two greased and sheated anchored monostrands. Specially designed anchorages were developed, which do not have any bearing plate or local zone spirals. The anchor head is

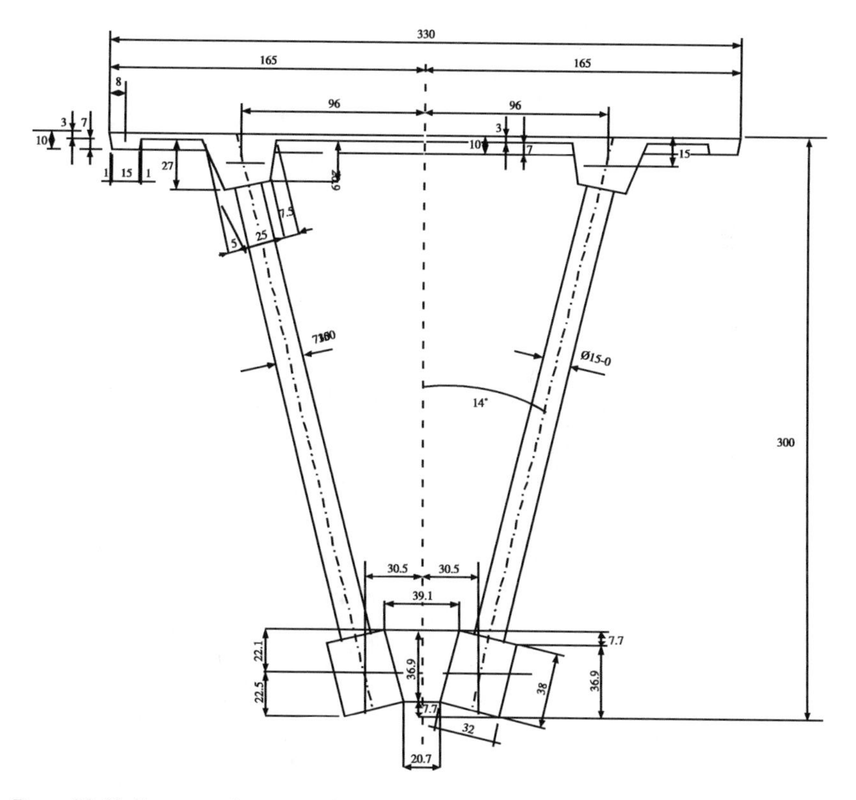

Figure 10.40 Transversal section of the superstructure (units: cm).

directly in contact with the RPC which is able to withstand the high compressive strength developed at the anchorage. Construction details have been previously published. After eight winters of service, the structure is behaving as expected.

10.14.3 Fabrication

10.14.3.1 Phase I: fabrication of the confined post-tensioned diagonals

Some 150-mm stainless steel tubes containing two post-tensionning cables were filled with a fibre-reinforced RPC. The mini anchors developed specifically for the project by VSL did not necessitate the use of anchors. The RPC was pressed (5 MPa) for 24 hours in order to eliminate the maximum of bubbles, to extract some mixing water and to compensate for some of the initial volumetric contraction. The top level of the RPC in the tube decrease by 100 mm during the pressing

Two days after their casting the diagonals were cured for two days at 90°C.

10.14.3.2 Phase II: construction of the deck and of the lower beam

Each of the six pre-cast elements was cast using usual pre-casting techniques. Each of the precast elements was used to cast the next element in order to have a perfect joint.

Figure 10.41 Fabrication of the diagonals.

Figure 10.42 Construction of a prefabricated element in the pre-cast plant.

10.14.3.3 Phase III: curing

The 90°C curing started 48 hours after the casting of the RPC. Polyethylene sheets were used to wrap each pre-cast element and create an enclosure in which hot vapour (90°C) could cure the RPC. This 90°C curing lasted 48 hours.

10.14.3.4 Phase IV: transportation to the site

The six pre-cast elements were transported by truck to the site without any particular problems. Each element was 10 metres long and weighed about 18 tonnes. The end elements weighed 24 tonnes.

10.14.4 Erection

10.14.4.1 Phase I: assembling the prefabricated elements

The assembling of the prefabricated elements was done on a stone bed that was temporarily built in the middle of the Magog river. The bridge was assembled first in two parts. The joints did not receive any special treatment (RPC against RPC) as can be seen in Figure 10.45.

Figure 10.43 Curing of a pre-cast element.

Figure 10.44 Transportation of the prefabricated elements.

Figure 10.45 Assembling the northern part of the bridge.

10.14.4.2 Phase II: installation of each half of the bridge

10.14.5 The past and the future

The pedestrian bridge of Sherbrooke is erected at the junction of the Magog River and the St Francis River close to two other bridges: one steel railway bridge built 100 years ago and a cast in place concrete. It is easy to compare the aesthetic performance of the three solutions. The photo was taken during spring thawing when Magog and St Francis rivers had reached their maximum level.

Figure 10.46 Assembling two prefabricated elements (dry joint).

Figure 10.47 Assembling the two halves of the bridge.

Figure 10.48 The completed bridge.

Figure 10.49 Confrontation of two technologies to cross the Magog River in Sherbrooke.

References

Acker, P., Ulm, F.-J. and Levy, M. (1997) 'Fire in the Channel Tunnel: Mechanical Analysis of the Concrete Damage', Concrete Canada – Technology Transfer Day, 1 October, Toronto, Ontario.

Aïtcin, P.-C. (1994) 'Durable Concrete-Current Practice and Future Trends', ACI SP-144 (*Concrete Technology: Past, Present, and Future*, edited by P.K. Mehta), pp. 83–104.

Aïtcin, P.-C., Neville, A.M. and Acker, P. (1997) 'Integrated View of Shrinkage Deformation', *Concrete International*, 19 (9): 35–41.

Aïtcin, P.-C., Pigeon, M., Pleau, R. and Gagné, R. (1998) 'Freezing and Thawing Durability of High Performance Concrete', Internal Symposium on High-Performance Concrete and Reactive Powder Concretes, Sherbrooke, August, Vol. IV, pp. 383–91.

Baalbaki, W. (1997) 'Experimental Analysis of Elastic Modules' Ph.D. thesis no. 1015, Université de Sherbrooke.

Baalbaki, W., Benmokrane, B., Chaallal, O. and Aïtcin, P.-C. (1991) 'Influence of Coarse Aggregate on Elastic Properties of High-Performance Concrete', *ACI Materials Journal*, 88 (5): 499–503.

Bache, H.H. (1981) 'Densified Cement/Ultra-fine Particle: Based Materials', *Second International Conference on Superplasticizers in Concrete*, Ottawa, Canada, 10–12 June.

Bonneau, O., Poulin, C., Dugat, J., Richard, P. and Aïtcin, P.-C. (1996) 'Reactive Powder Concrete from Theory to Practice', *Concrete International*, 18 (4): 47–9.

Breïtenbücker, R. (1996) 'High Strength Concrete C105 with Increased Fire Resistance Due to Propylene Fibers', *Utilization of High Strength/High Performance Concrete*, Paris: Ecole Nationale des Ponts et Chaussées, pp. 571–8.

Chan, S.Y.N., Peng. G.F. and Chan, J.K.W. (1996) 'Comparison between High Strength Concrete and Normal Strength Concrete Subjected to High Temperature', *Materials and Structures*, 29: 616–19.

Cheyrezy, M. and Beloul, M. (1999) 'Comportement des BHP au feu', French Civil Engineering Association Meeting, 9 March.

City of Montreal (1998) Standard Specification for High-Performance Concrete HPC (3VM–20).

Cook, W.D., Miao, B., Aïtcin, P.-C. and Mitchell, D. (1992) 'Thermal Stress in Large High-Strength Concrete Columns', *ACI Materials Journal*, 89 (1): 61–6.

Demorieux, J.-M. (1998) 'Le Comportement des BHP à hautes températures: État de la question et résultats expérimentaux', École Française du Béton et le Projet National BHP 2000, 24 and 25 November, Cachan, France.

Diederich, U., Spitzner, J., Sandvik, M., Kepp, B. and Gillen, M. (1993) 'The Behavior of High-Strength Lightweight Aggregate Concrete at Elevated Temperature', *High Strength Concrete*, Paris: Ecole Nationale des Ponts et Chaussées, pp. 1046–53.

Duval, R. and Hornain, H. (1992) 'La Durabilité du béton vis-à-vis des eaux aggressives', in J. Baron and J.-P. Ollivier (eds) *La Durabilité des bétons*, Paris: Presses de l'École Nationale des Ponts et Chaussées, pp. 376–85.

Felicetti, R., Gambavora, P.G., Rosati, G.P., Corsi, F. and Giannuzzi, G. (1996) 'Residual Strength of HSC Structural Elements Damaged by Hydrocarbon Fire or Impact Loading', *Utilization of High Strength/High Performance Concrete*, Paris: pp. 579–88.

Gagné, R., Lamothe, P. and Aïtcin, P.-C. (1993) 'Chloride-Ion Permeability of Different

Concretes', *Sixth International Conference on Durability of Building Materials Components*, Omiya, Japan, p. 1171–80.

Jensen, B.C. and Aarup, B. (1996) 'Fire Resistance of Fibre Reinforced Silica Fume Based Concrete', *Utilization of High Strength/High Performance Concrete*, Paris: pp. 551–60.

Jensen, J.J., Opheim, E. and Aune, R.B. (1996) 'Residual Strength of HSC Structural Elements Damaged by Hydrocarbon Fire on Impact Loading', in *Utilization of High Strength/High Performance Concrete*, Paris: pp. 589–98.

Jensen, O.M. and Hansen, P.F. (2001) 'Water-Entrained Cement-Based Materials, Part I', *Cement and Concrete Research*, 31 (4): 647–54.

Koylou, N. and England, G.L. (1996) 'The Effect of Elevated Temperature on the Moisture Migration and Spalling Behaviour of High Strength and Normal Concretes', ACI SP-167, pp. 263–8.

Kreijger, P.C. (1987) 'Ecological Properties of Building Materials', *Materials and Structures*, 20: 248–54.

Lachemi, M. and Aïtcin, P.-C. (1997) 'Influence of Ambient and Fresh Concrete Temperature on the Maximum Temperature and Thermal Gradient in a High-Performance Concrete Structure', *ACI Materials Journal*, 94 (2): 102–10.

Lachemi, M., Lessard, M. and Aïtcin, P.-C. (1996) 'Early-Age Temperature Development in a High-Performance Concrete Viaduct', ACI SP-167 (*High Strength Concrete: An International Perspective*, edited by J.A. Bickley), pp. 149–174.

Lessard, M. (1995) unpublished results.

Lessard, M., Dallaire, É, Blouin, D. and Aïtcin, P.-C. (1994) 'High-Performance Concrete Speeds Reconstruction of McDonald's', *Concrete International*, 16 (9): 47–50.

Mehta, P.K. and Monteiro, P. (1993) *Concrete: Microstructure, Properties, and Materials*, New York: McGraw-Hill.

Neck, U. (1999) Personal communication, March.

Neville, A.M. (1995) *Properties of Concrete*, 4th edn, London: Longman.

Nilsen, A.U. and Aïtcin, P.-C. (1992) 'Properties of High-Strength Concrete Containing Light-, Normal- and Heavyweight Aggregate', *Cement and Concrete Aggregates*, 14 (1): 8–12.

Noumowe, A.N., Clastres, P., Delvicki, G. and Costaz, J.-L. (1996) 'Thermal Stresses and Water Vapour Pressure of High-Performance Concrete at High Temperature', *Utilization of High Strength/High Performance Concrete*, Paris, pp. 561–570.

Phan, L.T. (1997) 'Fire Performance of High-Strength Concrete: A Report of the State-of-the-Art', Res. Rep. NISTIR 5934, NIST, Gaithersburg, Maryland, USA.

Richard, P. and Cheyrezy, M. (1993) Reactive Powder concrete With High Ductility and 200–800 MPa Compressive Strength, ACI SP 144 (*Concrete Technology: Past, Present, and Future*, edited by P.K. Mehta), pp. 507–18.

Sanjayan, G. and Stocks, L.J. (1993) 'Spalling of High-Strength Silica Fume Concrete in Fire', *ACI Materials Journal*, 90 (2, March–April): 170–3.

Shah, S.P., Weiss, J. and Yang, W. (1998) 'Shrinkage Cracking: Can It Be Prevented?', *Concrete International*, 20 (4): 51–5.

Weber, S. and Reinhardt, H.W. (1997) 'A New Generation of High-Performance Concrete: Concrete with Autogenous Curing', *Advanced Cement-Based Materials*, 6 (2): 59–68.

Chapter 11

The development of the cement and concrete industries within a sustainable development policy

11.1 Introduction

During the twentieth century, concrete became the most widely used material because it allows human beings to satisfy some of their fundamental needs (Aïtcin 1995). Concrete is a material that can be used to build houses, office buildings and urban infrastructures required by the development of modern societies as well as the transportation infrastructures necessary for the development of exchanges. This quasi-universal use of concrete is due to the intrinsic qualities of concrete: it is an inexpensive material; it resists water well; it does not rot; it does not rust; it does not burn; and it resists insect attacks. It is a material essentially made of local constituents; 80 per cent of it is water and aggregates. Moreover, at about $50 or € per tonne in 2000, cement is one of the least expensive industrial materials. Finally, the technology involved in making concrete is very simple and the initial investment is not very high.

If it is true that among living beings, only humans have succeeded in colonizing the entire planet, at the level of construction materials the same is true of concrete. Concrete is present everywhere that there are human beings. It should be remembered that in the year 2000, 1.5 billion tonnes of cement were manufactured and that each human being used 1 m^3 of concrete, weighing 2.5 tonnes.

Of course, this quasi-universal use of concrete has environmental consequences, as would the use of any other material (Kreijger 1987) (see Figures 11.1 and 11.2). It is essential to evaluate first the environmental impact of this huge amount of concrete and, second, to see what can be done to reduce it. The fabrication of 1 tonne of Portland cement implies the decarbonation of about 1 tonne of limestone at a temperature of between 700 and 900°C. This decarbonation of limestone is followed by the heating of the raw meal up to 1450°C.

Roughly, the production of 1 tonne of Portland cement generates the emission of 1 tonne of CO_2, to which NO_x emissions should be added. These nitrogen oxides are formed within the very hot flame (2500°C). Some SO_3 is also formed during the combustion of the sulphur contained in the fuel or in the raw materials. We have already seen that, usually, the sulphur is trapped within the

Figure 11.1

clinker, more specifically within the C_3A or in one of the different forms of alkali sulphates, but some SO_3 can also be found in the fumes within the chimney.

Presently, due to the efficiency of dedusting systems, practically no dust escapes from the chimney within the combustion gases.

The very high temperature reached within the flame certainly has a negative environmental impact, but it also has some positive consequences: cement kilns can be used to eliminate almost any organic toxic product. No organic molecule can withstand the temperature of a cement kiln's flame, therefore, cement kilns are currently used to eliminate PCBs (polychlorinated biphenyl), mad cows, varnishes, contaminated oil, etc. (Shinoda et al. 2000). Cement kilns can be used to burn domestic waste, scrap tyres, coals rich in sulphur otherwise unusable in any other industrial process.

Figure 11.2

Therefore, cement manufacturing presents two environmental aspects: one negative, the other positive. We will first see how to decrease the first aspect and, second, how to maximize the second one.

The making of concrete implies the use of large amounts of natural aggregates or crushed rock, some water and a small amount of admixtures.

In order to lower the ecological impact of concrete, it is essential to optimize the use of Portland cement to get a certain strength and workability (Gjørv 2000; Sakai and Banthia 2000; Sarja 2000; Sommerville 2000), in order increase the durability of concrete structures so that they will not have to be repaired or rebuilt as often. Industrial by-products could be used as concrete aggregates, and even crushed demolition concrete can be used as concrete aggregate (Tomosawa and Noguchi 2000; Toring 2000; Anonymous 2000).

In the following sections, we will see what can be done to lower the environmental impact of concrete and cements.

11.2 How to lower the environmental impact of concrete

It will be seen that there are several promising avenues to lowering the environmental impact of concrete. These avenues are easy to put into practice with the technology already available. Too much cement is wasted when making concrete and too much concrete is wasted when constructing concrete structures that are unable to withstand the environment in which they should fulfil these structural and economic functions for very long. It will be shown how it is possible to make more durable concrete with less cement.

11.2.1 Reducing the water/binder ratio

Concrete compressive strength and impermeability are essentially a function of the water/cement (W/C) or water/binder (W/B) ratio and not of the amount of cement used. A concrete having the same compressive strength, durability and workability can be made with less cement when using a water-reducer or a superplasticizer, or a more resistant and more durable concrete can be made when using the same amount of cement.

A water-reducer reduces by Δ W the amount of water W necessary to achieve the workability of a concrete having a given W/B ratio.

If the W/B ratio is kept constant

$$\frac{W}{B} = \frac{W - \Delta W}{B - \Delta B}$$

Therefore

$$\Delta B = \frac{1}{W/B} \Delta W$$

If we suppose that the most widely used W/B ratios today are between 0.50 and 0.80, the amount of binder that can be saved by using a water-reducer is:

$$1.5\ \Delta W \leq \Delta B \leq 2\ \Delta W$$

As a conventional lignosulphoante-based water-reducer reduces the amount of water necessary to achieve a given workability by about 8 per cent, the amount of cement that could be saved worldwide would be between 12 to 16 per cent.

As more than half the concrete that is presently made does not contain a water-reducer, this cement reduction should be applied to 800 million tonnes of cement. At a 13 per cent reduction rate, this means that in the year 2000, 100 million tonnes of cement were wasted because a water-reducer was not used systematically. This means that 16 million tonnes of fuel or 3 million tonnes of coal could have been saved and that 100 million tonnes of CO_2 emissions could have been avoided.

In order to eliminate such a waste, it would be so simple to force cement producers to systematically incorporate a water-reducer in their cement because *all* cements need to be dispersed by a water-reducer.

The beneficial action of water-reducers could be seen in another more positive way; 100 million tonnes of clinker or 135 millions tonnes of hydraulic binder, if a 35 per cent rate of substitution is applied to the clinker, it means that 400 million cubic metres of concrete could be made without any increase on the environmental impact of the cement industry.

The systematic incorporation of a water-reducer in hydraulic binders represents only a first step in the overall objective of reducing the environmental impact of concrete.

As a second step, let us consider the effect of an additional decrease in the W/C ratio when using a superplasticizer instead of a water-reducer. The use of a superplasticizer results in a reduction of 15 to 20 per cent of the amount of water in low W/B concrete. At the same time, the compressive strength of concrete increases from 25 to 75 Mpa, while the amount of cement used is increased by much less than a factor of 3. In a column that works in compression, this represents three times less concrete used to carry the same load, not to mention the longer durability of this concrete which will result in a much longer life for the concrete structure.

In order to go from a 25MPa concrete made without a water-reducer to a 75 MPa superplasticized concrete, the cement dosage has to be increased approximately from 300 kg of cement per cubic metre to 450 kg of cement. Therefore, when using a 75 MPa concrete to build a column instead of a 25 MPa concrete, the nominal load per kg of cement is multiplied by 2 (1.5 times more cement is used to get three times more strength). Therefore, the use of a 75 MPa concrete doubles the amount of columns that can be built with the same amount of cement. As far as the volume of aggregates used, it is reduced by 3 because three times less concrete has to be used to support the same load.

Of course, savings in cement and aggregates are less important in structural elements that are working in flexure such as beams, but they are significant.

The use of high-performance concrete having a lower W/B ratio than that of the conventional concrete presently used by the construction industry could have a tremendous environmental impact not to mention the increased service life of concrete structures, as will be seen.

11.2.2 Increasing the service life of concrete structures

The service life of a concrete structure depends on the severity of the environmental conditions in which this structure has to fulfil its structural function, on the adequacy of the concrete used to support these environmental conditions and to the care taken when placing and curing this concrete. Since it is difficult to modify local environmental conditions without extra expenses in order to protect the concrete from its natural environment or chemical attack it is better to use a durable concrete well adapted to its environment and to take precautions in its placing and curing in order to increase the service life of concrete structures (Gjorv 2000; Sakai and Banthia 2000; Sarja 2000).

Too often, concrete is still specified only on the basis of its compressive strength, without considering the environmental conditions in which it will have to fulfil its structural function or its placing conditions. As a consequence, many concretes are partially or totally unable to face their natural environment for a long time. The external parking garages built in Canada with 20 MPa concretes are an example of what should not be done.

From a structural point of view, it is possible to build a safe parking garage with a 20 MPa concrete, but such a concrete does not last very long when it has to face freezing and thawing cycles in saturated conditions in the presence of deicing salts that are dripping from cars always parked in the same place. The W/B ratio of a 20 MPa concrete is too high and cannot protect reinforcing steel against rusting.

It is very important that national codes be more demanding regarding durability when particular severe environmental conditions require a low W/B ratio. The designer will have to use the 'extra' MPa of this durable concrete adequately (Sommerville 2000).

Decreasing the W/B ratio is a necessary condition (but not sufficient) to increase the durability of concrete structures. It is not sufficient because a concrete that has a low W/B ratio can be wasted if it is not placed and cured correctly. In the field, it must be realized that placing and curing are as important, if not more so, than compressive strength upon delivery. It is necessary to establish very precise and clear specifications and to pay contractors specifically to improve the quality of the placing and curing, or to enforce severe penalties when placing and curing are not done properly. Of course, when specifications are very precise and clear, it is easy to verify that placing and curing are done properly. Paying contractors to place and cure concrete is a very profitable investment, a low initial investment that results in great long-term savings. The engineers of the City of Montreal, in Canada, have calculated that the cost of an adequate water curing varies from 0.1 to 1.0 per cent of the total initial cost of a concrete structure, at it has been seen in the previous chapter.

Presently, it is rather difficult to predict precisely the increase in service life that can be associated with good placing and curing practices, but no one will argue that a low W/B ratio concrete that is well placed and cured is more durable than a conventional concrete that is placed without any care and not cured at all (Shah et al. 2000).

The construction of concrete structures that last results in a decrease in concrete and cement consumption because the use of a durable concrete decreases the frequency at which this structure will have to be repaired or replaced (Sommerville 2000).

To convince cement and concrete producers that repairs and reconstruction are very costly from an overall point of view for their respective industry, it is only necessary to point out that very few dollars or euro are spent on materials for repair work; most of the cost of repairs is related to manpower or machinery. On the contrary, new construction involves the use of ten to twenty times more cement than repair work, as can be seen in Table 11.1. As infrastructure budgets are not elastic, more repair work means less new constructions.

11.2.3 Using concretes having a lower cement content

11.2.3.1 High-performance concrete having a low heat of hydration

Researchers from the Atomic Energy of Canada have patented a new type of concrete that has been called low-heat high-performance concrete (LHHPC) which contains only 100 kg of Type V cement, 100 kg of silica fume and 200 kg of finely ground silica filler. This concrete has a W/C+SF (C for cement, SF for silica fume) equal to 0.50 (the silica filler is not taken into account when calculating the W/B ratio). The use of such a concrete presents the great advantage of resulting of an increase of only 20°C in quasi-adiabatic conditions. In spite of its low cement dosage, the compressive strength of such a concrete was equal to 100 MPa. This is a particularly ecological concrete with outstanding performance. Unfortunately, there is not enough silica fume produced worldwide to generalize the use of such a concrete, unless silica fume is produced expressly for that purpose, which is not to be excluded.

Table 11.1 Relative importance of the cost of concrete and cement on the overall cost of different types of construction

Type of construction	*Cost of placing 1 m^3 of concrete*	*Cost of 1 m^3 of concrete*	*Relative cost (% of the total cost)*	
	(US$)	*(US$)*	*Concrete*	*Cement*
Building	600–900	45–60	6.7	2
Bridge	1 500–2 250	75–90	4	1
Repair of Bora-Bora wharf	27 000	?	?	0.1

11.2.3.2 *High volume fly ash and slag concretes*

CANMET researchers in Canada (Bisaillon et al. 1994; Ramachandran et al. 1998; Langley 1999; Mehta and Langley 2000) have developed the concept of high volume fly ash and slag concretes, a very interesting concept in a sustainable development context. Such a concrete not only decreases the environmental impact of concrete, but also adds value to industrial by-products of low commercial value. In such concretes, fly ash or slag represents 50 to 60 per cent of the binder. What sets this kind of concrete apart from a European concrete made with a European blended cement containing a high percentage of slag or fly ash is that their W/B ratio is drastically decreased by the use of a superplasticizer. The early compressive strength of such concretes is high enough not to bother contractors and their durability is excellent due to their low W/B ratio. Unfortunately, up to now very few demonstration projects have used this technology, but one day this technology will be used on a larger scale (Mehta and Langley 2000). A recent paper by Mehta and Manmohan (2006) shows that this technology is slowly taking off in the USA.

11.2.4 Recycling concrete

Recycling concrete at the end of its structural life to make new concrete results in virgin aggregate savings (Tomosawa 2000). However, it should be remembered that each time a material is recycled, its recycling results in a decrease in the performance of the new material, because the initially carefully selected materials used the first time have lost some of their original properties, have been contaminated or have been weakened during their service life (Torring 2000).

It is, however, encouraging to see that greater volumes of concrete are recycled every year, in spite of the fact that much more could be done. In terms of concrete recycling, high-performance concrete has the advantage of presenting a greater potential for recycling. In fact, due to their very low W/B ratio, they can be recycled two or three times before being used in road base.

Recycling concrete in urban areas presents several advantages because it is more and more difficult to obtain permits to open new quarries or extend existing ones. Moreover, it becomes more and more costly to dispose of construction waste (Anonymous 2000) and recycling concrete avoids the transportation of virgin aggregate over long distances, some aggregates represent 70 to 75 per cent of the mass of concrete. Recycling concrete reduces energy consumption, CO_2 emissions, and presents significant socio-economic advantages by reducing traffic.

11.3 The manufacturing of hydraulic binders presenting a more energy-efficient and ecological performance than present Portland cement

It will be seen below how it is sometimes easy, but also sometimes quite complex, to improve the energy efficiency and ecological performance of Portland cements in the context of sustainable development.

We will start by looking at what can be done to lower the energy requirement of Portland cement clinker before looking at the possibilities offered by the substitution of some Portland cement with mineral components at the cement plant or at the concrete plant. It will be seen that if the use of mineral components was generalized with the same degree of efficiency as it is in Belgium, the Netherlands and Germany, CO_2 emissions could be reduced by 30 per cent or that 30 per cent more of hydraulic binders could be produced with the same amount of clinker, equivalent to 450 million tonnes of Portland cement, which represents the production of 450 cement plants each producing 1 million tonnes of cement per year, without emitting a single extra tonne of CO_2 into the atmosphere except for the CO_2 emitted during the transportation of the mineral component to the cement plant.

11.3.1 Decreasing the energetic content of clinker

It must be recognized that cement companies have made significant efforts to reduce the amount of energy necessary to produce clinker following the two petroleum crises during the 1970s and 1980s. During these crises, the cost of a barrel of oil went from US$4.50 to almost US$30. This price increase had an immediate repercussion on the price of the other fuels. The dry process with a precalcination tower became the rule to produce clinker, and great efforts were made to limit heat losses and recover as much heat as possible during the cooling of the clinker. In a modern cement plant using a dry process, the fabrication of 1 kg of clinker now requires 2800 kJ, while it required 6000 kJ in a wet process plant.

However, in spite of these efforts, there are still areas where the situation can be improved. The following points will be considered in order:

- decreasing the clinkering temperature through the use of mineralizers;
- the fabrication of belite cement;
- the replacement of some limestone by another source of lime;
- the greater use of alternative fuels. This case does not really represent savings from an energy point of view but rather savings in virgin fuel.

11.3.1.1 *Decreasing the clinkering temperature through the use of mineralizers*

In Chapter 4, in which we studied the CaO–SiO_2–Al_2O_3 phase diagram, it was seen that the minimal temperature to form the C_3S and C_2S phases found in a clinker is 1450°C. Of course, this temperature corresponds to the formation of pure phases. However, it is well known that the presence of certain impurities in the crystalline network can substantially lower the formation temperature of one or several phases (Taylor 2000).

For a long time, researcher have tried to decrease the clinkering temperature by searching for specific ions that could be introduced into the raw meal without altering the hydraulic properties of the C_3S and C_2S. These ions should not substantially modify the process or the quality of the clinker. It could be dangerous to use toxic ions that could later be released in the hardened concrete (heavy metals for example). Moreover, the economic and ecological balance must be positive.

Presently, it seems that the two most promising mineralizers are sulphur and fluorine. When they are used in the right proportion, they can lower the clinkering temperature by 100 to 150°C (Marciano 2003). Several industrial tests have been conducted more or less secretly by certain companies. It is quite likely that the last technical difficulties will be solved and that many clinkers will be produced in this way in a few years.

A decrease in temperature of 100 to 150°C would result in a 2 or 3 per cent energy saving as well as a significant decrease in NO_x emissions.

11.3.2 Producing a belitic clinker

As already mentioned, it was to please contractors that cement producers started producing rapid hardening cement which allowed more rapid removal. Such cements are intended to accelerate the construction process. In order to reach this objective, cement producers have started producing cements that are richer in C_3S and C_3A and that are also ground more finely. Many cements found on the market have a C_3S content higher than 65 per cent. However, it is difficult to see the competitive advantage that these cements represent when all contractors have equal access to them. It will not be necessary to show the numerous technical drawbacks that these 'nervous' clinkers have on volumetric changes, early cracking, workability and, more generally, on the durability of concrete. In fact, to compensate for the 'nervousness' of the clinker, it is necessary to increase the water dosage to make a concrete having the desired workability. Any increase in water dosage results in a decrease in concrete durability. Moreover, these 'nervous' cements have a rheology that is difficult to control, so that too often it is necessary to retemper the concrete in the field before placing.

The answer to the request of contractors is more a decrease of the W/B ratio of the concrete and not an increase of the C_3S content and fineness of the cement. As efficient superplasticizers are presently available, it is definitely no longer necessary to manufacture cements having such a high C_3S and C_3A content in order to obtain high early strength. It has been well known for over 100 years that the strength of a concrete is related to its W/B ratio to a greater extent than to the 'strength' of the cement: the closer the cement grains in the fresh concrete, the stronger the concrete.

Of course, when it was not possible to bring cement particles closer to each other, it was necessary to use a greater water dosage to deflocculate these particles. Therefore, it was necessary to develop large amounts of C-S-H to obtain strength. High C_3S and C_3A contents were a prerequisite to obtain a high early strength. But now that superplasticizers are available, it is possible to bring the cement particles as close together as desired. It is therefore no longer necessary to form much 'glue' to get strength. This explains why research conducted by the US Corps of Engineers in Vicksburg and at the Université de Sherbrooke have shown that the best North American cements to make 200 MPa reactive powder concrete were ASTM Type V cements or the Type 20 M Canadian cement used by Hydro-Quebec for its mass concrete. These cements have low C_3S and C_3A contents. A reactive powder concrete can have a compressive strength equal to 50 MPa at 24 hours (O'Neil et al. 1996).

Therefore, the use of a C_2S-rich cement no longer means low early strength concrete. Of course, the manufacturing of a C_2S-rich cement implies the use of a lower amount of limestone and consequently the emission of less CO_2 and less fuel. The only cases where a belitic clinker is not so advantageous is when this clinker is blended with a mineral component because it does not produce as much lime as a C_3S rich clinker.

11.3.3 Using a source of lime other than limestone

Another interesting alternative to improve the ecological performance of clinker manufacturing by reducing the amount of CO_2 emissions, is to use a source of lime other than limestone. This would also result in energy savings because decarbonation of lime is a strongly endothermic reaction.

Crystallized slag can be used as a source of lime and to reduce the limestone content used in the raw meal (Table 11.2). In the theoretical case presented in Figure 11.3, it can be seen that 20 per cent of lime from the limestone can be saved when slag S and limestone L are used to produce clinker K instead of limestone L and clay C. Therefore, limestone replacement with a certain amount of crystallized slag represents a very interesting solution from an ecological point of view when slag is economically available, of course. Alinite cement produced from calcium chloride also perform better on an ecological basis.

Table 11.2 Diminution of CO_2 emission using a cristallized slag instead of a clay

	CaO content in limestone	*Quantity of limestone to make 1 tonne of clinker*	*Quantity of CO_2 producted by tonne of clinker during decarbonation*	*Diminution of CO_2 emission*	
				Decarbonation	*Total*
Cement made with clay	70%	1,25 t	0,55 t	Ref.	Ref.
Cement made with crystallized slag	50%	0,9 t	0,40 t	27%	10–15%

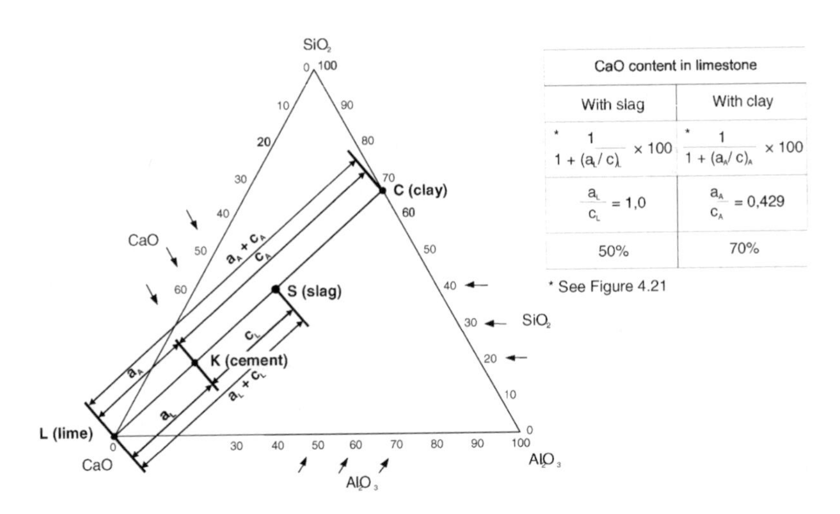

Figure 11.3 Comparison of the composition of a Portland cement made with a crystallized slag and a clay.

11.3.4 Use of mineral components

The replacement of a certain amount of clinker by a mineral component is a very efficient way of improving the energy and ecological efficiency of hydraulic binder processing. This replacement can be done at the cement plant when producing blended cement or at the concrete plant (Malhotra and Mehta 1996). In both cases, each kg of clinker or Portland cement replaced results in a saving of 3000 kJ and a decrease in CO_2 emissions of 1 kg. It is in that field that the most significant savings in energy and reductions in CO_2 emissions can be made.

Natural pozzolans, slag, fly ash, metakaolin, calcined clay, rice husk ash and even limestone and silica filler can be blended to Portland cement clinker.

However, the amount of clinker that can be substituted is somewhat limited by technological constraints. In the present state of the art, it seems that for general construction purposes a limit of 35 per cent is possible. In fact, the substitution of a certain amount of clinker by a mineral component has some drawbacks. As already mentioned, early strength is largely dependent on the amount of clinker present in the concrete. In certain types of construction or in a pre-cast plant, an excessively slow hardening cement can be uneconomical.

However, as previously mentioned, the situation can be improved in these cases by decreasing the W/B ratio through the use of a superplasticizer so that it is not necessary to develop as much C-S-H. Therefore, in a very near future, mineral components could replace 50 per cent of the clinker in a blended cement. On a global scale, this means that the worldwide hydraulic binder production could double with practically no additional fuel consumption and CO_2 emissions.

11.4 Would it be possible to eliminate Portland cement?

This could be perceived as a provocative question. The answer is why not, but it is not going to happen tomorrow. It is not excluded that less energy-demanding industrial processes could be developed to manufacture construction materials that harden in less than 24 hours. In Chapter 4 of their book *Natural Capitalism*, P. Hawken, A. Lovins, and L.H. Lovins (1999) question the future of biological processes based on biological models that show us that every day, some living organisms synthesize materials having mechanical properties far superior to those of the materials that we produce using a high amount of energy and generating considerable quantities of waste products.

According to Ernie Robertson of the Biomass Institute in Winnipeg, Canada (quoted in the above mentioned book), there at least three ways to transform limestone into a construction material. Limestone can be extracted from a quarry and cut into parallel sided blocks that could be piled up to build pyramids or cathedrals. This process is environmentally very clean but it is time consuming and not very efficient. Limestone can be blasted from a quarry, finely ground, and mixed with an appropriate amount of clay to produce Portland cement clinker. This is a very efficient way to create a very flexible construction material but from an energetic and environmental point of view, it is not a very efficient process. The third way is to give limestone to a hen and the next morning it is transformed into an egg shell that can be white or brown according to the desire of customers. This transformation is done at the body temperature of a hen.

Some shell fish even build stronger shells. They know how to select the necessary ions and molecules from sea water to build shells having an extraordinary strength and this is done at a much lower temperature than the temperature of a hen. The strength of abalone shells is still impressive for material researchers (Birchall and Kelly 1983). This shell is composed of very thin layers of calcium carbonate 0.2 m thick, separated by a larger layer of protein, as seen in Figure 11.4.

Therefore, why not dream that a construction material bio-industry will one day provide construction materials that can compete with concrete? Or that the cement industry could transform itself in a bio-industry, because after all, its role is to provide a construction material, whatever the process to make this construction material may be?

This is not for tomorrow, but which engineer among us, who started his engineering classes using a slide rule (like me), could have predicted that by the end of his or her active career, a cement plant would be controlled by a microscopic piece of silicon!

11.5 Conclusion

Concrete will remain the most widely used construction material for some time due to its intrinsic characteristics, its low cost and its versatility. Its use will not increase in industrialized countries, but it will increase tremendously in developing countries. It is unfortunate that in these countries people are not yet very concerned by ecological considerations.

The urbanization process, which is developing very rapidly at present in developing countries, is unavoidable because it is the only way to create wealth, as history teaches us. This urbanization will result in a drastic increase in the consumption of concrete with its concomitant increase in CO_2 emissions if

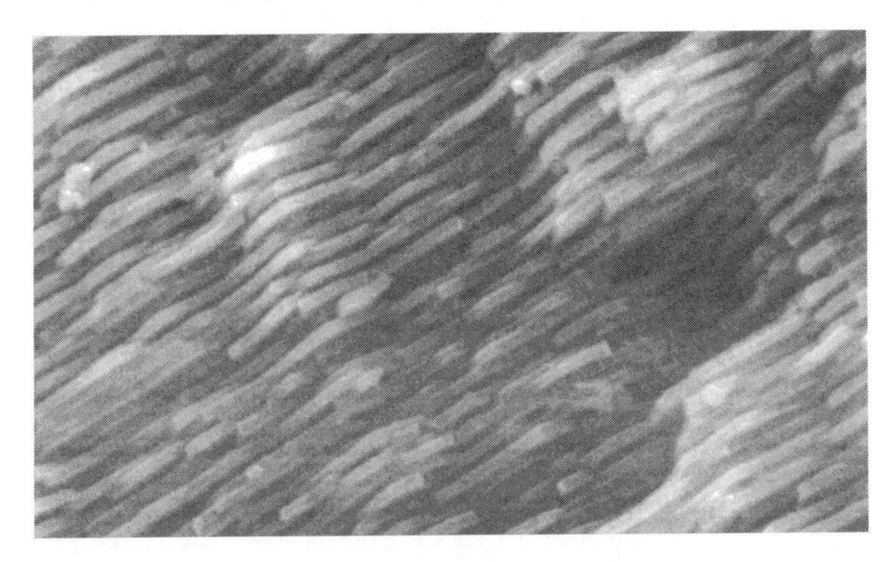

Figure 11.4 Microstructure of an abalone shell. We can clearly see layers of $CaCO_3$ 0.2 m thick, separated by a layer of protein (with the permission of ICI).

Table 11.3 CO_2 concentration in the air, in parts per million

	1000	*1200*	*1400*	*1600*	*1800*	*1900*	*1950*	*2000*
p.p.m	280*	283*	280*	275*	275*	295*	314*	375**

* from the air trapped in polar ice.
** direct measure at the top of Mauna Loa in Hawaii.

nothing is done to teach these people how to make more durable concrete using less clinker, because our forests and oceans are unable to absorb, all the CO_2 generated by human activities, as seen in Table 11.3, from data published by Ball (2000) and Doney (2006).

Unfortunately the know-how to implement such a technology is found in industrialized countries, where concrete consumption is decreasing.

It is our responsibility to teach these people how to make more durable concrete containing less cement, how to produce hydraulic binders having a lower energy content, and how to maximize concrete recycling. What a challenge!

References

Aïtcin, P.-C. (1994), Durable Concrete – Current Practice and Future Trends, ACI SP-144, pp. 85–104.

—— (1995) 'Concrete the Most Widely Used Construction Material', Adam Neville Symposium, Las Vegas.

Anonymous (2000) 'California to Begin Programs for Reusing Building Materials', *Engineering News Record*, 6 November, 245 (18): 19.

Ball, P. (2000) 'La Reforestation ne réduit pas forcément l'effet de serre', *Le Monde*, 10 November, p. 29.

Birchall, J.D. and Kelly, A. (1983) 'New Inorganic Materials', *Scientific American*, 248 (5, May): 104–15.

Bisaillon, A., Rivest, M. and Malhotra, V.M. (1994) Performance of High-Volume Fly Ash Concrete in Large Experimental Monoliths, *ACI Materials Journal*, 91 (2): 178–87.

Doney, S. (2006) 'L'Acidification des océans: l'écosystème menacé', *Pour la Science*, 343 (May): 58–65.

Gjørv, O.E. (2000) 'Controlled Service Life of Concrete Structures and Environmental Consciousness', in O.E. Gjøry and K. Sakai (eds) *Concrete Technology for a Sustainable Development in the 21st Century*, London: E & FN SPON, pp. 1–13.

Hawken, P., Lovins, A. and Lovins, L.H. (1999) *National Capitalism*, Boston, Mass.: Little, Brown and Co.

Kreijger, P.-C. (1987) 'Ecological Properties of Building Materials', *Matériaux et Structures*, 20: 248–54.

Langley, W.S. (1999) 'Practical Uses for High Volume Fly Ash Concrete Utilizing a Low Calcium Fly Ash', *Concrete Technology for Sustainable Development in the Twenty-First Century*, Hyderabad, India, November, pp. 65–96.

Malhotra, V.M. and Mehta, P.K. (1996) *Pozzolanic and Cementitious Materials*, Amsterdam: Gordon & Breach.

Malhotra, V.M. (2001) 'High-Performance, High-Volume Fly Ash Concrete for Sustainability', in A. Tagnit-Hamou, K. Khayat and R. Gagné (eds) *P.-C. Aïtcin Symposium on the Evolution of Concrete Technology*, December, Université de Sherbrooke, pp. 19–74.

Marciano, E. (2003) 'Sustainable Development in the Cement and Concrete Industries, Ph.D. thesis, no. 1452, Université de Sherbrooke.

Mehta, P.K. (1993), 'Concrete Technology at the Cross Roads – Problems and Opportunities', ACI SP-144, pp. 1–30.

Mehta, P.K. (2000) 'Concrete Technology for Sustainable Development: An Overview of Essential Elements', in O.E. Gjørv and K. Sakai (eds), *Concrete Technology for a Sustainable Development in the 21st Century*, London: E & FN SPON, pp. 83–94.

Mehta, P.K. (2001) 'Growth with Sustainability: A Great Challenge Confronting the Concrete Industry', in A. Tagnit-Hamou, K. Khayat and R. Gagné (eds), *P.-C. Aïtcin Symposium on the Evolution of Concrete Technology*, December, Université de Sherbrooke, pp. 89–98.

Mehta, P.K. and Langley, W.S. (2000) 'Monolith Foundation: Built to Last a "1000 Years" ', *Concrete International*, 22 (7, June): 27–32.

Mehta, P.K. and Manmohan, P. (2006) 'Sustainable High-Performance Concrete Structures, Concrete International, 28 (7): 37–42.

Neville, A. (2001) 'A Tribute to Aïtcin: A Challenge to Concretors', in A. Tagnit-Hamou, K. Khayat and R. Gagné (eds), *P.-C. Aïtcin Symposium on the Evolution of Concrete Technology*, December, Université de Sherbrooke, pp. 1–10.

O'Neil, E.F., Dauriac, C.E. and Gilliland, S.K. (1996) Development of Reactive Powder Concrete (RPC), Products in the United States Construction Market, ACI SP-167, pp. 249–61.

Ramachandran, V.S., Malhotra, V.M., Jolicoeur, C. and Spiratos, N. (1998) *Superplasticizers: Properties and Applications in Concrete*, Ottawa: Materials Technology Laboratory CANMET.

Sakai, K. and Banthia, N. (2000) 'Integrated Design of Concrete Structures and Technology Development', in E. Gjørv and K. Sakai (eds) *Concrete Technology for a Sustainable Development in the 21st Century*, Abingdon: E and FN Spon, pp. 14–26.

Sarja, A. (2000) 'Integrated Life Cycle Designs of Concrete Structures', in O. Gjørv and K. Sakai (eds) *Concrete Technology for a Sustainable Development in the 21st Century*, Abingdon: E and FN SPON, pp. 27–40.

Shah, S.P., Wang, K. and Weiss, W.J. (2000) 'Is High Strength Concrete Durable?' in O. Gjørv and K. Sakai (eds) *Concrete Technology for a Sustainable Development in the 21st Century*, Abingdon: E and FN SPON, pp. 102–14.

Sommerville, G. (2000) 'A Holistic Approach to Structural Durability Design', in O. Gjørv and K. Sakai (eds) *Concrete Technology for a Sustainable Development in the 21st Century*, Abingdon: E and FN SPON, pp. 41–56.

Taylor, H.F.N. (2000) *Cement Chemistry*, London: Thomas Telford.

Tomosawa, F. and Noguchi, T. (2000) 'New Technology for the Recycling of Concrete: Japanese Experience', in O. Gjørv and K. Sakai (eds) *Concrete Technology for a Sustainable Development in the 21st Century*, Abingdon: E and FN SPON, pp. 274–87.

Torring, M. (2000) 'Management of Concrete Demolition Waste', in O. Gjørv and K. Sakai (eds) *Concrete Technology for a Sustainable Development in the 21st Century*, Abingdon: E and FN SPON, pp. 321–31.

Chapter 12

Cements of yesterday and today, concretes of tomorrow

12.1 Introduction

Cement is still an essential material in making concrete, but in some modern concretes, it is no longer the most important material because these concretes are in fact composite materials. In a composite material, it is impossible to decide which is the most important material, because by its nature, a composite material has properties that are always much better than the simple addition of each component's individual properties. In the fable of the blind and the lame, it is impossible to decide who, between the blind and the lame, is the most important character.

Modern concrete is more than simply a mixture of cement, water and aggregates. More and more frequently, modern concrete contains mineral components having very specific characteristics that give specific properties to concrete and also chemical admixtures which have even more specific effects. Modern concrete is becoming a very complex chemical material where mineral products and amorphous products, and not just ground clinker and calcium sulphate, interact with organic molecules or polymers. These are specially developed to highlight certain characteristics of concrete or correct certain deficiencies of current cements because current cements can present some deficiencies in some of their applications.

It would be pretentious to believe that nothing else will be discovered in the field of concrete. The science of concrete is only beginning to develop and it should be expected that in the years to come, new types of concretes that will better fulfil different socio-economic needs will be developed.

The development of different types of concrete will not necessarily result in an increase in the number of cement types to be produced, but it will require that the quality of the cement be much more consistent than at present. In the future, cements will have to fulfil tighter specifications.

The development of numerous high-tech concretes will also not necessarily result in an increase in the overall consumption of cement or binder used in a cubic metre of concrete because cement and binders will be used more and more efficiently: in the best-case scenario, in developed countries, it will be possible to make more concrete with the same amount of binder.

The binders of tomorrow will contain less clinker so that the cement industry will become the hydraulic binder industry, an industry that will market fine powders that harden when mixed with water. Interestingly, this increasing use of mineral components other than ground clinker will help the cement industry to fulfil some of its objectives in a sustainable development perspective, that will in future be imposed by governments. It is already very important that today's cement industry highlight this new role.

12.2 Concrete: the most widely used construction material in the world

It has been seen in Chapter 3 how Portland cement became, during the twentieth century, the most widely used construction material. In 1900, total world production of cement was about 10 million tonnes and according to CEMBUREAU, in 1998 it was 1.6 billion tonnes. If we suppose that, on average, 250 kg of cement are used to produce 1 m^3 of concrete, in 1900, only 40 million m^3 of concrete were used, while in 1997, the amount produced was about 6.4 billion m^3. This is a little more than 1 m^3 of concrete per person per year, or more than 2.5 tonnes of concrete per person per year. Only fresh water is used in larger amounts, and this is very often because it is wasted.

This stupendous increase in cement and concrete consumption can be linked to the steady urbanization process that most societies experienced and to the globalization of trade. However, if we look closely at Figure 12.1 where the progression of the amount of cement produced in the world is presented, it is seen that it is during the second half of the twentieth century that the consumption of cement started to increase at a very rapid pace. Of course, it

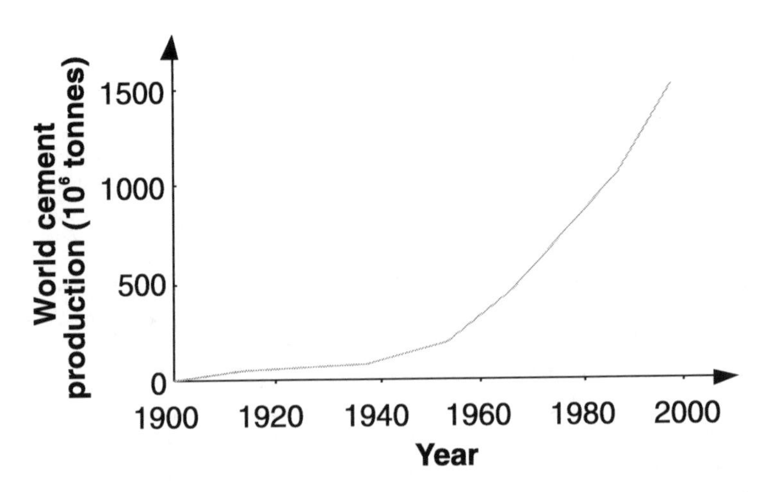

Figure 12.1 World production of cement during the twentieth century, according to CEMBUREAU.

was necessary to rebuild Europe and Japan after the last war which was particularly devastating in terms of infrastructure, but the peaceful period that followed this war can be characterized by a strong trend toward urbanization in many countries and an increase in the standard of living.

Historians and sociologists teach us that a society gets richer and increases its standard of living when it becomes urbanized. The development of a city always results in a considerable increase in infrastructure needs and, consequently, in an increased cement consumption: a house, a school, a hospital, a theatre, a restaurant, a sports centre, a water and sewerage network, a sewage and water treatment plant are always built using concrete. It is therefore quite normal that this drastic increase in cement production occurred in the world during the second half of the twentieth century.

In Chapter 3, it has been possible to establish a direct relationship between the consumption of cement and the gross national product per inhabitant, but also that cement consumption no longer increases when the standard of living reaches a certain level. Several reasons can be put forward: the urbanization process has reached saturation; the major parts of the infrastructure needs have been built and technological progress results in better technical uses of concrete, as with any other material, so that it is possible to satisfy any socio-economic need with less and less material. Presently, in industrialized countries, each material is facing a saturated market, and only maintenance, replacement and the natural progression of the market are the driving forces for its use.

The only cement markets that should experience a spectacular expansion in the years to come will be in developing countries, and this raises further questions.

12.3 Progress achieved by the cement industry in recent years

In recent years, the cement industry has achieved significant progress, specifically in the field of processes and energy savings. However, here too, the last word has not been said in spite of the fact that, from a thermodynamic point of view, the $CaO–SiO_2–Al_2O_3–Fe_2O_3$ phase diagram still governs Portland cement manufacturing. It is already possible to decrease significantly the temperature in cement kilns through a better control of the use of so-called mineralizers, as shown in the previous chapter (Taylor 1997).

With time, Portland cement processing has become more and more complex, and more and more technical. The processes are less robust and require the use of powerful computing facilities to run expert systems which very often require highly skilled personnel to maintain them.

Moreover, very interesting progress has been achieved in the field of sustainable development, and some cement plants are already safely eliminating numerous pollutants or industrial wastes (Uchikawa 1996). Sometimes, some cement plants even produce cement at a negative cost because they are paid to eliminate pollutants, which means that they start making money before having sold a single tonne of cement!

There are, however, some fields in which the evolution of the cement industry has not been so good from a technological point of view. This aspect will be developed later on in this chapter.

12.4 The emergence of a science of concrete

During these past thirty years a science of concrete which is now attracting 'pure' scientists has developed. Concrete is both the fruit of a simple technology and a complex science which is beginning to be mastered, but not in all of its details. In fact, the hardening of a modern concrete results from reactions between amorphous or mineral products, water, more or less complex organic molecules, and in some cases, with mineral salts. Presently, concrete studies are done using quite sophisticated observation and measurement techniques, so that we continue to improve our understanding and control of concrete technology. We are even able to develop new uses for concrete in fields that were difficult to foresee some years ago.

For example, as passionate as I was in 1970 when I started to take interest in concretes known at that time as high-strength concretes, I was far from imagining that in 2000 the highest building in the world would be a concrete high-rise building built in Malaysia (the Petronas Towers). I was also very far from realizing that in 1998, the deepest offshore platform would be built in concrete in Norway (the Troll platform) and that this structure could be taller than the Eiffel Tower. I was far from imagining that in 1998, I could make a reactive powder concrete stronger in compression than ordinary structural steel. But it must be recognized that this spectacular progress in the field of concrete is essentially due to progress achieved in the field of admixtures rather than progress realized in the field of cement manufacturing, as will be seen in the next section.

12.4.1 Recent progress achieved in the field of chemical admixtures

The idea of adding admixtures to concrete is not new. We have seen in Chapter 7 that Roman masons added egg white or blood to their concrete (Venuat 1984). The recent discovery of the beneficial effects of some organic molecules on very specific properties of concrete has often been quite fortuitous, but it can now be explained scientifically.

For a long time, the technology of chemical admixtures has been a field reserved to a few companies within which secrets were jealously kept and, in my opinion, this is a field that has for too long not drawn enough interest from cement companies. It must also be recognized that admixture companies have always been, and are still today, very clever at presenting themselves as a complementary industry to the cement industry, an industry essentially linked to the world of concrete and not to the world of cement.

I am personally convinced that it is this desire not to face cement companies that made them select the expression 'water-reducer' rather than 'cement-reducer' to describe the organic molecules that are used by concrete producers to make a concrete of a given workability with less water but also with less cement. Of course, since the work of Feret and Abrams (Neville 1995), it is well known that it is not the amount of cement that is used in a concrete that influences its strength and durability but rather the ratio between the amount of water and cement that is used. Feret's law, formulated in 1896, is still valid even in the case of high-performance concrete that does not contain enough water to hydrate all the cement particles (Aïtcin 1998).

Unfortunately, admixture companies have been quite successful in deepening the mystery surrounding the use of admixtures by creating and developing an unnecessarily complex and peculiar terminology. Chemical admixtures are not such mysterious products; their action is dictated by the complex laws of physics, chemistry and thermodynamics. As seen in Chapter 7, for V. Dodson (1990), there are only four types of admixtures:

1. those that disperse cement particles;
2. those that modify the kinetics of hydration;
3. those that react with one of the sub-products of hydration reaction;
4. those that have only a 'physical' action.

For many years, to keep their market, chemical admixture companies were forced to limit the price of their admixtures to be able to demonstrate to concrete producers that, from an economic point of view, it was in the end more profitable to use an admixture rather than increase the amount of cement in order to achieve a given compressive strength. Of course, each time such a calculation has to be made it is realized that, in spite of what concrete producers are saying, Portland cement is not such an expensive material. Therefore, in its first phase of development, the admixtures industry began taking advantage of the interesting properties of some industrial by-products. This phase corresponds to the marketing of vinsol resin as an air-entraining agent and lignosulphonates as water reducers, both being by-products of the pulp and paper industry.

The almost simultaneous development in Japan and Germany (Hattori 1998; Meyer 1987), at the end of the 1960s and the beginning of the 1970s, of the exceptional dispersing properties of certain synthetic polymers, known presently as superplasticizers, created a great change in the development of the admixture business. For the first time, quite pure products were especially manufactured as concrete admixtures, products more costly than lignosulphonates and much more efficient. In fact, when using a lignosulphonate-based water-reducer, it is possible, in the best case, to reduce the amount of mixing water by 8 per cent to 10 per cent (and of course cement by the same amount) when making a concrete having a 100 mm slump (the minimum to ensure an efficient placement in the field with vibrators). Suddenly, using a superplasticizer, it was possible to make 200 mm slump concrete. This is the kind of concrete contractors have been waiting for all these years. Even in the case of high-performance concrete with a very low water/cement (w/c) ratio, it is now possible to make 200 mm or 250 mm slump concretes that do not contain enough water to hydrate all the cement grains and do not need to be vibrated to be placed.

Fifteen years have been necessary to see superplasticizers enter the concrete market significantly and it must be admitted that they are still not used to their full potential nor as often as they should be.

The exceptional properties of superplasticizers were the basis for the development of the science of admixtures which very rapidly made possible the understanding of the limits of lignosulphonates and the superiority of superplasticizers. In fact, we now know why it is quite difficult to introduce more than 1 litre of a lignosulphonate water reducer into 1 cubic metre of a concrete without seeing the negative secondary effects of the impurities contained in the lignosulphonate, although it is possible to use superplasticizers with dosages up to 15 litres in high-performance concrete, and even to 40 litres in reactive powder concretes (Bonneau et al. 1996).

However, when high superplasticizer dosages began to be used to decrease the w/c ratio in high-performance concrete, it was realized that, in some cases, some cements and some superplasticizers were incompatible and it was no longer possible to take out the water reducer and add more cement and water to solve the problem (Aïtcin et al. 1994). It became necessary to understand why this phenomenon occurred only with some cements and some superplasticizers, but not with others.

The very recent science of admixtures has progressed quite rapidly, but advances have still to be made, because in some cases, we do not yet understand well enough all the interaction mechanisms governing the mixture of Portland cement and superplasticizers. However, we know how to make 150 MPa concrete on an industrial basis. At such a level of strength, it is the coarse aggregate that becomes the weakest link in concrete: it is only necessary to take out coarse aggregate to be able to increase concrete compressive strength and make reactive powder concrete having a compressive strength of 200 MPa; it is only necessary to confine this reactive powder concrete in thin-walled stainless steel tubes to see the compressive strength increased to 375 MPa; and it is only necessary to replace the sand by a metallic powder to see the compressive strength of concrete increased to 800 MPa (Richard and Cheyrezy 1994).

A 1 000 MPa concrete (1 GPa) is no longer a dream; it will become a reality in this century.

12.4.2 Progress achieved in observing the microstructure and in understanding the nanostructure of concretes

Several other significant achievements have been made in the field of concrete technology due to the use of new very sophisticated scientific apparatus to observe concrete microstructure, and even more recently concrete nanostructure. A scanning electron microscope (SEM) is not only an instrument that makes beautiful pictures; it is an instrument that allows us to better understand what concrete is in its more detailed structure. The use of SEMs has been a key factor in the emergence of these new sciences of concrete and admixtures, as shown in Chapter 6. It is possible now to better observe and understand the effect of a particular admixture on concrete microstructure. However, we must not forget that a good visual observation or a detailed observation under an optical microscope, under polarized or natural light, are still as necessary for the study of modern concrete. Using these sophisticated observation tools, it has been possible to master and direct the evolution of concrete technology rather than continue to progress by trial and error to find organic molecules with useful properties when included in concrete. Macrostructural properties are intimately linked to microstructural properties.

The first uses of magnetic resonance to study the nanostructure of concrete are very encouraging and result in giant steps in the understanding of the true nature of the calcium silicate hydrates formed during cement hydration and that are still identified with the very vague notation C-S-H because we have nothing better to describe them with. We are beginning to understand better what we already know quite well in the case of the detailed structure of the silicon tetrahedron and aluminum octahedron or the magnesium octahedron in kaolin and chrysotile asbestos, two close cousins of C-S-H. Kaolin and chrysotile asbestos are, respectively, aluminium silicate hydrate and magnesium silicate hydrate.

As seen in Chapter 6, the aluminum ion which has a 0.50 Å radius in an aluminum octahedron in kaolinite fits quite well with the plane composed of silicate tetrahedra. Contrary to this, the magnesium ion, which is a little larger at 0.78 Å, forces the octahedral planes to curve and results in the formation of a fiber: asbestos fibres. As the calcium ion is much larger, 1.06 Å, it does not allow the formation of large planar or fibrous structures but rather a structure where there are small amounts of silica tetrahedra and alumina octahedra that can only develop as very small crystals.

Finally, due to the progress made with the atomic force microscope, it is now possible to explain why, when the C/S ratio of C-S-H is low, the intrinsic strength of C-S-H increases and how some admixtures interact at the nanometer level (Nonat 2005).

12.5 Cements of yesterday and today

12.5.1 *Evolution of their characteristics*

Generally speaking, the cements of yesterday were not as fine and did not contain as high an amount of C_3S as today's cements, but these are not the only differences. Others, more hidden difference, have important technological impacts. Before discussing these hidden differences, the first two differences will be reviewed.

As stated previously in Chapter 11, it is supposedly from the pressure of contractors that some cement companies decided to increase the fineness and the C_3S content of their cements in order to allow contractors to remove formwork more rapidly and therefore increase their competitiveness. Personally, I cannot accept such an argument because I don't see how the competitiveness of contractors is improved by the fact that all the cement companies are offering contractors cements that have an increased fineness and C_3S content. This attitude has even had a negative effect on the cement industry, because with these new cements, it is possible to achieve a higher twenty-eight-day compressive strength using less cement, but this higher twenty-eight-day compressive strength is achieved at the expense of long-term durability. For example, at one time, the state of California was forced to put an end to what was known in California as the 'Blaine war'. Cement producers should rather ask themselves seriously what the advantages and the disadvantages are in increasing the specific surface and the C_3S content of a cement when making concrete. There are many other, much less dangerous ways, to increase initial compressive strengths of a concrete.

Of course, in some cases, these increases in the fineness and C_3S are acceptable, for example in the case of a structural concrete used in a mild environment, but in many cases these increases in the fineness and C_3A content have catastrophic results when concrete is exposed to harsh environmental conditions.

In fact, when a finer cement richer in C_3S is used, it is possible to reach a higher twenty-eight-day compressive strength with a higher W/C ratio. According to Wischers (1984), in 1960 in England, a 30–35 MPa concrete could be made using 350 kg/m^3 of cement and a W/C ratio of 0.45. In 1985, the same structural concrete could be made using only 250 kg/m^3 of cement and a W/C ratio of 0.60. For the designer who is making structural calculations, these two concretes are equivalent. However, from a microstructural point of view, the porosity and the permeability of these two concretes are completely different. A concrete having a 0.60 W/C ratio will carbonate more rapidly than a concrete having a W/C ratio of 0.45 and its durability to sea water, freezing and thawing and deicing salts will also be not as good.

As soon as a concrete is exposed to a severe environment, the key factor that conditions its durability is its W/C ratio and not its compressive strength. It must be admitted that at an equal fineness and an equal content of C_3S, twenty-eight-day compressive strength depends on the W/C ratio, but this is no longer the case when the fineness and the C_3S content are different.

Therefore, the good old cements of yesterday which were coarser and less rich in C_3S were used to make concretes whose compressive strength continued to increase after twenty-eight days, while our modern efficient cements have achieved almost all the strength they can at twenty-eight days. For the designer and the cement producer, these cements are equivalent, but for owners concerned with maintenance costs and life cycle, the good old concretes were finally much stronger and more resistant than their twenty-eight-day compressive strength indicated.

12.5.2 Standards

Portland cement is a very complex mineral product composed of at least five principal mineral phases C_3S, C_2S, C_3A, C_4AF and calcium sulphate, and made from very simple raw materials that also contain small amounts of oxides other than SiO_2, CaO, Al_2O_3 and Fe_2O_3 which result in what are called the minor cement phases. These minor phases are quite different from one cement to another. Taking into account these differences in the composition of cement, it is absolutely fundamental that a given Portland cement satisfy the requirements of different standards, so that it will be possible to make a concrete that has predictable characteristics and be adaptable to precise needs.

The tests performed on mortars or pure pastes with a W/C ratio of about 0.50, which is a quasi-universal way of testing cement in the world, have for a long time been very safe from a W/C point of view because most industrial concretes had a W/C ratio higher than 0.50, but this is no longer true. The rheology of low W/C ratio concretes is no longer dictated by the amount of water used to make them or by the shape of coarse aggregates but rather by what is now called the compatibility between the cement and the admixtures used. This raises very important questions: Are the standards presently used in the world to test cement well adapted to the real world of concrete? Are we testing the right properties?

My answer is yes, if the long term objective of the cement industry is to base its development on the use of a commodity product having a 20 MPa compressive strength. My answer is no, if the long-term objective of the cement industry is to transform some of the concrete that will be used into smart concretes that will be able to face the competition of other construction materials and keep its share of the construction market.

12.5.3 Cement admixture/compatibility

According to concrete producers and contractors, compatibility problems between water-reducers (based on lignosulphonates) and cement occurred episodically some time ago, although not as often as with today's superplasticizers. However, as seen in Chapter 8, scientific documentation of these compatibility problems is quite rare. It seems that the phenomenon has not received all the necessary attention from admixture manufacturers and cement producers. Moreover, each time a solution was found in the field to solve the problem (very often, it consisted of the replacement or the omission of the admixture), no one was interested any longer in knowing what had happened. In the rare cases found in the literature (Dodson and Hayden 1989; Ranc 1990), it seems that one frequent cause was the presence of a high amount of anhydrite in the calcium sulphate. Of course, a cement producer can satisfy the SO_3 content of present standards by adding either gypsum or a mixture of gypsum and anhydrite, but it seems that when the amount of anhydrite is too high, and a lignosulphonate is used, the rate of solubility of sulphate ions was drastically reduced (Ranc 1990).

The frequency of incompatibility problems has increased drastically with the use of polysulphonate superplasticizers in high-performance concretes having a low W/C ratio or a low water/binder (W/B) ratio; that is, a W/B ratio much lower than that used in the sacrosanct standards. As in such cases it is no longer possible to solve the problem by taking out the superplasticizer, it became necessary to study the problem in more detail. Of course, presently, this problem is not yet important for the cement industry. It is much more important for the admixture industry because high-performance concrete (HPC) is a very promising market; several litres of superplasticizer are used in each m^3 of HPC. This situation, which is not a general one for any cement and superplasticizer, has attracted the attention of admixture companies and some university researchers who are trying to understand it from a fundamental point of view.

When polycarboxylate superplasticizers started to be used to replace polysulphonate superplasticizer in incompatible and non robust combinations, they were found to be very efficient, but as they were tried with other combinations, they were also found to be incompatible with some cements. As a general rule it can be said that the more compatible and robust a polysulphonate-cement combination is, the more incompatible it is with a polycarboxylate superplasticizer and vice versa.

This strong desire to understand and solve compatibility problems is, as far as I am concerned, one of the reasons for the rise in the science of admixtures.

Although we still do not understand fully how superplasticizers interact with all the cement and the sulphate phases in Portland cement, we have already found some practical solutions to solve incompatibility problems, in many cases. The double introduction method is one of these methods. It consists of adding the superplasticizer in two doses: the first at the beginning of the mixing, and the second at the end of the mixing, or just before placing the concrete in the field. We can also mention the addition of a very small amount of retarder or the addition of some sodium sulphate with polysulphonate superplasticizers or calcium chloride with polycarboxylates. However, there are still cement/superplasticizer combination problems that are not solved.

In fact, if the SO_3 content of today's and yesterday's cements have not changed, this is not true for the SO_3 content of the clinker. Not long ago, the SO_3 content of clinkers was usually in the order of 0.5 per cent, but it can be now as high as 1.5 per cent in some cases or even higher (values as high as 2.5 per cent have been reported). As the maximum amount of SO_3 permitted by standards in a cement is still 3.5 per cent, cement companies are, in some cases, limited in the amount of calcium sulphate they can add to their clinker, specifically when they are adding iron sulphate to complex hexavalent chromium which is responsible for mason scabies.

If, from a purely chemical point of view, the SO_3 content of modern cements is the same, it is not certain that, from the solubility rate of SO_4 ions, it is the same when the amount of SO_3 indicated in the chemical analysis is coming from an alkali sulphate, or whether it is dissolved in the C_2S in the clinker, or whether the SO_3 is combined in one form of the calcium sulphate that has been added to the clinker during its grinding.

The recent results obtained at the University of Sherbrooke by S. Jiang et al. (1999) seem to demonstrate, from a rheological point of view, that for many cements that will be superplasticized with a polysulphonate, there is an optimal amount of soluble alkalis. This ideal alkali soluble content is not reached with some modern cements because, in order to please some agencies specifying the use of cements having a low alkali content in order to avoid potential, or very often imaginary, alkali-aggregate reactions, some cement companies are selling cement with an unnecessarily low alkali content.

Another problem that could cause trouble in the years to come is the influence of the SO_3 content in delayed ettringite formation (DEF). A number of researchers who were working in the field of alkali-aggregate reaction, which no longer seems to interest as many people, are switching to DEF. Already, many papers have been written on the subject, intensive courses have been given, and books written on this subject.

12.6 Concretes of yesterday and today

12.6.1 A commodity product or a niche product

For a long time, the concrete industry and designers have produced and specified a universal concrete, good enough to be used under any circumstances, whose compressive strength was usually between 15 and 25 MPa. In some countries, with the years, it has been possible to see a slight increase in concrete compressive strength so that presently, in some developed countries, the concrete used for structural purposes has a compressive strength between 25 and 35 MPa. This does not mean that 20 MPa concrete is no longer used: there are numerous applications where a designer does need a concrete with a compressive strength higher than 20 MPa (footings, basements, etc.), but more and more structures are built with a slightly stronger concrete.

However, during the 1970s, concrete having a higher strength (40 to 50 MPa) began to be specified for columns in high-rise building, because slender columns offered more architectural possibilities and more renting space (Albinger and Moreno 1991). Over the years, the name of these initial high-*strength* concretes has been changed to high-*performance* concrete because it was realized that these concretes have more than simply a high strength. These concretes started to be used outdoors and faced more severe environments such as off-shore platforms, bridges, roads, etc. Little by little, it was realized that the market for this concrete was not only the high-strength market, but also more generally the market for durable concrete, that represents more or less one third of the present market for concrete.

As seen in Chapter 11, it has also been realized very recently that this type of concrete is more ecologically friendly, in the present state of technology, than usual concrete because it is possible to support a given structural load with less cement and of course, in some cases, one-third of the amount of aggregates necessary to make a normal strength concrete. Moreover, the life cycle of HPC can be estimated to be two or three times that of usual concrete. In addition, HPC can be recycled two or three times before being transformed into a road base aggregate when structures have reached the end of their life.

HPCs, which are simply concretes with a low W/C or W/B ratio, are economical concretes on an initial cost basis, because it is possible to build an equivalent structure with less formwork, less concrete to be placed and less reinforcing steel. The Quebec Ministry of Transportation has calculated that the initial cost of a 50 to 60 MPa concrete bridge is 8 per cent less than that of a 35 MPa concrete without taking into consideration the increase in the life cycle of the bridge (Coulombe and Ouellet 1994).

The acceptance of HPC is slow, but it is progressing constantly and this progression will continue because designers and owners will realize the value and durability of this concrete. Of course, HPCs are not a panacea that will stop the development of all other kinds of concrete. HPC has its limitations, but it is durable concrete that will allow designers and architects to go beyond the limits of present concrete.

Along with these developments in the field of HPC, it has been possible to see recently other high-tech concretes finding some niche markets, such as fibre-reinforced concrete, roller-compacted concrete and reactive powder concrete. All these concretes are designed for small but lucrative markets where competition is limited. Only serious companies are involved in these markets, because they are the ones at the cutting edge of technology. When they are able to provide such high-tech concretes they are able also to supply the large amounts of regular concrete that must be used to complete the building of the infrastructure.

It is obvious that this short list of special concretes is not exhaustive and will grow very rapidly with time because it will be more and more interesting to offer contractors more elegant, more durable and more profitable solutions than the one that can be put into practice using 20 to 30 MPa concrete. The development of BOOT (Build Own Operate and Transfer) projects will undoubtedly accelerate this trend towards the formulation of niche concretes and, interestingly enough, sometimes it might be discovered that some of these concrete could benefit additional markets for which they were not initially conceived. For example, one can mention the case of self-compacting concrete which was developed in Japan to facilitate the placing of concrete in congested structural elements, and is now starting to be used as 'silent concrete' or 'noiseless' concrete, because it can be cast any time of the day or night without disturbing the neighbourhood since it does not need to be vibrated. When such a concrete is used in a precast plant, this precast plant can be less noisy than a discotheque.

12.6.2 Strength or durability

The design of concrete structures is done by structural engineers, engineers who know only one thing about the material, its twenty-eight-day compressive strength, and who are not very familiar with concrete durability.

The knowledge of the twenty-eight-day compressive strength of a concrete is, of course, fundamental to make calculations that will allow the construction of a safe structure, but it is necessary to ensure that this concrete will maintain its mechanical strength during the whole life of the structure. Unfortunately, many examples show us concretes that had an adequate initial twenty-eight-day compressive strength, but have lost most of their functionality because they faced an environment for which they were not designed or because they were not placed or cured properly. There is no need to go beyond such examples to find the reason for the very poor image that concrete has with the public: it is only necessary to look at the very poor appearance of much of our present infrastructures or the numerous repair works that are consuming so much time and so many dollars. It is a pity to have to demolish so much infrastructures that have reached only half of their intended life cycle, not to mention the enormous socio-economic costs associated with these repairs (deviations, traffic jams, loss of time, pollution, etc.).

The cement industry is paying a very high price for these errors: instead of taking advantage of the very lean budgets allocated for the construction of new infrastructure that would use cement, these budgets are used to do repair work that consumes very little material but a lot of labour. Contractors don't care, because in both cases, the volume of the work to be done, in terms of dollars, is the same.

Several codes now emphasize concrete durability rather than strength when selecting the concrete to be used to build a structure. It is about time!

The problems of external parking garages built in Canada with 20 MPa concrete caused a great deal of trouble to the cement and concrete industry due to their poor freezing and thawing and scaling resistance. Accelerated carbonation of 20 MPa concrete in Europe is also costing cement companies a lot, just as sea water attack is doing in the Middle East.

Moreover, a more rapid degradation of infrastructure and limestone monuments in large cities due to atmospheric pollution in highly urbanized areas is foreseeable. In fact, hidden inside a porous 20 MPa concrete is a whole world of microscopic life. Bacteria, germs, moss, and lichens are prospering in concrete pores. The thiobacillus ferroxidans bacteria need calcium sulphate to develop. This calcium sulphate is present as the result of the attack of limestone or concrete by acid rain, but these bacteria then proliferate, producing sulphuric acid as a by-product, which attacks concrete and limestone to make new calcium sulphate.

HPC, which has a much lower porosity, better withstands the effect of pollution, and this will be one of the reasons for its growing use in the future, simply because it is less porous. As far as the additional MPa characteristic of high-performance concrete are concerned, designers will have to learn how to use them efficiently.

Having worked in a BOOT project, I have seen that when a contractor has to maintain a structure for twenty-five or thirty years and guarantee a 100-year life cycle, he does not hesitate very long about the quality of the concrete to be used: he uses HPC.

12.6.3 The race for more MPa

Although concrete compressive strength is not its essential characteristic, because it is its durability that it is more important, it must be admitted that these two characteristics are intimately linked. The overwhelming importance of compressive strength in the codes and the ease with which it is measured can explain why the increase in compressive strength of concrete has been, in a certain way, a constant preoccupation. Periodically, some researchers announce that they have succeeded in making a concrete with a very high compressive strength, but it must be admitted that all these efforts translate into very few industrial applications, except perhaps the DSP (an acronym for 'Densified Cement/Ultra-Fine Particle-Based Material') concept developed by H.H. Bache (1981) and the reactive powder concept (RPC) developed by Pierre Richard (Richard and Cheyrezy 1994) that was used for the first time during the construction of the Sherbrooke pedestrian bridge (Aïtcin 1995).

Every day, my students are making concretes of 400 MPa, using appropriate, simple mechanical and thermal treatments. Pierre Richard, for his part, was able to make an 800 MPa concrete using a metallic powder, so that a 1 000 MPa reactive powder concrete (1 GPa) is no longer a utopian concrete. What is the future of such a concrete three times stronger in compression than steel? No one knows, but I have no doubt that in this century, this type of concrete could be made and surely used. The importance of the compactness of the aggregate skeleton, of appropriate thermal treatments and the benefits of confinement were well known. The creative work of Pierre Richard has been to transform these well-known technologies into a simple and usable concept.

Of course, the price of 1 m^3 of RPC is frightening to many engineers who still compare this price to the price of 1 m^3 of ordinary concrete or to HPC. However, it should be realized that the unit price of reactive powder concrete must not be calculated in relation to 1 m^3, but rather to 1 tonne of material, because RPC has to be compared to steel and not to ordinary, or high-performance, concrete.

When the Sherbrooke pedestrian bridge was built, the cost of the materials used to make the RPC could be estimated at roughly US\$1 000/m^3. Presently, the same RPC could be produced for US\$750/m^3, and in the near future, it could perhaps be possible to make it for US\$600 to \$650/m^3. The cost of the materials used to make an RPC can be split roughly into two equal parts, the cost of the powders and the cost of the fibres.

However, even at US\$1 000/m^3, that is US\$400/t, in some applications where durability and compressive strength are keys factors, RPC can compete with structural steel which cost US\$1 200 to \$1 500/t in year 2000.

Moreover, in some applications, RPC should not compete with steel only, but also with pig iron, aluminium and even wood.

12.7 The concrete of tomorrow in a sustainable development perspective

As seen in the previous chapter, the concept of sustainable development is not a passing fad but rather a policy that will last and become more important because the great ones of this world breathe the same polluted air as all of us. It is no longer possible in the northern hemisphere to live in the fashionable west ends of large cities to benefit from pure air, because these areas are situated east of another urbanized centre and are receiving their pollution. Because the presidents and prime ministers of the G8 countries breathe the same polluted air as everyone else, sustainable development concerns them as much as the rest of us; this is why sustainable development will survive the departure of current world leaders.

Is it absolutely necessary to satisfy, at any price, the greed of shareholders, or is it better to preserve this heritage for our children and grandchildren, and the children and grandchildren of these shareholders?

When it is realized that only one-third of the world has the advantage of a high standard of living and that this one-third is not interested in going back to the living conditions of the good old days, while two-thirds of the world has only one thing in mind, that is to enjoy the same standard of living as the richest countries, it is obvious that it is urgent that a sustainable development policy be enforced all over the planet in order to avoid repeating the same errors that resulted in the present situation. The application of such a policy will not be easy, because it is always others who are polluting and wasting more than us.

The cement and concrete industry has no choice: it must add this direction to its development and its other constraints. I have no doubt that the cement and concrete industries will succeed in making this change. The cement industry, or rather the hydraulic binders industry, which will become the name of the cement industry, will be a GREEN industry. There is no choice, there is no point in complaining or fighting back, it is better to face, as soon as possible, this new situation and transform the cement and concrete industry into in a GREEN industry and proclaim this loud and clear to the public. In the end, there are many ways of doing this and this commitment must be known to the public.

The concrete of tomorrow will be more durable and will be developed to satisfy socio-economic needs with the least environmental impact. The cost of a project in the future will have to incorporate not only the present economic costs that we are used to calculating now, but also social and environmental costs ranging from the extraction of the raw materials, to their utilization, including their elimination at the end of the life cycle of the structures.

12.7.1 The ecological impact of concrete

It is not satisfactory to say that the energy content of 1 m^3 of concrete is negligible, because at 6.4 billions times a small energy content, it is no longer negligible. There are many ways of decreasing CO_2 and NO_x emissions and of decreasing the amount of aggregates needed to make concrete structures. It will also be necessary to learn how to incorporate ecological and socio-economic costs when evaluating a given project.

When the importance of such costs is taken into account, it will be found that HPC and RPC are not such costly materials compared to 20 to 30 MPa concrete, because these materials are more durable, can very easily be recycled several times before ending up as granular road base material and require the use of less material.

This century will see cement companies developing alternative binders which are more environmentally friendly from a sustainable development point of view. This is why the use of mineral components, which has not been promoted very strongly during the past thirty years, will be exploited more seriously. It is no longer necessary to return to the Roman texts to be convinced of the need to use natural pozzolans. The cement and concrete industry will begin controlling the artificial pozzolans market (metakaolin, rice husk ash, diatomaceous earths, amorphous silica, calcined clay or shale). It is no longer true that the blending of any mineral component decreases the early compressive strength of concrete, because since it is possible to use superplasticizers, the initial compressive strength of a concrete is no longer controlled only by the amount of C_3S and the fineness of the cement, but also by the density of the hydrated cement paste. This density is a function of the W/B ratio and/or of the W/C ratio. In a very dense system (with a very low W/B ratio), it is not necessary to have too much glue (C-S-H) to obtain the necessary MPas to allow a contractor to rapidly remove the formwork of a structure. But at present, such a solution is not always practical or economical, so that there is a practical limit above which the use of pure Portland cement is less expensive when a high early compressive strength is necessary.

Until recently, when Portland cement was compared to the other principal binders still in use in the industry (gypsum and lime), it could be said that gypsum and lime were more ecologically friendly. Gypsum and lime are used in a cyclic way due to the chemical reactions involved, whereas the use of Portland cement is linear (Aïtcin 1995). But if the time scale is changed, this conclusion is no longer valid. In fact, if the time scale is increased to a geological scale, all concrete will one day end up as a mixture of limestone, clay, iron oxide and sand which are the stable form of calcium, aluminum, iron and silicon ions in our environment on earth. This is what we learn from observing nature.

12.7.2 The binders of tomorrow

I cannot pretend to read the future in tea leaves, but I can make some predictions. It is not too risky to do so when you are sixty-eight years old. In any case, I am prepared to take the credit for all the right predictions and the blame for the wrong ones.

The binders of tomorrow will contain less and less ground clinker; they will not necessarily have such a high C_3S content; they will be made with more and more alternative fuels. They will have to meet tighter standard requirements and they will need to be more and more consistent in their properties, because the clinker content of blended cements will be lower. The binders of tomorrow will be more and more compatible with more and more complex admixtures, and their use will result in making more durable concrete rather than simply stronger concrete.

This is only the beginning of the list the qualities the binders of tomorrow will have, not to mention that these binders will also have to be not too expensive. This is the challenge for the cement industry!

12.7.3 The admixtures of tomorrow

Admixtures will be more and more numerous, and they will more often be made specifically for the concrete. They will be more and more pure, more and more specific, and more and more precise in their action. It will become more and more difficult to blame the admixture in the case of incompatibility and it will be necessary for cement producers to provide a list of compatible and incompatible admixtures to their customers in order to specify the right admixture as well as the admixture that will be unsuitable.

Admixtures are becoming an essential component in making concrete. This is not a new constraint and neither is it a very interesting constraint, because the marriage of cement and admixture has to be seen as a technological opportunity that very few cement producers are presently realizing. The marriage of organic and mineral chemistry, and the chemistry of amorphous materials and colloidal materials, is the secret to the success of concrete in the next century.

12.7.4 The concrete of tomorrow

The concrete of tomorrow will be GREEN, GREEN, AND GREEN. Concrete will have a low W/B ratio, it will be more durable and it will have various characteristics that will be quite different from one another, for use in different applications. The time is over when concrete could be considered a low-priced commodity product; the time has come for concrete 'à la carte'.

Concrete and cement producers have to realize that they can make more profit by selling small amounts of concrete 'à la carte' than by selling a cheap commodity product. Contractors and owners have to realize that what is important is not the cost of 1 m^3 of concrete, but rather the cost of 1 MPa or one year of life cycle of a structure. When the cement and concrete industry, as well as contractors and owners, realize that, then the construction industry will have made a great step forward.

I am personally convinced that the greater use of BOOT projects will force contractors to put into practice, and owners to accept, a revolution in the construction industry.

The concrete producer of tomorrow will have to know how to deal with all the different types of products offered by cement and admixture producers, to provide contractors with concrete that will be more high-tech and more economical, not in terms of the cost of 1 m^3, but in terms of performance.

The concrete industry of tomorrow will have to continue to produce a commodity product, but also to produce niche concretes with a high added value.

12.8 The development of the concrete industry and the cement industry in the twenty-first century

I have presented my personal thoughts on the past history of cement and concrete and my predictions for the future. As you will have noted, I am not totally objective in this field. My love of concrete is great, but it is not without limits, because I know concrete too well not to realize that it is not the 'only' material of the future. It is simply a marvellous material, flexible in its composition, ecological when we take care, and a material still full of unexplored possibilities or even unexploited possibilities.

The years to come will not be more difficult than the present or the past ones; they will simply be different, with new challenges to face.

Cement and concrete will remain, at least during the first half of the twenty-first century, the most widely used construction materials in the world, although this future concrete could be quite different from that used today. The concrete of the later part of this century will be a concrete having a balanced ecological content; concrete will be a material made and used to serve human beings, and not simply to maximize the short-term profit of some shareholders.

The competitiveness of the cement industry will be a function of the speed at which these changes are made. This is quite a difficult task for a heavy industry that is still too fragmented and an industry that for so long rested on its laurels, an industry that for too long had only to wait to pick up orders, an industry that has for too long shown very little interest in concrete, which is its only market.

As we enter a world where the greatest wealth will be knowledge, and after leaving the university world, I would like to take the opportunity to pass along a very strong message: it will be vital for the cement industry to take more interest in universities during this century. Not all university professors are dreamers! This cooperation is vital for the sake of the industry in order to train the engineers who will make the cement industry more competitive, but also, and this is more important, to give the desire to future engineers, administrators and architects who are studying in universities, and who one day will become decision-makers, to think of concrete when they have to select the material they will use for a construction project.

It is at the University of Toulouse in France that I fell in love with concrete, because two professors made me share their passion for concrete. It is at the University of Sherbrooke that I tried to transmit this passion for concrete and cement to my students and, very modestly, I think this has been the biggest achievement in my career as a professor.

12.9 Conclusion

After reading this presentation on the cements of yesterday and today and on the concrete of tomorrow, I would be very happy if only one idea remained: each time one of the parameters in the process of cement-making is modified for any reason, on top of the two legitimate questions that have to be answered about the economic and acceptance standard consequences of this move, it will be necessary to raise another question: what will be the result of this modification on the quality of the concrete made with this new cement? Quality means durability more than strength. Cement is made to make profit, but also to ensure the competitiveness of concrete as a universal construction material.

It is totally wrong to think that cement is solely a material that has to fulfil standards on mortars that are more and more outdated and very far from the real world of concrete. What can be made with cement except concrete? This is the challenge of the cement and concrete industry for the twenty-first century. It is a stimulating challenge.

References

Aïtcin, P.-C. (1995) Concrete, the most widely used construction materials, ACI SP-154, pp. 257–66.

—— (1998) *High-Performance Concrete*, London: E & FN SPON.

Aïtcin, P.-C., Jolicoeur, C. and MacGregor, J. (1994) 'Superplasticizers: How They Work and Why They Occasionally Don't', *Concrete International*, 16 (5): 45–52.

Albinger, J. and Moreno, J. (1991) 'High Strength Concrete: Chicago Style', *Concrete Construction*, 29 (3): 241–5.

Bache, H.H. (1981) 'Densified Cement/Ultra-Fine Particle-Based Materials', *Second International Conference on Superplasticizers in Concrete*, Ottawa, Ontario, Canada.

Bonneau, O., Poulin, C. Dugat, J., Richard, P. and Aïtcin, P.-C. (1996) 'Reactive Powder Concretes: From Theory to Practice', *Concrete International*, 18 (4): 47–9.

Coulombe, L.-G. and Ouellet, C. (1994) 'The Montée St-Rémi Overpass Crossing Autoroute 50 in Mirabel: The Savings Achieved by Using HPC', *Concrete Canada Newsletter*, 2 (1): 2

Dodson, V.H. (1990) *Concrete Admixtures*, New York: Van Nostrand Reinhold.

Dodson, V.H. and Hayden, T. (1989) 'Another Look at the Portland Cement/Chemical Admixture Incompatibility Problem', *Cement, Concrete, and Aggregates*, 11 (1): 52–6.

Hattori, K. (1978) Experience with Mighty superplasticizer in Japan, ACI SP-62, pp. 37–66.

Jiang, S.P., Kim, B.-G. and Aïtcin, P.-C. (1999) 'Importance of Adequate Alkali Content to Ensure Cement/Superplasticizer Compatibility', *Cement and Concrete Research*, 29 (1): 71–8.

Meyer, A. (1987) Experience in the use of superplasticizer in Germany, ACI SP-62, 1987, pp. 21–36.

Neville, A.M. (1995) *Properties of Concrete*, London: Longman.

Nonat, A. (2005) 'The Structure of C-S-H', *Cement Wapno Beton*, 2, (March): 65–73.

Ranc, R. (1990) 'Interaction entre les réducteurs d'eau-plastifiants et les ciments', *Ciments, Bétons, Plâtres, Chaux*, 782: 19–20.

Richard, P. and Cheyrezy, M. (1994) Reactive Powder Concrete with high ductility and 200–800 MPa compressive strength, ACI SP-144, pp. 507–518.

Scheubel, B. and Nachtwey, W. (1997) 'Development of Cement Technology and its Influence on the Refractory Kiln Lining', Refra Kolloquium, Berlin, pp. 25–43.

Taylor, H.F.W. (1997) *Cement Chemistry*, London: Thomas Telford.

Uchikawa, H. (1996) 'Cement and Concrete Industry Orienting toward Environmental Load Reduction and Waste Recycling', IVPAC Conference, Seoul, pp. 117–49.

Venuat, M. (1984) *Adjuvants et traitements*, n.p.

Wischers, G. (1984) 'The Impact of the Quality of Concrete Construction on the Cement Market', Report to Holderbank group, No. DIR 84/8448/4.

Chapter 13

My vision of clinkers and binders

13.1 Introduction

During my entire career as a concrete researcher, I had no choice but to use the cements that were manufactured by cement producers. Many of these cements did not fulfil my needs exactly and sometimes they did not have exactly the characteristics I was looking for. Several times, I had to import very specific cements from as far as Europe, the USA, Mexico or Chile, fly ashes from Australia and slags from Belgium in order to obtain the kind of hydraulic binder I needed for my research work. These cements and/or binders had specific characteristics that could be used to pinpoint their influence and make my demonstrations more convincing. Observing the behaviour of alls the cements I have used during my career has given me a clear idea of what an ideal clinker should be.

In fact, as it will be seen, it is possible to make all the concretes presently used in the construction industry, from unshrinkable fill of 1 to 2 MPa, to reactive powder concrete (RPC) that could soon reach 1000 Mpa, with only two types of clinker: a grey and a white clinker. In this chapter, I will describe the characteristics the grey clinker should have.

13.2 My vision of clinkers and binders

I will start by putting cement producers' minds at ease: this clinker is not revolutionary, it is a clinker already manufactured by some cement plants that don't know that they are making what is for me an ideal clinker. It is essentially a Type V clinker used to make sulphate resistant Portland cements whose particular characteristics will be detailed later.

In fact, when someone dedicates his career to improving the overall quality of concrete or to developing more resistant and more durable concretes, the same problem must be faced: the excessively high C_3A content of today's Portland cement. I am convinced that C_3A is poison to concrete; for rheology, poison for polysulphonate superplasticizers, and poison for durability (sulphate attack, sea water attack, delay ettringite formation and freeze-thaw cycles).

Unfortunately, it is a necessary poison because it is quite difficult and uneconomical to totally eliminate C_3A from clinker. Therefore, as it is very difficult to make a clinker that does not contain any C_3A, let's make one with the minimum amount of C_3A.

As to favour the transformation of the C_2S produced before the clinkerization process into C_3S, it seems to be necessary to make a clinker having 14 to 16 per cent of interstitial phase. The ideal clinker will maintain this amount of interstitial phase. As its C_3A content will be low, its C_4AF content will be high, it will be a dark-coloured clinker.

I am not like Dick Burrows who is trying to introduce a Type VI clinker in ASTM standards very close to my ideal clinker. It is not my intention of creating a new clinker type, the ideal clinker will be basically an ASTM Type V clinker.

The second essential characteristic of an ideal clinker is that it should not contain an excessively high C_3S content. Now that it is easy to get cement grains as close to each other as desired through the use of superplasticizer, it is not necessary to develop high amounts of C-S-H to obtain a certain level of early and/or long-term compressive strength and compactness (durability). It is only necessary to lower the W/C or W/B ratios to get the desired strength. Of course, to *increase early compressive strength* it is always possible to increase *slightly* Blaine fineness and/or to increase *slightly* the initial temperature of concrete to accelerate hydration reaction. The ideal clinker should have a *maximum* content of 50 to 55 per cent of C_3S.

The third very important characteristic of the ideal clinker is a *minimum amount of soluble alkalis* of 0.4 per cent. In fact, in the low W/C or W/B ratio concretes that will be used more frequently in the future, alkali sulphates provide very rapidly, much more rapidly than the different forms of calcium sulphate that are added to clinker during its grinding, the necessary SO_4^{2-} ions necessary to form the ettringite that blocks the hydration of the C_3A and of the interstitial phase.

It is well documented that the lack of compatibility and robustness of most of the Portland cement/polysulphonate combinations is due to an unbalanced amount of reactive C_3A and SO_4^{2-} ions in the interstitial solution in fresh concrete (presently, on a cost/property, basis polysulphonates are the most efficient superplasticizers and the only ones that allow an easy adjustment of the entrained air content).

In order to compensate the lack of rapidly soluble alkali sulphates, it is always possible to add a small amount of sodium sulphate (Kim et al. 2000a) to improve compatibility and robustness. Sodium sulphate does not work with polycarboxylate, it rather deteriorates the situation. It is better to add calcium chloride which provides Ca^{2+} ions that favour the steric fixation of the polycarboxylate chains on the surface of the cement particles (Yamada et al. 2000; Nawa et al. 2001).

Moreover, experience shows that the cements that are the least compatible and robust with polysulphonate superplasticizers (polynaphthalene, polymelamine and lignosulphonates) are white and low alkali cements. Usually, white cements have high C_3A and low alkali contents. Recent studies have also demonstrated that these cements react poorly with some polycarboxylate superplasticizers (Yamada et al. 2000; Nawa et al. 2001).

To improve the compatibility and robustness of cement/superplasticizer compatibility, it is always possible to delay as much as possible the introduction of a polysulphonate superplasticizer or to sacrifice some of the lignosulphonate water reducer or a part of the polynaphthalene or polymelamine superplasticizer in order to avoid, as much as possible, a direct reaction between the SO_3^- terminations of these superplasticizers and C_3A (Baalbaki 1998).

Of course, those who fear an alkali/aggregate reaction could add an amount of sodium sulphate that will respect the limit of 0.6 per cent Na_2O equivalent that qualifies a cement as being a low alkali cement. Personally, I have always thought that the best way to avoid any alkali/aggregate reaction is to use an accelerated test on concrete samples to avoid the use of alkali reactive aggregates, and when *all* the aggregates available in a particular area are potentially reactive, the best way to avoid any alkali/aggregate reaction is to use a blended cement containing slag or fly ash. Rapidly soluble alkalis activate slag and fly ash concrete and are able to trap most of the alkalis in the initial C-S-H that is formed (Duchesne and Bérubé 2001).

Personally, I think that in spite of the very abundant literature on the subject, alkali/aggregate reaction is not the essential cause of concrete deterioration, rheological problems occurring when concrete is delivered on the site constitute the *major* source of lack of durability of concrete in the year 2000 rather than alkali/aggregate reaction or delayed ettringite formation that are a marginal cause of lack of durability of some concretes.

Table 13.1 presents the characteristics of the ideal clinker.

13.3 The ideal Portland cement

The ideal Portland cement is a cement that does not need to be particularly fine, because it is much better from a cracking and durability point of view to bring the cement particles closer to each other because it is not necessary to produce enough C-S-H to get the desired strength and durability. For usual applications, I would impose a *maximum Blaine fineness* of 350 m^2/kg.

The same SO_3 maximum content that is presently fixed at 2.3 per cent for ASTM Type V cement that have a C_3A content lower than 5 per cent could be enforced.

As far as the compressive strength of standard mortar cubes, rather than fixing minimum strength requirements, I would instead fix maximum strength requirements. This requirement is somewhat redundant, because I have already fixed a maximum on C_3A, C_3S, and fineness values.

Moreover, I would add an essential requirement that is presently lacking in

Table 13.1 The characteristics of the clinker and Type V Portland cement of my dreams

		Specifications	
		Ideal clinker	*Type V cement*
C_3A content		< 5%	< 5%
C_3S + C_3A content		60 à 65 %	–
SO_3 content		< 2.3 %	< 2.3 %
Soluble alkalis (Na_2O équiv)		> 0.4 %	–
MgO content	%	< 6.0	< 6.0
L.O.I.	%	< 0.75	< 0.75
Insoluble residues	%	< 3.0	< 3.0
(C_4AF + 2 C_3A)	%	< 25	< 25
Fineness	m^2/kg	300 < Blaine < 350	> 280
Air content	%	< 12	< 12
Initial flow	%*	> 100 %	–
Loss of workability at 1h 30	%*	< 20 %	–
Setting time (Vicat)	initial	–	> 45 min
	final	–	> 75 min

* Measured using the mini-cone slump test on a 0.42 W/C ratio paste.

all cement standards, it is a test that could ensure that initial cement rheology is not altered during the first hour and a half that follow W/C mixing.

Present initial and final setting times are inappropriate tests to control the initial rheology of a cement during which it must be controlled. Initial and final setting time are, indirectly, tests that control the rate of hardening of a particular cement, not its rheology. When cements are setting, they have already been placed in the forms for 2 to 5 hours.

13.4 Perverse effects of C_3A

13.4.1 Perverse effects of C_3A on the rheology of concrete

It is well known that C_3A is the most reactive phase of a Portland cement and that the addition of some form of calcium sulphate is the simplest and most economical way to neutralize early hydration of C_3A and of the interstitial phase.

In the absence of calcium sulphate, C_3A hydrates nearly instantaneously to form hydrogarnets. On the contrary, when sulphate ions are available, the C_3A is transformed into ettringite which slows down considerably the further hydration of the C_3A at the surface of the ground clinker grains. It is only when portlandite starts to precipitate at the end of the dormant period that C_3A hydration starts to form more ettringite again. When the calcium sulphate added to the cement is exhausted, SO_4^{2-} ions are becoming quite rare, ettringite becomes a source of calcium sulphate and is transformed into monosulphoaluminate. When there are no more SO_4^{2-} ions provided by ettringite, the remaining C_3A hydrates as hydrogarnet, but when these two minerals are formed, it is a long time since the concrete has been placed in the forms and it has already hardened. In principle, in a concrete that has just finished hardening (one to three days) some monosulphoaluminate, some ettringite and some hydrogarnets can be found. Usually, microscopic observation shows us only ettringite because the photos of an hydrated cement paste look better when they contain beautiful needles of ettringite.

When a Portland cement has not been well sulphated, it can present a flash set phenomenon (undersulphated) because C_3A is hydrating instantaneously as hydrogarnet, or a false set phenomenon; during grinding too much gypsum has been transformed into hemihydrate and the conversion of this hemihydrate into gypsum causes a temporary loss of initial workability. In such a case, an increase of the mixing time restores the desired workability, but in the case of a flash set, it worsens the situation.

13.4.2 Perverse effects of C_3A on the compatibility and robustness of polysulphonate-based superplasticizers and water reducers

During its grinding, numerous electrical charges appear on the fractured surface of cement particles, essentially negative charges on C_3S and C_2S crystals and positive charges on the surface of C_3A and C_4AF crystals. As a result of these numerous electrical charges present at their surface, cement particles flocculate very rapidly when they come into contact with a liquid as polar as water. Consequently, *all cements* need to be dispersed in order to improve their performance, otherwise cement particles will flocculate and Portland cement will lose some of its binding potential. In order to avoid cement particle flocculation, it is only necessary to neutralize the superficial electrical charges with organic molecules known as water-reducers or superplasticizers. For fifty years, lignosulphonate water-reducers have been used very successfully to deflocculate cement particles.

Carboxylates have also been used for the same purpose, but to a lesser extent because some are slightly more expensive and because they delay somewhat further the achievement of early strength.

Polysulphonate and polycarboxylate molecules are presently extensively used to lower the W/C or W/B ratio in order to produce high-performance concrete. Polysulphonates and polycarboxylates are very efficient dispersing admixtures, but as presently the cost/performance ratio of polycarboxylate is usually higher than that of polynaphthalene sulphonate, polysulphonates are the most widely used superplasticizers in ready-mixed concretes. It is only with cement having a high C_3A and C_3S content and a low soluble alkali content that in some precast applications, where air entrainment is not a prerequisite, some polycarboxylates are cost-effective when compared to polynaphthalene superplasticizers.

When cement particles are not allowed to flocculate, not only are cement particles well dispersed within the mass of concrete but all the water trapped within the cement flocs acts as a lubricant, therefore, improving concrete workability. The same workability can be obtained with less initial mixing water or a higher slump can be obtained with a lower amount of mixing water.

Water is no longer the only means of controlling concrete rheology, polysulphonates and polycarboxylates offer much more flexibility and possibilities.

Of course Portland cement hydration reaction is altered by the presence of these organic molecules that are coating cement particles, and to a certain extent, can act as a barrier to water molecules. It becomes a matter of equilibrium between the ionic species that are found in the interstitial solution of the fresh concrete.

In some cases, it has been shown that the SO_3^- terminations of polysulphonate superplasticizers react with the C_3A to form an organomineral component resembling ettringite, but that does not crystallize into beautiful ettringite needles but rather like an 'amorphous' material. When too many polysulphonate molecules react in that way with the C_3A, a more or less rapid slump loss is observed depending on the number of superplasticizer molecules that are lost.

13.4.3 The perverse effects of C_3A on the durability of concrete

Each time damaged Portland cement concretes are observed under an electron microscope, it is always surprising to see large amounts of well-crystallized ettringite needles. These ettringite crystals can be found in cracks, in the transition zone, or even, in air bubbles. It is well known, but not well understood (Mehta 1973), that when ettringite is formed in the presence of high concentrated lime, which is the case of Portland cement concrete, it crystallizes in an unstable and expansive form. On the contrary, when it forms in the presence of a reduced amount of lime, like in the supersulphated cements it crystallizes in a very stable form and not in an expansive form (Gebauer et al. 2005).

The observation of a fractured surface of a specimen of high-performance concrete (W/B ratio equal to 0.35, spacing factor of 180 μm) has shown that when this specimen failed after 1960 cycles of freezing and thawing according to Procedure A of ASTM C666 Standard (Freezing and Thawing in Water) it was full of ettringite. The inside of some air bubble had been invaded by beautiful ettringite needles. At least, this is proof that there is a movement of the interstitial solution towards air bubbles during freezing and thawing cycles.

When certain cracked concretes that still are in relatively good shape are also observed under an electron microscope, ettringite crystals are often found in these cracks. This ettringite is termed secondary ettringite because it crystallized after the formation of the crack.

The durability of concrete to sea water atttack can also be linked to the C_3A content of the cement.

Finally, the not-very-well-understood phenomenon called delayed ettringite formation (DEF) is also caused by the presence of C_3A in Portland cement.

Therefore, the less C_3A a clinker contains, the better it is.

13.5 Making concrete with an ASTM Type V cement

The range of concretes presently used by the concrete industry has widened significantly during the past years following the use of very specific admixtures that have become more and more efficient in their effects. High-performance concretes, self-compacting concretes, high performance roller-compacted concretes, fibre-reinforced concrete, RPCs and underwater concretes have been developed and are increasingly being used. Is it possible to make all these 'smart' concretes with a Portland cement and hydraulic binders made with a Type V Portland cement clinker.

Of course, Type V Portland cement applications, that in the USA represent 3 per cent of the cement market, are satisfied and a maximum Blaine fineness of 300 m^2/kg could be sufficient for such a cement.

If a Type V clinker is ground to a fineness of 350 m^2/kg, it can be used for most applications presently using a Type I or II cement. The concrete will be somewhat darker and slightly less resistant at 24 hours, but it will be less prone to cracking due to plastic shrinkage, autogenous shrinkage, and drying shrinkage if appropriately water-cured. In hot countries, such a clinker is ideal since it would be very easy to control its rheology because of its controlled 'reactivity'.

Concrete users who find that the compressive strength of these concretes made with clinkers having a C_3S and C_3A content lower than those of the cements they are used to using would only have to lower slightly their W/C or W/B ratio or to increase slightly the initial temperature of their concrete. As this cement has a low C_3A and Na_2O equivalent contents, it is compatible with any *good quality* polysulphonate superplasticizer, and, moreover, it requires a low amount of superplasticizer to decrease significantly the amount of mixing water and increase significantly the initial compressive strength. Polymelamine superplasticizers usually increase initial compressive strength slightly more than polysulphonates and polycarboxylates.

As far as self compacting concrete are concerned the use of a Type V clinker is very advantageous because its lower 'reactivity' and not very high Blaine fineness mean it requires less superplasticizer to reach a flowing state.

As far as reactive powder concretes are concerned, the different tests conducted by the US Army Corps of Engineers and the Université de Sherbrooke have shown that Type V cements were the ideal and most economical cements to produce such concretes. The Sherbrooke passerelle was built with a special cement used by Hydro-Québec to build its dams that has a maximum C_3A content lower than 3 per cent.

The presence of rapidly soluble alkali sulphate in the ideal clinker its also very advantageous because it limits the early consumption of superplasticizer molecules.

It is only in cold weather that a Type V cement could present too low an early strength, but for winter concreting it is always possible to increase its fineness, to increase the initial temperature of the concrete and/or to lower slightly the W/C or W/B ratio using a polymelamine superplasticizer and to use insulated forms.

In all these cases, a Type V clinker presents the great advantage of making a concrete whose initial rheology can be kept under control during the entire time necessary to place the concrete. It is no longer necessary to add water in the field to restore an adequate slump to place the concrete easily in the forms.

What a giant step in the direction of concrete durability!

For mass concrete, a Type V concrete is ideal, especially if it is blended with a slag, a fly ash or a pozzolan. The blending can be adjusted to the desired maximum temperature or twenty-eight-day compressive strength.

Moreover, such a cement that has enough rapidly soluble alkalis will be compatible with almost all Portland and blended cements and very economical with polynaphthalene superplasticizers.

Presently, as mineral components are available almost everywhere, *it will no longer be necessary to make a special low alkali cement,* because if the only aggregates available in a given area are potentially reactive, the best way to get rid of an alkali/aggregate reaction is to use a blended cement. *Therefore, Type V clinker could be used with potentially reactive aggregates if it is blended with a mineral component.*

It is only in architectural applications, where a white clinker is a must, that a second type of clinker will be needed that should look like that used by Aalborg cement in Denmark to manufacture its white cement with a low C_3A content.

13.6 Conclusion

The universal ideal clinker can be used, with some slight adjustments, to make any kind of concrete, if taking advantage of the numerous possibilities offered by mineral components and admixtures.

The universal ideal clinker facilitates the control of the rheology of concrete very easily and gives contractors enough time to place concrete very easily, in any case, much more easily than the present C_3S and C_3A rich clinkers. Adding water in the field would no longer be necessary, so that in general, concrete durability will be increased and much less Portland cement will be wasted.

Finally, the ideal clinker is very efficient and economical from an admixture point of view.

References

Baalbaki, M. (1998) 'Influence of the Interactions of Cement and Superplasticizer on the Properties of Concrete: Influence of the Introduction Mode', Ph.D. thesis, Université de Sherbrooke.

Duchesne, J. and Bérubé, M.-A. (2001) 'Long-Term Effectiveness of Supplementary Cementing Materials Against Alkali Reaction', *Cement and Concrete Research*, 31 (7): 1057–63.

Gebauer, J., Ko, S.-C., Lerat, A. and Roumain, J.-C. (2005) 'Experience with a New Cement for Special Applications', Second International Symposium on Non-Traditional Cement and Concrete, pp. 277–83.

Kim, B.-G., Jiang, S.P. and Aïtcin, P.-C. (2000a) 'Effect of Sodium Sulphate Addition on the Properties of Cement Pastes Containing Different Molecular Weight PNS Superplasticizer', *Sixth CANMET/ACI International Conference on Superplasticizers and Other Chemical Admixtures in Concrete*, Nice, October, pp. 485–504.

—— (2000b) 'Slump Improvement Mechanism of Alkalis in PNS Superplasticized Cement Pastes', *Materials and Structures*, 33 (230): 363–9.

Mehta, P.K. (1973) 'Mechanism Associated with Ettringite Formation', *Cement and Concrete Research*, 3 (1): 1–6.

Nawa, T., Uematsu, C. and Ohnuma, H. (2001) 'Influence of the Type of Cement on the Dispersing Action of Polycarboxylate-Based Superplasticizer', CONSEC 01, Vancouver, Vol. II, pp. 1441–8.

Yamada, K., Ogawa, S. and Hanehara, S. (2000) Working mechanism of poly-beta-naphthalene sulphonate and polycarboxylate superplasticizer types from the point of cement characteristics, ACI SP-195, pp. 367–382.

Appendix I

How Vicat prepared his artificial lime (Mary 1862)

Free translation by the author

The operation that will be described here is an actual synthesis that binds intimately, through the action of fire, the essential principles that an analysis separates in the case of hydraulic limes. This operation consists, in transforming the lime that is to be modified into a fine powder in a dry shelter, to knead it with some water and a certain proportion of grey or brown clay, or simply with a clay used to make bricks and to shape the obtained paste as bowls that are left to dry before being fired at a suitable temperature.

Of course, when mastering the proportions of lime and clay it is possible to give the desired degree of energy to the artificial lime and to equal or beat the best natural limes.

The best limes can contain 0.2 part of clay for 1.00 of lime; less performing ones can contain 0.15, 0.10 parts, even 0.06 is sufficient for those that already have some hydraulic properties.

When the clay proportion goes up to 0.33 to 0.40 the lime produced does not burn out, but after firing it is easily pulverized and give, when mixed with water, a paste that harden very rapidly under water.

When a clayey soil contains some gravel or limestone debris it must be cleaned by washing it in a large water basin; the clayey soil is agitated with a scraper: the clay slurry that is formed is diverted in a second basin where it is mixed with an appropriate amount of lime to make bowls. With some experience it is easy to find the best proportion of the mix. In spite of the fact that the material is soaked, the handling is done in a much easier way than with any other mixing process.

It is not true that a clay burned separately and thereafter mixed with common lime, in the previous proportions, gives the same results than when the clay and the lime are mixed before firing. Firing modifies both constituents of the mixture and gives rise to a new compound having new properties. This fact becomes palpable when comparing, for example, the colour of an artificial lime (made from a ferrugious clay) that borders on yellowish pale green, to that of a common lime that is ground with a some red cement. Moreover, there is a great difference in the manner in which these compounds behave under water.

In order to clear all doubts about transforming (on a large scale) common lime into hydraulic lime, an approximate breakdown of what it would cost to produce one cubic metre of such lime in Souillac, the place where I lived, is presented in the following.

A square common wood kiln can contain 59 m^3 of materials; 17 m^3 of natural limestone can be used to build the vault on which 42 m^3 of bowls or prisms of artificial lime can be placed, the mix proportion is 1.00 part of slacked lime for 0.20 part of clay that is measured on a dry powder basis . . .

Appendix II

AD 1824 . . . . .No. 5022

Artificial stone: Aspdin's specification

TO ALL TO WHOM THESE PRESENTS SHALL COME, I, Joseph Aspdin, of Leeds, in the county of York, Bricklayer, send greeting.

WHEREAS His present most Excellent majesty King George the Fourth, by His Letters Patent under the Great Seal of Great Britain, bearing date at Westminster, the twenty-first day of October, in the 6th year of His reign, did, for Himself, His heirs and successors, give and grant unto me, the said Joseph Aspdin, his especial license, that I, the said Joseph Aspdin, my exors, adniors, and assigns, or such others as I, the said Joseph Aspdin, my exors, adniors, and assigns, should at any time agree with, and no others, from time to time and at all times during the term of years therein expressed, should and lawfully might make, use, exercise, and vend, within England, Wales, and the Town of Berwick-upon-Tweed, my Invention of '**An Improvement in the Modes of Producing an Artificial Stone**;' in which said Letters Patent there is contained a proviso obliging me, the said Joseph Aspdin, by an instrument in writing under my hand and seal, particularly to describe and ascertain the nature of my said Invention, and in what manner the same is to be performed, and to cause the same to be inrolled in His Majesty's High Court of Chancery within two calendar months next and immediately after the date of the said in part recited Letters Patent (as in and by the same), reference being thereunto had, will more fully and at large appear.

NOW KNOW YE, that in compliance with the said proviso, I, the said Joseph Aspdin, do hereby declare the nature of my said Invention, and the manner in which the same is to be performed, are particularly described and ascertained in the following description thereof (that is to say): –

My method of making a cement or artificial stone for stuccoing buildings, waterworks, cisterns, or any other purpose to which it may be applicable (and which I call Portland cement) is as follows: – **I take a specific quantity of limestone***, such as that generally used for making or repairing roads, and I take it from the roads after it is reduced to a puddle or powder; but if I cannot

* Aspdin does not indicate the proportions of the various ingredients used nor the temperature at which the mix must be calcined.

procure a sufficient quantity of the above from the roads, I obtain the limestone itself, and I cause the puddle or powder, or the limestone, as the case may be, to be calcined. **I then take a specific quantity of argillacious earth or clay***, and mix them with water to a state approaching impalpability, either by manuel labour or machinery. After this proceeding I put the above mixture into a slip pan for evaporation, either by the heat of the sun or by submitting it to the action of fire or steam conveyed in flues or pipes under or near the pan till the water is entirely evaporated. Then I break the said mixture into suitable lumps, and calcine* them in a furnace similar to a lime kiln till the carbonic acid is entirely expelled. The mixture so calcined is to be ground, beat, or rolled to a fine powder, and is then in a fit state for making **cement or artificial stone**. This powder is to be mixed with a sufficient quantity of water to bring it into the consistency of mortar, and thus applied to the purposes wanted.

In witness whereof, I, the said Joseph Aspdin, have hereunto set my hand and seal, this Fifteenth day of December, in the year of our Lord One thousand eight hundred and twenty-four.

JOSEPH (l.s..) ASPDIN

AND BE IT REMEMBERED, that on the Fifteenth day of December, in the year of our Lord 1824, the aforesaid Joseph Aspdin came before our said Lord the King in His Chancery, and acknowledged the Specification afore-said, and all and every thing therein contained and specified, in form above written. And also the specification aforesaid was stamped according to the tenor of the Statute made for that purpose.

Inrolled the Eighteenth day of December, in the year of our Lord One thousand eight hundred and twenty-four.

LONDON : Printed by George Edward Eyre and William Spottiswoode, Printers to the Queen's most Excellent Majesty. 1857.

Appendix III

And if the first North American natural cement was made in 1676 in Montreal

In 1991 the research group on concrete of the Université de Sherbrooke was asked by the Fine Art Division of the City of Montreal to develop a mortar similar to the one used in 1676 to build the foundations of the first fortification wall of Montreal. These foundations have been unearthed and can be seen in the Pointe-à-Caillère Museum in the old Montreal quarter. In order to restore these foundations the architects did not want to use a modern mortar that would have been aesthetically ugly beside of the old mortars that were still in pretty good shape in some part of this foundation.

Samples of the original mortar were observed through optical and electron microscopy to identify its composition. The mortar was made with a natural sand, most probably taken from the riverside of the St Lawrence river. It was a sand rich in dark-coloured particles that are no longer found in the fluvioglacial sands that are presently used to make concrete and mortar in Montreal area.

Surprisingly the binder used was what we would call presently a natural cement, some of its C_2S was not yet fully hydrated when it was observed.

As this mortar had been for centuries in contact with a dark beige clayey soil it was necessary to incorporate yellow and brown pigments in the reconstituted mortar. A black sand was introduced in the mix to contamine the modern mortar sand available. Different beige mortars were prepared and submitted to the architect for his final selection and finally 2000 mortar sacks of 20 kg having different colours were prepared.

At a first glance it was somewhat surprising to find that the mortar used to join the large limestone blocks constituting the foundation of the first fortification wall of Montreal was an hydraulic lime or even a natural cement. However, after reflection this is not so surprising because most of the limestone found on Montreal Island has in fact almost the chemical composition of Portland cement. Although this kind or limestone needed a quite high temperature to be burned this was not a problem because wood was abundant.

For many years two cement plants operated on the island of Montreal. These plants were only adding 5 per cent of corrective materials to produce Portland cement. The last operating cement plant built on the side of Metropolitan highway, the main East–West through city highway, closed its operations in 1987.

Appendix IV

The SAL and KOSH treatments

The SAL method (salicylic acid/methanol treatment)

(Hjorth-Leuven (1971) *Cement and Concrete Research*, 1 (1): 27–40)

The SAL treatment dissolves the silicate phases C_3S and C_2S so that it is easy to identify the minerals present in the interstitial solution: sulphates, calcite, MgO, and other minor constituents.

1 Dissolve 30 g of salicylic acid in 150 ml of methanol.
2 Add 5 g of Portland cement in the solution.
3 Agitate for 2 hours.
4 Filter.
5 Wash the residue with methanol.
6 Dry the residue at 90°C.
7 Grind the residue to form a very fine powder.

Kosh treatment (KOSH [Potassium hydroxide/ sucros])

(Gutteridge (1979) *Cement and Concrete Research*, 9 (3))

This treatment dissolves the interstitial phase.

1 Dissolve 30 g of KOH and 30 g of sucrose in 300 ml of distilled water.
2 Heat the mixture up to 95°C.
3 Add 9 g of cement.
4 Agitate for 1 minute.
5 Filter (filter paper No 4).
6 Wash the residue with 50 ml of distilled water.
7 Wash the residue with 100 ml of methanol.
8 Dry at 60°C.

Appendix V

Determination of the Bogue potential composition

1. The amount of free lime (CaO) is subtracted from the total amount of lime to get the actual amount of combined lime.
2. It is assumed that all the SO_3 present in the Portland cement is combined exclusively with lime under the form of calcium sulphate $CaSO_4$. The amount of CaO that is combined in the calcium sulphate is calculated and deducted from the total amount of combined CaO.
3. It is assumed that all the Fe_2O_3 is combined into C_4AF. The amount of lime and alumina combined with Fe_2O_3 to form C_4AF is subtracted from the remaining amount of lime and the initial amounts of Al_2O_3 found by the chemical analysis.
4. It is assumed that all the remaining Al_2O_3 is combined into C_3A. The amount of lime combined with alumina to form C_3A is subtracted from the remaining amount of lime.
5. Finally the lime and silica are split between C_3S and C_2S.

N.B. In reviewing all the hypotheses made during this calculation, it is found that it is very optimistic to give the Bogue potential composition with a precision greater than 1 per cent.

It is better to write that the Bogue potential composition of a particular Portland cement is: 50 per cent of C_3S, 24 per cent of C_2S, 7 per cent of C_3A, and 9 per cent of C_4AF, rather than 49.8 per cent of C_3S, 24.3 per cent of C_2S, 7.25 per cent of C_3A and 8.85 per cent of C_4AF.

Example: a Portland cement has the following chemical composition:

Silica (SiO_2)	20.5%	Magnesium oxide (MgO)	2.38%
Alumina (Al_2O_3)	4.6%	Sulphuric anhydride (SO_3)	3.1%
Iron oxide (Fe_2O_3)	2.9%	Loss on ignition (LOI)	1.69%
Total lime (CaO)	63.2%	Insoluble residue	0.64%
Free lime (CaO)	0.9%		

What is the potential Bogue composition of this Portland cement?

Calculations

1. The amount of free lime is subtracted from the total lime content to find the amount of lime that is actually combined

$$63.2 - 0.9 = 62.3\% \text{ of CaO}$$

This lime is combined into calcium sulphate, the two silicate phases, C_3A and C_4AF.

2. Calcium sulphate content (Hypothesis 2)

It is assumed that the calcium sulphate present in the Portland is in the form of gypsum $CaSO_4 \cdot 2\,H_2O$

The chemical formula for gypsum is

$$\underbrace{Ca\,SO_4 \cdot 2\,H_2O}_{172} = \underbrace{CaO}_{56} + \underbrace{SO_3}_{80} + \underbrace{H_2O}_{36}$$

In gypsum, 80 g of SO_3 are combined to 56 g of lime to give 172 g of gypsum, so as we have 3.1 per cent of SO_3 in the cement, the gypsum content of the Portland cement is $\frac{172 \times 3.1}{80} = 6.7\%$

The amount of lime that is combined to the 3.1 per cent of SO_3 is $\frac{56}{80} \times 3.1 =$ 2.2%

The amount of lime that remains in the minerals composing Portland cement clinker is $62.3 - 2.2 = 60.1\%$

This amount of lime will be split between C_3S, C_2S, C_3A, and C_4AF.

The oxides that are combined in C_3S, C_2S, Al2O3, and Fe_2O_3 are 60.1 per cent of CaO, 20.5 per cent of SiO_2, 4.6 per cent of Al_2O_3, and 2.9 per cent of Fe_2O_3.

3. C_4AF content

C_4Af is the only phase that contains the 2.9 per cent of Fe_2O_3

$$\underbrace{C_4}_{224\,g} + \underbrace{A}_{102\,g} + \underbrace{F}_{160\,g} = \underbrace{C_4AF}_{486\,g}$$

In C_4AF 160 g of Fe_2O_3 are combined in 486 g of C_4AF

$$C_4 \cdot A \cdot F \text{ content is therefore } \frac{486}{160} \times 2.9 = 8.7\%$$

$$\text{The amount of alumina combined in } C_4AF \text{ is } \frac{102}{160} \times 2.9 = 1.8\% \text{ of } Al_2O_3$$

$$\text{The amount of lime combined in } C_4AF \text{ is } \frac{224}{160} \times 2.9 = 4.0\% \text{ of CaO}$$

At this stage of the calculation the remaining oxide contents are

Lime $60.1 - 4.0 = 56.1\%$ of CaO
Alumina $4.6 - 1.8 = 2.8\%$ of Al_2O_3

And still 20.5% of SiO_2

4. C_3A content

The 2.8 per cent of alumina are combined in C_3A

$$\underbrace{C_3}_{163\text{ g}} + \underbrace{A}_{102\text{ g}} = \underbrace{C_2A}_{270\text{ g}}$$

102 g of alumina are combined to give 270 g of C_3A. The amount of C_3A is therefore

$$\frac{270}{102} \times 2.8 = 7.4\% \text{ of } C_3A.$$

$$\text{The amount of lime that is combined into } C_3A \text{ is } \frac{168}{102} \times 2.8 = 4.6\% \text{ of CaO.}$$

Therefore the amount of lime that is not yet combined is

$56.1 - 4.6 = 51.5\%$

The amount of uncombined SiO_2 is still 20.5 per cent.

5. C_2S and C_3S content

$$\underbrace{C_2}_{102\text{ g}} + \underbrace{S}_{60\text{ g}} = \underbrace{C_2S}_{172\text{ g}} \quad \text{and} \quad \underbrace{C_3}_{168\text{ g}} + \underbrace{S}_{60\text{ g}} = \underbrace{C_3S}_{228\text{ g}}$$

We will assume that the S content of C_2S is x per cent and that the S content of C_3S is y per cent.

Each gram of S is combined with $\frac{112}{60} = 1.87$ g of C in C_2S

Each gram of S is combined with $\frac{168}{60} = 2.8$ g of C in C_3S

Therefore $x + y = 20.5$ (S)
$1.87x + 28y = 51.5$ (CaO)

solving the system gives

$x = 6.36\%$ and $y = 14.16\%$

The amounts of C_2S and C_3S are calculated as follows

$$C_2S = \frac{172}{60} \times 6.36 = 18.2\%$$

$$C_3S = \frac{228}{60} \times 4.16 = 53.9\%$$

6. *Final result*

When the phase contents are rounded out to the closest 1 per cent, the potential Bogue composition is

C_3S	54%	
C_2S	18%	
C_3A	7%	16%
C_4AF	9%	
Gypsum	7%	

Instead of repeating all these calculations each time, the following formulas give the Bogue composition directly

$$C_3S = 4.0710\ CaO - 7.6024\ SiO_2 - 1.42397\ Fe_2O_3 - 6.7187\ Al_2O_3$$

$$C_2S = 8.6024\ SiO_2 + 1.0785\ Fe_2O_3 + 5.0683\ Al_2O_3 - 3.0710\ CaO$$

$$\text{or} = 2.8675\ SiO_2 - 0.7544\ C_3S$$

$$C_3A = 2.6504\ Al_2O_3 - 1.6920\ Fe_2O_3$$

$$C_4AF = 3.0432\ Fe_2O_3$$

Appendix VI

Example of a very simple binary diagram

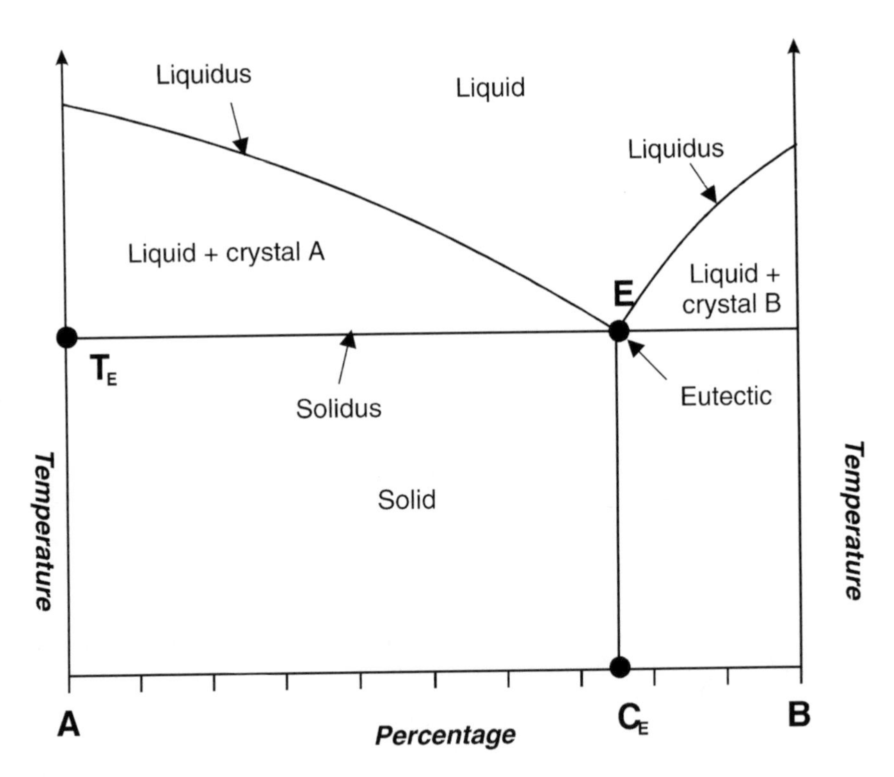

Figure A6.1 Example of a very simple phase diagram (also called fusibility diagram) of two pure materials A and B.

Appendix VII

Ternary diagrams

Trilinear representation (Lievin 1935a and 1935b; Aïtcin 1983)

A. Geometrical background: Rooseboom triangle

In an equilateral triangle the sum of the distance between any point situated inside the triangle is equal to the length of the height of this triangle (Figure A7.1).

If h is assumed to be equal to 100, $h_1 + h_2 + h_3 = 100$ where h_1, h_2 and h_3 represents the amounts of the three components in the mixture. For example h_2 represent the amount of A in the mixture, h_3 the amount of B, and h_1 the amount of C.

It could possible to graduate the 3 perpendicular heights, 10 per cent by 10 per cent and to plot each of the internal points to this system of reference. However, in order to free the centre of the ternary diagram, a system of parallel lines is drawn and it is their intersection with the 3 sides of the triangle that are graduated. Therefore, the sides are graduated in percentage of S, C and A, as shown in Figure A7.2. In this figure, small arrows indicate what parallel to draw to find the content of each component.

B. Representative points of the three basic oxides

The points representing the three basic oxides are, of course, the three summits of the triangle (Figure A7.2).

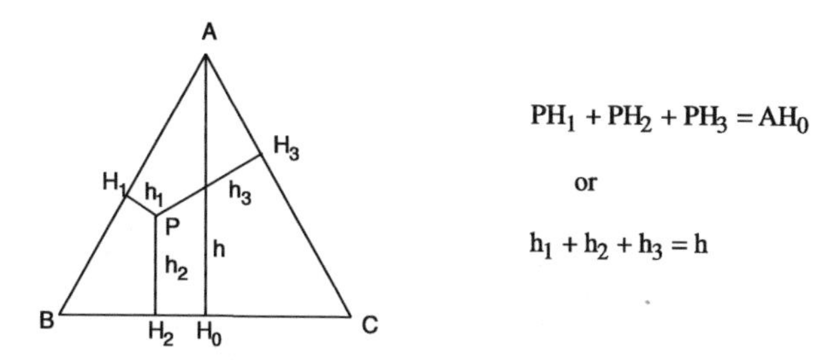

Figure A7.1 Principle of the method of the triangle.

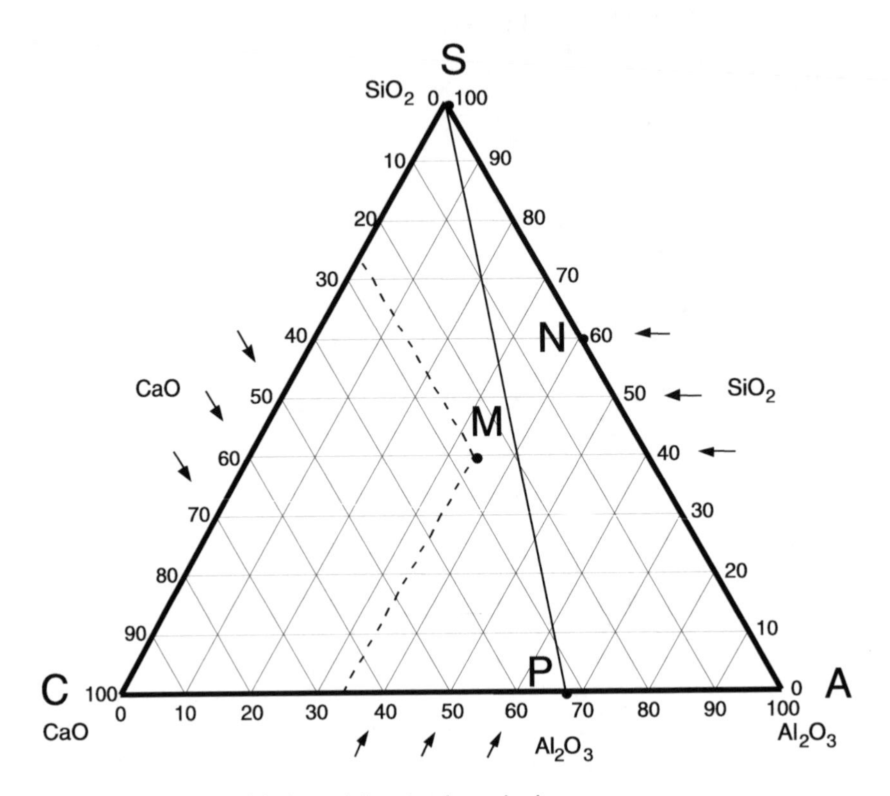

Figure A7.2 Example of the use of the triangle method.

C. Binary mixes

One component is missing in the mix, all the points representing such mixes are situated on the opposite side of the missing oxide. For example, point P in Figure A7.2 represents a mix composed of 2/3 Al_2O_3 and 1/3 CaO.

What is the composition of the mix represented by point N on Figure A7.2? N is located on the SA side of the triangle, therefore it corresponds to a mix that does not contain any CaO. N represent a mix composed of 60 per cent SiO_2 and 40 per cent Al_2O_3.

D. Ternary mixes

A mix is composed of 40 per cent SiO_2, 25 per cent lime, 35 per cent Al_2O_3, where is the point M that represents this mix?

M is obtained, for example, by drawing an horizontal line passing by point 40 per cent on the right side (SiO_2) of the triangle and by a parallel to the left side of the triangle passing by point 35 per cent (Al_2O_3) on the lower horizontal side of the triangle. It can be verified that point M could also be obtained as the intersection of the horizontal line passing by point 40 per cent on the right side (SiO_2) and a parallel the right side (CaO) of the triangle passing by point 25 per cent.

On Figure A7.2, let us draw a line joining the summit S to the previous point P on the CA side. We have already seen that point P corresponds to a mix containing 2/3 Al_2O_3 and 1/3 CaO. Let us look at where this line intersects the horizontal lines corresponding to mixtures containing 10, 40 and 70 per cent SiO_2.

In the case of a mix containing 10 per cent SiO_2, the intersection defined a point containing 60 per cent Al_2O_3 and 30 per cent CaO. The Al_2O_3 content is equal to twice the CaO content. It is seen that it is the same for a mix containing 40 per cent SiO_2, in this case the percentage of Al_2O_3 is 40 per cent and the percentage of CaO is equal to 20 per cent. In the case of the horizontal line passing by the SiO_2 point equal to 70 per cent, the percentage of Al_2O_3 is equal to 20 per cent and the percentage of CaO is 10 per cent. In all these particular cases, the amount of Al_2O_3 in the mix is twice as great as the amount of CaO.

This rule can be generalized: any line joining a summit to a given point on the opposite side represents mixtures having the same proportions of the two other components.

Representative point of a mix containing three components.

Any mix of the three basic components is represented by a point inside the triangle. Let us represent the three following mixes A_1, B_1, C_1 whose composition are given in the following table.

	SiO_2	CaO	Al_2O_3
A_1	8%	39%	53%
B_1	90%	5%	5%
C_1	40%	50%	10%

Points A_1, B_1, C_1 are represented in Figure A7.3 where they define the triangle $A_1B_1C_1$. It is possible to demonstrate that any point P located inside the $A_1B_1C_1$ triangle can be obtained by mixing A_1, B_1, C_1 in a particular proportion and reciprocally that all the A_1, B_1 and C_1 mixtures are contained in triangle $A_1B_1C_1$. In figure A7.4, we will see, without demonstrating them, three formulas that can be used to calculate the proportions of A_1, B_1 and C_1 that should be mixed to obtain a mix corresponding to point P.

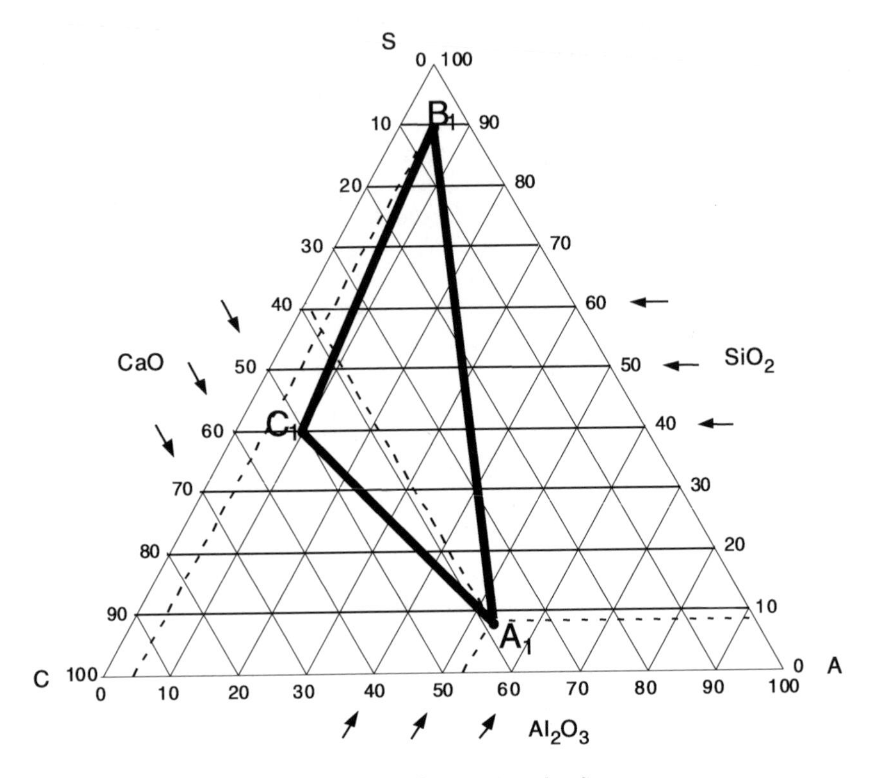

Figure A7.3 Representation of A_1, B_1 and C_1 in a triangular diagram.

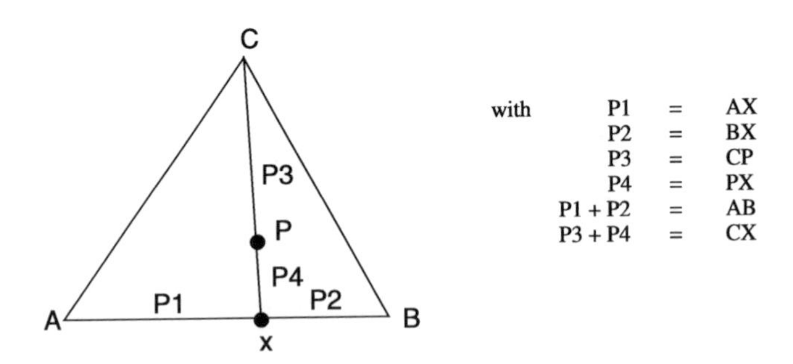

Figure A7.4 Determination of the proportion of a combined material when using the triangle method (general formulas).

$$X_A = \frac{P_2}{P_1 + P_2} \times \frac{P_3}{P_3 + P_4} \qquad X_A = \frac{1}{1 + \frac{P_1}{P_2}} \times \frac{1}{1 + \frac{P_3}{P_4}}$$

$$X_B = \frac{P_1}{P_1 + P_2} \times \frac{P_3}{P_3 + P_4} \qquad X_B = \frac{1}{1 + \frac{P_2}{P_2}} \times \frac{1}{1 + \frac{P_3}{P_4}}$$

$$X_C = \frac{P_4}{P_3 + P_4} \qquad X_C = \frac{1}{1 + \frac{P_3}{P_4}}$$

X_A, X_B X_C represent the decimal fraction of A, B and C that have to be mixed to obtain a mix represented by P. It is easy to verify that $X_A + X_B + X_C = 1$.

In Figure A7.4, the formulas do not depend on the scale of the figure

In what proportions do the previous points A_1, B_1 and C_1 have to be mixed to obtain a point P containing 60 per cent SiO_2, 20 per cent Al_2O_3 and 20 per cent CaO?

First, it is essential to check that point P is within the $A_1B_1C_1$ triangle. As shown in Figure A7.5, P is within the triangle, therefore it can be obtained by mixing A_1, B_1 and C_1. The calculations are shown in Figure A7.5.

E. Case of three aligned points

Figure A7.6 shows how to calculate the proportions of A and C that must be mixed to obtain a mixture B within the segment AC. Of course the closer B is from C, the higher its content in C, and the closer to point A, the higher its content in A. If B is just in the middle of AC, it can be obtained by mixing in equal proportions A and C. More generally, the so-called proportion rule is applied to calculate the amounts of A and C that must be mixed to obtain B. The formulas are presented in Figure A7.6.

References

Aïtcin, P.-C. (1983) *Technologie des granulats*, Quebec: Éditions du Griffon d'Argile.

Liévin, A. (1935a) 'Propriétés du diagramme ternaire appliquées à l'industrie des ciments' Part I, *Revue des Matériaux de Construction et de Travaux Publics*, 309 (June): 137–41.

Liévin, A. (1935b) 'Propriétés du diagramme ternaire appliquées à l'industrie des ciments', *Revue des Matériaux de Construction et de Travaux Publics*, 310 (July): 166–73.

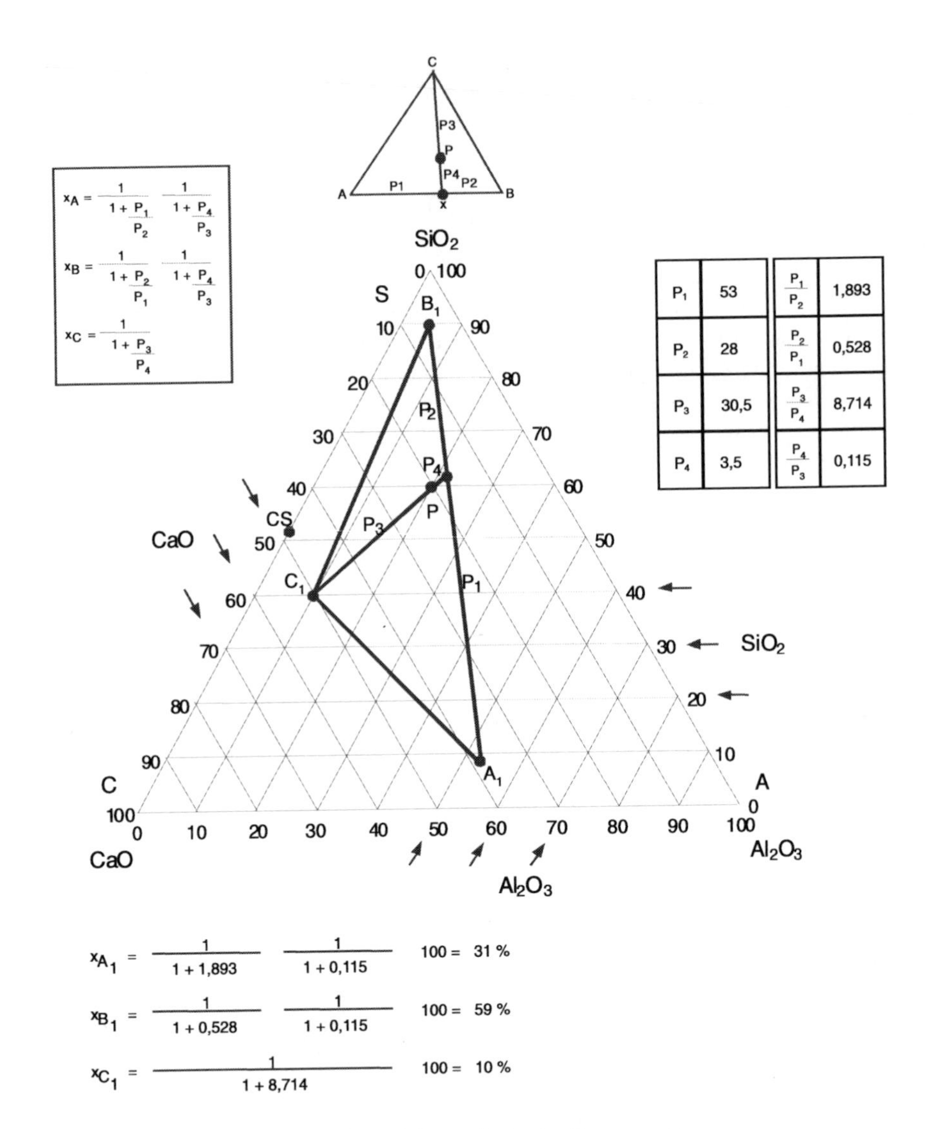

$$x_A = \frac{1}{1+\frac{P_1}{P_2}} \quad \frac{1}{1+\frac{P_4}{P_3}}$$

$$x_B = \frac{1}{1+\frac{P_2}{P_1}} \quad \frac{1}{1+\frac{P_4}{P_3}}$$

$$x_C = \frac{1}{1+\frac{P_3}{P_4}}$$

$$x_{A_1} = \frac{1}{1+1{,}893} \quad \frac{1}{1+0{,}115} \quad 100 = 31\ \%$$

$$x_{B_1} = \frac{1}{1+0{,}528} \quad \frac{1}{1+0{,}115} \quad 100 = 59\ \%$$

$$x_{C_1} = \frac{1}{1+8{,}714} \quad 100 = 10\ \%$$

Figure A7.5 Proportions of A_1, B_1 and C_1 to obtain the P composition.

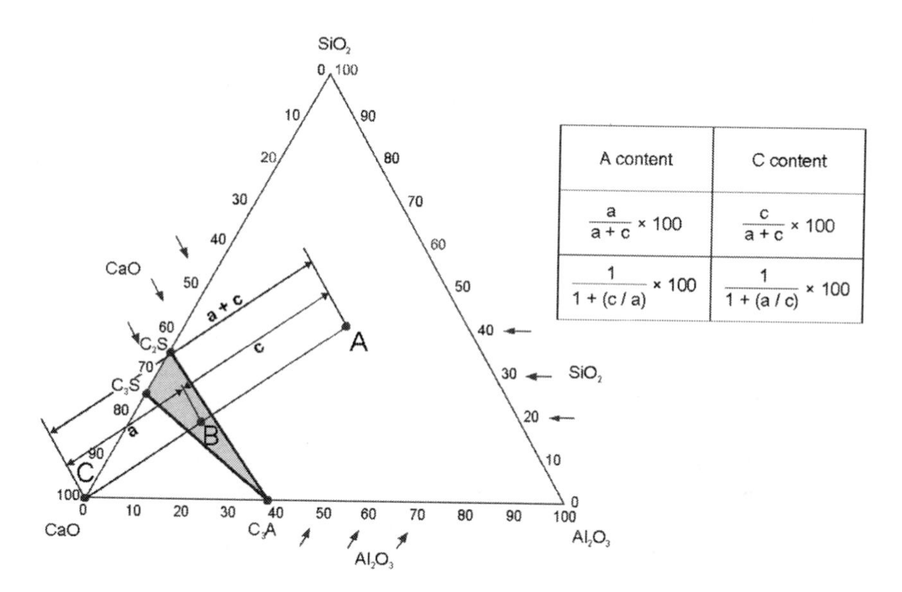

A content	C content
$\frac{a}{a+c} \times 100$	$\frac{c}{a+c} \times 100$
$\frac{1}{1+(c/a)} \times 100$	$\frac{1}{1+(a/c)} \times 100$

Figure A7.6 Proportions of A and C to obtain composition B.

Appendix VIII

Ternary phase diagrams

Binary and ternary phase diagrams are very useful because they can be used to predict the minerals that are formed when heating a mixture of 2 or 3 oxides. Binary and ternary diagrams are used in ceramics, metallurgy, and in the cement industry. In the case of Portland cement, more specifically, the knowledge of $CaO–SiO_2$ and $CaO–SiO_2–Al_2O_3$ phase diagrams is very helpful in understanding the formation of the principal phases found in a clinker. Although phase diagrams are obtained by mixing pure phases that are later submitted to very precise conditions of heating and cooling, they brought a lot of very useful informations, in spite of the fact that the actual conditions in a cement kiln are far from the ones used to establish phase diagrams.

A ternary phase diagram seems very complicated at first glance because there are so many numerical values and lines. But after learning how to look at them, they are not so complicated:

1 The numerical values found inside the diagram indicate the temperature at which the *first crystal* is formed from the melted mix.
2 *Isotherms* link points for which the temperature at which the first crystal appears is the same. This temperature is indicated directly on the isotherm. When isotherms have been obtained experimentally, they are drawn in a full line, when they have been estimated there are drawn as dotted lines.
3 The bold lines (sometimes bold dotted lines) are *boundary lines*. These lines delimit zones where the *primary phase*, that crystallizes first from the melted mix, is the same. In the case of Figure A8.1, the *crystallization zone* of gehlenite has been shaded. Along boundary lines, arrows indicate the variation of the crystallization temperature. Figure A8.2 represents all the crystallization zones found in the ternary phase diagram $CaO–SiO_2–Al_2O_3$.
4 There are also a certain number of straight lines linking the points representing the composition of minerals that can be formed when mixing two or three of the three basic oxides. These lines are called *alkemade lines*, they delimit a triangle called the *compatibility triangle* in which the minerals located at the three summits exist in the solid mix after cooling. In

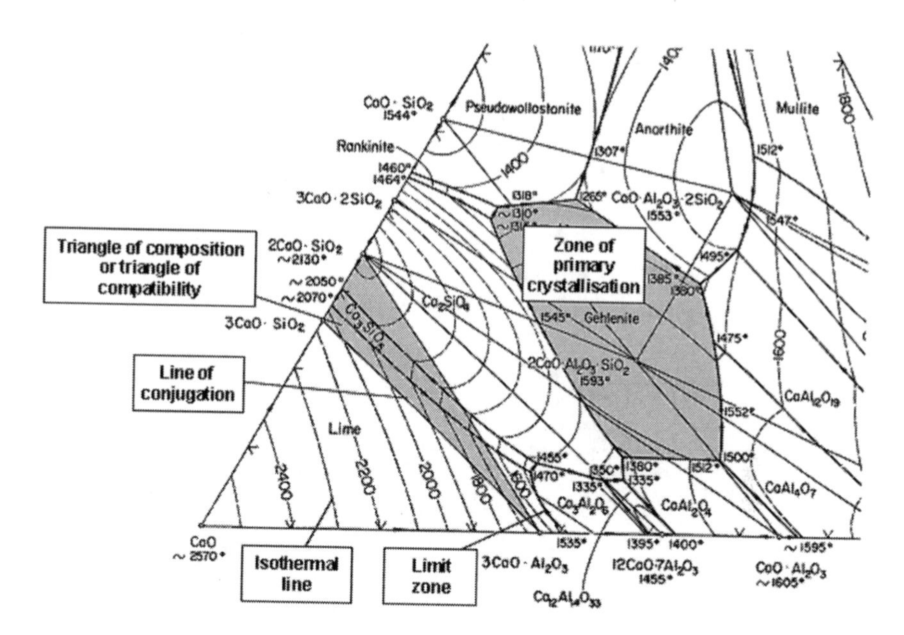

Figure A8.1 Scale up of a zone of the ternary diagram $CaO–SiO_2–Al_2O_3$ where the zone of primary crystallization of gehlenite and the compatibility zone of the bicalcium and tricalcium silciate and the tricalcium aluminate is shown *(adapted from Figure A8.3)*.

Figure A8.1, the $C_3S–C_2S–C_3A$ *compatibility triangle* has been shaded. Any composition represented by a point within this triangle is a mixture of C_3S, C_2S and C_3A and reciprocally any mixture of C_2S, C_3S and C_3A is represented by a point within this *compatibility triangle*; therefore the points representative of all Portland cements must be located in this triangle.

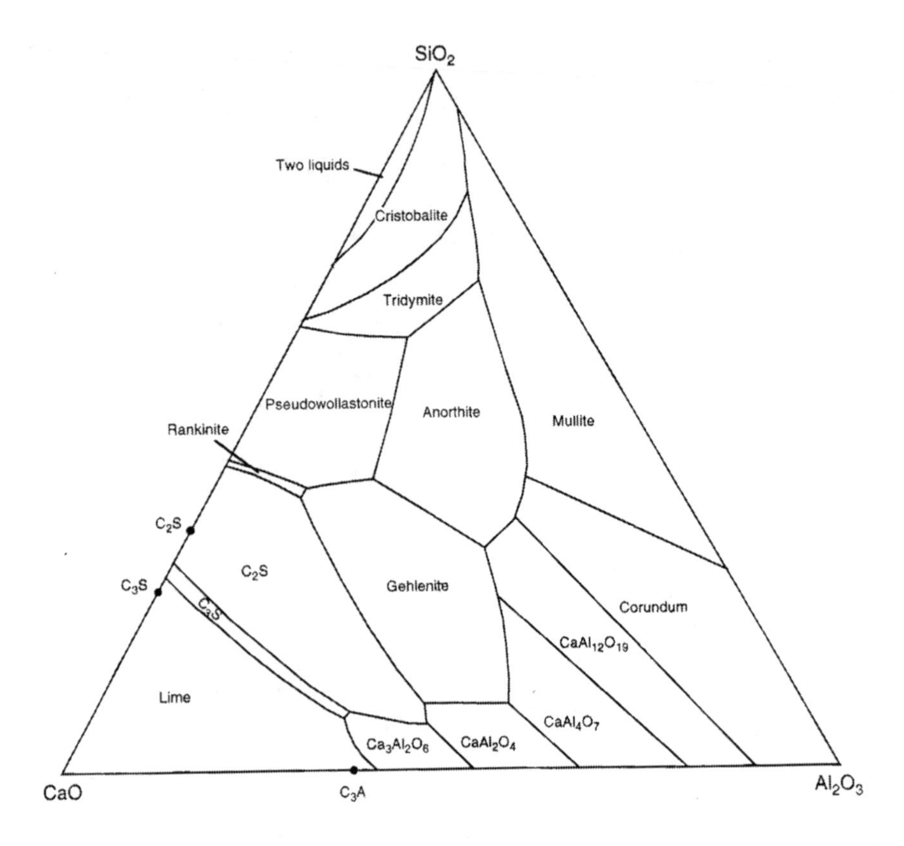

Figure A8.2 Limit lines and crystallization zones in the diagram $CaO–SiO_2–Al_2O_3$ *(adapted from figure A8.3).*

$CaO–Al_2O_3–SiO_2$

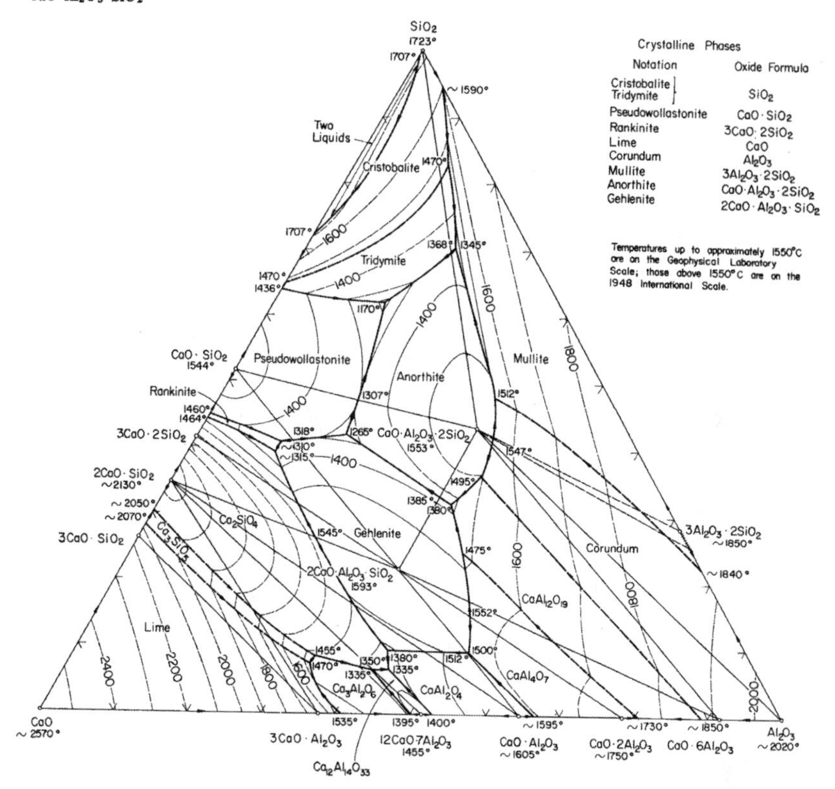

Figure A8.3 $CaO–Al_2O_3–SiO_2$ phase diagram (with the permission of the American Ceramic Society).

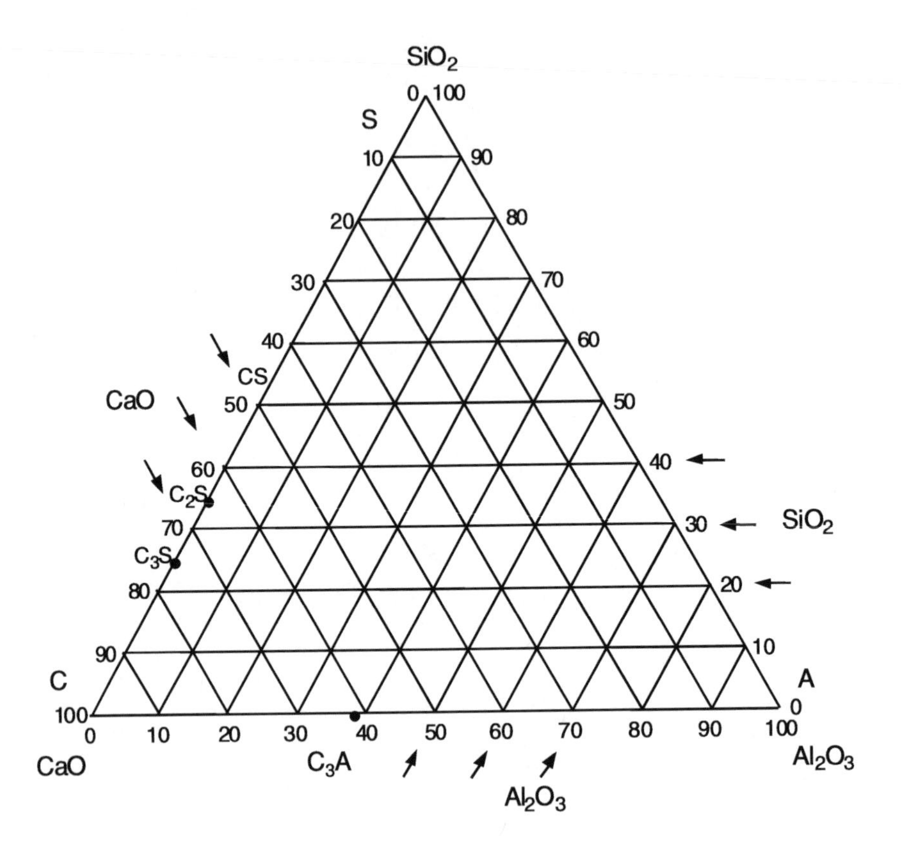

Figure A8.4 Typical exercise sheet.

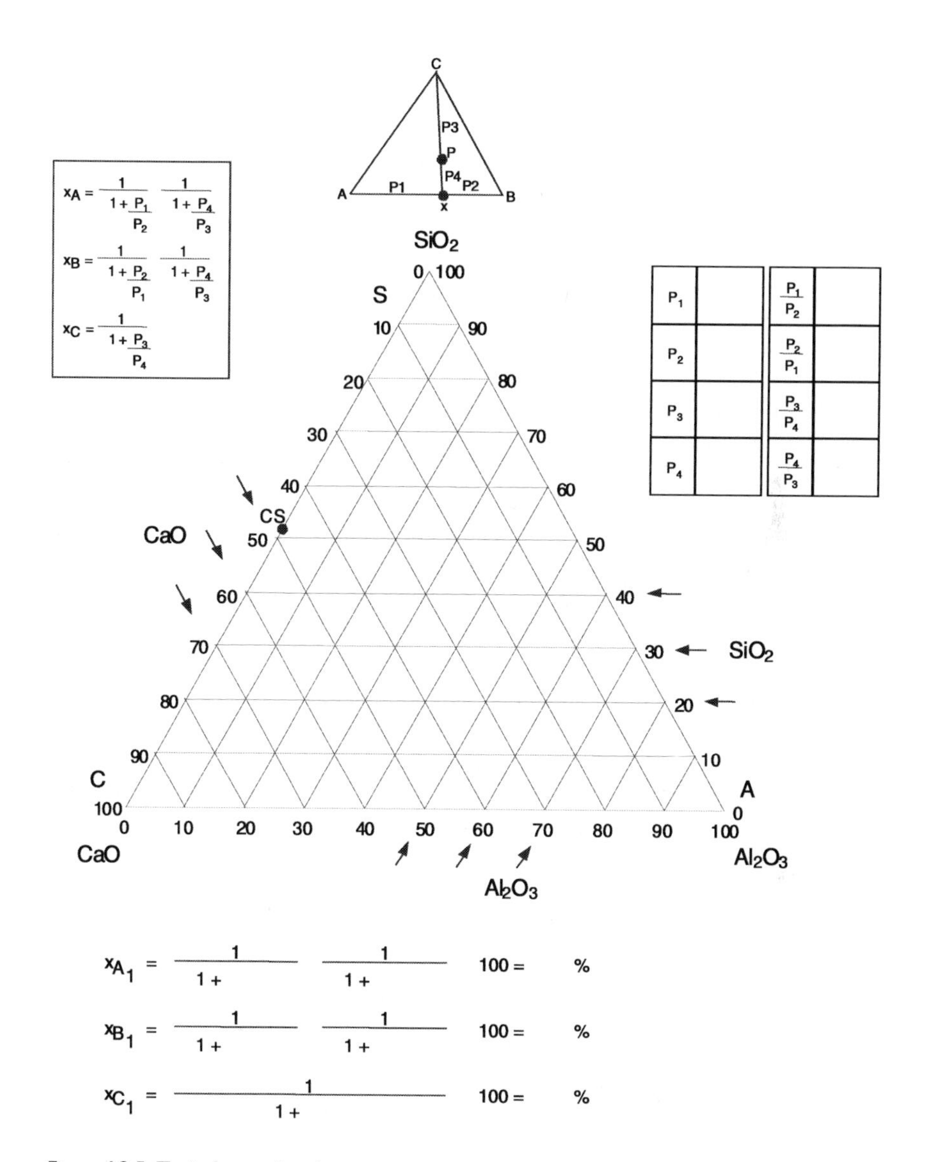

Figure A8.5 Typical exercise sheet.

Appendix IX

Influence of the alkalis on the nature and morphology of hydration products

The few alkali sulphates that are found in a Portland cement play an important role as far as the nature and the morphology of the first hydrates formed when the cement particles enter in contact with water is concerned, because these sulphates are highly soluble. In fact they are much more soluble than the different forms of the calcium sulphate that are found in a Portland cement after its grinding (even the hemihydrate form).*

In Figure A9.1 it is possible to see some syngenite that has been formed. The syngenite is a double sulphate of calcium and potassium that is soluble. This syngenite is formed by the side of short prisms of ettringite in a cement rich in alkalis. In a cement poor in alkalis only short prisms of ettringite are seen. After 2 hours of hydration in both cases ettringite and syngenite crystals can be observed.

After 5 to 6 hours of hydration it is possible to observe the growth of C-S-H crystals in the low alkali cement (Figure A9.1, bottom). In the high alkali cement syngenite starts to dissolve to produce some secondary gypsum.

After 10 hours of hydration the amount of C-S-H formed has increased, syngenite has disappeared and has been replaced by crystals of secondary gypsum. After 10 hours of hydration ettringite appears as short prisms in the high alkali cement.

On Figure A9.2 it is possible to see the growth of the C-S-H and ettringite.

Figure A9.3 and represents the sequential formation of the different hydrates that has been observed in an environmental microscope. It is possible to observe the non-negligible role of the syngenite during the first hours that follow the beginning of hydration. Moreover it is also possible to notice its disposition around 10 hours later.

Figure A9.4 represents the aspect of a hydrated cement paste observed in an environmental microscope 5 and 6 hours after the beginning of hydration. After 5 hours it is possible to see large crystals of portlandite (Figure A9.4a).

* *The figures have been reproduced with the permission of the authors, J. Stark, B. Möser, and A. Eckart*

10 minutes

Low alkali content

High alkali content

2 hours

Figure A9.1 Comparison of the microstructure of a region rich in C_3A in two hydrating cement pastes (10 min and 2 h), one with a low alkali content and the other a high alkali content. With permission of the Institut für Baustoffkunde, Bauhaus-Universität Weimar.

Between 5 and 24 hours it is possible to observe the growth of C-S-H crystals (Figure A9.4b).

On Figure A9.4c, it is possible to see the first C-S-H crystals that begin to interlock and provide some mechanical strength to the hydrated cement paste.

Figure A9.5 represents the hydration products in a cement rich in aluminates and sulphates. It is possible to see ettringite needles bridging the gap between a clinker grain and the hydration products formed outside the particles by dissolution and precipitation.

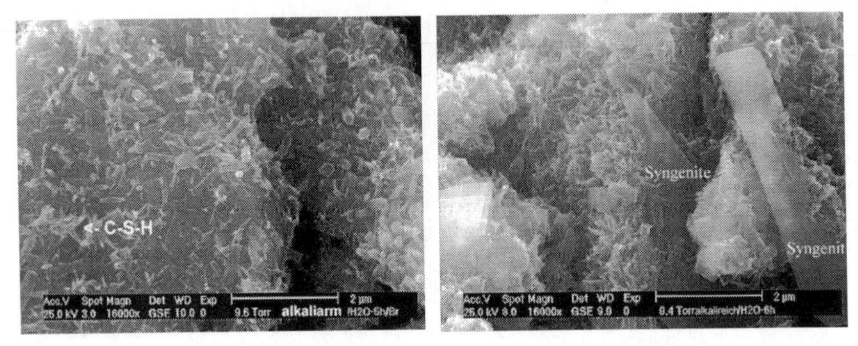

5 to 6 hours

Low alkali content

High alkali content

10 hours

Figure A9.2 Comparison of the microstructure of a region rich in C_3S in two cement pastes hydrating (at 5 to 6 hours and 10 hours), one with a low alkali content and the other with a high alkali content. With permission of the Institut für Baustoffkunde, Bauhaus-Universität Weimar.

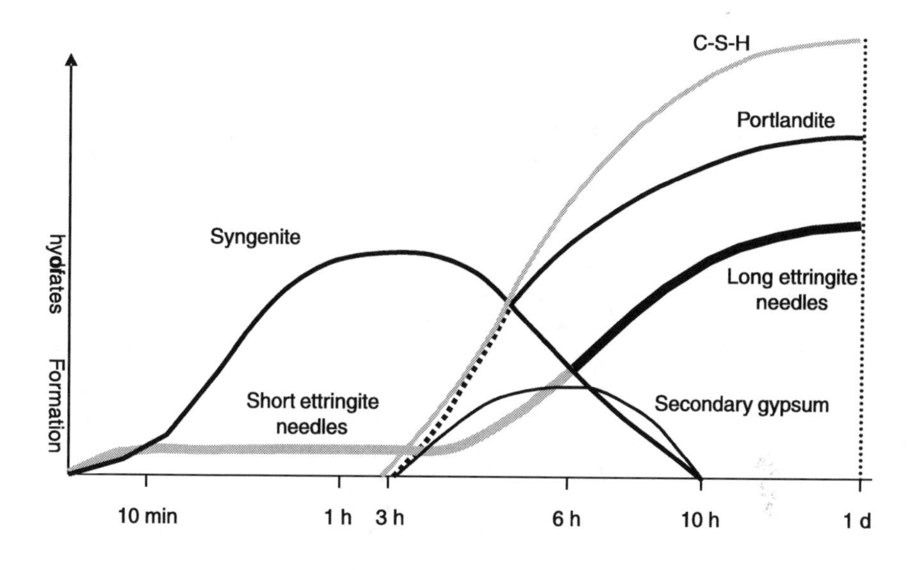

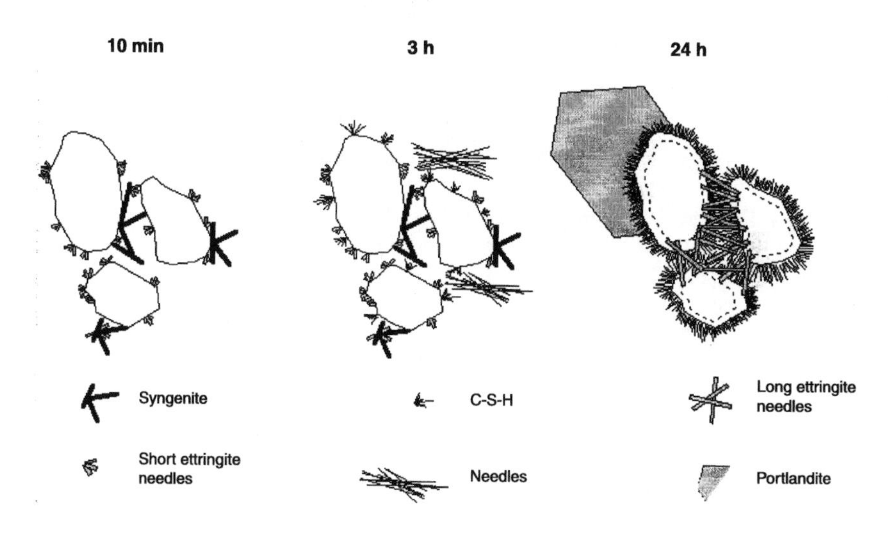

Figure A9.3 Schematic representation of the appearance of different kinds of minerals in respect to hydration time. With permission of the Institut für Baustoffkunde, Bauhaus-Universität Weimar.

(a) (b)

after 5 hours of hydration — after 24 hours of hydration

(c)

Figure A9.4 Evolution of the microstructure of hydrated cement paste. With permission of the Institut für Baustoffkunde, Bauhaus-Universität Weimar.

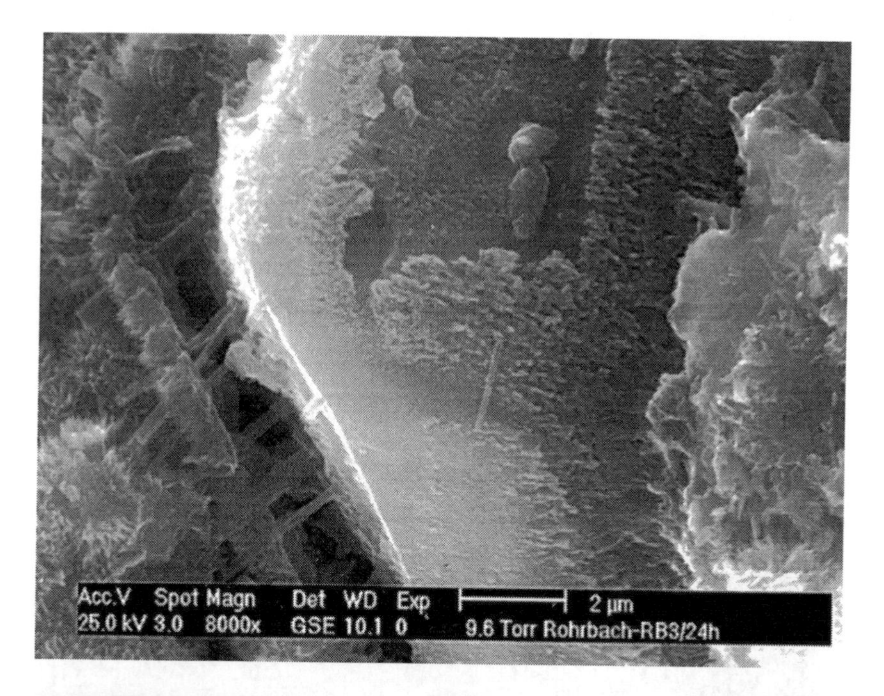

Figure A9.5 Hydration after 24 hours of the aluminate phase in a cement rich in sulphate. Ettringite needles that have grown along their c axis have traversed the space, between the cement grains and the newly formed hydrates, that is still filled with interstitial solution during all the hydration process. With permission of Institut für Baustoffkunde, Bauhaus-Universität Weimar.

(a)

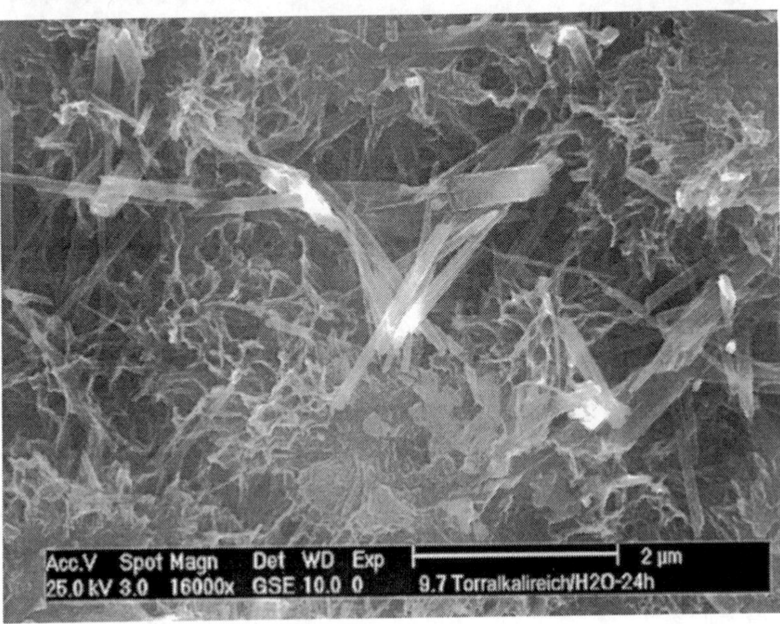

(b)

Figure A9.6 Comparison of the microstructure of an area rich in C3S in two hydrating cement pastes after 24 hours, one with a low alkali content (a), the other with high alkali content (b). With permission of the Institut für Baustoffkunde, Bauhaus-Universtät Weimar.

Appendix X

Colouring concrete

1. Pigments

Pigments are essentially very fine particles of different metallic oxides. These oxides are practically the same as those once used by prehistoric men to decorate their caves. For a long time, the only non-metallic pigment used was carbon black. Organic pigments have recently been developed but it does not seem that they are used in large quantities. They are based on copper and phthalocyamine complexes.

Green pigments contain chromium (α Cr_2 O_3 or Cr O (OH)), blue pigments contain cobalt or lapis lazuli (ultramarine blue Na_2 Al_6 S_6 O_{24} Si). All of these pigments cost much more than the more widely used iron oxide pigments. Moreover, blue pigments are not very stable. They have a tendency to get paler with time due to a change in the valence of cobalt. Blue and green pigments are normally used only with white Portland cements.

Yellow, brown, red, and black pigments are iron oxide-based. Scrap iron is attacked by concentrated hydrochloric acid under reducing conditions to form limonite (Kroone and Blakey 1968). Limonite and goethite are yellow iron oxides. When undergoing oxidizing conditions, rust is formed ($Fe(OH)_3$). If $Fe(OH)_3$ is heated in oxidizing conditions, red ferric oxide (Fe_2O_3) is formed, but if the heating occurs in rather strong reducing conditions, black magnetic oxide (Fe_3O_4) is formed instead. If magnetic oxide is heated further, it is transformed into red iron oxide (Fe_2O_3). Various shades of red and brown can then be obtained by varying the nature and length of the firing conditions.

2. Using pigments

Pigments can be used either with grey or white Portland cements. Of course with white Portland cement, pigments are more effective and produce nicer colors. Pigment dosage is almost always given as a percentage of the mass of Portland cement; a 3 to 4 per cent dosage is usually sufficient to colour concrete. A dosage higher than 10 per cent is almost never used. Of course, the greater the pigment dosage, the more intense the color (Venuat 1984; Bensted 1998).

When using pigments great care must be taken to avoid any source of contamination in order to produce the most stable and uniformly coloured concrete. Before producing any colored concrete, it is absolutely necessary to clean all the mixing equipment and particularly the mixer and its agitating system.

When buying coloured concrete elements, such as bricks and paving stones, it is always better to order 5 to 10 per cent more than actually needed, because if the initial purchase was a little bit short, it is not certain that exactly the same shade can be found again. Colour variations can be due to so many causes, such as a variation in the colour of the cement, a slight variation in the colour of the pigment and/or its grain size distribution, a variation in the water/binder ratio, a variation in the colour or cleanliness of the sand, a variation in curing conditions, etc.

Pigments do not modify compressive strength and other mechanical properties of concrete. As mentioned in the previous chapter, Detwiler and Metha (1989) were able to show that carbon black could be as efficient as silica fine in increasing concrete compressive strength. However, pigments cannot be used to replace silica fume because they are much more expensive.

3. Efflorescence

When dealing with coloured concrete, it is necessary to talk about efflorescence, because efflorescence can spoil the visual appearance of coloured concrete elements. The whitish stains or deposits that appear on the surface of concrete can be very unappealing. When the deposit is thick enough it can even hide the original colour. When the efflorescence that appears on a concrete surface is uniform, it is not as bad as when it is localized in very specific spots. Efflorescence also develops in grey concrete, but because of the natural pale grey colour of concrete, it is usually not as visible (Neville 2002a and 2002b).

In concrete plants producing bricks and paving blocks, efflorescences appear during very specific periods of the year, in the spring and in the fall, in very particular hygrometric conditions.

Efflorescence has been classified in two different classes: primary and secondary efflorescence (Kresse 1983 and 1989). Primary efflorescence develops when concrete is still in its plastic state and continues to develop during the hardening process. This phenomenon appears when a thin film of water exists on the surface of the concrete (light rain, fog, dew or condensation). Calcium ions migrate from the inside to the surface and when the film of water evaporates, calcium ions are left on the concrete surface where they react with carbon dioxide to form a calcium carbonate deposit.

Secondary efflorescence is produced much later and can reach maximum intensity one year after the fabrication of the concrete elements. Usually, it should have been washed out by rain a year after its appearance. It is believed that secondary efflorescence is caused by external water that temporarily penetrates concrete before being evaporated when drier conditions prevail. This is the only difference between the two types of efflorescence, which are both due to the migration of internal Ca^{2+} ions to the surface of the concrete.

Some efflorescence can also be found at the base of concrete walls in contact with a soil when superficial water, or that of the water table, contains dissolved ions that can move up in the concrete due to osmotic pressures. Of course, in such cases, the chemical nature of the efflorescence depends on the chemical nature of the ions present in the water. Usually, this kind of efflorescence is found in poorly consolidated concretes or in concretes having a high water/binder ratio.

3.1 Mechanism of formation of efflorescence

Much research has been done to find the major causes of the appearance of efflorescence, mainly in Germany, because for a long time, Germany was the largest producer of pigments (Kresse 1983 and 1989; Deichsel 1982). The French organization CERIB did also some work in this field during the 1970s (Rabot et al. 1970; Dutruel and Guyader 1974 and 1977; Anonymous 1977).

If a freshly cast concrete surface dries very rapidly, there are very few chances that any efflorescence will appear on its surface. In such a case efflorescence rather appears on the surface of the capillaries inside the concrete. Consequently, when the weather is hot and dry or when there are strong drafts, there is little likehood of seeing primary efflorescence appearing on the surface of the concrete. However, later on, the risk of seeing secondary efflorescence appear on concrete surface is increased, because many calcium ions have been transported close to the surface of the concrete within the capillaries (Neville 2002a and 2002b).

On the contrary, when the surface of the fresh concrete dries slowly and when a film of water remains long enough on the concrete surface, the chances of seeing primary efflorescence are increased, because the film of water remains on the concrete surface long enough to get saturated in Ca^{2+} ions. Later on, when the film of water evaporates, Ca^{2+} ions will be carbonated. Of course, when efflorescence is developed just at the opening of the capillaries they can close theses capillaries and therefore stop the occurence of any further efflorescence.

3.2 How to decrease the risk of efflorescence

In order to get rid of efflorescence, or at least decrease the possibility of its formation, it is necessary to close the surface of the concrete to avoid any exchange of water with the external environment. However, in such a case, if one part of the surface is not well closed efflorescence will be concentrated in the unprotected spots.

In order to avoid or at least diminish the appearance of primary efflorescence, several measures can be taken. Concrete permeability and sorptivity can be decreased by a reduction of the water/binder ratio by increasing the volume of the paste by using a finer cement, or using a blended cement containing a pozzolanic addition that will fix some of the lime released by the hydration of the silicate phases. However, it is very optimistic to hope that a pozzolan could initially fix all of the lime released during hydration of the silicate phase so that the risk of seeing primary and secondary phases is totally eliminated (Neville 2002a and 2002b).

On horizontal surfaces, it is very important to decrease the amount of bleeding water by increasing the amount of fine particles or by incorporating colloidal agents which will retain water inside the concrete. Everything that can be done to limit the movement of the calcium ions should be done. It is also important to try to protect the surface against any rapid cooling to avoid any condensation. Finally the concrete surface can be sealed with siloxanes, silanes or acrylates.

3.4 How to get rid of efflorescences

It is quite easy to get rid of efflorescence once it has appeared: it is only necessary to dissolve it with a 10 per cent solution of muriatic acid. Calcium carbonate is transformed into the more soluble bicarbonate form. Thereafter, it is necessary to wash the concrete surface with clear water. Light sandblasting or pressurized water can also be used to eliminate old efflorescence.

In a plant making coloured concrete bricks and paving blocks, it is better to stop the production of coloured elements where the hygroscopic conditions prevailing favour the formation of efflorescence and concentrate on these days on the production of grey elements.

References

Anonymous (1977) 'Les efflorescences', *Le Bâtiment-Bâtir*, 5: 43–6.

Bensted, J. (1998) 'Special Cements', in P.C. Hewlett (ed.) *Lea's Chemistry of Cement and Concrete*, 4th edn, London: Arnold, pp. 779–835.

Deichsel, T. (1982) 'Efflorescence-origins, Causes, Counter-measures', *Betonwerk + Fertigteil-Technik*, 10: 590–7.

Detwiler, R.J. and Mehta, P.K. (1989) 'Chemical and Physical Effects of Condensed Silica Fume', in *Concrete, Supplementary Paper at the Third International CANMET/ACI Conference on Fly Ash, Silica Fume, Slag and Natural Pozzolans in Concrete*, Trondheim, pp. 295–306.

Dutruel, F. and Guyader, R. (1974) *Évolution de la teinte d'un béton et étude des mécanismes qui s'y rattachent*. Technical publication No. 11 by Centre d'Études et de Recherche de l'Industrie du Béton Manufacturé.

—— (1977) 'Influence de la carbonatation associée à l'étuvage et des conditions de stockage sur la teinte d'un béton', *Revue des Matériaux de Construction*, 704 (January): 27–30.

Kresse, P. (1983) 'Studies on the Phenomenon of Efflorescence on Concrete and Asbestos Cement', *Betonwerk + Fertigteil-Technik*, 9.

—— (1989) 'Coloured Concrete and its Enemy: Efflorescence', *Chemistry and Industry*, February 20: 93–5.

Kroone, B. and Blakey, F.A. (1968) 'Some Aspects of Pigmentation of Concrete', *Constructional Review*, 41 (7): 24–8.

Neville, A.M. (2002a) 'Efflorescence: Surface Blemish or Internal Problem, Part I, The Knowledge', *Concrete International*, 24 (8): 86–90.

—— (2002b) 'Efflorescence: Surface Blemish or Internal Problem, Part II, Situation in Practice', *Concrete International*, 24 (9): 85–8.

Rabot, R., Coulon, C. and Hamel, J. (1970) 'Étude des efflorescences', *Annales de l'Institut Technique du Bâtiment et des Travaux Publics*, 23rd year, July–August, No. 271–2, pp. 70–84.

Venuat, M. (1984) *Adjuvants et traitements*, n.p.

Appendix XI

Relevant ASTM standards

Admixtures

C 260 Spec. for Air-Entraining Admixtures for Concrete
C 311 Tests for Sampling and Testing Fly Ash or Natural Pozzolans for Use as a Mineral Admixture in Portland-Cement Concrete
C 494 Spec. for Chemical Admixtures for Concrete
C 618 Spec. for Coal Fly Ash and Raw or Calcined Natural Pozzolan for Use as a Mineral Admixture in Portland Cement Concrete
C 979 Spec. for Pigments for Integrally Colored Concrete
C 1017 Spec. for Chemical Admixtures for Use in Producing Flowing Concrete

Aggregate

C 29 Test for Unit Weight and Voids in Aggregate
C 33 Spec. for Concrete Aggregates
C 40 Test for Organic Impurities in Fine Aggregates for Concrete
C 70 Test for Surface Moisture in Fine Aggregate
C 87 Test for Effect of Organic Impurities in Fine Aggregate on Strength of Mortar
C 88 Test for Soundness of Aggregates by Use of Sodium Sulphate or Magnesium Sulphate
C 117 Test for Materials Finer than 75-μm (No. 200) Sieve in Mineral Aggregates by Washing
C 123 Test for Lightweight Pieces in Aggregate
C 131 Test for Resistance to Degradation of Small-Size Coarse Aggregate by Abrasion and Impact in the Los Angeles Machine
C 227 Test for Potential Alkali Reactivity of Cement-Aggregate Combinations (Mortar-Bar Method)
C 289 Test for Potential Alkali-Silica Reactivity of Aggregates (Chemical Method)

C 294 Descriptive Nomenclature for Constituents of Natural Mineral Aggregates
C 295 Guide for Petrographic Examination of Aggregates for Concrete
C 330 Spec. for Lightweight Aggregates for Structural Concrete
C 331 Spec. for Lightweight Aggregates for Concrete Masonry Units
C 332 Spec. for Lightweight Aggregates for Insulating Concrete
C 441 Test for Effectiveness of Mineral Admixtures of Ground Blast-Furnace Slag in Preventing Excessive Expansion of Concrete Due to the Alkali-Silica Reaction
C 566 Test for Total Moisture Content of Aggregate by Drying
C 586 Test for Potential Alkali Reactivity of Carbonate Rocks for Concrete Aggregate (Rock Cylinder Method)
C 682 Evaluation of Frost Resistance of Coarse Aggregates in Air-Entrained Concrete by Critical Dilation Procedures
C 1105 Test for Length Change of Concrete Due to Alkali-Carbonate Rock Reaction
C 1137 Test for Degradation of Fine Aggregate Due to Attrition
E 11 Spec. for Wire-Cloth Sieves for Testing Purposes

Cement

C 91 Spec. for Masonry Cement
C 109 Test for Compressive Strength of Hydraulic Cement Mortars (Using 2-in. or 50-mm Cube Specimens)
C 115 Test for Fineness of Portland Cement by the Turbidimeter
C 150 Spec. for Portland Cement
C 151 Test for Autoclave Expansion of Portland Cement
C 186 Test for Heat of Hydration of Hydraulic Cement
C 191 Test for Time of Setting of Hydraulic Cement by Vicat Needle
C 204 Test for Fineness of Hydraulic Cement by Air Permeability Apparatus
C 230 Spec. for Flow Table for Use in Tests of Hydraulic Cement
C 243 Test for Bleeding of Cement Pastes and Mortars
C 266 Test for Time of Setting of Hydraulic Cement Paste by Gillmore Needles
C 430 Test for Fineness of Hydraulic Cement by the 45-pm (No. 325) Sieve
C 452 Test for Potential Expansion of Portland Cement Mortars Exposed to Sulphate
C 595 Spec. for Blended Hydraulic Cements
C 845 Spec. for Expansive Hydraulic Cement
C 917 Test for Evaluation of Cement Strength Uniformity from a Single Source
C 989 Spec. for Ground Granulated Blast-Furnace Slag for Use in Concrete and Mortars

C 1012 Test for Length Change of Hydraulic-Cement Mortars Exposed to a Sulphate Solution
C 1157 Performance Specification for Blended Hydraulic Cement
C 1240 Spec. for Silica Fume for Use in Hydraulic-Cement Concrete and Mortar

Concrete

C 31 Making and Curing Concrete Test Specimens in the Field
C 39 Test for Compressive Strength of Cylindrical Concrete Specimens
C 42 Test for Obtaining and Testing Drilled Cores and Sawed Beams of Concrete
C 78 Test for Flexural Strength of Concrete (Using Simple Beam with Third-Point Loading)
C 94 Spec. for Ready-Mixed Concrete
C 116 Test for Compressive Strength of Concrete Using Portions of Beams Broken in Flexure
C 125 Terminology Relating to Concrete and Concrete Aggregates
C 138 Test for Unit Weight, Yield, and Air Content (Gravimetric) of Concrete
C 143 Test for Slump of Hydraulic Cement Concrete
C 156 Test for Water Retention by Concrete Curing Materials
C 157 Test for Length Change of Hardened Hydraulic-Cement Mortar and Concrete
C 171 Spec. for Sheet Materials for Curing Concrete
C 173 Test for Air Content of Freshly Mixed Concrete by the Volumetric Method
C 192 Making and Curing Concrete Test Specimens in the Laboratory
C 215 Test for Fundamental Transverse, Longitudinal, and Torsional Frequencies of Concrete Specimens
C 231 Test for Air Content of Freshly Mixed Concrete by the Pressure Method
C 232 Test for Bleeding of Concrete
C 293 Test for Flexural Strength of Concrete (Using Simple Beams with Centre-Point Loading)
C 309 Spec. for Liquid Membrane-Forming Compounds for Curing Concrete
C 260 Test for Ball Penetration in Freshly Mixed Hydraulic Cement Concrete
C 403 Test Time of Setting of Concrete Mixtures by Penetration Resistance
C 418 Test for Abrasion Resistance of Concrete by Sandblasting
C 457 Test for Microscopical Determiaation of Parameters of the Air-Void System in Hardened Concrete
C 469 Test for Static Modulus of Elasticity and Poisson's Ratio of Concrete in Compression

C 470 Spec. for Moulds for Forming Concrete Test Cylinders Vertically

C 496 Test for Splitting Tensile Strength of Cylindrical Concrete Specimens

C 512 Test for Creep of Concrete in Compression

C 531 Standard Test Method for Linear Shrinkage and Coefficient of Thermal Expansion of Chemical-Resistant Mortars, Grouts, and Monolithic Surfacings

C 567 Test for Unit Weight of Structural Lightweight Concrete

C 597 Test for Pulse Velocity Through Concrete

C 617 Capping Cylindrical Concrete Specimens

C 642 Test for Specific Gravity, Absorption, and Voids in Hardened Concrete

C 666 Test for Resistance of Concrete to Rapid Freezing and Thawing

C 671 Test for Critical Dilation of Concrete Specimens Subjected to Freezing

C 672 Test for Scaling Resistance of Concrete Surfaces Exposed to Deicing Chemicals

C 684 Making, Accelerated Curing, and Testing of Concrete Compression Test Specimens

C 685 Spec. for Concrete Made by Volumetric Batching and Continuous Mixing

C 779 Test for Abrasion Resistance of Horizontal Concrete Surfaces

C 803 Test for Penetration Resistance of Hardened Concrete

C 805 Test for Rebound Number of Hardened Concrete

C B56 Petrographic Examination of Hardened Concrete

C 873 Test for Compressive Strength of Concrete Cylinders Cast in Place in Cylindrical Moulds

C 878 Test for Restrained Expansion of Shrinkage-Compensating Concrele

C 900 Test for Pullout Strength of Hardened Concrete

C 918 Test for Developing Early-Age Compression Test Values and Projecting Later-Age Strengths

C 944 Test for Abrasion Resistance of Concrete or Mortar Surfaces by the Rotating-Cutter Method

C 1038 Test for Expansion of Portland Cement Bars Stored in Water

C 1074 Estimating Concrete Strength by the Maturity Method

C 1078 Test for Determining the Cement Content of Freshly Mixed Concrete

C 1079 Test for Determining the Water Content of Freshly Mixed Concrete

C 1084 Test for Portland-Cement Content of Hardened Hydraulic-Cement Concrete

C 1138 Test for Abrasion Resistance of Concrete (Underwater Method)

C 1150 Test for the Break-Off Number of Concrete

C 1152 Test for Acid-Soluble Chloride in Mortar and Concrete

C 1202 Test for Electrical Indication of Concrete's Ability to Resist Chloride Ion Penetration

Index

abalone shell 409, 410
Abram's law 334, 417
abrasion resistance 350
absolute volume contraction 148
absorptivity effect of 'airplugs' 186
accelerator 251
calcium chloride 252
calcium formate 252
calcium nitrite 252
triethanolamine 252
acceptance
criteria 58, 241
standards 135
ACI journals and meetings 61, 62
ACI/CANMET 62
ackermanite 276
activation
of slag 279, 282, 283
with alkalis 282, 283
with calcium sulphate 282, 283
with lime 282, 283
activity index of slag 284
addition of gypsum 36
additives *see* cementitious materials
adiabatic calorimetry 216
admixtures 22, 206, 413, 416
addition to cement 264
alteration of hydration reaction 206
different types 210–11, 417
dispersing agents 213
electrostatic repulsion 217
history 207
lignosulphonate 417
of tomorrow 429
physical action 206
progresses achieved 416
reactive sites 215, 216, 219
relative cost in concrete 264
science of admixtures 208
secondary effects 418
surface absorption 217
terminology 209, 210, 212
use of synthetic polymers 209
vinsol resin 259, 417
AF
current values 116
influence on the viscosity of the interstitial phase 116
afm *see* monosulphoaluminate
aft *see* ettringite
aggregate skeleton
effect on microcracking 186
aggressive ions
penetration 350
air bubbles 258
effect on the properties of the fresh concrete 208
air entrained Portland cement 318
air entraining agents 207, 258
beneficial effect 258
difference with foaming agents 258
effect on permeability 258
mode of action 259
spacing factor 258
'air plugs' 176, 186, 258
'a la carte' blended cement 306
'a la carte' concrete 430
alkali aggregate reaction 25, 320, 440
effect of cementitious materials 309
alkali cycle 123
alkali sulphates 73, 116, 119, 123, 140, 192, 228, 466
soluble 228, 230
alkaline activation 282, 327
alkemade lime 460

alinite 329, 407
alite 61, 72, 74, 109, 124
proportion 87
shape 125
alternative fuels 36, 112, 405
aluminium powder 257
alumino-ferritic modulus (AF) 69
current values 116
influence on the viscosity of the interstitial phase 116
aluminous cement 314, 320
alumina content 321
chemical composition 320
conversion reaction 323
failures 324
heat development 323
phase composition 320, 321
refractory application 321
uses 323
American Portland Cement Association 59
amorphous solids 274
slag 276
analcine 17
anhydrite 123, 228, 327
addition 422
as a contaminant of gypsum 130
natural 156, 158, 228
solubility 134, 156
anhydrous phase 147
anti freeze admixtures 262
aphtitalite 123, 228
apparent volume increase 172, 173, 194
Aqueduct of Segovia 19, 28
arcanite 123, 228
Arche de la Défense 358
Arrhenius law 187, 188
artificial lime 90, 442
artificial pozzolan 287, 428
artificial stone 444
asbestos 150
ash content of the coal 117
Aspdin's specification 444
ASTM 59, 212
ATILH 59
atomic force microscope 419
autoactivation of hydration reaction 188
autogenous shrinkage 174, 177, 188, 333, 341
as a consequence of chemical shrinkage 174
comparison with drying shrinkage 174
control of 345
definition 174
discovery 174
effect on concrete 175, 176
influence of curing conditions 342
influence of w/b 175
role of 'air plugs' 176

Baalbaki model 335
bag of cement 105
automatic bagger 107, 108
automatic roto-packer 108
production in North America 106
ball mill 35, 103
control of the temperature 133, 134
diaphragm 104
efficiency 102
first ball mill 35
grinding under water 114
retention time 134
barrel of cement 58, 92, 105, 138
bauxite 322
beach pebbles 102
belinite 329
belite 61, 72, 74, 86, 109, 118
nest 130
picture 165, 166
proportion 87
secondary 125
shape 124, 126
belitic cement 405
composition 406
production 406
big bag 105, 138
binary diagram 80, 452, 454
example 452
binary mixes 454
binder of my dreams 433
binders of tomorrow 414, 429
Blaine fineness 133, 136, 137, 316, 434
influence on cube strength 137
influence on the fresh properties 133
influence on the hardened properties 133
blast furnace 276
blast furnace slag *see* slag
bleeding 147
blended cement 52, 109, 144, 273, 306, 314, 320, 440
addition of cementitious material 306
commercialization 53
heat development 308

history of use 52
standardization 309
Blue Circle 54, 56
boat transportation 105, 111
Bogue composition 67, 136, 448
calculation 70
clinker 76
method of calculation 448–51
potential 74, 76
BOOT projects 51, 350, 426, 430
boundary lines 460
Britannic cement 34
brucite 68, 136
buff cement 315
build-ups 123
burlap sacks 105
burner 109
bypass 123

C_3A 74–8, 87, 398, 420, 461
a necessary poison 434
acceleration of C_3S hydration 151
as a poison to concrete 433
average content 136
control of hydration 153
disadvantage 141
hydration 134, 171, 192
Na_2O content 78
negative effect 420
perverse effect on robustness 437
perverse effect on rheology 436
prehydration 134
C_4AF 74–7, 87
heat evolution 171
Ca–langbeinite 123, 228
Calcifrit 327
characteristics 327
chemical composition 327
production 327
ternary cement 328
calcination zone 117
calcined clay 27, 273, 274, 287, 299
metakaolin 286, 287, 292, 200, 300, 408
calcined shale 273, 287, 299
calciumaluminate 320
calcium carbonate
as a contaminant of gypsum 130
cycle 26
calcium chloride
addition to a superplasticizer 250
addition to polycarboxylates 434
as an accelerator 251
attack of reinforcing bars 251
limitation 251
maximum dosage 251
mode of action 251
prohibition 252
calcium formate 252
calcium langbeinite *see* Ca-langbeinite
calcium monosulphoaluminate *see* monosulphoaluminate
calcium nitrate 252
calcium silicate hydrate 28
calcium sulphates 69, 74, 102, 106, 280, 282, 69, 74, 102, 106, 280, 282, 436
addition 128, 422
as a retarder 251
attack by thiobacillus ferroxydans 426
cocktail 158
crucial role during hydration 228
different types found in Portland cement 156, 157, 228
influence on cube strength 137
optimum content 230, 231
solubility 130, 158
synthetic calcium sulphate 228
true nature in Portland cement 156
wastes 133, 159
calcium sulphoaluminate cement 314, 325
applications 326
hydration 325
manufacturing 325
phasic composition 325
Candlot salt *see* ettringite 153
CANMET 404
$CaO–SiO_2$ phase diagram 81
$CaO–SiO_2–Al_2O_3$ phase diagram 80, 83, 84, 274, 406
eutectic composition 279
use by the aluminous cement industry 83, 84
use by the glass industry 83, 84
use by the metallurgic industry 83, 84
use by the refractory industry 83, 84
$CaO–SiO_2–Al_2O_3–Fe_2O_3$ 415
capillary water 180
carbonation 420
carbon black 207
carbon dioxide 86, 142
emission 86
carboxylates 437
celite 61, 72
cellular concrete 257
CEMBUREAU 60

cement
  C_3S content 420
  complexity 421
  evolution of the characteristics 420
  fineness 420
  minor phases 421
  world production 1
cement/admixture compatibility 422
cement consumption 44
  as a function of birth rate 48
  as a function of GDP 45, 49, 50
  exportations 35, 58
  factors influencing 50
  importations 55
  market 54
  per capita, per country 46
  periodicity 54, 56
cement fabrication
  clay content 33
cement industry
  as a green industry 428
  distribution 55
  profitability 55
  structuration 54
cementitious additions *see* cementitious materials
cementitious materials 273, 274, 275
  effect on alkali aggregate reaction 309
  effect on concrete properties 308
  effect on durability of concrete 309
  effect on early strength 308
  effect on long term strength 308
  effect on shrinkage 309
  effect on water demand 307
  importance of water 308
  mode of introduction 307
  use in high-performance concrete 334
  use of admixtures 308
cementitious system 327
  no clinker binder 327
cement kiln
  different types 109
  long kiln 121, 122
  production 101, 102
  schematic representation 117, 118
  short kiln 119, 120
  with a precalcinator 109, 117
  with a pre-heater 109, 117
  without pre-heating and precalcinator tower 101
cement kiln dust (CKD) 112
  reintroduction 123, 133
cement marketing 57
cement/polysulphonate compatibility 229, 230, 231
  compatible combination 234, 238
  influence of a retarder 236, 241
  influence of cement type 234, 240
  influence of the intersticial phase 228
  of low alkali cement 230
  of white Portland cement 230
cement production
  by continent 47
  by country 45
  concentration 55
  flowing trend 50
  progression during the twentieth century 44–6
cement/superplasticizer
  compatibility 69
  incompatibility 418
  robustness 70
cement tanks 138
CEMEX 54, 56
cenosphere 289
CERIB 59
CERILH 59
chalk 90
  quarry 112
chemical analysis 88
  clinker 135
  importance 135
  raw meal 135
chemical attack
  in a marine environment 348, 349
chemical composition 315
  evolution 140
  of fly ashes 290
  of natural pozzolans 302, 303
  of principal cementitious materials 275
  of silica fume 298
  of slag 279
  Portland cement 64
  raw meal 76
  simplified notation 74
chemical contraction 177, 180, 185, 186
chemically linked water 176
chemical shrinkage 174
  restriction by the aggregate 174
chemical trigger 190
chlorine cycle 122
chlorine ions
  effect on reinforcing steel 349

chromium
complexation of hexauvalent chromium 160
chrysotile asbestos 150, 196, 420
'Ciment du jour' 306
Ciment Fondu 322
CIMENT FRANÇAIS 56
citric acid 318
CKD 133, 143, 327
Class C fly ash 292
definition 292
Class F fly ash 292
definition 292
classification of admixtures
ASTM 212
Dodson classification 212
Mindess and Young classification 212
clay 85, 86, 90
calcined 27
content 33
optimum proportion 33
proportion to make artificial lime 443
clay deposit 113
clinker 70, 74, 90
characteristics 436
chemical analysis 70
degree of crystalinity 75
free lime content 75
hardness 135
morphology 72, 124–31
nervous 141
of my dreams 433
outside 128
phase composition 70
porosity 123, 135
quenching 72, 109, 119
reproducibility 58
storage 109, 128
clinkerization process 75
clinkering temperature 88
clinkering zone 72, 88
closed system 176
cluster of free lime 114, 131
CO_2
concentration in air 410, 411
decrease due to water reducer 206
emission 112
per tonne of clinker 142, 143
reduction 143
world wide 143
coal
ash content 116
clean 160
rich in sulphur 116, 140
coal equivalent 142
coalescence of the bubbles 259
coarse aggregate 418
compressive strength 334
effect on elastic modulus 335
influence on shrinkage 343
cohesive forces
origin 198, 199
cold ground clinker 135
cold slag 280
colloïdal agents 232, 260, 318
different types 261
use in concrete 261
colour
of slag 276
of concrete made with blended cement 308
of silica fume 295
of slag concrete 308
commodity product 423
compatibility 229, 434
of type V cement 439
problems 208, 209
compatibility triangle 460, 461
for C_3S, C_2S and C_3A 461
compatible combination 238
composite material 334, 413
compressive strength 167, 425
concrete
as a composite material 334, 413
average consumption 12, 415
confined concrete 418
cost 428
cover 350, 351
durability 420, 440
ecological impact 428
effect of air entrained bubbles on rheology 259
effect of the cementitious materials 307
environmental impact 428
heat development 308
high-tech concrete 413, 424
history 16
industry of the twenty-first century 430
macrostructural properties 419
microstructural properties 419
microstructure 419, 420
mixer 38
modern concrete 36, 37
nanostructure 420

of tomorrow 430
origin of shrinkage 194
permeability as a function of the w/c ratio 347
resistance to freezing and thawing 438
specialized applications 2, 424
strength and weaknesses 10, 11, 12, 406
concrete bacillus 154
condensation 221
conditionning of Portland cement 138
Confederation Bridge 350, 371, 373, 374
confined reactive powder concrete 386
Congress of Cement Chemistry 62
connectivity of the pore system 348
consolidation of the cement industry 51, 55
construction of a cement plant 114
consumption
of concrete 12, 413
of superplasticizer 159
progression of consumption 414
control of autogenous shrinkage 186
importance for durabaility 186
control of the rheology 438
control room 113
conversion reaction of aluminous cement 323
cooling zone 117, 119
cooperation with universities 431
corn syrup as a retarder 256
corrective materials 114
corrosion inhibitors 263
cost of a cement plant 51, 112
cristobalite 297
crushers 114
crystal distortions 77
crystallization zone 460, 462
in the $CaO–SiO_2–Al_2O_3$ phase diagram 460, 462
crystal reactivity 77
crystalline
network 78
influence on rheology 78
solid 78
crystallized slag 277
as a source of lime 407
crystallographic phases 76
C/S ratio
of C-S-H 163, 197
CSA specification for the spacing factor 351, 352
C_2S 74, 76, 77, 81, 83, 85, 461
crystals 161
reaction with pozzolans 285
stability at high temperature 1
C_3S 74, 76, 77, 81, 83, 85, 461
average content 136
crystals 161
disadvantage 141
maximum content 434
negative effect 420
optimum content 434
prehydratation 136
reaction with pozzolans 285
C-S-H 74, 148, 193, 283, 407
as seen in a SEM 161, 164, 165
C/S ratio 163, 197
crystallites 161, 196
diffusion 170
first layer 161
growth of crystals 466, 467
hydration of C_3S 148, 161
lamellar character 197, 198
precipitation 161
produced by cementitious materials 309
produced by pozzolans 285
quasi crystallites 200
structural models 152, 195–8
structure 419
synthesis of the present knowledge 197
'vitreous' aspect 164, 165
weakening effect of sulphate ions 171
cube strength 69, 137, 160
curing 350
critical period 343
effect on 0.35 w/c concrete 345
fogging 344
initial water curing 343
of high-performance concrete 344
ponding 344
prewetted geotextile 344
ways to cure high-performance concrete 344
wet burlap 344
curing condition
importance on volumetric changes 185
curing of concrete 308, 309
influence of curing condition on volumetric changes 173
membrane 343
of high-performance concrete 344
paying contractor 310, 346, 347

cycle
  of calcium carbonate 26
  of gypsum 23

decarbonation 118, 397
  degree of decarbonation 98
  limestone 86, 109
decreasing the energetic content of clinker 405
  alternative fuels 405
  belitic cement 405
  decreasing the clinkering temperature 405
  other source of lime 405
  use of mineralizers 405
dedusting system 112
DEF *see* delayed ettringite formation
defoamer 220
defoaming agent
  tributylphosphate 256, 257
degraded protein 263
degree of hydration 180, 181, 188
degree of sulphatization 69
dehydrated gypsum 134, 135, 228
delayed addition of a superplasticizer 250, 435
delayed ettringite formation 423, 433, 439
development of modern concrete 36, 37
diatomaceus earth 303
  photo 304
  X-ray diffractogram 286
dicalcium silicate 70, 74
DICKERHOFF 92
differential thermal analysis 297
diffusion process 84, 170
discovery of binders 6
  first hydraulic binder 33
  gypsum 20
  hydraulic binder 33
  lime 16
  natural cement 33
  Portland cement 30, 33
  Pozzolans 6, 27
dispersing agents 213
  chemical effect 215, 216
  mode of action 214, 216
  physical effect 214
dispersion of cements 437
dissolution rate 228, 229
  of calcium sulphate 233
Dodson classification of admixtures 212
dormant period 167, 169, 189
double introduction method 422
dry process 36, 94
drying shrinkage 175, 345
  of high-performance concrete 342
  of ordinary concrete 342
DSP concrete 426
durability 356, 425, 427
  as a function of the w/b ratio 181
  importance of autogenous shrinkage 186
  in marine environment 348–51
  influence of admixtures 206
  influence of porosity 347
  of high-performance concrete 347
  role of C_3A 433
durability factor 351
Dusseldorf Airport fire 355

early strength
  how to increase it 249
ecological binders 405
ecological concrete development 399
ecological impact of Portland cement 39, 143
economical aspect
  of reconstruction 403
  of repair 403
EDAX 161, 164
Eddystone lighthouse 31, 32
efflorescence 314, 474
  decreasing the risk 475
  formation 474
  getting rid of 476
  primary efflorescence 474
  secondary efflorescence 474
egg shell 409
electrical condictivity 147, 152, 190, 191, 193
  variation during hydration 152
  variation with time 167
electrostatic precipitator 143
electrostatic repulsion 217
elimination of organic toxic products 398, 415
  contamined oil 398
  domestic wastes 398
  industrial wastes 398
  mad cow 398
  PCB 398
  pollutants 415
  scrap tires 398
  varnishes 398

energy efficient binder 405
entrained air bubbles
  correction for a coarse sand 258
  effect on compressive strength 259
  effect on crack propagation 258
  effect on freezing and thawing 260
  effect on rheology 258–60
  factor influencing 260
  in high-performance concrete 260
entrained water 346
entrapped air 260
environmental conditions
  impact on concrete 397
  influence on criteria 332
  influence on service life 402
environmental SEM 151
equivalent time 188
ettringite 128, 137, 153, 163, 283, 326, 436, 437, 438, 466
  clusters in concrete 157
  first layer 170
  formation 189–92
  freezing and thawing 156
  influence of lime 326
  number of water molecules 153
  unit cell 155
  secondary ettringite 156
eutectic 80
  composition 279, 452
expert system 112, 415
exportation of cement 35, 58
external parking garage 425
external source of water 175, 184, 185, 342
  role in decreasing autogenous shrinkage 175, 342

factorial experimental planning 232
false set 130, 134, 137, 152, 158, 437
fat lime 25
felite 61, 72
Fe_2O_3 86
Feret's law 334, 417
ferroaluminous calcium phase 84
FICEM 60
fillers 275, 305, 306
  effect of silica fume 295
  rate of substitution 305
final grinding 114, 134, 142
  influence of press rolls 142
  influence on the mechanical strength 133
  influence on the rheology 133
fire resistance
  fire in the Channel Tunnel 353
  high-performance concrete 352
  spalling of concrete 355
firing conditions 75
firing process 90
  one step 90
  two steps 90
firing zone 117, 119, 121
first crystal 460
flash set 137, 152, 158, 437
floating silos 105
flocculation of cement particles 213–16, 437
fluidized bed 144
fluorine
  as a mineralizer 406
  in regulated set cement 317
fluxing agents 83, 278
  addition to slag 278
  alumina 83
fly ash 287, 288, 289, 408
  beneficiated 292
  carbon content 287, 292
  chemical composition 290
  commercialization 53
  effect of carbon content on admixtures 287
  effect of grinding 293
  formation 288, 289
  history of use 53
  micro fly ash 292
  picture 289, 291
  super fly ash 292
  unburned carbon 287
  X-ray diffractogram 290
foaming agents 263
fogging 344
formation of ettringitte in high concentration of lime 438
formation of portlandite 190
free lime 75, 88, 119, 131
  clusters 114
free water 176
freeze-thaw cycles 433
  durability of deicing salt 309
  durability of slag concrete 309
freezing and thawing resistance 207, 231, 351, 426

of high-performance concrete 351
testing 351
fuels
clean fuel 160
cost 116
gas 116
heavy fuel 116
organic wastes 116
petroleum coke 116
pulverized coal 116
rich in sulphur 116, 130, 140
type 116
fusibility diagrams *see* binary diagram 452

gas 116
gas permeability 347
gaseous phase 180, 182, 185
gehlenite 276, 461
gel water 176, 180
geopolymer 17
geotextile 344
glass 78
globalization
market 55
of the cement industry 51
of the concrete industry 51
global shrinkage 194
grain size distribution and shape 141
grappier 34
Green concrete 430
grinding
advantages and disadvantages 142
aids 104, 106, 316
closed circuit 147
facility 138
open circuit 141
operation 56
grinding mill
ball mill 102
history 102
horizontal mill stones 102
pendular mill stones 102
press role 104
gypsum 74, 134, 228, 325
addition 36, 130, 422
board production 23
controlling C_3A hydration 153
cycle 23, 429
dehydration 23, 158
discovery 20, 22
hot clinker 158
in Portland cement 158
low grade 130
optimization 171
partial dehydration 158
role played during hydration 152
solubility 130
solubility rate 158
strength and weakness 23, 24
gypsum board production 23, 24

Hadrian's Wall 19
hard-burnt lime 136
heat development 167
heat of evolution of C_3A 170
heat of evolution of C_3S 169
heat of evolution of Portland cement 169
heat of hydration 168
heavy fuel 116
HEIDELBERG 54
hemihydrate 130, 134, 156, 437
solubility rate 158
Herculaneum 28
hessian *see* wet burlap 344
Hibernia offshore platform 373, 375, 376
high alkali cement 472
high-performance concrete 418, 422, 424
aggregate compressive strength 334
autogenous shrinkage 333
bridges 362–74
chloride ion permeability 348
critical period for curing 344
CSA requirement for the spacing factor 351
curing 344
definition 333
durability 356
ecological aspects 332, 356, 424, 426, 428
economical aspects 424
fire resistance 352
importance of water content 335
recycling 357
roller compacting concrete 379, 381, 382
self-compacting 378, 380
self-desiccation 333, 341
shrinkage 341
spacing factor 351
the future 356
the making 335

ways to cure it 344
with a low heat of hydration 403
high-strength concrete 333, 347, 424, 426
high-tech concrete 413, 424
high volume fly ash concrete 404
high volume slag concrete 404
history
of cement 39
of concrete 36–8
of gypsum 20, 22, 23
of hydraulic lime 34, 39
of lime 25, 33
of natural cement 34
of Portland cement 30, 44
of pozzolans 27–9
HOLCIM 54, 55
homogeinization 92
by air 114
of calcium sulphate 159
requirements 114
hot clinker 158
hot slag 280
HUMBOLDT WEDAG 117
hump 78, 280
hydrated calcium silicate 74
hydrated phases 147
hydration
autoactivation 188
degree of hydration 180, 181, 188
dormant period 189
influence of the temperature 187, 188
minimum amount of water 184
of pozzolamic cement 286
of slag 282, 283
role of portlandite 190
sequential description 188–95
hydration of C_2S 148
heat evolution 171
morphology 165, 166
hydration of C_3S 148
morphology 163
progression by diffusion process 170
progression by dissolution/precipitation 170
hydration reaction 146, 176
absolute volume decrease 174
apparent volume increase 174
chemically linked water 176
complexity 201
creation of mechanical strength 194
different approaches 151
experimental data 176
free water 176
gel water 176
heat released 193
indirect observation 166
influence of admixtures 206
influence of polysulphonate 226
Jensen and Hansen model 177–85
laws governing hydration 153
modeling 200
retardation due to polysulphonates 226
slow down period 193
type of C-S-H 176
volumetric contraction 174
w/c needed 176
hydraulic binders 52, 314
hydraulic cement 34
hydraulic lime 34, 92
hydrogarnet 153, 156, 436, 437

ideal Portland cement 143, 435
maximum Blaine fineness 435
maximum cube strength 435
SO_3 content 435
importance of curing conditions 185
impure phases 72, 75
incompatibility 422
infrared spectrograph 244
initial heat of hydration
influence on the fineness 133
initial rheology 440
initial setting time 191
insulated forms 440
insulating concrete 257
interaction cement/ polysulphonates 224, 225, 226
complexity 225
interstitial liquid 94
interstitial phase 72, 73, 94, 118, 119, 124, 127, 146, 315, 434
content 80
crystallinity 70, 75, 124, 127
morphology 73
porosity 75, 119, 124, 127
vitreous 124, 127
ionic
diameter 77
inclusions 76
radius 77, 78
substitution 76, 77
isothermal conditions 177
isothermal shrinkage 178

isotherms 460
ITALCIMENTI 56

Jensen and Hansen model 177–85
Joigny Bridge 364, 366

kaolinite 150, 292, 419
kieselghur *see* diatomaceus earth
kiln dust 122
Köln aqueduct 20
KOSH treatment 70, 71, 447

LAFARGE 34, 54, 59
Lambot boat 37
latexes
  different types 262
L/D ratio 119, 121
lean lime 25
Le Chatelier
  contraction 149
  experiment 172–4
  work 149
level of substitution 284
life cycle 332, 350
lightweight aggregate from slag 276
lignite 287
lignosulphonate 94, 218, 219, 259, 400, 417
  as an entraining agent 259
  as retarder 25
  as superplasticizer 243
  calcium salt 220
  consistency 219, 220
  effect of impurities 220
  limitation in the dosage 221
  slurry preparation 94
  solid content 219, 220
  sugar 220
  surface active agent 220
  water reducer 437
lime
  cluster of free lime 131
  cycle 26, 429
  discovery 16
  fat lime 25
  free lime 119, 131
  history 25
  hydraulic lime 34
  lean lime 25
  mortar 25
  preparation 25, 90
  production 26
  slacked lime 25
  strength and weakness 26
  uncarbonated 26
  uncombined 119
  use 25, 33, 34
lime saturation factor (LAF) 69
lime and silica phase diagram 61
limestone 85, 90
  containing clay 33
  decarbonation 86, 109
  firing 26
  impure 31, 113
  marl 34
  of Portland island 33
  optimum proportion 33
  production of cement 33
  quarry 113
  transformation into a construction material 409
limestone filler 273, 408
  history of use 53
limestone monuments
  degradation 426
limit lines
  in the CaO–SiO_2–AlO_3 phase diagram A 462
liquid phase 80, 88
liquidus 80, 280
LO I (loss of ignition) 68
LONE STAR 329
long term durability 420
long term strength
  effect of cementitious material 308
low alkali cement 136, 319, 472
  alkali content 319
  definition 319
  white cement 315
LSF
  calculation 116
  current values 116
  variation 116
lunar concrete 12

macrocracks 186
magnetic resonance 196, 419
main oxides 64
Manual of ConcretePractice 60
manufacturing of cement 83
  ferroaluminous calcium phase 84
  final grinding 106
  fluidized bed 144
  negative aspects 399

positive aspects 399
preparation of the raw meal 106
process 106
production of clinker 106
role of Fe_2O 83
marine environment
brucite 349
C_3A influence 349
carbonation 349
factors in fluencing marine durability 348, 349
mechanism of attack 349, 350
monochloroaluminate 349
sulphate attack 349
market durability 424
marl 34
clayed marl 113
Marsh cone method 232, 234, 237
masonry cement 318
composition 318
manufacturing 318
percentage of cement consumption 318
mathematical models 151
maturity meters 188
maximum temperature 308
McDonald's restaurant entrance 375, 377, 378
mechanical strength 194
influence of the fineness 133
melilite 276, 280
menisci 182
role in autogenous shrinkage 174
metakaolin 287, 292, 299, 300, 408
X-ray diffractogram 286
metallic oxides
as pigments 473
use with grey or white cement 473
metallic powder 418
metallurgical coke 276
Metropolitan Opera House 58
microcement 320
composition 320
microcracks 186, 194
micro structure
in a high alkali cement 472
in a low alkali cement 472
of high-performance concrete 337
of high w/c concrete 336
mid-range water-reducer 218
mill certificate 64
Mindess and Young classification of admixtures 212
mineral components 408, 413
decrease of CO_2 emission 408
use in Belgium 405
use in Netherlands 405
mineralized admixtures *see* cementitious materials
mineralizer 405, 406, 415
mineral phases 76
mineral wool 276
minimum amount of water 184, 185
minislump method 232–6
minor components 146
minor oxides 68
minor phases 421
monosulphoaluminate 129, 171, 154, 437
crystals 61
Montée St-Rémi viaduct 370, 371
morphology of hydration product 471
morphology of the phases
of clinker 72, 73
of Portland cement 146
of the intersticial phase 73
mortar
lime mortar 25, 27
pozzolanic mortar 27
muriatic acid 476

Na_2O equivalent 66
nanostructure 419
natural anhydrite 158, 228
natural cement 33, 34, 92
natural pozzolan 273, 287, 300, 408
basic reaction 300, 303
chemical analysis 302, 303
definition 300
effect on CO_2 emission 303
occurrence 302
X-ray diffractogram 302
nervous cement 141, 407
influence of C_3A 141
influence of C_3S 141
influence of fineness 141
neutralization of a PNS 224
New York Stock Exchange 59
niche product 423
NOx 143
emission 112, 397, 406
noiseless concrete 261
Normandie Bridge 370, 372

NOVAPb 328
nuclear magnetic resonance 151

oil well cement 316
  Blaine fineness 316
  characteristics 316
  different types 316
optical microscope observation 419
optical microscopy 75
optimum C_3S content 434
organic wastes 116
osmotic pressure 182
other constituents *see* cementitious material 272
other types of cement 328
  oxysulphate magnesium 328
  Sorel cement 328
oxide composition 66

Pantheon 6, 9, 17, 28–30
paper bag 138
Passerelle of Sherbrooke 439
PCB 398
pH
  increase of pH 167
percolation point *see* structuration point 168, 186
periclase 68, 76, 119, 136
perlite 304, 305
  photo 305
permeability
  effect of air bubbles 258
  effect of pozzolan 286
  effect of slag 284
  influence of "air plugs" 258
perverse effect of C_3A
  on compatibility and robustness of cement/superplasticizer combination 437
  on durability 438
  on rheology 436
pet coke (petroleum coke) 106, 140
Petronas Towers 360, 362, 416
phase diagram 80
  $CaO–SiO_2–Al_2O_3$ 80
  general information 80
  lime-silica 61
  ternary 61
phase transformation 80
phasic composition 69, 76, 140
phillipsite 17
phosphates as retarders 254
pigglet farm 377–9
pig iron 276
pigments 314, 473
  maximum dosage 473
plaster of Paris 22
plerosphere 289
PNS *see* polynaphtalene
polyacrylates 231
  dosage 231
polycarboxylate 231, 422, 437
  addition of sodium sulphate 246
  characteristics 245, 246
  configuration 245, 246
  dosage 231
  entrainment of air 231, 246
  incompatibility 246
  influence of the grafted chaine 246
  robustness 246
  use in the present industry 247
polymelamine 219, 229, 241
  characteristics 241, 242
  reason to use it 242
polynaphthalene 219, 222, 229
  characteristics 242, 243
  condensation 223
  efficiency 223
  neutralization 224
  polymerization process 223
  reasons to use it 243
  sulphonation 223
  use in gypsum board 222
polysulphonate 68, 134, 224, 225, 226, 315, 422, 437, 468
  action on inert powder 225
  action on pure phases 225
  dispersive properties 208
  hydration in presence of 225, 226
  interaction with cement 226
  reactivity with white cement 315
  retarding action 226
Pompeii 28
ponding 344
Pont du Gard 18, 28
porosity
  influence on durability 347
  of the cement paste 195
  of the intersticial phase 75
Portland cement
  Aspdin's patent 444, 445
  chemical composition 64
  conformity to standard 69

cube strength 69
discovery 30
ecological impact 39, 142
fabrication 35
heat of hydration 68
history 30–6
hydration 146
market price 113
phase composition 70
practical properties 68
precalcination 36
production 34
proto-Portland cement 35
sulphate resisting 68
white 473
X-ray diffractogram 72
portlandite 161, 162, 257, 436, 470
as a chemical trigger 190
crystals 193
crystals as seen in a SEM 162
formation 190
precipitation 170, 191
Portneuf Bridge 368, 369
potassium aluminate as a set accelerator 254
pozzofume 292
pozzolan 6, 27, 31, 285
activation 292
ASTM definition 285
vitreous silica content 285
X-ray diffractogram 302
pozzolanic materials 6, 27, 52, 275
ASTM classes 290
chemical composition 302, 303
different types 287
effect on permeability 285
effect on reinforcing bar passivation 285
history 27, 33
pozzolanic reaction 285, 309, 310
pozzolanic reactivity 285, 287
schematic representation 285
use 28, 29
water demand 308
precalcination 36
precalcinator 112, 117, 119
pre-heater 98, 109, 112, 117
prehomogeneization hall 94, 114
preparation of the raw meal 109
press-roll 104, 134, 142
pressure of crystallization 194
prestress concrete 38
primary phases 460
processing of raw materials 90
production of cement 35
cost 53, 106
energy costs 106
financial costs 106
first French cement plant 35
first production 35
first US cement plant 35
linear process 36
professional associations 58, 59, 60
protoconcrete 16
pulverized coal 116, 287
Puzzoli 28
Pyrament 18, 329
Pyramids 17
pyrite 132
pyroclastic rock 300

quality control of Portland cement 135
importance 135
quaternary cement 53, 285, 307
quartenary phase diagram 83
quartz 85, 86
quenching
of fly ash particles 287
of slag 276, 279
of the clinker 72, 86, 88, 109, 119, 316
quenched slag *see* slag 279
quick lime 90

railway car 109, 138
rapid chloride-ion permeability 347
correlation with water permeability 347, 348
raw meal 68, 76, 85, 114, 121, 282
chemical analysis 135
chemical transformation 88
degree of decarbonation 94, 98
fineness 114
firing 94
homogeneity 92, 114
homogeneization 94, 98, 99, 100
reaction wall of the University of Sherbrooke 378, 380
reactive aggregates 435
reactive powder concrete 213, 357, 379–87, 402, 407, 418, 439
concept 383–6
ductility 385, 386
ecological advantage 428

failure surface 385
microstructure 384
scale factor 380
thermal treatment 385
type of cement 407
typical composition 384
reactive sites 216, 219
reconstruction
economical aspect 403
recycling concrete 357, 404
reduction of the w/b
effect on the ecological performance 400
refractories 83
regulated set cement 317
effect of citric acid 318
fabrication 318
uses 318
regulation of the cement industry 58
reinforced bridge 38
reinforced concrete 38
repair
economical aspect 403
residence time 118, 120
in a ball mill 142
resistance to freezing and thawing 438
restriction of shrinkage 186
retarders 254
action on C_3A 255
action on C_3S 255
corn syrup 256
lignosulphonates 255
mode of action 255
other types 256
phosphates 254
sodium gluconate 256
sugar 254
use with a superplasticizer 236, 250
zinc oxide 254, 256
retarder addition 422
retention time
in the ball mill 134
rheological behaviour 147
different types 239
influence on cement fineness 239
influence on cement type 240
reactivity of C_3A 78
ways to improve it 249
rheological problem 435
direct cost 160
rheology
control 137
influence of the fineness 133
rhyolite as a source of perlite 304
rice husk ash 287, 299, 300
effect on water demand 300
SEM photo 301
production 299, 300
X-ray diffractogram 286
Riecke law 194
RILEM 60
meetings 62
robustness 229, 231, 246, 422, 434
influence of colloidal agents 232
roller-compacted concrete 318
high performance 379, 381, 382
Roman
cement 33
concrete 28
network of roads 21
Rooseboom triangle 83
Rosin-Rammler number 133, 142
rotary kiln
first rotary kiln 35, 94
routine tests 135
rules of crystallization 76

sacks of cement 138
burlap 107
three-ply paper bags 138
weight 138
sacrificial superplasticizer 435
sacrificial water reducer 435
Sagrada Familia cathedral 11, 12, 359
salicylic acid treatment (SAL method) 70, 71, 447
salt of carboxylic acid 256
as retarder 256
SAP (superabsorbant polymers) 346
saturated lightweight sand
substitution to normal sand 346
saturation point 232, 238
scale factor for reactive powder concrete 380
scaling resistance 426
scanner electron microscope (SEM) 151
influence of w/b 163
observation of fractured surface 151
science of admixtures 208, 418, 422
science of concrete 332, 413
scientific literature and meetings 61, 62
Scotia Plaza 364

sea water attack 433, 439
self-compacting concrete 261
secondary constituents *see* cementitious materials
secondary ettringite 156
segregation 147, 232
self-desiccation 184, 333, 341
  control 149
  of the paste 182, 183
semi-dry process 94, 97
semi-wet process 94, 95
serpentine 149
service life 356, 401
  cost of water curing 402
  decreasing the w/b 402
  increasing the service life 402
  influence of environmental conditions 402
  specifications 402
set accelerator 253
  for shotcrete 253
  potassium aluminate 254
  sodium silicate 254
setting 167
  final setting 168
  initial setting 168
  Vicat 168
shell calcination 33
Sherbrooke pedestrian bikeway 387–94, 427
  characteristics 387
  cost 427
  erection 389, 391
  fabrication 388–90
  transportation 389, 390
shipping of Portland cement 138
shrinkage
  of a cement paste 173
  of high-performance concrete 341
  of ordinary concrete 343
  reducing admixtures 261, 346
shrinkage compensating cement 317
  different types 317
  US production 317
  uses 317
silent concrete 261, 378, 379
silica filler 408
silica fume 273, 287, 292, 293
  average size of particles 293, 295
  chemical composition 298
  effect on concrete microstructure 297, 298
  effect on concrete properties 297
  formation 293–5
  from zirconium 293
  history of use 53
  marketing 298, 299
  pozzolanic effect 296
  silicon content 295
  unburned carbon 293
  white silica fume 295
  X-ray diffractogram 286, 297
silica modulus SM 69
silicate phase 94
  content 80
siliceous filler 273
silicoaluminous fly ash 273
  typical chemical composition 290
silicocalcareous 292
silicocalcic fly ash 273
  typical chemical composition 290
silicocalcium phase 84
silicon content of silica fume 295
silos 114
simplified chemical notation 74
Siporex® 257
skyscrapers 358, 360, 363
slaked lime 25, 136
slag 273–6, 408
  activation 279–83
  blended cement 283
  chemical composition 279
  cold slag 280
  colour 276
  different types 279
  effect on concrete properties 283
  fineness 284
  first slag used 274
  hot slag 280
  hydration 282
  molten slag 276
  production 279
  quenching 276, 279
  use in the raw meal 282
  X-ray diffractogram 79, 282
slag cement
  activation 52
  commercialization 52
  history 52
slump loss 147
  with a PNS 240
slump retention 232
slurry 92
  homogeneization 116

SM current values 116
small bridges 364
'smart'concretes 439
SO_2 112
SO_3
content 131, 140, 160, 231, 422, 423
emission 140, 397
limitation 231
limits 140
maximum amount 140
of clinker 160
of modern clinker 140
variation in clinker 160
SO_4^{2}/AlO_2 227
influence on setting of cement 227
socio-economic criteria 332
sodium abietate 259
sodium gluconate 256
sodium nitrite 262
as an anti-freeze admixture
sodium silicate 254
as a set accelerator 253
sodium sulphate 422
addition to a superplasticizer 434
solid gel 177–81
solidus 80, 280
soluble alkali 319, 423, 437
minimum amount 434
optimum content 319, 423
sulphates 440
soluble anhydrite 22, 156
Sorel cement 328
spacing factor 351, 352
spalling of concrete under a fire 355
special concretes 424
special Portland cement 314
air entrained 314
buff cement 315
chemical composition 314, 315
coloured 315, 316
low alkali 319
microcement 320
oil well cement 316
regulated set cement 317
shrinkage compensating cement 317
white cement 315
specialized applications 2
specific surface area 136
Blaine fineness 136
Standards 421
limitations 421
Statue of Liberty 58
steel balls 102
steric repulsion 218
storage of the clinker 128
structuration point 152, 168, 171
sugar as a retarder 254
sulphate attack 349, 433
sulphoaluminate 140, 257, 285
crystals 78
sulphocalcareous fly ash 292
typical chemical composition 290, 292
sulphonated hydrocarbons as air entraining agent 259
sulphonation 223
sulphur
as a mineralizer 406
coal and fuel rich in sulphur 116, 140
cycle 123
superabsorbant polymers (SAP) 346
super fly ash 292
superplasticizer 134, 401, 417
architecture 247
average molecular weight 248
beneficial action on sustainability 401
composition of commercial product 221, 247
control of the quality 247
different types 221
dosage 248
history 222
parameters influencing the dosage 248
powder or liquid 249
reaction with the intersticial phase 226
rheoligical efficiency 248
role of C_3A 433
saturation point 248, 249
selection 240
supersulphated cement 52, 327, 438
supplementary cementitious materials *see* cementitious materials
surface absorption 217
sustainable development 144, 332, 397, 414, 427
swelling
due to brucite 136
growth of portlandite crystals 174
of a cement paste 173
of concrete 344
syngenite 466
synthetic calcium sulphate 228

Taiheyio Cement 54

technical date sheet of superplasticizer 224
temperature
gradient 340
influence of the cement dosage 339, 340
influence on the compressive strength 339, 340
on high-performance concrete 335–40
influence of the kinetics of hydration 187
rise 147, 340
ternary blended cement 53, 285, 307
ternary diagrams 453–9
Rooseboom triangle 453
ternary phase diagram 61, 460–5
$CaO–SiO_2–Al_2O_3$ 461–5
ternary mixes 454–7
case of three aligned points 457, 459
proportions 455–9
tetracalcium ferroaluminate 70, 74
thaumasite 349
thermal activation 188
thermal gradients 188
thermal shrinkage 194
thiobacillus ferroxydans 426
tobermorite 197
transition zone 117–22, 297, 299, 334
transportation
by barrel 92, 138
by boat and barges 57, 105
by helicopter 140
by plane 140
by railway car 105, 138
by truck 105
cost 57, 111
trass 287
tributylphosphate as a defoaming agent 256
tricalcium aluminate 70, 74
tricalcium silicate 34, 70, 74
triethanolamine (TEA) 252, 253, 257
mode of action 252
Tripoli *see* diatomaceous earth
tuff 27
Two Union Square 365
Type V
cement 350
applications 439
clinker 433
Type VI clinker 434

unburned carbon 308
effect on air entrained admixture 308
effect on the bubble system 308
in fly ash 287, 308
in silica fume 293, 295
universal clinker 440
use of admixtures 209
use of cementitious material
effect on CO_2 emission 274

Van Breughel model 201
variation of SO_3 content in clinker 160
Vernet schematic representation of hydration 188
vertical kiln 90, 94
Vicat 54 442
artificial lime 442
needles 137
vinsol resin 259, 417
vitreous silica 285, 300
volatiles
alkalis 116
circulation 121, 123
volcanic ash 287
volcanic tuff 287
volumetric contraction 148
during hydration 174
volumetric stability
due to coarse aggregate 343
volumetric variations 147, 178, 194
absolute volume decrease 173
apparent volume increase 173
during hardening 186
during hydration 151, 167
effect of coarse aggregate content 186
growth of C-S-H 174
growth of portlandite 174
influence of curing 173
isothermal conditions 186
with and without an external source of water 186
VOTORANTIN 56

Waldorf-Astoria Hotel 58
water
addition 440
chemically linked water 176
free water 176
gel water 176, 180
percolation 186
role when making concrete 147
water/binder ratio 333

water/cement ratio 333
water curing
  cost 402
water demand
  effect of cementitious material 308, 309
water reducer 400, 401, 417
  beneficial action on sustainability 401
  beneficial effect 264
  decrease of CO_2 emission 206
  different types 210
  introduction in cement 401
  lignosulphonate 417
  saving in cement 206
water repellant admixture 261
Water Tower Place 361
wet burlap 344
wet process 94, 112
white Portland cement 68, 315
  alkali content 315
  C_3A content 315
  compatibility with polysulphonate 229, 230, 315
  iron oxide content 315
  intersticial phase 315
  polycarboxylate 315
  reactivity 315
  reactivity with polysulphonate 315
white rust *see* zinc oxide

X-ray diffractogram
  of a crystallized fly ash 291
  of a diatomaceous earth 286
  of a glass 78
  of a vitreous slag 79
  of fly ashes 290
  of hydrating paste 161
  of metakaolin 286
  of natural pozzolan 302
  of Portland cement 72
  of pozzolans 286
  of rice husk ash 286
  of silica fume 286, 297
  of slag 79, 282

yeelimite 325
Ytong® 257

zeolite 287, 300
zeolitic phases 17
zinc oxide
  as a retarder 254, 256